# Fetal and Newborn Cardiovascular Physiology

**Donald Henry Barron, Ph.D.**

# Fetal and Newborn Cardiovascular Physiology

## Volume 1
## Developmental Aspects

Proceedings of a Symposium to honor Donald H. Barron held in conjunction with the fall meeting of the American Physiological Society, 11 to 14 August 1976, Bryn Mawr, Pennsylvania

Edited by
Lawrence D. Longo &
Daniel D. Reneau

**Garland STPM Press**
**New York & London**

Copyright © 1978 by Garland Publishing, Inc.

All rights reserved. No part of this work covered by the copyright hereon may be reproduced or used in any form or by any means — graphic, electronic, or mechanical, including photocopying, recording, taping, or information storage and retrieval systems — without permission of the publisher.

15  14  13  12  11  10  9  8  7  6  5  4  3  2  1

**Library of Congress Cataloging in Publication Data**
Main entry under title:

Fetal and newborn cardiovascular physiology.

    Includes bibliographies and indexes.
    CONTENTS: v. 1. Developmental aspects. — v. 2. Fetal and newborn circulation.
    1. Fetus — Physiology — Congresses.   2. Infants (Newborn) — Physiology — Congresses.   3. Cardiovascular system — Growth — Congresses.   4. Blood — Circulation — Congresses.   5. Physiology, Experimental — Congresses.   6. Barron, Donald Henry, 1905-   I. Barron, Donald Henry, 1905-   II. Longo, Lawrence D. III. Reneau, Daniel D.
[DNLM:   1. Cardiovascular system — Physiology — Congresses.   2. Infant, Newborn — Physiology — Congresses.   3. Fetus — Physiology — Congresses.
    WQ210 F419 1976]
    RG600.F45      612.6'401'1      77-20591

   ISBN 0-8240-7012-7 (v. 1)
   ISBN 0-8240-7013-5 (v. 2)

Printed in the United States of America

# Contents

Contents of Volume 2 . . . . . . . . . . . . . . . . . . . . ix

Acknowledgments . . . . . . . . . . . . . . . . . . . . . . xi

Contributors to Volumes 1 & 2 . . . . . . . . . . . . . xiii

Preface . . . . . . . . . . . . . . . . . . . . . . . . . . . xvii

An Appreciation of Donald Henry Barron . . . . . . . . xxvii
    Lawrence D. Longo

To Donald Henry Barron . . . . . . . . . . . . . . . . . . . xxxv
    Giacomo Meschia

## HISTORICAL CONSIDERATIONS

1. A History of Fetal Respiration: From Harvey's
   Question (1651) to Zweifel's Answer (1876). . . . . . .1
       Donald H. Barron

2. Many Slender Threads: An Essay on Progress in
   Perinatal Research . . . . . . . . . . . . . . . . . .33
       S. R. M. Reynolds

## DEVELOPMENTAL ASPECTS OF THE CARDIOVASCULAR SYSTEM

3. Ontogenesis of the Autonomic Control of Cardiovascular Functions in the Sheep . . . . . . . . . . 47
       Nicholas S. Assali, Charles R. Brinkman III
       James R. Woods, Jr., Bahis S. Nuwayhid, and
       Adrian Dandavino

4. Postnatal Maturation of the Central Neural Cardiovascular Regulatory System . . . . . . . . . . . . 93
   *Phyllis M. Gootman, Nancy M. Buckley, and Norman Gootman*

5. Development of Fetal Cardiovascular Responses to Alpha-Adrenergic Agonists . . . . . . . . . . . . 153
   *G. R. Van Petten, W. H. Harris, and G. J. Mears*

6. Reactivity of Fetal Vascular Smooth Muscle to Sympathetic Nerve Stimulation and Vasoactive Agents . . . . . . . . . . . . . . . . . . . . . .167
   *Che Su, B. L. Pegram, John A. Bevan, Nicholas A. Assali, and Charles R. Brinkman III*

7. The Electrical Properties of Embryonic Chick Cardiac Cells . . . . . . . . . . . . . . . . . . . 191
   *Nick Sperelakis and Michael J. McLean*

8. Development of Electrical Activity in Embryonic Myocardial Cells . . . . . . . . . . . . . . . . .237
   *Melvyn Lieberman, C. Russell Horres, Joyce E. Purdy, and Linda R. Halperin*

9. Metabolic Maturation in the Fetal Mouse Heart . . . 257
   *Kern Wildenthal*

10. Glycolytic Control Mechanisms in Cardiac Muscle from Fetal Rhesus Monkeys . . . . . . . . . . . . 271
    *Clarissa H. Beatty, Rose Mary Bocek, and Martha K. Young*

CARDIAC OUTPUT AND ITS CONTROL

11. Venous Return and Control of Fetal Cardiac Output . . . . . . . . . . . . . . . . . . . . . . .299
    *Raymond D. Gilbert*

12. Water Transfer Across the Placenta: Hydrostatic and Osmotic Forces and the Control of Fetal Cardiac Output . . . . . . . . . . . . . . . . . .317
    *Gordon D. Power, Philip J. Roos, and Lawrence D. Longo*

# Contents of Volume 2

Umbilical Blood Flow

Oxygenation and the Circulation

Hypoxia and the Circulation

Placental Exchange and the Circulation

# Acknowledgments

The Symposium was aided by support from:

    The National Foundation--March of Dimes, New York, New York.

    The Frost Foundation, Shreveport, Louisiana.

    The Ruston Hospital Corporation, Ruston, Louisiana.

We thank Mrs. Linda Moore and Mrs. Helen Little for editorial assistance.

# Contributing Authors to Volumes 1 and 2

Robert M. Abrams
William W. Allen
Endla K. Anday
Nicholas S. Assali
J. Ayromlooi
June N. Barker
Donald H. Barron
Frederick C. Battaglia
Clarissa H. Beatty
John A. Bevan
Amrutha Bhakthavathsalan
John M. Bissonnette
Rose Mary Bocek
Robert Boyd
Susan C. Brennan
Charles R. Brinkman III
Sonya Brotman
Nancy M. Buckley
Laurence I. Burd
B. Burns
L. Allan Butler
James M. Cameron, Jr.
Sidney Cassin

Donald Caton
Ronald A. Chez
James F. Clapp III
Herbert E. Cohn
David R. Cook
Adrian Dandavino
Geoffrey S. Dawes
Maria Delivoria-Papadopoulos
Karel J. deNeef
Barry Dvorchik
Richard A. Ehrenkranz
Robert Elsner
Wilhelm Erdmann
J. Job Faber
James D. Ferguson
William F. Friedman
Takashi Fuchigami
E. O. Fuller
P. M. Galletti
Ronald F. Gautieri
D. T. Gibbons
Raymond D. Gilbert
Norman Gootman

Phyllis M. Gootman
Hiroshi Goto
Eric J. Guilbeau
Gail H. Gurtner
Linda R. Halperin
Lewis A. Hamilton, Jr.
Eric Harinck
W. H. Harris
D. Jane Henderson
Robert Holland
Robert J. Hollister
C. Russell Horres
Benjamin T. Jackson
Jos R. C. Jansen
F. Johnson
M. Douglas Jones, Jr.
Stanley E. Kirkpatrick
Savitri P. Kumar
Wolfgang Künzel
Charles Leffler
James A. Lemons
Melvyn Lieberman
Maida Liu
Lawrence D. Longo
I. Lysak
Leon I. Mann
John W. Manning
J. G. Maylie
Margaret K. McLaughlin
Michael J. McLean
G. J. Mears

Leena Mela
Giacomo Meschia
James Metcalfe
Leonard D. Miller
Etsuro K. Motoyama
Joan C. Mott
Jerome W. H. Niswonger
M. Notelovitz
Donald O. Nutter
Bahij S. Nuwayhid
Gary K. Oakes
Kirk D. Pagel
Julian T. Parer
Marilyn Paul
B. L. Pegram
Nancy Peress
Martin L. Pernoll
George J. Piasecki
Gordon G. Power
Joyce E. Purdy
Edward J. Quilligan
John H. G. Rankin
Daniel D. Reneau
S. R. M. Reynolds
Philip J. Roos
S. David Rubenstein
Richard L. Schreiner
Roger E. Sheldon
Ian A. Silver
Michael A. Simmons
Nick Sperelakis

*Contributors*

T. E. Stacey
Ann R. Stark
Che Su
Hazel Szeto
Kent L. Thornburg
Paola S. Timiras
M. E. Towell
Thomas N. Tulenko
Thom Tyler
Motoaki Umezu
J. H. van Bemmel
Cornelis J. van Nie
G. R. Van Petten
Adrian Versprille

A. F. L. Veth
Adrian M. Walker
Richard Wallis
R. H. T. Ward
A. P. Weedon
Howard C. Wieland
Charles J. Wilcox
Kern Wildenthal
Wayne W. Wolstenholme
James R. Woods, Jr.
J. T. M. Wright
James F. Wyatt
Martha K. Young

# Preface

In his monumental description of the circulation of the blood, *Exercitatio anatomica de motu cordis et sanguinis in animalibus* (Francofurti, sumpt. G. Fitzeri, 1628), William Harvey first described the circulation of the fetus as well as that of the adult. He used the example of the fetal heart to bolster his argument that the blood passes from the right to the left heart through the pulmonary vascular bed rather than through invisible pores in the septum of the heart. Harvey observed that

> . . . there is absolute identity between what happens in the human embryo and what happens in others, in which the unions in question are not in process of abolition. Hence, the heart, by its movement, transfers blood very freely from the vena cava, through both ventricular conduits, into the great artery [aorta]. The

right ventricle receives blood from the
auricle and then drives it forward through
the artery-like vein  pulmonary artery  and
its offshoot, the so-called artery-like channel [ductus arteriosus], into the great artery.  The left ventricle, in like manner,
simultaneously receives blood, that has been
directed from the vena cava, by a different
route, through the oval opening [foramen
ovale], by means of the auricular movement,
and by its tension and contraction it drives
this blood through the root of the aorta into
the same great artery.  Thus, in the embryo,
while the lungs are idle and devoid of activity or movement, as though they did not exist,
Nature uses the two ventricles of the heart
as one for the transmission of the blood.

Of course, the unique anatomical features of the fetal circulation had been described long before Harvey's time. For instance, Galen first described the foramen ovale and the ductus arteriosus (*Opera omnia.* Ediderunt Andreas Asulanus et J. B. Opizo. Venetiis in aedibus Aldi . . . , 1525), and Vesalius apparently first described the ductus venosus in a work published posthumously (*Anatomicarum Gabrielis Falloppii observationum examen* . . . , Venetiis, Apud Franciscum de Franciscis, Senensem, 1564).

Modern knowledge of the fetal circulation, however, originated to a great extent with two groups of investigators. In 1927 Huggett determined that the oxygen affinity of blood of the fetal goat differed from that of its mother. Although we know that Huggett's oxyhemoglobin dissociation curves were incorrect, his conclusions stimulated other workers to determine these interrelations under various circumstances and tease out the mechanisms of this phenomenon. A decade later, Sir Joseph Barcroft, Donald H. Barron, Alfred E. Barclay, and Kenneth J. Franklin used cineangiography to demonstrate for the first time the circulation of the intact fetus delivered by Cesarean section. Using this technique, these workers also first demonstrated the time of functional closure of the ductus arteriosus.

Several previous monographs have been devoted to the circulation of the fetus and/or the newborn. These include *The Foetal Circulation and Cardiovascular System, and the Changes They Undergo at Birth* by Barclay, Franklin, and

Prichard (1944); Sir Joseph Barcroft's, *Researches on Prenatal Life* (1946); the work of Lind, Stern, and Wegelius, *Human Foetal and Neonatal Circulation* (1964; an essentially new and much expanded edition, edited by Walsh, Meyer, and Lind, appeared in 1974); Cassel's *The Ductus Arteriosus* (1973), and *The Heart and Circulation in the Newborn and Infant* (1966); and most recently, Rudolph's *Congenital Diseases of the Heart* (1974). Each of these volumes presents a reasonably complete account of the major morphologic features of the heart and great vessels and the physiologic aspects of the fetal and/or newborn circulation. In addition, several of these works review *in extenso* congenital heart disease and other clinical aspects of the circulation.

During the past decade an increasing number of investigators have attempted to elucidate the ontogeny of the control mechanisms during development of the circulation in the fetus and newborn. They have probed such areas of investigation as the development of baroreceptors and chemoreceptors in autonomic control, the control of cardiac output and distribution of blood flow to the peripheral tissues, the control of blood flow through the ductus arteriosus, the factors that determine the mean blood pressure or blood pressure set point, and the effects of hypoxia in the fetus and newborn infant.

The titles of the papers in this monograph reflect some of the recent progress in understanding these and related problems. Because of the interdisciplinary nature of this field of research, and the fact that many individuals working in related areas present their work at different meetings of different societies and so may be unaware of one another's work, we organized a symposium at which investigators not only could present their newest and exciting work, but could interact and share ideas. This symposium was held 11 to 14 August 1976 in Bryn Mawr, Pennsylvania, in conjunction with the fall meeting of the American Physiological Society. About 90 investigators from 11 countries gathered to test their ideas on one another.

Because of his contributions to this field, the conference and these volumes are dedicated to Donald H. Barron. A high point of the symposium was a banquet Thursday evening, 13 August. Donald Barron spoke on "From Harvey's Question (1651) to Zweiffel's Answer (1876)."

Following this presentation, a number of Dr. Barron's friends and former colleagues reviewed his contributions to science and his influence on their lives. Dr. Barron's paper, and some reminiscences by Samuel R. M. Reynolds, constitute the introductory section to this volume.

The next section considers some developmental aspects of the fetal and neonatal cardiovascular system. Assali and his colleagues present an overview of the fetal response to autonomic sympathetic and parasympathetic pharmacologic agents and the changes that occur in the fetal responses during maturation. Gootman et al. investigate postnatal maturation by stimulation of afferent nerves or the central nervous system directly, in addition to examining the interactions of these systems and the effects on them of pharmacologic agents. Van Petten et al. further explore the time course of vascular responses to adrenergic drugs, the development of presynaptic nerves, and the development of the receptor-effector system in the fetus and newborn. Su et al. document the difference in the developmental pattern of the adrenergic neuro-effector synapse in various blood vessels of the fetus. Sperelakis and McLean note the striking changes that occur in the electrical properties of myocardial cells during embryonic development, while Lieberman et al. review some problems associated with electrophysiologic studies of these embryonic myocardial cells. Some unique features of the metabolic maturation of the fetal heart are reported by Wildenthal and by Beatty and her colleagues.

In the following section several workers consider different aspects of the control of cardiac output. Gilbert discusses venous return and role of mean systemic pressure and vascular compliance in the control of cardiac output. Power et al. hypothesize on the role of the relative concentrations of carbon dioxide and bicarbonate ion, amino acids, and glucose in regulating placental transcapillary water exchange and fetal blood volume, and thus cardiac output; and Longo et al. consider the interrelations of blood volume and extracellular fluid volumes and the role of blood volume in the regulation of fetal cardiac output. Kirkpatrick and Friedman examine the problem of whether the Frank-Starling relation operates in the fetus and discuss the role of the changing myocardial fiber length in determining the cardiac output. Maylie et al. suggest that ultrastructural development of the myocardial T-tubular system and sarcoplasmic reticulum determines the force-frequency relations of the developing heart. Versprille et al. present a morphologic analysis of the developing heart that correlates ventricular geometry with function. Mott reviews the development of the renin-angiotensin system in the fetus and its role in regulation of extracellular fluid volume and vasomotor tone.

An exciting development--both from the standpoint of understanding the control of blood flow through the ductus arteriosus and great vessels of the fetal heart, and its implications in the treatment of newborn infants with certain congenital heart defects--is the discovery of the role of prostaglandins and related hormones in the fetal and neonatal circulation. Cassin et al. analyze the role of prostaglandins in the control of the developing pulmonary circulation, and Friedman and Kirkpatrick present the therapeutic uses of prostaglandins in closure of the ductus arteriosus. Both M. A. Heymann (University of California, San Francisco) and F. Coceani (Hospital for Sick Children, Toronto, Canada) also participated in this session of the symposium; however, their discussions are not included. Finally, Rankin presents an interesting hypothesis on the role of prostaglandins in regulating maternal and fetal placental blood flows.

The first volume closes with some theoretical considerations. Although physics and mathematics have strong mathematical underpinnings, too often biologists have been content to gather data and perform experiments with little appreciation of the theoretical aspects of a given problem. Fortunately, mathematical approaches are being used increasingly in an attempt to understand certain aspects of the circulation and respiratory gas transport in the fetus and newborn. Cameron et al. present a thorough mathematical analysis using a deterministic, lumped parameter formulation of the fetal circulation. Allen and his colleagues develop a somewhat different approach for studying the time course and magnitude of changes of oxygen levels in response to hypoxia. Butler et al. present a theoretical consideration of maternal and fetal placental blood flows during uterine contractions and the implications on transplacental oxygen exchange. Both Veth and van Bemmel, and Gibbons et al. present a mathematical basis for interpreting some of the changes in fetal heart rates observed during labor and delivery.

Volume II considers several aspects of the peripheral circulation in the fetus and newborn infant. Numerous questions relate to the regulation of the fetal umbilical circulation and the control of the fraction of cardiac output perfusing the placenta during development from a minute embryo, in which almost all of the output from the heart goes to the placenta, to a near-term fetus, in which this fraction decreases to about 50% of total cardiac output. This control probably involves not only adrenergic agents as discussed by Chez et al. and Tulenko, but also the respiratory gases in blood as reported by Motoyama et al. Reynolds compares the fetal umbilical circulation to a pulsometer pump, while Bissonnette examines the problem of recruitment versus distension and the regulation of volume of the placental capillaries.

Obviously the ultimate purpose of the evolving circulation is to deliver adequate oxygen and other nutrients to, and catabolites from, the cells of developing tissues. Mela et al. review some unique features of fetal mitochondria during the perinatal period, such as changes in the concentration of cytochromes, their turnover rate and rate of respiration. Using microelectrodes, Silver compares the oxygen tensions and the pattern of electrical discharge during normoxia and hypoxia in several areas of the fetal and adult brain. In turn, Erdmann presents the oxygen tension response of these cells when anesthetic agents are administered to the mother. In contrast to the brain and other organs, the fetal liver is supplied with blood of widely varying oxygen tension. The left physiologic lobe is perfused with blood (from the umbilical vein) with a relatively high oxygen tension, while the right physiologic lobe receives blood (from the portal vein) at a relatively low oxygen tension. Dvorchik compares the distribution of some components of the mixed-function oxidase system with the fetal hepatic circulation and hepatic oxygenation. Caton et al. demonstrate that factors other than just fetal mass determine the rate of oxygen consumption by the fetus. Barker analyzes the role of blood flow in determining the genesis and growth of the cerebral microcirculation during development. Abrams et al. use recently developed thermal techniques to calculate fetal cerebral blood flow under a variety of conditions. Schreiner and his colleagues study the preferential utilization of various substrates by the fetus of fed and fasting sheep.

Manifestly, hypoxia can affect the circulation of the fetus and newborn infant; however, one cannot extrapolate from hypoxic effects in the adult to the qualitative or quantitative effects in the fetus or newborn infant. Dawes reviews some of the problems associated with circulatory studies in the chronically catheterized lamb fetus. Several caveats he notes in working with these preparations concern cyclical variations in fetal heart rates, blood gases, breathing movements, and lability due to noise or maternal stress. Parer demonstrates that fetal oxygen consumption is a function of arterial oxygen tension, and Cohn et al. present data on the role of the autonomic nervous system and the redistribution of blood flows during hypoxia. Longo and co-workers compare the fetal circulatory response to hypoxia induced by breathing low oxygen mixtures and that associated with carbon monoxide. The effects of partial umbilical cord compression (Towell and Lysak) or complete occlusion (Künzel et al.) are reviewed. Mann et al. analyze the fetal electroencephalographic and metabolic response to graded hypoxia, while Brotman et al. present the effects of prenatal or postnatal hypoxia on brain biogenic amines and other neurotransmitters. Ferguson and his colleagues describe the redistribution of blood flow in newborn lambs following exchange transfusion; and Anday et al. contrast the cardiovascular response of low birth weight infants to exchange transfusion with fresh packed erythrocytes to that with whole blood. Pernoll et al. examine the effects of maternal exercise on maternal and fetal heart rates and blood pressure. Elsner makes a fascinating comparison of the fetal cardiovascular responses in the deep-water diving mammal (seal) and terrestrial animal (sheep). Quilligan explores the problem of the extent to which fetal heart rates reflect fetal oxygenation.

Of course, a vital link in the chain of an adequate supply of oxygen and other nutrients for the fetal circulation to transport is that of placental exchange. Fuller and her co-workers demonstrate that the elusive perfused uterine preparation for experimental studies is indeed a reality. Gurtner and Burns discuss the hypothesis of carrier-mediated respiratory gas exchange in the placenta and present evidence that fulfills certain of their criteria for facilitated transport. Thornburg and Faber clarify the role of the various cells layers of the placental membrane as resistances to diffusion, while Boyd et al. present data

on the placental permeability to a number of solutes and
discuss the implications of these permeabilities on the
fluxes of various ions and solutes. Holland measures the
reaction rate of carbon dioxide with the hemoglobin in
both maternal and fetal erythrocytes, and discusses the
implications of these reaction rates on placental respiratory gas exchange. Finally, Gautieri and Wolstenholme
discuss the effects of various drugs on uterine and umbilical blood flow in the placental vasculature.

Perhaps several caveats should be noted. The overriding purpose of these volumes is to present some of the
latest and best studies that are being carried out in various laboratories around the world toward an understanding
of important physiologic problems in the circulation of the
developing fetus and newborn. Little is presented on the
anatomy and morphology of the fetal circulation or the
specific changes that occur during birth. These topics are
dealt with *in extenso* in other monographs and texts. Of
necessity, most of the experiments reported in this monograph were performed in lambs, goats, monkeys, chicks, or
other species. Obviously, the functional demands on the
developing human circulation differ from those in various
animals. For instance, the blood flow to the brain of a
human near-term fetus or infant will be several times that
of the blood flow to the lamb. On the other hand, there
are many aspects of the circulation in both that are comparable, and these should not be overlooked. To a certain
extent, the major divisions of the monograph are somewhat
arbitrary. Several of the manuscripts could have been included in sections other than those in which they appear.
However, we have attempted to group the papers so that each
section will be as logically coherent as possible. Much
of the data presented here is, as in most of science, reductionist and analytic in approach; that is, systems are
dissected into smaller and finer bits and pieces. Unfortunately, we seem to pay too little attention to the other
possible approach, that is, to synthesizing the bits and
pieces again into an understandable system and integrated
whole.

Finally, many of the studies are concerned with an
understanding of fundamental biologic problems and may have
little apparent relevance to clinical problems or disease.
We are reminded of the words of Severinus (seventh century)

who wrote: "Go my sons, buy stout shoes, climb the mountains, search . . . the deep recesses of the earth . . . In this way and in no other will you arrive at a knowledge of nature and the properties of things." The path to deep understanding and insight winds through the jagged peaks of science rather than across the broad plains of technology. While some governmental administrations might wish to avoid this tortuous path by short cuts, there is, unfortunately, no "yellow brick road" to some scientific "Oz." Rather, our understanding will increase only as first-rate pioneers continue to explore the frontiers. Since the time of William Harvey it has become increasingly apparent that development of the heart during the perinatal period has a profound effect not only on the heart itself, but on the person as a whole.

Lawrence D. Longo

# An Appreciation of Donald Henry Barron

Lawrence D. Longo

To describe in a few words the place which belongs to Donald Barron as a scientist can only end in a gesture. A brief scientific biography must content us at present.

Donald Henry Barron was born on 9 April 1905 on a farm near Flandreau, South Dakota. His father was a farmer and his mother a teacher. After graduating from Carleton College in 1928 he moved to Iowa State College, where one year later he received a Master of Science degree in plant physiology, also performing work which led to his first publication, a paper on the effects of mold growth on the temperature of stored oats, wheat, and barley (1). This training in plant physiology constituted Dr. Barron's only formal training in physiology.

He then moved to Yale as an assistant in zoology. Here he studied under Ross Granville Harrison and Harvey Burr Ferris and did his thesis work with John Spangler Nicholas.

During this time he studied the effects of cyanide on embryos (2) and showed that the results of a given efferent nerve impulse depend on the muscle innervated rather than on specific differences in the impulses themselves (3). Following his Ph.D. in zoology in 1932, he moved to the Department of Anatomy at Albany Medical College, where he remained for a year. During this time he studied structural changes in the anterior horn cells following lesions of the central nervous system or the spinal cord (4).

This interest in the nervous system was strengthened through reading Sir Charles Scott Sherrington's *Integrative Action of the Nervous System*, and an association with John Farquhar Fulton, newly arrived at Yale from Oxford. In 1933, on the advice of Fulton, Donald Barron went to Berne as a National Research Council Fellow and spent six months with Professor Leon Asher. In spring of the following year, he moved to the laboratory of Edgar Douglas Adrian (later Lord Adrian) at Cambridge University and soon began work with Sir Bryan Harold Cabot Matthews, a master experimentalist, on the potentials in the spinal cord. They were not long in obtaining the first records of impulse activity in the spinal tract (5,6). He remained at Cambridge and continued to collaborate with Matthews for the next six years.

During the summer of 1934, he met Joseph Barcroft (as yet untitled); Barcroft had suffered a minor injury and Barron went to his aid. Later, in a chance encounter during afternoon tea, Barcroft mentioned to Matthews that he had just arranged the purchase of fifty ewes to be stocked for his experiments the coming winter. To Barron's question whether he proposed to study the functional development of the nervous system, Barcroft replied that he knew nothing of such a possibility; he invited Barron to bring him up to date on the subject.

Barron told him about the studies of G. E. Coghill on *Amblystoma* and the generalization he had drawn regarding the development of overt behavior, and added that this application to the development of the reactivity in mammals had been questioned by William F. Windle but supported by Angulo y Gonzalez and D. Hooker. Barcroft then asked Barron to join him during the coming winter to study the question, using his sheep as subjects, and, a few days later, invited him to his office to discuss the possibilities of looking into the functional development of the nervous system. Barcroft then obtained a special grant

from the Rockefeller Foundation to support him and the
work for a year. Following a brief visit to Albany, Barron
returned to Cambridge in October. He and Barcroft did
their first experiment on 9 November 1934. It was on a
fetus of 46 days' gestational age which, fortunately for
the future of the project, was very active, "respiring"
spontaneously--that is to say, there were rhythmic contractions of the diaphragm. Barcroft was excited by what he
saw--and his interest in the study of fetal activity grew
from that time (7). He made films of human fetal activity
in the months prior to his death.

In the spring of 1935 Barcroft invited representatives
of the two schools of thought current in America on the
subject to join him and Barron in their studies the next
year (1935-36). In the meantime, Barron was appointed to
a vacant demonstratorship in the Cambridge School of Anatomy under the newly appointed Professor H. A. Harris.
Accordingly, invitations were sent to Windle and to Coghill's pupil Angulo y Gonzalez, as Coghill's health prevented his making the trip. Although Angulo y Gonzalez
declined, Windle accepted the invitation, and in the winter
of 1935-36 the group attempted to determine whether the
first fetal movements were local reflexes or mass movements.
Their conclusions were far from satisfactory. All agreed
that the somatic respiratory movements appeared before the
central nervous system mechanism was functional (8,9), but
the interpretations as to the character of the movements
were "as numerous as the observers."

Although Barron continued to collaborate with Matthews
on studies on the spinal cord, he became increasingly involved in other aspects of Barcroft's fetal studies and
more frequently in his company and under his influence.
This proved to be a very productive period for both Barron
and Barcroft. In fact, they helped create the emerging
discipline of fetal physiology. A collaborative study with
Alfred E. Barclay and Kenneth J. Franklin of Oxford resulted in the first serial radiographs of closure of the ductus
arteriosus in the newborn (10) and in the now classic paper
on the first radiographic demonstration of the circulation
in the fetal heart and great vessels (11). This latter
work confirmed the fact that blood from the superior vena
cava flows largely into the right atrium, right ventricle,
pulmonary artery, ductus arteriosus, and to the descending
aorta; while that from the inferior vena cava flows largely
through the foramen ovale, left atrium, left ventricle,

aorta, and the brachiocephalic artery. In a review of
the changes of the fetal circulation at birth (12), Donald
Barron summarized this work of several years and displayed
his historical perspective, relating his own work to the
previous studies of Sabatier (Memoire sur les organs de la
circulation de sang les foetus, *Mem. Acad. Roy. Sci.*
(Paris) 198:1774) and Wolff (*Novi Comment. Acad. Sci. Imp.
Petropolit* 20:357, 1776).

With the outbreak of World War II, Dr. Barron returned to the United States, where, in the Department of Zoology at the University of Missouri, he continued to study
the development of neurons in the spinal cord.

In 1943 (in what I believe was a stroke of genius)
John Fulton, with the insight so characteristic of him,
asked Donald Barron to join the Department of Physiology
at Yale. At New Haven, Barron established fetal physiology
in America as a field in its own right. This he did both
by important scientific studies and by attracting students
that were to carry on his work in many centers in America
and Europe. Dr. Barron had the great gift of attracting
young colleagues, aiding them in discovery, and developing
their potential for research.

In the early 1950s Dr. Barron, in original studies
of placental oxygen exchange, measured for the first time
what we now call the placental diffusing capacity (13,14).
He published a perhaps little appreciated but important
study on the effects of carbon dioxide and bicarbonate ion
on blood osmotic pressure (15,16) and of the electrical
potential difference across the placental membranes (17).
He was also fascinated by the problem of fetal oxygenation
during pregnancy at high altitudes. He asked a penetrating
question: Because at sea level the fetal arterial blood
oxygen tensions are only 20 to 30 torr (1/5 to 1/3 maternal
values), what are the blood gas values at high altitude
where ambient and arterial oxygen tensions are so much
lower? As Joseph Barcroft had done before him, he headed
an expedition to Cerro de Pasco in the Peruvian Andes with
collaborators such as André Hellegers, William Huckabee,
Giacomo Meschia, James Metcalfe, and Harry Prystowsky.
This adventure resulted in a series of landmark papers on
maternal and fetal adaptation to high altitude, including
those on the oxygen supply to the fetal llama (18) and
sheep (21), the fetal oxyhemoglobin dissociation curves
(20), growth rates and organ weights (22), and blood volumes (19). These studies showed that oxygen partial

pressures in the arterial blood of fetuses at high altitude were little different from those at sea level, despite the much-reduced partial pressures of oxygen in the atmosphere and maternal blood. This work raised numerous questions regarding the mechanisms accounting for these compensatory adaptations, questions which remain unanswered.

Upon returning to New Haven, Donald Barron and his colleagues continued to work on problems relating to placental transfer and fetal circulation. In what Dr. Barron perhaps regards as a rather minor report of the technique that he and Giacomo Meschia developed for chronically catheterizing fetal blood vessels (23), he almost literally opened up a new field of study: the fetus under relatively physiologic conditions. Much of the work offered in this present volume represents an outgrowth of these initial studies.

During the New Haven years Dr. Barron accepted numerous academic responsibilities, including those of Assistant Dean of the School of Medicine from 1945 to 1948 and Acting Chairman of the Department of Physiology in 1964. In addition, he served as Managing Editor of the *Journal of Comparative Neurology* and by his participation in the Human Embryology and Development Study Section of the National Institutes of Health, profoundly influenced the development of fetal and neonatal physiology in America.

In 1969 Donald Barron moved to Florida, not to retire, but to assume the first endowed chair at the University of Florida, the J. Wayne Reitz Professorship of Reproductive Biology and Medicine, a chair which he still fills. Despite reaching the scriptural age two years ago, he does not fit the picture that one associates with threescore and ten. A model scientist, he continues to carry out productive research and stimulate students and fellows. In honor of his enormous contribution to reproductive physiology, in 1975 he was elected an honorary member of the Royal Society of Medicine.

The physiology of the fetus has been an almost unexplored wilderness. To a great extent, the work that has been done is a credit to Donald Barron, not only because he penetrated that wilderness himself, but because he cut a path for others to follow. In the obituary for Sir Joseph Barcroft that Dr. Barron published in *Science* (24), he wrote:

Differences in the bloods of fetuses and their mothers and their apparent functional advantages led Barcroft to study the respiratory function of the blood in the fetus. The readiness to follow an interest "beyond the visible horizons," so characteristic of the man, served to develop the whole field of fetal physiology, including the functions of the placenta, the circulatory change at birth, and the functional development of the nervous system.... All of these investigations were based upon simple ideas and questions and carried through with direct methods and simple techniques. They were not designed to gather details except insofar as they were essential for the development of method in technique; they were designed to reveal principles of function or integrations. Once these were at hand, Barcroft moved on to expand his chosen field of interest, leaving the intricacies to be explored by those with special knowledge.

To a large extent, these words apply as well to Donald Barron as to his mentor, Sir Joseph. Were Donald Barron a British subject, I have no doubt that we would be addressing him as Lord New Haven or at least Sir Donald. But since he is our man, we are honored to pay him tribute for his great contribution through both his own distinguished scientific studies and his inspiring leadership of fellow investigators in a field where new concepts are anxiously awaited by the world.

## REFERENCES

1. Gilman, J.C. and Barron, D.H. 1930. Effect of molds on temperature of stored grain. *Plant Physiol.* 5: 565-573.

2. Barron, D.H. 1931. Imbibition in disintegration. *Proc. Soc. Exp. Biol.* 26:1019-1020.

3. ----. 1932. Muscle response to foreign innervation. *Proc. Soc. Exp. Biol.* 29:184-186.

4. ----. 1933. Structural changes in anterior horn cells following central lesions. *Proc. Soc. Exp. Biol.* 30:1327-1329.

5. Barron, D. H. and Matthews, B. H. C. 1935. Conduction in the spinal cord. *J. Physiol.* 84:9P-11P. London.

6. ----. 1935. Intermittent conduction in the spinal cord. *J. Physiol.* 85:73-103. London.

7. Barcroft, J.; Barron, D. H.; and Matthews, B. H. C. 1936. The genesis of respiratory movements in the sheep. *J. Physiol.* 86:29P. London.

8. Barcroft, J.; Barron, D. H.; and Windle, W. F. 1936. Some observations on genesis of somatic movements in sheep embryos. *J. Physiol.* 87:73-78.

9. Barcroft, J. and Barron, D. H. 1937. Movements in midfoetal life in the sheep embryo. *J. Physiol.* 91:329-351.

10. Barclay, A. E.; Barcroft, J.; Barron, D. H.; and Franklin, K. J. 1938. X-ray studies of the closing of the ductus arteriosus. *Brit. J. Radiol.* 11:570-585.

11. Barclay, A. E.; Barcroft, J.; Barron, D. H.; and Franklin, K. J. 1939. A radiographic demonstration of the circulation through the heart in the adult and in the foetus, and the identification of the ductus arteriosus. *Brit. J. Radiol.* 12:505-518.

12. Barron, D. H. 1944. The changes in the fetal circulation at birth. *Physiol. Rev.* 24:277-295.

13. ----. 1946. The oxygen pressure gradient between the maternal and fetal blood in pregnant sheep. *Yale J. Biol. Med.* 19:23-27.

14. ----. 1952. Some aspects of the transfer of oxygen across the syndesmochorial placenta of the sheep. *Yale J. Biol. Med.* 24:169-190.

15. Meschia, G. and Barron, D. H. 1956. The effect of $CO_2$ and $O_2$ content of the blood on the freezing point of the plasma. *Quart. J. Exp. Physiol.* 41:180-194.

16. ----. 1956. Freezing point depression of arterial and venous plasmas in vivo. *Yale J. Biol. Med.* 29:54-59.

17. Meschia, G.; Wolkoff, A. S.; and Barron, D. H. 1958. Difference in electrical potential across the placenta of goats. *Proc. Nat. Acad. Sci.* 44:483-485.

18. Meschia, G.; Prystowsky, H.; Hellegers, A.; Huckabee, W.; Metcalfe, J.; and Barron, D. H. 1960. Observations on the oxygen supply to the fetal llama. *Quart. J. Exp. Physiol.* 45:284-291.

19. Prystowsky, H.; Hellegers, A.; Meschia, G.; Metcalfe, J.; Huckabee, W.; and Barron, D. H. 1960. The blood volume of fetuses carried by ewes at high altitudes. *Quart. J. Exp. Physiol.* 45:292-297.

20. Meschia, G.; Hellegers, A.; Prystowsky, H.; Huckabee, W.; Metcalfe, J.; and Barron, D. H. 1961. Oxygen dissociation curves of the bloods of adult and fetal sheep at high altitude. *Quart. J. Exp. Physiol.* 46:156-160.

21. Metcalfe, J.; Meschia, G.; Hellegers, A.; Prystowsky, H.; Huckabee, W.; and Barron, D. H. 1962. Observations on the placental exchange of the respiratory gases in pregnant ewes at high altitude. *Quart. J. Exp. Physiol.* 47:74-92.

22. ----. 1962. Observations on the growth rates and organ weights of fetal sheep at altitude and sea level. *Quart. J. Exp. Physiol.* 47:305-313.

23. Meschia, G.; Cotter, J. R.; Breathnach, C. S.; and Barron, D. H. 1965. The hemoglobin, oxygen, carbon dioxide and hydrogen ion concentrations in the umbilical bloods of sheep and goats as sampled via indwelling plastic catheters. *Quart. J. Exp. Physiol.* 50:185-195.

24. Barron, D. H. 1947. Sir Joseph Barcroft: 1872-1947. *Science* 106:160-161.

# To Donald Henry Barron
Giacomo Meschia

　　　　Among the investigators of Fetal Physiology, none
have contributed more to its formation in a systematic
body of knowledge than Donald H. Barron. His studies on
the physiology of intrauterine life began forty years ago
under the guidance of Joseph Barcroft. Their collaboration initiated because of a common interest in the development of the nervous system and brought Barron to focus
his attention on the environment in which the mammalian
nervous system develops. We can trace back to the days
of this collaboration the origin of a set of questions
about the physiology of the fetus which have helped greatly in giving to this field of investigation a direction
and a basic theme. How much oxygen does the normal fetus
require? What structural characteristics of the placenta
are relevant to its function as the respiratory organ of
the fetus? In what ways does the respiratory function of
fetal blood differ from that of maternal blood? What are
the functional correlates of histological differences

among placentae of different species? What are the
substrates of fetal metabolism? How is fetal growth integrated
with placental growth and the regulation of
uterine blood flow? Does the margin of safety of the
fetus with respect to the supply of oxygen and metabolic
substrates deteriorate in the course of gestation? How
do the uterine circulation, the placenta, and the fetus
react to acute or chronic hypoxia? After his appointment
to the Yale faculty in 1943, Barron proceeded to develop
the methodology and conceptual framework that was needed
in order to answer the above questions. It was a difficult
task, carried out with what might have seemed to most
people inadequate financial and technical assistance. In
the age of "big science," it was one man's effort in the
venerable tradition of "small science." His success has
been remarkable. Most of the questions originally posed
have been answered or are in the process of being answered
by application of the methods which he pioneered. Prominent
among these is the use of sheep with chronically indwelling
catheters in the uterine and fetal circulations,
which began in Barron's laboratory about seventeen years
ago. Such use has revolutionized the approach to the
study of intrauterine life and has contributed to making
fetal research one of the most active areas of investigation
in mammalian physiology.

# Historical Considerations

# 1

# A History of Fetal Respiration:
## From Harvey's Question (1651) to Zweifel's Answer (1876)

Donald H. Barron

Department of Obstetrics and Gynecology
University of Florida
School of Medicine
Gainesville, Florida

Of the landmarks in the development of our understanding of life *in utero*, none appears more important than the demonstration a hundred years ago--by Paul Zweifel, a young Swiss obstetrician, that the fetus respires via the placenta. His demonstration gave the final answer to a question raised by William Harvey in 1651 in his *Exercitationes de Generatione Animalium* (Fig. 1):

> How does it happen that the foetus continues in its mother's womb after the seventh month? Seeing that when expelled after this epoch, not only does it breathe, but without respiration cannot survive one little hour; whilst, as I have before stated, if it remains *in utero*, it lives in health and vigour more than two months longer without the aid of respiration at all.

Fig. 1. Title page of the London edition of Harvey's De Generatione, which appeared in 1651.

It established the placenta as the fetal "lungs." It paved the way to meaningful studies of the mechanisms involved in the initiation of pulmonary ventilation at birth and of the metabolism of the fetus *in utero*. It opened the modern era of fetal physiology.

When it was first raised, Harvey's Question, as it came to be known, aroused the interest of both the philosophers and the experimentalists, but in the absence of: (1) any clear concept of the phenomena embraced by the word "respiration" and (2) the knowledge that the air of the atmosphere is a mixture of gases, each with different chemical properties, no satisfactory answer would have been possible before 1774, when Priestley's discovery (15) of what he called "dephlogisticated air"--oxygen--demonstrated that it was essential to both combustion and respiration, and suggested both were "phlogistic" processes.

The inferences which Priestley drew from his experiments about the nature of "phlogistic" processes and about respiration proved to be in error. It is Lavoisier (6,7) to whom we are indebted for recognizing that oxygen plays an active role in respiration and combustion, not a passive one as Priestley assumed; that it was not a menstruum into which phlogiston was discharged, but a gas that combined chemically with carbon to form carbon dioxide.

But Priestley's demonstrations that "dephlogisticated" air was as necessary for the survival of a mouse as it was for the burning of a candle and that dark blood brightened to scarlet when exposed to "dephlogisticated" air--and only to that specific component of common air--certainly opened the way for the development of our understanding of respiratory processes. Scarcely less important was his recognition that the change in the color of blood in its passage through the ventilated lung was a fundamental aspect of respiration.

Soon the change of dark venous blood to scarlet arterial was accepted as evidence that it had been exposed to oxygen. Quite independently, it appears, Blumenbach (2) in Germany, and Hunter (5) in England inferred, from the change in the color of the blood en route through the allantoic circulation, that it served to aerate the blood of the chick *in ovo*.

As a result of these advances, the answer to Harvey's question appeared to be at hand. Writing in 1796, only twenty-two years after the discovery of oxygen, Erasmus Darwin (3), grandfather of Charles, put forward the view that the fetus *in utero* respired via the placenta (Fig. 2), stating his reasons as follows:

> First...the basis of atmospherical air, called oxygene, is received by the blood through the membranes of the lungs; and that by this addition the colour of the blood is changed from a dark to a light red. Secondly, that water possesses oxygene also as a part of its composition and contains air likewise in its pores; whence the blood of fish receives oxygene from the water, or from the air it contains, by means of their gills, in the same manner as the blood is oxygenated in the lungs of air-breathing animals; it changes its colour at the same time from a dark to a light red in the vessels of their gills, which constitute a pulmonary organ adapted to the medium in which they live. Thirdly, that the placenta consists of arteries carrying the blood to its extremities, and a vein bringing it back, resembling exactly in structure the lungs and gills above mentioned; and that the blood changes its colour from a dark to a light red in passing through these vessels.
>
> This analogy between the lungs and the gills of animals, and the placenta of the fetus, extends through a great variety of other circumstances; thus air-breathing creatures and fish can live but a few minutes without air or water or when they are confined in such air or water, as has been spoiled by their own respiration; the same thing happens to the fetus, which, as soon as the placenta is separated from the uterus, must either expand its lungs, and receive air, or die. Hence from the structure as well as the use of the placenta, it appears to be a respiratory organ, like the gills of the fish, by which the blood in the fetus becomes oxygenated.

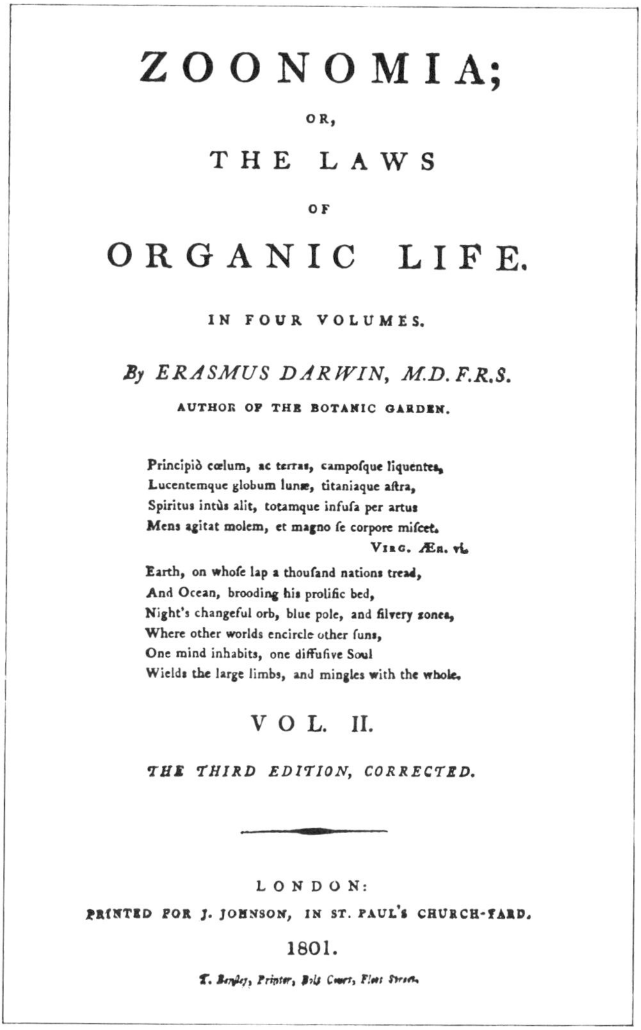

ZOONOMIA;

OR,

THE LAWS

OF

ORGANIC LIFE.

IN FOUR VOLUMES.

*By ERASMUS DARWIN, M.D. F.R.S.*

AUTHOR OF THE BOTANIC GARDEN.

Principiò cœlum, ac terras, campoſque liquentes,
Lucentemque globum lunæ, titaniaque aſtra,
Spiritus intùs alit, totamque infuſa per artus
Mens agitat molem, et magno ſe corpore miſcet.
VIRG. ÆN. vi.

Earth, on whoſe lap a thouſand nations tread,
And Ocean, brooding his prolific bed,
Night's changeful orb, blue pole, and ſilvery zones,
Where other worlds encircle other ſuns,
One mind inhabits, one diffuſive Soul
Wields the large limbs, and mingles with the whole.

VOL. II.

*THE THIRD EDITION, CORRECTED.*

LONDON:

PRINTED FOR J. JOHNSON, IN ST. PAUL'S CHURCH-YARD.

1801.

*T. Bensley, Printer, Bolt Court, Fleet Street.*

Fig. 2. Title page of volume 2 (3rd edition) of Zoonomia, Erasmus Darwin's most important scientific work. Published originally in 1794-1796, the work is best known for the statement of his views on generation.

Darwin's view was based upon several assumptions that appear to have been widely accepted at the time although their broader implications may not have been so fully recognized by those who did; among them were that oxygen enters the blood during its passage through the lungs and changes its color from dark venous to bright arterial and that oxygen is transported by the blood to the placenta and there enters the fetal circulation. At that time these were assumptions. All that had been established by experiment about respiration was the disappearance of oxygen from the inspired air and the concomitant appearance of a similar amount of carbon dioxide in the air expired and the simultaneous release of a quantity of heat which correlated fairly closely with the amount of carbon dioxide exhaled. Priestley had shown that dark blood brightens when exposed to oxygen and that the volume of oxygen diminished as the blood brightened, but no one had shown that the oxygen content of the blood increased as it did so.

The weakest, and in some respects the most important link in Darwin's chain of reasoning was the assumption that the fetal blood changed from dark venous to bright arterial in passing through the placenta--that it took up oxygen there. It was the most important, for, if correct, the validity of the other assumptions involved was virtually established.

On this question, opinion at that time appears to have been about equally divided; some claimed to have seen a color difference on the two sides of the umbilical circulation, others that they observed none. But of those who interested themselves in the question none appears to have considered the broader implications of their conclusion for an understanding of the phenomena of respiration. Their interest appears to have been centered on the fetus, not in the analysis of the nature of respiratory mechanisms.

Of the claims both pro and con, none appears to have been supported by relevant details of the circumstances in which the observations were made--the condition of either the mother or the fetus. For the most part they appear to have been based on incidental or chance observations-- made without particular attention to the existing circumstances and in the absence of any appreciation that the color of the umbilical venous blood would depend on them.

Representative of these claims is one by the Danish obstetrician, Paul Scheel (16), who states in a footnote in his treatise, *"Liquore Amnii Asperai Arteriae Foetuum Humanorum"*: "The blood carried to the fetus via the umbilical vein is a little brighter than that which returns to the placenta via the arteries, but no redder than the venous blood of an adult who enjoys perfect respiration." This he claimed "to have seen a hundred times."

One of the earliest, if not the first, to attempt to answer the question by a series of planned experiments was G. F. Schüz, a pupil of F. H. J. Autenrieth, Professor of Medicine at Tübingen. Autenrieth and Schüz (19), interested primarily in the relation of respiration to the genesis of animal heat, appear to have questioned the premise that "all living things respire--change dephlogisticated arterial into phlogisticated venous blood"--and to have regarded the evidence in support of the generalization as inconclusive. So they were moved to investigate an example that was in question: "Does the placenta really guarantee the fetus an alternative for the lungs?" If it did, they reasoned, the fetus should be a source of heat (Fig. 3).

From the observations on the color of the blood in the umbilical vessels of "near term" fetuses of rabbits and cats, delivered by cesarean section, Schüz concluded: "The blood which is sent to the placenta via the umbilical arteries is not changed, and it truly appears to be the same so far as its color goes, when it comes back from the placenta." But when exposed to air and when the fetus began to breathe: "...deeply and slowly in an irregular manner, the blood in the abdominal aorta took on a color very similar to that of the arterial blood of the mother." On the basis of these observations, he concluded: "the blood of the fetus is of the same dark color everywhere, not because it can't be transformed into arterial blood, but oxygen, the agent necessary for that transformation, is lacking."

To answer the question "Does the fetus produce heat?" Schüz compared the rate of cooling of fetal kittens and of fetal rabbits delivered by cesarean section with the umbilical circulation intact, with that of litter mates detached from the placenta and killed. The temperature was determined "by means of a thermometer thrust into the abdomen of the fetus." As the rate of cooling appeared to be the same in the two series, Schüz concluded the fetus produced no heat, that it was warmed by the mother until birth when it moved to a "higher order of life" in which it did consume oxygen and release heat.

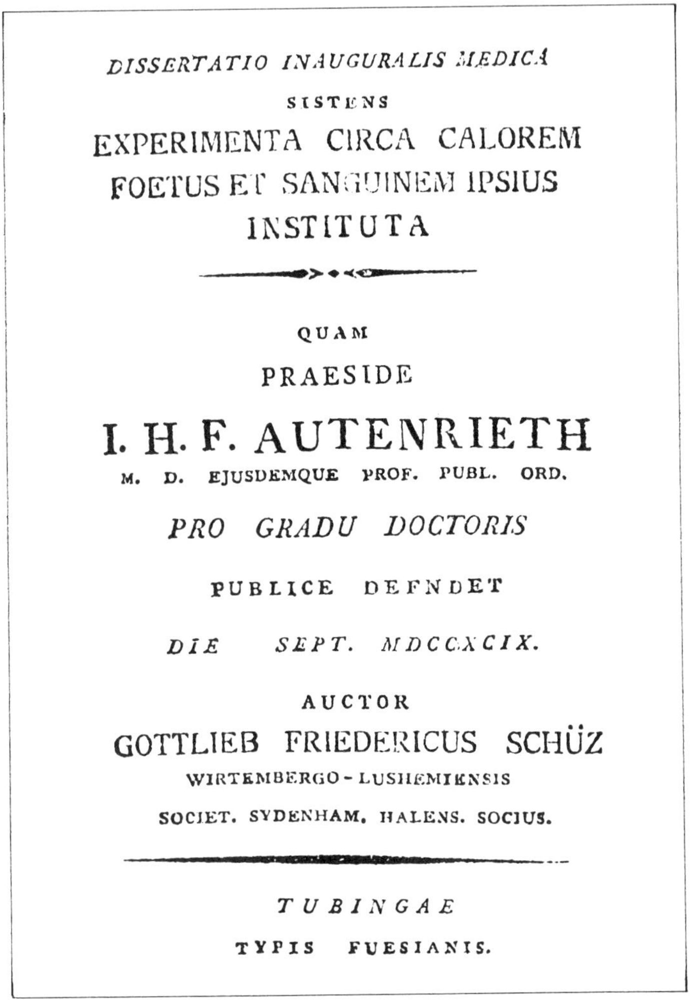

*DISSERTATIO INAUGURALIS MEDICA*

SISTENS

EXPERIMENTA CIRCA CALOREM
FOETUS ET SANGUINEM IPSIUS
INSTITUTA

QUAM

PRAESIDE

# I. H. F. AUTENRIETH

M. D. EJUSDEMQUE PROF. PUBL. ORD.

*PRO GRADU DOCTORIS*

PUBLICE DEFNDET

*DIE   SEPT.   MDCCXCIX.*

AUCTOR

GOTTLIEB FRIEDERICUS SCHÜZ

WIRTEMBERGO - LUSHEMIENSIS

SOCIET. SYDENHAM. HALENS. SOCIUS.

*TUBINGAE*

TYPIS FUESIANIS.

Fig. 3. The title page of G. F. Schuz' dissertation.

Although his experiments did not justify his conclusions, Schüz' inference that the fetus *in utero* produced no heat went uncontested by experimentalists for fifty years. But obstetricians who in their practice witnessed the regular transition at birth from a dependence for survival on the umbilical circulation to one on pulmonary ventilation were not convinced by his conclusion that the fetal blood was not oxygenated in the placenta.

A persistent difference of opinion was undoubtedly the stimulus that prompted the medical faculty of the University of Bonn to set as a prize task for their students in 1820: "The determination through observation and experiment on living animals if the fetus breathes whilst it remains *in utero* and enclosed in its membranes."

The prize, a gold medal, was won by a first-year student, Johannes Müller, who within a few years established himself as the leading figure in German physiology, numbering among his pupils such well-known names as Helmholtz, Du Bois-Reymond, Schwann, Virchow, and Henle (Fig. 4).

As others before him, Müller sought an answer to the question by comparing the color of the blood on the two sides of the umbilical circulation after delivery of the cord by cesarean section. His observations are set forth at length in a thesis entitled *"De Respiratione Foetus"* (11), but his comments in a review, *"Zür Physiologie des Fötus,"* that appeared the next year (12) are less speculative and more informative.

There Müller states that most embryologists denied the existence of a color difference; but he regarded any opinion, whether pro or con, as unreliable which was based --as most were--on observations of small animals:

> I have, through much experience, learned that such investigations can never contribute anything to deciding the question. If the vessel was opened as quickly as possible; even if the embryo was well developed, so little blood always flowed from the open vessel I could form no opinion. I need not mention I place no value whatsoever on observation on blood within vessels. In such a delicate investigation as the present, the blood from the different vessels must be compared simultaneously against a white background, something

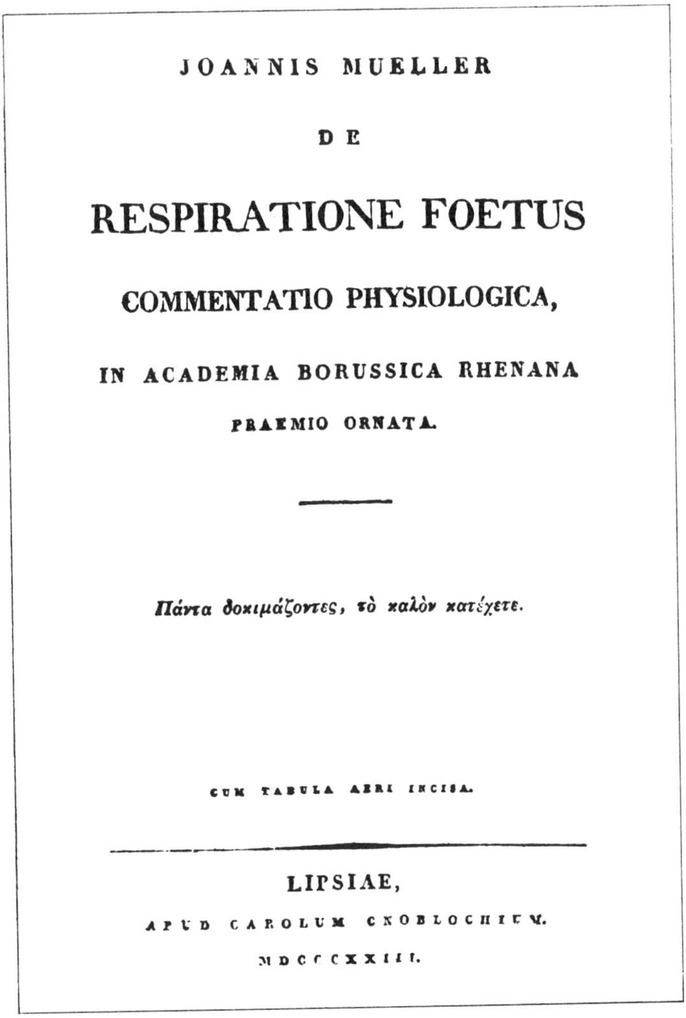

JOANNIS MUELLER

DE

# RESPIRATIONE FOETUS

COMMENTATIO PHYSIOLOGICA,

IN ACADEMIA BORUSSICA RHENANA

PRAEMIO ORNATA.

Πάντα δοκιμάζοντες, τὸ καλὸν κατέχετε.

CUM TABULA AERI INCISA.

LIPSIAE,
APUD CAROLUM CNOBLOCHIUM.
MDCCCXXIII.

Fig. 4. Title page of Johannes Muller's prize thesis. (Courtesy of the Historical Library of the Yale University School of Medicine)

of paper or ivory and finally held against
the light. How difficult this is with such
a limited quantity of blood is apparent to
everyone.

The only opinion Müller did not question was one based on observations of large animals. In the horse and pig, Joerg (1815) claimed to have seen a color difference on the two sides of the umbilical circulation; and for reasons not stated, Müller says that on the basis of that claim he was "fully prepared for a decisive experiment on a pregnant sheep which according to the shepherd would deliver within eight days."

After opening the abdomen and the uterus, Müller and his associates fixed their attention on the color of the umbilical vessels:

The larger umbilical vessels and their branches
had the same bluish color, but in the smaller
and smallest branches of the chorion the difference was unmistakable.... The arterial
(umbilical venous) blood in the vessels of the
chorion was not as bright as the arterial of
the mother; the venous blood of the latter,
darker than the venous (umbilical arterial) of
the chorion. I opened an umbilical vein at
once, the blood that flowed out was not as
bright as the arterial of the adult but much
brighter than its venous blood. I collected
two ounces and thirteen grains. Boiled oil
was poured over the collected blood to exclude
the air of the atmosphere.... When I opened
the artery the blood spurted out. The wit-
nesses of the experiment were as convinced as
I, that this blood was much darker than that
of the umbilical vein.... The blood of the
umbilical vein appeared to us much brighter
than that of the maternal jugular vein. The
latter, however, was about as dark as the
blood in the umbilical artery.

As evidence of the state of the preparation under observation, Müller offered "the mother and the fetus were both alive at the end of the experiment. The latter moved so actively that it had to be held by two persons."

Although this experiment appears to be the only one in which Müller saw a color difference in the umbilical vessels, he concluded in his thesis that the fetus did "respire *in utero*" and turned his attention to the estimation of its "respiratory needs." This he did by determining, as an index of such needs, the "survival time," or the length of time the fetus or neonate continued to make gasping movements without access to air. Müller found, as Le Gallois (8) had earlier, that so judged, the survival time of the fetus (rabbit) decreased as gestation advanced; that it fell sharply at birth and continued to fall in the neonatal period. These observations appear to be a part of the basis for his conclusion that the "respiratory needs" of the fetus were quite low, that they increased slowly as gestation advanced and rose significantly at birth. It is a view that was to find favor with many half a century later.

After the publication of his thesis, Müller's interests in respiration appear to have centered on its chemical aspects--the genesis of the carbon dioxide in the expired air and the respiratory function of the blood. Was carbon dioxide present in the blood when it entered the lungs or produced there by the action of the inspired air on it? Did blood serve simply for the transport of oxygen to, and carbon dioxide from, the tissue capillaries or was it directly involved in carbon dioxide production? Many were asking the same questions.

Müller failed, as did others, to demonstrate, with the techniques then in use, the presence of significant amounts of either oxygen or carbon dioxide in blood:

> If one gradually heats the blood of an adult to $200°F$ ($74.6°R$), that is, above the coagulation point of the proteins, in a vessel connected to a gas collecting tube, no air is given off from the blood--neither oxygen or carbon dioxide. The air passing over it is the unaltered air of the atmosphere contained in the vessel and collecting tube. H. Davy, in an early experiment of this kind, must have been mistaken for he believed he observed the development of air and many others have fallen under the same delusion. As I have heated blood from the umbilical vein of a near-term sheep in every way, the result could not be otherwise.

Müller was equally unsuccessful in changing the color of blood by subjecting it to reduced pressure:

> In human venous blood, quite fresh and still liquid, I could not observe the slightest brightening under the air pump, nor the slightest darkening of bright red blood. The brightening of blood through respiration cannot be due to the exhalation of carbon dioxide present in it; to the contrary the bright red color of arterial blood must be due either to the removal of a part of the carbon that combines, in respiration, with the atmospheric oxygen to form carbon dioxide or to the binding of a part of the oxygen with the blood corpuscles.

These failures must have raised doubts in Müller's mind about the assumption that oxygen was transported to the maternal side of the placenta for transfer there to the fetal blood, and questions about the chemical basis of the change in the color of the blood in the tissue capillaries.

In any event, the results of his own experiments with blood and the demonstration by his pupil, Theodore Schwann (20), that the hen's egg cannot be incubated successfully without access to oxygen appear to be the basis for the conclusion Müller expressed in his *Handbuch der Physiologie des Menschen* (13), i.e. that "mammalian fetus does not respire in the usual sense of the word" and to suggest that "the process is obviated by the connection of the fetus with the mother."

Müller went further and completely disavowed his conclusion of ten years earlier:

> The process in the placenta which replaces, or renders respiration unnecessary in the mammalian fetus must be of a very special nature, for there is in man and the mammals no observable difference in color between the blood in the umbilical artery and that in the umbilical vein.... Earlier, I... could never detect a difference in rabbit, guinea pig and kitten fetuses. And these small animals are quite as good, indeed, even better suited for observations than

> larger animals. In the same period, when
> as a student I was interested in every-
> thing, I thought I saw a difference at the
> vivisection of a "near term" sheep; others
> present believed they too saw one, and Joerg
> claims to have observed a difference in the
> chorion of the horse...but all my later ob-
> servations are in agreement with those I
> made on small animals.

But some of these later observations appear to have been made in quite different circumstances than the one on the "near term" ewe for Müller continued:

> Many ewes are slaughtered at Bonn; so during
> part of the winter, embryos of sheep and even
> of cows can be easily obtained, together with
> the uterus, and often whilst they are still
> warm. During the winter months such parts
> were brought to me regularly for anatomical
> purposes and I have never observed a distinct
> difference in the color of the blood of the
> umbilical vein and the arteries.

Müller's expectation that examination of the color of the blood in the umbilical vessels in these circumstances would throw light on the question of respiration by the fetus illustrates the confusion current at the time concerning the role of respiration in animal survival—the ends served by, and the sites of, the formation of carbon dioxide in the body.

The confusion is further illustrated by the attempts which were made to explain why the eggs of birds required a continuous supply of oxygen for their successful incuba-tion, whereas embryos of viviparous animals developed in the genital tract without apparent access to air. One of the first to offer an explanation for this seeming differ-ence between the embryos of the ovipara and the vivipara was the comparative anatomist and embryologist T. L. W. Bischoff (Fig. 5). His explanation is of special interest, for like Müller, he too appears to have believed at one stage in his career that the fetus *in utero* did respire. Writing in 1842, Bischoff acknowledged:

> Several years ago in Berlin, I too, investi-
> gated the question (the color of the blood in

Fig. 5. Theodore L.H. Bischoff (1807-1882) from Munchener Medizinische Wochenschrift 80 (1933). Reproduced from a copy in the picture archives of the Institute of the History of Medicine, Vienna, with permission of the Director, Professor Erna Lesky.

the umbilical vessels) by observations on human fetuses and I must say, that I as well as many others who did not know the purpose of the studies frequently recognized a distinct color difference, if I collected the blood from the sectioned artery and vein in different watch glasses as quickly as possible after the delivery of the child.

But when Müller, his teacher, once the leading proponent of a placental respiration, disavowed his earlier conclusion, Bischoff had doubts about his own observations; he could not be certain the bloods examined were obtained before the baby had taken a breath.

Bischoff's views on the nature of respiration in 1842 (1) reflected those of his close friend and colleague Justus von Liebig, who held that carbon dioxide formation and heat production in the animal body were basically excretory processes necessitated by the nature of the materials with which it was nourished. As plants used only the simplest elements in their growth and maintenance, they excreted nothing and produced no heat. But animals lived directly, or indirectly, on plants and so were obliged to excrete large amounts of carbon. This they did by oxidizing and exhaling it.

Respiration also served to eliminate the carbon and hydrogen in the breakdown products of organ function. They were thought to be removed from the blood by the liver, sequestered in the gall, then reabsorbed by the blood, carried to the lungs, and there combined with oxygen to form carbon dioxide and water.

As Bischoff accepted the conclusion of Autenrieth and Schüz that the fetus produced no heat, the absence of a respiration *in utero* was, he reasoned, in complete accord with Liebig's concept and lent it support:

> The fetus takes in no food by mouth with which to maintain respiration. It takes from the maternal blood only those nitrogen containing constituents required for the formation of its organs. Further, the increase in the fetal mass is so great relative to the functional activities of its organs that the breakdown products they form must be very limited in amount.... Certainly the nitrogenous materials removed from the blood by the Wolffian body and the kidney which are stored in the allantois and amniotic fluid must be small compared with the adult. The breakdown products containing carbon are removed by the blood from the liver and as this is the only avenue of their excretion, the liver is larger relatively, than it is in the adult in whom the greater part of the carbon is eliminated in the lungs as carbon dioxide. The excretory products (of the liver) accumulate in the gut as meconium; they are not reabsorbed and hence not available for burning. As there is nothing to burn, the fetus produces no heat.

Put simply, in Bischoff's view, the mammalian fetus lived as an organ of the mother; it took from her blood only those elements which had been acted upon by her respiration. Hence it had no need to respire.

Circumstances, Bischoff reasoned, were quite different for an embryo that depended upon a yolkstore laid down in the ovary for the materials utilized in its growth and development. Before the yolk could be utilized by the embryo it had to be converted into blood. The conversion required oxygen but differed, in Bischoff's view, from the transformation of orally ingested foodstuffs into blood by the adult in that no heat was released. Yolk had been partially transformed by maternal processes. The transformation into blood was completed by the embryo.

Oxygen consumption, carbon dioxide formation, and heat production Bischoff believed to be related processes; their presence or absence in a living system depended upon the nature of the materials utilized in its maintenance.

Order began to emerge from confusion concerning the nature of respiration and the origin of the carbon dioxide in the expired air during the late eighteen thirties, aided by the development of improved methods for the extraction of gases from blood by Gustav Magnus (10) and stimulated by Theodore Schwann's (21) enunciation of the "cell theory." At the time both men were at the University of Berlin, Magnus as professor of chemistry, Schwann as Müller's assistant.

Magnus developed his methods to determine, as he put it,

> Whether the carbon dioxide (in the expired air) was initially formed in the lungs by the oxidation of a part of the carbon in the venous blood to carbon dioxide by the oxygen in the air, or if the venous blood contained the carbon dioxide already formed when it arrived in the lungs, so it was eliminated by the presence of the air of the atmosphere; whether the inspired oxygen was immediately exhaled as carbon dioxide, or was carried by the blood throughout the body and used in oxidation not only in the lungs but in the body generally.

By exposing freshly drawn blood to reduced pressure over mercury, Magnus established the presence of significant amounts of carbon dioxide and of oxygen in it; and that the proportions of the two gases were different in arterial and in venous blood. That difference led him to conclude:

> As the ratio of carbon dioxide to oxygen is
> not the same in both blood types, the greater
> quantity of carbon dioxide in the venous can
> only develop during the circulation, either
> produced in the blood or absorbed by it. They
> make it very probable that the oxygen inspired
> in the lungs is absorbed by the blood and dis-
> tributed throughout the body so that it serves
> in the so called capillary vessels for oxida-
> tion and probably in carbon dioxide formation.

Magnus offered neither evidence nor opinion on whether the carbon dioxide gained by the blood in its circulation was produced in or absorbed by it.

That some isolated tissues--brain, lungs, skin, muscle--took up oxygen and gave off carbon dioxide in air and continued to release carbon dioxide in nitrogen had been demonstrated by Spallanzani a few years before his death in 1799, but his observations, first published posthumously by his friend Jean Senebier in 1803 and 1807 (23, 24), were either overlooked, ignored by, or unknown to those who centered their attention on the role of blood in respiration.

The enunciation of the "cell theory" quickened interest in cells not only as structural, but as functional and metabolic units (Fig. 6). Without citing evidence in support of his claim--he may have assumed none was needed-- Schwann (21) (Fig. 7) wrote:

> Gaseous, or loosely bound oxygen or carbon dio-
> xide, is certainly essential to the metabolic
> phenomena of the cell. Oxygen disappears and
> carbon dioxide is formed as a result (in animal
> cells), or carbon dioxide disappears and oxygen
> is released (in plant cells). The universality
> of respiration rests on these fundamental re-
> quirements of cells.

Those who believed as Schwann, that gaseous oxygen was essential for the metabolism of animal cells assumed the blood served only as a transport system for the respiratory gases, that they diffused in and out of the blood in the lungs and in the tissues in accordance with their concentration gradients. But there were a number of obstacles to the concept that respiration took place solely

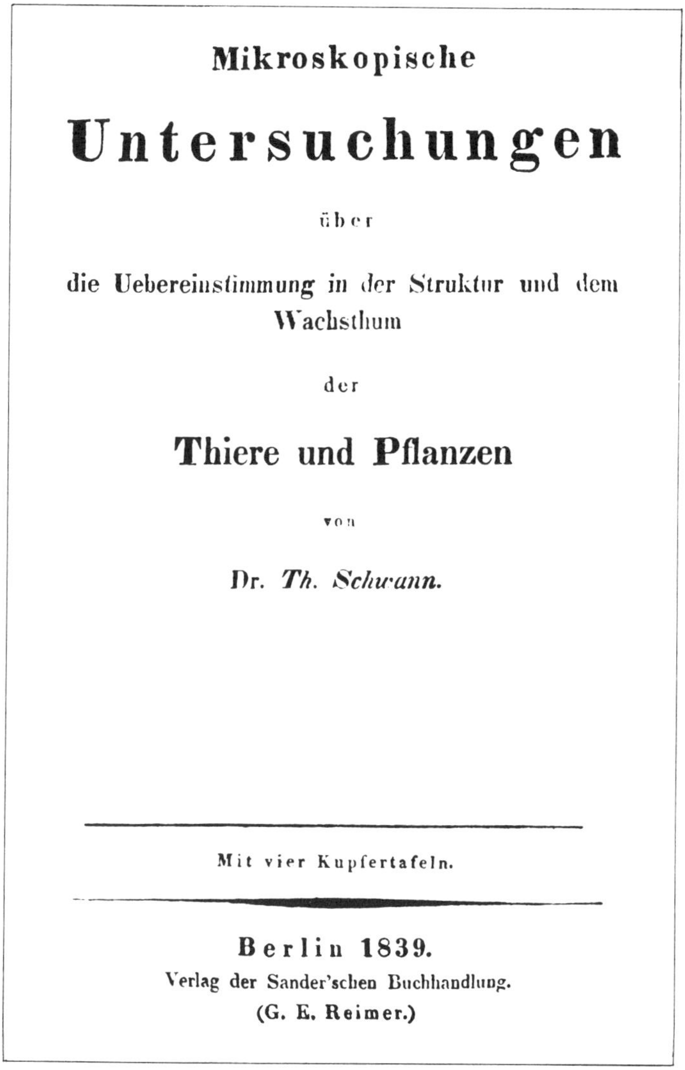

## Mikroskopische
# Untersuchungen

über

die Uebereinstimmung in der Struktur und dem Wachsthum

der

## Thiere und Pflanzen

von

Dr. *Th. Schwann.*

Mit vier Kupfertafeln.

**Berlin 1839.**
Verlag der Sander'schen Buchhandlung.
(G. E. Reimer.)

Fig. 6. Title page of Schwann's monograph, in which he set forth the "cell theory." Printed with the permission of Professor Wilhelm Auerswald, Secretary of the Gesellschaft der Ärtze in Wien.

Fig. 7. Theodor Schwann (1810-1882), from the collection of the picture archives of the Institute of the History of Medicine, Vienna, with permission of the Director, Professor Erna Lesky.

in the tissues. Among them were: (1) the observation that freshly drawn blood kept at room temperature over mercury slowly darkened; (2) the inability of investigators to demonstrate the presence of oxygen in lymph and tissue fluids; and (3) the slow rate at which oxygen is lost from blood exposed to an inert gas or reduced pressure, compared with its rapid disappearance from blood in its passage through tissue capillaries.

To bring these observations into accord with Magnus' demonstration, many--among them Carl Ludwig and Claude Bernard--adopted what came to known as the Chemical Theory of Respiration, i.e., that the by-products of tissue metabolism diffused from the tissues into the capillary blood and were burned there.

The view that the primary seat of oxidation was in the cells--the Absorption Theory of Respiration--was difficult to reconcile with Müller's conclusion that the umbilical blood did not brighten in passing through the placenta--that the fetal tissues did not respire. But the Chemical Theory offered a simple explanation for his statement that its relation to the maternal circulation obviated a fetal need to respire. Moreover, it offered an acceptable alternative for the argument advanced by Litzmann (1846) that "oxygen must be given up inside the placenta (to the fetal blood) for an extraordinary abundance of oxygenated blood is carried through the large uterine arteries into an equally wide capillary bed, whereas oxygen poor blood leaves the uterus." One need only assume that the by-products of fetal metabolism were carried by the umbilical circulation, to the placenta, there entered the maternal blood and oxidized.

Support for the view that respiration took place in cells grew gradually from a variety of sources--through demonstrations that oxygen was taken up by isolated "blood-free" frog muscle (9), by the cells of insects exposed directly to the atmosphere of the air by the tracheal system, by germinating seeds, etc. (25). But these demonstrations did not exclude the possibility that capillary blood was the principle site of respiration in vertebrates, or establish that oxygen diffused from capillary blood into tissues.

The first possibility was rendered unlikely by the failure of investigators to find in the blood of asphyxiated animals the increased amounts of the by-products of tissue metabolism which the theory required; the second was given indirect support by the observations of Schwartz (22) on the blood of infants asphyxiated during birth.

Schwartz, as many obstetricians, was an ardent proponent of the theory of tissue respiration. He reasoned:

> We know with certainty from the investigations of Du Bois Reymond and G. Liebig that a muscle requires free oxygen dissolved in its fluids for its activity. It follows, *a priori* that it cannot be otherwise in the fetus. The fundamental chemical and physical processes in the fetus and in the newborn, as their anatomical organization, must be similar.

His failure and that of others to observe a color difference in the blood on the two sides of the umbilical circulation, Schwartz explained: "We never see truly normal fetal blood. Every normal birth modifies the metabolism of the fetus and finally shuts off the placental exchange entirely even before the delivery from the birth canal is complete."

Opening the uterus to inspect the cord, Schwartz maintained, so compromised the circulation on the two sides of the placenta, and the exchange of the respiratory gases across it, that the color difference between the arterial and venous bloods in the umbilical circulation disappeared. Then, as the fetus continued to take oxygen from, and add carbon dioxide to the blood, it became progressively darker.

In support of his view, Schwartz pointed out that:

In those infants that are born normally and without any decrease in their vitality, the blood in the arteries and veins always has the same dark color--that of the venous blood of the adult. If, however, the fetus comes into the world in a condition of decreased vitality, apneic, or dying, the dark color and the dissolution of the blood obtained from the fetal end of the umbilical cord are both increased quite independently of the course of the labor and always in relation to its decreased vitality. If the fetus dies during the delivery, whatever the cause, one always finds the blood very dark and thin.... These facts are in complete parallel with the findings in the asphyxiation of adults. They have their significance in the observation, recently confirmed by Brucke, that the decrease in the free oxygen and the increase of carbon dioxide decreases the bright color and the coagulation properties of the blood.

Ten years later (14) the distinguished physiologist, Edward Pflüger, wrote of these observations: "This is the most important passage that is to be found in the entire literature concerning the existence of a placental respiration," and he quoted it verbatim for the benefit of his readers.

But he continued, "One assumes without hesitation that it is permissible to carry over to the unborn, facts which have been established for the newborn, forgetting however, that the latter lives in entirely different circumstances. In fact, it is easy to see that a need for a respiration by the fetus worthy of note scarcely exists." And, he pointed out "The only (fetal) muscle that works more, perhaps than that of the adult, is the heart. A limited amount of work must be done. But it has not been established that it involves the use of free oxygen.... One has no justification for making the genesis of muscle contraction dependent on the presence of free oxygen." A number of investigators had shown that an isolated muscle contracted in an oxygen-free atmosphere. Pflüger continued:

> The disappearance of oxygen from blood flowing through a contracting muscle, observed by Bernard and by Ludwig and Sczelkow, finds its simplest and most obvious explanation in this: that the acids which are formed (during the action of the muscle) diffuse into the blood and, as all acids, are at once firmly bound by oxygen. It is possible (in the case of the fetus) that as the oxidizable breakdown products are limited in amount they could cross the placenta and be burned there initially.

The suggestion of von Baerensprung (26) and others that fetal tissues are a source of heat and by inference the site of oxidative processes, Pflüger argued, proved nothing. Even if the fetus produced no heat it would be at the same temperature as the mother; hence a very limited heat production by the fetus would suffice to elevate its temperature. But he added: "This heat source is not of necessity linked to the presence of free oxygen, as any muscle can demonstrate."

Finally, Pflüger conceded that Schwartz might be justified in his contention that "we never see normal fetal blood" and should therefore draw no inferences about a fetal respiration from the color of the vessels in the cord, but he used bold type (see Fig. 8) to indicate the great emphasis he placed on the testimony of a, "large number of excellent observers, that when inspected whilst the fetus was in the amniotic fluid and still in the most complete placental union with the mother, the blood of the umbilical artery had the same color as that of the umbilical vein."

He insisted that even a very limited placental exchange would result in a significant color difference and concluded "that if the fetus did have a respiration, its oxygen consumption must be infinitesimally small compared with that of the adult." That conclusion was soon to be challenged by obstetricians, viz. B. S. Schultz and Adolph Gusserow (Fig. 9).

für spricht. Obenan für die Beurtheilung steht die von einer grossen Zahl ausgezeichneter Beobachter bezeugte Thatsache, dass bei der Betrachtung des Nabelstranges eines Fötus im Fruchtwasser, der noch in vollkommenster Placentar-Verbindung mit dem lebendigen mütterlichen Organismus steht, **das Blut der Nabelarterien dieselbe Farbe besitzt, wie das der Nabelvene.** So berichten Emmert, Autenrieth, Schütz, Haller, Hunter, Osiander, Bichabt, Scheel, Magendie, E. H. Weber, Schwartz und Andere (s. Schwartz a. a. O. p. 46.). Einige behaupten allerdings bei menschlichen Embryonen einen schwachen Unterschied bemerkt zu haben, indem das Blut der

Fig. 8. A portion of page 62 of Pflüger's 1868 paper, in which he used bold type to emphasize his belief that the bloods on the two sides of the umbilical circulation were of the same color.

During his training years, Gusserow had spent some time in Virchow's laboratory. While there, his views concerning the metabolism of the fetus and its relation to that of the mother were undoubtedly influenced by his association with Felix Hoppe-Seyler, one of the early proponents of cellular respiration, and regarded by many as the founder of Physiological Chemistry. Gusserow (4) agreed with Pflüger that the organs of the fetus, except for the heart and the kidneys, were functionally inactive or relatively so, but he insisted with Schultz (17,18) that the oxygen requirements of the fetus were equal, or nearly so, to those of the neonate because of its rapid rate of growth--an aspect which Pflüger and others appear to have overlooked in their reasoning.

Gusserow, "a good operator, an excellent teacher with an attractive outgoing personality," was in his early thirties when appointed the Director of the Frauenklinik at Zurich in 1867, after less than a year as Professor of Obstetrics at Utrecht. In Zurich he was soon heavily involved in the planning of a new Lying-in Hospital, and he was obliged to pursue his research interests by encouraging his assistants to put his views to test and arranging facilities for them to do so. Thus it was that G. Wurster (28) reexamined the question: "Is the fetus a source of heat and so by inference its tissues the site of oxidation?"

There was at the time indirect, but no direct evidence of a heat production by the fetus. Estimated by means of a thermometer introduced into the peritoneal cavity through a small incision in the body wall, von Baerensprung (26) found "the uterus and the pelvic cavity of the nonpregnant dog and rabbit to be a little less warm than the abdomen, whereas in the pregnant state the uterus was warmer than the pelvis, the pelvis warmer than the abdomen." The pregnant uterus was almost a degree warmer than the non-

Fig. 9. Adolph Gusserow (1816-1906), from Arch. Gynack., vol. 78.

pregnant. These results and his demonstration that the incubated hen's egg produces heat were the basis for von Baerensprung's conclusion: "The concept that the fetus produces a heat of its own and adds it to the heat imparted by the mother appears to be correct." His conclusion was given added support by Wurster's (28) observation that the anal temperature of the newborn infant, taken immediately after delivery--and when possible before the cord was tied--was higher, in the majority of his cases, than the vaginal temperature of the mother.

Before any further advance was made in testing his views, Gusserow was called to the chair of Obstetrics at the University of Strassburg, reorganized by Prussia after the annexation of Alsace and Lorraine at the end of the Franco-Prussian War. Other young men who had distinguished themselves as teachers and investigators were drawn from Austria and all parts of Germany to newly created chairs. Among them was Gusserow's friend Hoppe-Seyler; he was called to one in physiological chemistry, the first to be established in that discipline.

Gusserow's assistant at the time, Paul Zweifel (Fig. 10), a native of Zurich, went to Strassburg with him. As facilities for research in the Frauenklinik were limited, Gusserow arranged for Zweifel to work in the newly established Institute for Physiological Chemistry. There, with Hoppe-Seyler's guidance, he investigated the chemistry of meconium, the action of ergot, and the placental transfer of chloroform.

One of Hoppe-Seyler's chief interests was in the chemistry of haemoglobin--he gave the compound its name-- and he was especially interested to establish whether or not the spectral bands of oxyhaemoglobin could be observed in blood within the circulation. To answer this question he persuaded Zweifel to examine, spectroscopically, blood sequestered in the umbilical cord at the birth of a child.

In doing so Zweifel (30) discovered "the absorption bands of oxyhaemoglobin were very clearly visible and specifically in cases in which I clamped the cord at delivery before the child had taken a breath.... It is very interesting that these bands in fetal blood which is enclosed in the umbilical cord by double clamping, persist for a very long time.... It is absolutely clear, therefore, that the spectroscopic demonstration of oxygen in the blood of the fetus is as precise as one can wish." And he reasoned "with the complete isolation of the fetus from atmospheric air, the blood of the mother can be the only source of oxygen and thus a direct proof of the (placental) transfer of oxygen is provided. It must pass from the one blood to the other through the epithelium of the chorionic villi and it is of interest to learn how fast such a movement can take place between qualitatively different blood types." But the demonstration of the presence of oxygen in the fetal blood did not establish that it was being utilized by the fetal tissue. A positive answer to that question required the demonstration of a difference in the color of the blood on the two sides of the umbilical circulation.

Either through his own experience, or through the comments of Schwartz, Zweifel appears to have been persuaded that the failure of earlier investigators to observe the color difference was due to changes in the circulation

Fig. 10. Paul Zweifel (1848-1927), from Arch. Gynack., vol. 131.

on the two sides of the placenta set up by the contractions of the uterus when it was opened. In his study:

> Rabbits were used exclusively.... I believe the most important advantage to be that one is certain to avoid the placenta which is always located on the mesometrial side of the uterus. Then, to avoid the influence of atmospheric air and quite specifically the cooling of the uterus which always induces strong contractions and as a consequence premature respiratory movements of the fetus, I decided to carry out the experiment under water or a warmed neutral fluid.

The pregnant rabbit fastened on a frame was immersed after tracheotomy and laparotomy up to the neck in a saline bath held at 38°C. At his first experiment Zweifel obtained the answer he sought:

> After the animal had been quiet for some time a place on the uterus was quickly opened with two forceps by raising a fold and tearing through it. The stretched membranes of the fetus appeared in the opening with two chorionic vessels running next to each other. The membranes were opened, the fetus brought carefully out of the uterus and the vessels observed directly. Here two darkly colored umbilical arteries and one bright red colored vessel (the umbilical vein) are very clearly recognized. The pulsation of one umbilical artery is visible at the umbilicus....
>
> 10:20 A.M. Air closed off (maternal airway).
>
> 10:22 A.M. Very marked dyspnoea of mother. The color difference persists in the umbilical vessels....
>
> 10:23 A.M. The blood of the umbilical vein becomes darker, asphyxial cramps of the mother. Breathing movements of the fetus begin but soon cease.
>
> 10:23:25 A.M. The blood of the umbilical vein is no longer different in color from

> the blood of the umbilical artery. The (maternal) airway free again. The animal takes over its own respiration.
>
> 10:24 A.M. The blood of the umbilical vein begins to become a bright red. The little animal still remains quiet.
>
> 10:25:15 A.M. The first respiratory movement of the fetus begins after bright red blood had streamed in for many seconds and from then on they occurred in rapid succession. With the same fetus the same experiment was repeated.

With this experiment and others of its kind, Zweifel answered Harvey's question finally and unequivocally. But his study did more than that; it provided the first visual evidence in support of the view that the difference between the oxygen tension in the capillaries and in the tissues is adequate to account for the disappearance of the oxygen from blood whilst it is en route through the peripheral circulation.

Within a year Zweifel's observation was confirmed by Pflüger's colleague, Nathan Zuntz (29). Thereafter, interest focused on the rate of oxygen consumption by the fetus and the relative oxygen requirements of the fetus and the neonate--an aspect still under investigation.

## ACKNOWLEDGMENTS

The collection and organization of the material which forms the substance of this review was aided by a grant from the Association for the Aid of Crippled Children, now the Foundation for Child Development. The work was carried out, for the most part in the libraries of the Gesellschaft der Ärtze and the Institut für Geschichte der Medizin in Vienna. For the privilege of working in them I am most grateful to Professor Wilhelm Auerswald, Secretary of the Gesellschaft, and to Professor Erna Lesky, Director of the Institut.

## REFERENCES

1. Bischoff, T. L. W. 1842. *Entwicklungsgeschichte des Säugethiere und des Menschen*. Leipzig.

2. Blumenbach, J. F. 1786. *Institutiones Physiologicae*. Göttingen.

3. Darwin, E. 1796. *Zoonomia; or the Laws of Organic Life*. 4 vols. London.

4. Gusserow, A. 1872. Zur Lehre vom Stoffwechsel des Foetus. *Arch. Gynaek*. 3:24-270.

5. Hunter, J. 1794. *Treatise on blood, inflammation and gun-shot woulds to which is prefixed a short account of the author's life by Everard Home*. London.

6. Lavoisier, A. L. 1777. Expériences sur la respiration des animeaux et sur les changements qui arrivent à l'air en passant par leur poumon. *Oeuvres de Lavoisier*, vol. 2, pp. 318-333. Paris Imprimerie Imperiale.

8. Le Gallois, M. 1812. *Expériences sur le Principe de la Vie*. Paris.

9. Liebig, G. 1850. Ueber die Respiration der Muskeln. *Arch. Anat. Physiol*. pp. 393-416, Leipzig.

10. Magnus, G. 1837. Ueber die im Blute enthaltenen Gase, Sauerstoff, Stickstoff, und Kohlensäure. *Ann. Phys. Chem*. 40:583-606, Leipzig.

11. Müller, J. 1823. *De Respiratione Foetus*. Leipzig.

12. ----. 1824. Zur Physiologie des Fötus. *Anthrop*. 2:423-483, Leipzig.

13. ----. 1834. *Handbuch der Physiologie des Menschen*. Coblenz.

14. Pflüger, E. 1868. Ueber die Ursache der Athembewegungen, sowie der Dyspnöe und Apnöe. *Pflüg. Arch. ges. Physiol*. 1:61-106.

15. Priestley, J. 1790. *Experiments and Observations on Different Kinds of Air.* 3 vols. Birmingham.

16. Scheel, P. 1798. *Liquore Amnii Asperae Arteriae Foetuum Humanorum.* Copenhagen.

17. Schultz, B. S. 1868. Die Placentarrespiration der Fötus. Jena. *Z. Med. Naturw.* 4:541-552.

18. ----. 1871. *Der Scheintod Neugebornen.* Jena.

19. Schüz, G. F. 1799. *Experimenta Circa Colorem Foetus et Sanguinem Ipsius Instituta.* Tübingen.

20. Schwann, T. 1834. *De Necessitate Aeris Atmosphaerici ad Evolutionem Pulli in Ovo Incubito.* Berlin.

21. ----. 1839. *Mikroskopische Untersuchungen über die Uebereinstimmung in der Struktur und dem Wachstum der Thiere und Pflanzen.* Berlin.

22. Schwartz, H. 1858. *Die vorzeitigen Athembewegungen.* Leipzig.

23. Senebier, J. 1807. *Rapports de l'air avec les êtres organisés ou Traités de l'action du poumon et de la peau des animeaux sur l'air, comme de celle des plantes sur ce fluide. Tirés des Journaux d'observation et expériences de Lazare Spallanzani. Avec quelques Mémoires de l'Editeur sur ces Matières, par Jean Senebier,* 3 vols. Geneva.

24. Spallanzani, L. 1803. *Mémoires sur la respiration, par Lazare Spallanzani, traduit en Français, d'apres son manuscrit inédit par Jean Senebier.* Geneva.

25. Traube, M. 1861. Ueber die Beziehung der Respiration zur Muskelthätigkeit und die Bedeutung der Respiration überhaupt. *Virchows Arch.* 21:386-414.

26. von Baerensprung, F. 1851. Untersuchungen über die Temperaturverhältnisse des Foetus und des erwachsenen Menschen im gesunden und kranken Zustande. *Arch. Anat. Physiol. Wissen. Med.*, pp. 126-175.

27. Willis, R. 1847. *The Works of William Harvey, M.D.* Translated from the Latin with a life of the author. Sydenham Society, London.

28. Wurster, G. 1869. Ueber die Eigenwarme der Neugebornen. *Berl. klin. Wschr.* 37:393-395.

29. Zuntz, N. 1877. Ueber die Respiration des Säugethierfoetus. *Pflüg. Arch. Ges. Physiol.* 14:605-627.

30. Zweifel, P. 1876. Die Respiration des Fötus. *Arch. Gynaek.* 9:291-305.

# 2

## Many Slender Threads:
## An Essay on Progress in Perinatal Research

### S.R.M. Reynolds
Department of Anatomy
University of Illinois
College of Medicine
Chicago, Illinois

One of the ineluctable consequences of growth in any field of science is that subjects of inquiry once established tend to give birth to sub-subjects and that these subjects once established will in turn undergo further mitotic division (8).

Establishment of a subject of inquiry almost always begins with an attempt to find an answer to a relatively simple question; the knowledgeable investigator does not think of solving the gamut of problems of a scientific discipline. Change enters into the sequelae which an investigation may open up. Other workers pick up the threads that the original investigator may not have recognized. It is inevitable, therefore, that the fabric of a subject is woven of many slender threads, and in the ultimate pattern --when it is known--many of the first threads have been

lost to sight as others of greater strength and importance
supplant them.

It has been said that ninety per cent of the world's
accumulated knowledge is the product of work and thought
in the twentieth century. Although there are small roots
that may be seen in the history of the ancients and the
Middle Ages, it is true that the flowering process was slow.
The exceedingly rapid developments of the past fifty years
are related directly to the exploitation of a wide range
of technologies that were undreamed of before this time.
As I see it, many investigators tend to use some of the
modern methods and equipment as technologists rather than
as expert professional masters of original thought. This
results in the slender threads being pulled beyond their
limits of endurance. Many of us stand to be led astray by
complexities that are beyond our ken. I may speak of this
because I, myself, have been both perpetrator and victor of
such circumstances.

The scholar honored by this volume, Professor Donald
H. Barron, has reviewed what men understood of embryonic
and fetal respiratory demands from the time of William
Harvey in 1628 to that of Paul Zweifel, near the turn of
the last century. Harvey was a teacher, physiologist, and
clinician; Zweifel was a gynecologist whose intellect and
curiosity touched several broad areas of reproductive biology. In both instances, their studies, separated by nearly three centuries, were made long before the field of
study and the medical subspecialty of perinatology could
possibly have been imagined, but was fundamental to this
very recently recognized field of study and medical practice.

Historically, Western civilization developed from
many earlier ones. Some 2,500 years ago, Hippocratic physicians (3) wrote that

> when the child is big and the mother can no
> longer continue to provide him with enough
> nourishment, he becomes agitated, breaks
> through the membranes and incontinently passes
> into the external world free of any bonds....
> Those that have the least food for the fetus
> come quickest to birth and vice versa. And
> that is all that I have to say upon the sub-
> ject.

There is truth in this historic view, but like most such statements, it is incomplete. It was observed very early that a small amount of yolk is contained within its sac just before the birth of a chick; there is still some food available for the chick at the time of hatching. It is equally true, as Hammond and others in the present century showed, that the offspring outgrow their confining space--shell or uterus--so that regardless of the intricate and varied mechanisms related to birth, Nature provides its own solution among the several species. The alternative to birth at the required time is death *in utero*.

In more recent times, the late eighteenth century, the first definitive account of the placenta was given by the brothers William and John Hunter, physicians with inquiring minds of scientists. To them goes credit for establishing that the maternal and fetal circulations are separate entities (5). Over 1,500 years before the Hunters, Galen (2) described the ductus arteriosus and showed that normally it is present in the fetus, but not in the adult. In like manner, Vesalius, in a posthumous work (7), described the ductus venosus as a unique fetal structure. The Dutch physician-anatomist, Hoboken (4), noted what are now called the nodes of Hoboken in the umbilical arteries. As with so many other aspects of developmental physiology and anatomy, it is only within the past thirty years that the mechanism of their origin has been described. It remained for Barclay, Franklin, and Prichard in the 1950s to describe the specialized sphincter which serves as a vital valve in the ductus venosus, but it was not until later that its real function was defined. This perspective brings us to very recent times indeed concerning the fragile threads of ideas of human contact, of thinking, and of communication that imparted to the field of fetal physiology such vitality as exists today. Some will call it fetal physiology, some developmental biology, and still others will call it neonatal or perinatal physiology. All of these terms refer to the same living and growing phenomena. The use of a term shows the perspective of the user.

It will not be amiss to anticipate, at this point, what is historically significant and will be described below. The honoree of this volume, Professor Donald H. Barron, began his scientific studies in zoology at Yale University after sound and broad preparation at Carlton College. At Yale, he came under the tutelage of Professor J. S. Nicholas, where he wrote his thesis on the role of nerves in the development of skeletal musculature. His interests in comparative neurology were lasting. Later he became the editor of *The Journal of Comparative Neurology*, long after he had contributed substantially to the field of fetal physiology and trained many present-day leaders in the field of developmental studies.

By 1932, Dr. Barron was a Fellow of the National Research Council in Berne and Cambridge. Although he published three papers from his sojourn in Berne, it was his going to Cambridge that opened up the flood gates of a lifetime dedication to fetal physiology and to training others in the field. Brian Matthews was the first neurophysiologist to whom Barron attached himself there, but it was only a year later when Dr. Barron associated himself with Sir Joseph Barcroft. This collaboration led to a publication on the initiation of respiration at birth. During his stay in England almost all his subsequent publications were with Barcroft, and five years later, upon his return to the United States, there appeared from time to time other papers with British collaborators. The last was perhaps the most significant. This was an assemblage of the data and views of Barcroft that resulted, in 1946 in England, and America, in the final publication under Barcroft's name, but with credit to Barron, of *Researches on Prenatal Life* (1). This is still a current volume since it is in process of republication at the present time. This monograph summarizes most of the work of Barcroft, Barron, Barclay, and many others; it contains a breadth of information which is highly valuable, although it would be wrong to assume that many ideas have not changed, or been added to, or that the original text is without some errors. But what publication will stand without change and alteration for more than thirty years in these days of rapidly moving research?

What has been said above is totally inadequate testimony to the role that Barcroft played in only one of his remarkably diverse interests in science. It is sufficient, however, to suggest that Barron was one of many notable beneficiaries of Barcroft's genius and academic and scientific power, as well as a testimonial to Barcroft of a very long and productive career as a physiologist, teacher, and scientific statesman. It is amusing to recall that when the editors of the *Journal of Physiology* adopted many years ago the policy of requiring that the listing of authors on papers be in alphabetical order, without consideration of rank or seniority, Barcroft is reputed to have made sure that none of his future collaborators possessed names starting with A.

From 1932, when Barcroft published his first paper on fetal respiration, 76 additional papers appeared before his death in which Barron was a co-author. In all, Sir Joseph published 331 titles, of which 213 were published on a range of subjects in physiology. Although his earlier interests were wide, most bore heavily on the respiratory functions of the blood, respiratory mechanisms in animals and man, and the qualities and properties of hemoglobin. The major part of Barcroft's studies after 1932 dealt with the many characteristics of fetal physiology. Considering what has just been said, it is clearly apparent that after substantial contributions to his earlier fields of interest, the slender thread that led Barcroft into the new field of fetal physiology was a chancy and slender one at best. His basic interest in respiratory physiology carried over into the new field, but it was the work of another investigator that led him into so rich a new field as fetal physiology.

Modern fetal physiology began with the work of a relatively unknown physiologist, who, in turn, was influenced by chance by a senior obstetrician of an inquiring mind. This junior worker was Arthur St. George McCarthy Huggett, who was a Beit Memorial Fellow in Physiology at St. Thomas' Hospital Medical School. The obstetrician was Dr. John Fairbairn. Huggett published with Sir John Mellanby, his chief at St. Thomas, a half-dozen papers which led him directly to an analysis of the sensitivity of the respiratory center of fetal goats and lambs. This was in 1926 and from this time on Huggett seldom deviated from studying various aspects of fetal and placental physiology in his nearly 100 publications before his death in 1968. Huggett adopted a nineteenth century method of working on large fetal mammals still attached to the placenta. Mother and fetus were immersed in a saline bath as Preyer (6) had done forty years before, but the method had remained unexploited. Huggett soon became Professor of Physiology at a small and undeveloped Department of Physiology at St. Mary's Hospital Medical School of the University of London. It is interesting to read the Obituary Notice of Professor Huggett in the *Proceedings of the Royal Society*, of which he was a Fellow, that in a joint publication with R. H. Elliot, and F. G.

Hall, an American physiologist then working in Barcroft's laboratory in Cambridge, and Professor Huggett in 1933, then at Leeds, acknowledged Barcroft's "personal aid" in the determination of the blood volume of fetal lambs. Clearly, chance and circumstance among several scholars having widely different ideas and experiences focused attention on an object uniquely suited to physiological investigation and which was relatively unexplored. The slender threads of chance were stretched thin with respect to fetal studies in the late 1920s and 30s.

In 1953, few investigators worked on fetal physiology or developmental biology. Most were in the field of zoology; almost none worked in the medical sciences. There was no vehicle for communications between them except through established zoological, anatomical, and physiological journals. A small number knew each other personally and visited one another or corresponded by mail. A few hardy souls undertook to publish in mimeographed form a *Fetal News Letter* which was circulated on an informal basis among friends having related interests. After World War II, there was forthcoming a burst of scientific effort that was generated by the National Institutes of Health through a wide variety of grants, fellowships, and traineeships. The funds followed the interests of the applicants; but those in fetal physiology and development were few. Such applications as did arrive for consideration were reviewed by a committee in the Division of General Medical Sciences. It was not until the mid-1960s that the number and quality of requests in developmental biology reached the point that a new National Institute for Child Health and Human Development was organized. The development of priorities and programs to be stressed has required years to be defined, so that in my opinion this Institute has made less impact than the longer acting force of the earlier committees of the National Institutes of Health. Throughout many of these years since World War II, money came from the governments, especially in the United States and the United Kingdom. Students were free to move into any field of research with which they had contact or knowledge. It is of interest, therefore, to look at the manner by which impetus was given to fetal studies by one man, Huggett. From that, so much has followed that we cannot narrate the whole nor even the sequence of events that transpired to make fetal physiology was it is today.

To put the matter in perspective, I met "Hugo" (Huggett) in 1950 at a meeting of the Physiological Society in London. At dinner, I mentioned to him that, in the light of the brilliant studies of Barcroft and Barron, and of Barclay, Franklin, and Prichard (to which Barcroft contributed so much) on the physiology of fetal lambs, and the circulatory changes at birth, and the merited attention that these recent studies were attracting, Hugo's initial role in doing studies on large fetal mammals in 1926 was overlooked. Hugo's studies failed to receive the credit which they deserved. As so often happens, it is the latest study that gains attention, and not the basic one on which later studies rest. Hugo was pleased by my remark. He called to his wife, at my other side, and said: "Esther, Esther, did you year that?" His wife, the late Professor Esther Killick, was Professor of Physiology at the Royal Free Hospital for Women. Obviously, Hugo saw the field of fetology come to life with many workers, and his own original and continuing contributions were, I knew and he felt, overlooked or overwhelmed by the more spectacular achievements of cineangiography and cardiovascular physiology, in contrast to his then-current studies of fructose and glucose passage and production in the placenta.

The next time I saw Hugo was when he came with Professor Amoroso to my hotel room in London just before New Year's Eve, 1950. I was ill with a respiratory infection. They brought me the sure cure, Amo's choice, in the form of a supremely fine port wine from a vintner who, I was told, was licensed in the old City of London by Royal Charter, operating on an hereditary basis going back for many centuries. The circumstances, all together, were such that I recovered well enough to get back to the Nuffield Institute where, with Drs. Prichard and Ardran and joined by Dr. Geoffrey Dawes, the new young director of Nuffield, recognized for his special talents and his initiation into fetal physiology, we engaged in a season of work on fetal lambs.

In the spring, Hugo invited me to visit him at St. Mary's, which was with pleasure and to my profit. He said that before he retired he would have one more sabbatical leave; his desire was to come to the Department of Embryology of the Carnegie Institution of Washington in Baltimore where I was a staff member at the time. He wanted to study the transfer of radioactive fructose and glucose across the monkey placenta between fetus and mother. He wanted to see to what extent, if at all, he could extrapolate his conclusions from studies in goats and sheep (syndesmochorial placentas) to the primate with hemochorial placenta. Hugo came to Baltimore with his wife and two daughters for the year 1953-54. That year, Dr. William M. Paul, just through his residency training in Toronto (where he later became Professor and Head of Obstetrics and Gynecology) and Dr. Vittorio Danesino, from Naples and now Professor of Obstetrics and Gynecology and the University of Sassari, were with me. As things turned out, for a time all of our work revolved around fructose and glucose in monkeys.

That year ten pregnant monkeys were assigned to us from about double that number available in the laboratory. Huggett was convinced that the operative procedures could be carried out as he had done in the sheep and goat. This was laparotomy of the mother in a warm saline bath and delivery of the fetus with ready access to the fetal and maternal blood vessels. Dr. Paul and I were convinced, one from surgical experience in women, the other from operations on pregnant monkeys, that the uterus simplex would not permit this type of experimentation. Arguments continued while skills in the appropriate carbohydrate determinations were acquired in our laboratory under the tutelage of Hugo. We enlisted the assistance of Dr. Francis P. Chinard, now head of the Department of Medicine at the New Jersey Medical College, and Dr. Theodore Enns, now at the Scripps Institute, La Jolla.

The time came for the first experiment. This was determined by the condition of the pregnant monkeys and our election to do the experiments in the last trimester of pregnancy. Hugo prevailed, and the first fetus was lost immediately upon incision of the uterus and prompt delivery of the placenta.

Our next experiment was scheduled within a week and a wholly new approach simply had to be developed. Without a solution, the value of Hugo's sabbatical year would be lost and the time of a good many of us wasted. I cannot say when or how the idea of locating the interplacental blood vessels for catheterization came to me. It would be nice to say it all came clear in a flash when I awoke one night from a deep sleep. The late Professor John Fulton at Yale had a Dictaphone by his bedside to capture such inspirations as they came to him. For myself, I can only say that I went home one evening without the solution; I returned to the laboratory the next morning and told Dr. Paul of the idea.

We talked of the possibilities. We first tried to visualize the interplacental blood vessels in the second monkey using a child's cystoscope, inserted into the amniotic cavity through a point incision with a strong purse-string suture around it. William Paul's surgical skills were indispensable. We failed, however, to identify any vessel satisfactorily, owing to the small size of the field visualized. We worked in a darkened room. Although we failed in that procedure, we noted that the uterus, illuminated from inside, looked something like a lighted jack-o'-lantern in the dark. Dr. Paul closed the stab incision and we used two pen-type flashlights against the uterus and saw the pulsating interplacental arteries, as well as their matching veins. A convenient spot was identified, and marked by an Allis clamp over the vessel to be catheterized on the visceral peritoneum, and Dr. Paul, with great skill, completed the operation. The rest is history. For the first time, to my knowledge, the fetal circulation was reached and made available *in utero* without loss of amniotic fluid or opening the amniotic sac.

Hugo and entourage then performed his first fructose-carbohydrate experiment in a primate. When we finished, we connected the catheter to a Statham strain gauge and recorded for the first time fetal arterial blood pressures with the fetus *in utero*. Drs. George Corner, Elizabeth Ramsey, and everyone we could find in the Old Hunterian Building, in which the Carnegie Laboratory was then situated at Johns Hopkins, were asked to come behold "this first."

Within a few weeks after publication, Dr. A. A. Plentl wrote to me saying that he and his group, then including Drs. Donald Hutchinson and Emmanuel Friedman, used the procedure without difficulty in their studies which previously had lacked access to the fetal fluid compartment. I knew then that the method would find many applications. Hugo did nine successful experiments on monkeys that year.

I have often thought of the nature of the circumstances that enabled me to think of the interplacental vessels in the monkey when the need arose. I had not, myself, then worked on monkeys except in isolated studies. However, as a post-doctoral fellow with Carl Hartman in 1931-32, one simply had to know, to see, and talk about what he could learn, and observe in monkeys, or any of the variety of other things going on at Carnegie at that busy time. Then, while I was a staff member from 1940 to 1956, the same spill-over of awareness and learning came from Drs. Elizabeth Ramsey, George Streeter, George Corner, George Bartelmez, and other superb technicians and fellows. It was surely at this time, in laboratories and at the renowned Carnegie lunch table, where the food was thin and the talk was thick, that one learned all kinds of useful or interesting things. I am sure that the environment, chance though it was for conveying needed information for a specific problem, generated a reservoir of unrecognized "wisdom" which could be tapped for use when necessary.

A final work, for fetologists, may be said about their debt to Professor Huggett. In 1953, many workers around the world studied fetal physiology. Except for isolated instances, there were not many such investigators who knew each other personally, as mentioned above, except through reading the literature, exchange of reprints, or rare letter exchanges existing among fetologists as a means of communication. In the fall of 1953, Hugo said one morning, "Sam, we should have a week-long conference of European and American physiologists who work on the fetus. We should have twenty Europeans and twenty Americans." I suggested that he make up a list of which foreign workers should participate in such a meeting. It was assumed that American workers could find sources of travel money, an assumption that proved only too true. The Division of General Medical Sciences, NIH, and the Association for Crippled Children each gave $10,000 to the Cold Spring Harbor Biological Conference for its Symposium that year. Twenty-one scientists from Great Britain, Scotland, Ireland, France, Sweden, Belgium, and Finland came, expenses paid. One hundred sixteen North Americans came; it was hoped to limit the Conference to sixty at most, but come they did. This Conference was the start of much international cooperative research, especially between fetologists in Sweden, Great Britain, France, and the United States. The activating spirit in this, virtually unrecognized, was Huggett. It was not until the year of the death of the man who perhaps was among the first of what is now a world of fetologists to use large mammals that I wrote of his contribution. We see from this the complexity and the importance of the of the pattern of intellectual and technological interdependence that is woven, even unconsciously, of many slender threads; they comprise the ways of the warp and woof of science. Together they make the fabric in which the pattern of knowledge of the fetus is discernible.

At the time Hugo generated the need for *in vivo* studies of primate fetuses, the ground was ripe for stimulating general interest on the part of investigators in many fields of study. In addition to the names of individuals mentioned above, scores of investigators were at work in the field of developmental biology in areas that impinged on fetal physiology. This is shown by the sequelae to the Cold Spring Harbor Conference. The Association for Aid to Crippled Children organized a three-day meeting in New York in which "bench scientists" studying fetal physiology in this country and Canada, as well as the United Kingdom and Sweden, maternal and child public health agencies, sociologists, and scholars from related clinical fields were brought together in order to generate communication which hopefully would contribute to understanding. The correlated presentations and discussions were published by the Association and widely distributed here and abroad.

Within a year the Josiah Macy, Jr., Foundation sponsored the first of two conferences, the second commencing three years later. Each group met for two days each year for five years at the Nassau Inn, Princeton, New Jersey. The personnel for both study groups consisted of about eighteen individuals; to these were added each year four to six other investigators from here and abroad. The first conference was on gestation under the direction of the late George Wislocki, the other under Clement A. Smith, Professor of Anatomy at Harvard and Professor of Pediatrics, respectively. The informal and informative proceedings were published so that a far larger audience might be aware of the breadth and scope of fetal studies, which now were no longer limited or parochial. The range and quality of thinking in this area was recognized to be far broader in scope than had been envisaged at Cold Spring Harbor. Both Dr. Wislocki, and Dr. Smith had been at Cold Spring Harbor and saw the value of communication, but thought the time had come for sustained communications against a larger backdrop. Time proved them to be correct as the numbers of investigators studying fetuses and support for them increased rapidly.

In the early 1960s there were so many investigators attracted to fetal studies that two relatively significant events were little noticed, although they made an impact on the quantity and the quality of the research in subsequent years. One was a meeting held at the Commodore Hotel in New York City under the aegis of the American Society of Electrical Engineers. Within this organization were some individuals who were interested in biological applications of the newer and still developing electronic techniques; they were, in a sense, influencing what was to become a new medical specialty, that of biomedical engineering. The second meeting held at Stowe, Vermont, was organized by Dr. Jerold Lucey, Professor of Pediatrics, and included more than thirty individuals who were working then on the physiology of the fetal primate. The purpose was to discuss, compare, and criticize the many procedures and experimental designs then being studied on a face-to-face basis.

The investigators invited to form a panel at the New York meetings (as I recall, since no record was kept) were Drs. Barron, Caldeyro-Barcia of Montevideo, Dawes of Oxford, Hon, Hellman, and myself. Several of the group took the positive view that modern technology made it possible to monitor and evaluate the clinical status of human fetuses *in utero*; the rest held that more could and should be learned by experimental methods at hand or to be developed about the kind of changing cardiovascular physiology of the fetus in comparison with that of adults. It was sensed by me and some others, I feel sure, that this meeting was but a prelude to commercial exploitation of applied electronics through developing cardiotachometers and uterine activity recorders which would have wide appeal to obstetricians and pediatricians. This is exactly what happened.

Over the last dozen to fifteen years there have been numerous well-organized conferences relating to fetal physiology in North and South American and in Europe. Each has had a focus on some specialized aspect of fetal, placental, uterine, or other aspect of fetology. It is no news today that the American College of Obstetricians and Gynecologists has recognized for several years a subspecialty within their organization, that of Maternal-Fetal Medicine. A Perinatal Research Society exists, and journals in this field have come into existence. These and other activities, agencies, and organizations now offer a wide outlet for the student of fetology. This was not true two decades ago, and looking back one may observe what great strides have been made since Hugo suggested that it was time that students in this field should come together for a week in order to know one another, to exchange ideas, and realize that it was time to get a move on.

This work testifies to what has been accomplished by only some of the students of the fetal circulation in very recent years. It proves the point made by Dr. Weech in the quotation at the beginning of this chapter: Knowledge grows upon knowledge, and specialization begets specialization, and it is on the spinning wheels standing in the pools of ignorance where these many slender threads are spun.

## REFERENCES

1. Barcroft, Jr. 1946. *Researches on pre-natal life.* Oxford: Blackwell.

2. Galen. 1525. *Opera omnia.* Ediderunt Andreas Asulanus et J. B. Opizo. 5 vols. *In aedibus Aldi, et Andreae Asulani soceri.* Venetiis.

3. Hippocrates. 1849. *The genuine works of Hippocrates.* Translated from the Greek, with a preliminary discourse and annotations by Francis Adams. 2 vols. London: Sydenham Society.

4. Hoboken, N. 1669. *Anatomia secundinae humanae,...*Ad Nicolaum Tulpium adscriptum. Cum annexo s. spicilegio epistolarum, rem potissimum genertoriam referentium. Trajecti ad Rhenum, J. Ribbius.

5. Hunter, J. 1786. *Observations on certain parts of the animal oeconomy.* London.

6. Preyer, T. W. 1885. *Spezielle Physiologie des Embryo.* Leipzig.

7. Vesalius, A. 1564. *Examen observationum Falloppii.* Venetiis.

8. Weech, A. A. 1970. *Physiology of the perinatal period,* vol. 1. Uwe Stuve, New York: Appleton-Century-Crofts.

# Developmental Aspects of the Cardiovascular System

# 3

## Ontogenesis of the Autonomic Control of Cardiovascular Functions in the Sheep

N.S. Assali, C.R. Brinkman III, R. Woods, Jr.,
A. Dandavino, and B. Nuwayhid
Department of Obstetrics and Gynecology
UCLA School of Medicine
Los Angeles, California

The normal growth of the fetus during intrauterine life, its ability to withstand the stress of labor and delivery, and its healthy development during the neonatal period depend to a great extent upon the integrity of its circulatory system.

During the fetal state, this system has the major responsibility of transporting the essential elements across the placenta to the fetal tissues and of returning to the placenta the fetal catabolic products to be eliminated. After birth, the circulatory system plays a major role in adjusting the newborn to the external environment and in securing its normal development.

Morphologically and physiologically, the fetal circulation differs considerably from the neonatal circulation (1,2,3). The salient features unique to the fetus consist of the following:

1. The umbilico-placental vascular bed, which represents a vast network of blood vessels without neural supply; this network absorbs a major fraction of the cardiac output and assists in maintaining a relatively low fetal systemic vascular resistance and pressure.

2. The pulmonary vascular bed, with its high resistance and pressure and a very reduced blood flow.

3. The relatively fast heart rate and a high cardiac output per unit of weight.

4. The various vascular shunts, which play important hemodynamic roles.

5. The regional vascular beds, which have high vascular resistance and reduced blood flows.

After birth, the expansion of the lungs and the initiation of breathing bring about a profound fall in the pulmonary vascular resistance and pressure and a marked increase in blood flow. Elimination of the umbilico-placental bed results in a rise in systemic resistance and pressure, and redistribution of cardiac output. The vascular shunts close during this transitional period and, shortly thereafter, the cardiovascular system becomes fully adjusted to the external environment.

Although many of the physical and physiological factors that play important roles in fetal and neonatal hemodynamics have been extensively studied, the role of the autonomic nervous system has not been adequately assessed.

This chapter deals with data on the sympathetic and parasympathetic control of the fetal and neonatal cardiovascular functions. The autonomic control exerted in the nonpregnant adult sheep was used as a standard reference to which the fetal and neonatal results were compared.

## METHODS AND MATERIALS

For the purpose of methodology and investigative approach, this study was divided into the following two subprojects:

## Fetal Studies

The studies of the autonomic control of the cardiovascular functions in the fetus were carried out by *in vivo* and *in vitro* methods. We shall confine our presentation to the *in vivo* studies, since Dr. Su will present the *in vitro* data; these consisted of observations on the development of the sympathetic and parasympathetic tones and reflexes controlling the resting heart rate, vascular pressures, and blood flows in the great vessels of fetal lambs weighing 300-5800 g (65-165 days' gestation). We also investigated the patterns of reactivity of the cardiovascular functions to exogenous administration of autonomic agonists in the same group of animals. Although the methodology has been described in detail elsewhere (4,5,6), a brief account is given here.

Time-dated pregnant ewes were obtained from our supplier. Confirmation of the gestational length and fetal size was obtained radiologically. We arbitrarily classified the maturity of the fetal lambs on the basis of their weight as follows: Immature--300-1500 g; premature--1600-2600 g; and mature--those weighing more than 2700 g.

An attempt was made to carry out most of the studies in chronically instrumented fetuses in the unanesthetized condition. When, however, an experimental protocol proved to be unsuitable or difficult to carry out in chronically instrumented animals, we did not hesitate to perform acute experiments with the mother under spinal anesthesia. Previous studies from our laboratories have shown that the fetal circulatory functions in acute experiments carried out under spinal anesthesia were not significantly different from those observed in the chronically instrumented fetus (3).

Blood flow rates in the great vessels, including the ductus arteriosus, were measured with electromagnetic flowmeters; pulmonary and systemic arterial pressures were monitored with implanted catheters connected to matched Statham strain gauges; and heart rate was obtained from the pressure pulse using a cardiotachometer. Blood respiratory gases and pH were analyzed at frequent intervals by methods used currently in our laboratories.

To test the development of autonomic control of the resting circulatory functions throughout fetal development, the following procedures were used:

1. Stimulation of the afferent vagal pathways by intraatrial (right) or intravenous injections of veratridine (Bezold-Jarisch reflex), as well as by a right atrial stretching.

2. Abolition of total parasympathetic tone by bilateral cervical vagotomy and by muscarinic cholinergic blockade with atropine.

3. Inhibition of total neural transmission by ganglinic blockade with trimethaphan and stimulation of transmission in the autonomic ganglia with 1,1-dimethyl-4-phenylpiperazinium iodide (DMPP).

4. Alpha-adrenergic receptor blockade with phenoxybenzamine and beta adrenergic blockade with propranolol.

The responses of the various cardiovascular functions to stimulation of autonomic receptors during fetal development were studied as follows:

1. The alpha-adrenergic receptors were stimulated with progressively increasing doses of l-norepinephrine given in single bolus injections. For a given fetal age, a dose-response relationship curve was constructed using the changes in the systemic arterial pressure as the target parameter. The norepinephrine dosage that induced half of the maximal arterial pressure change (ED50) was calculated from the curves of the immature, premature, and term lambs. This parameter provides a relatively reliable information on receptor maturation.

In addition to l-norepinephrine, the effects of dl-epinephrine were tested in the same manner in some animals.

2. The beta-adrenergic receptors were stimulated with intravenous administration of progressively increasing doses of dl-isoproterenol in a manner similar to that used in the norepinephrine testing.

3. The cholinergic receptors were stimulated with intravenous administration of progressivley increasing doses of acetylcholine. A dose-response relationship curve was constructed for each fetal age group, using the changes in the pulmonary artery pressure as the target parameter. The $ED_{50}$ for acetylcholine was calculated from the curve of each age group to give information about the degree of cholinergic receptor maturation.

Since, in the fetus, the pulmonary and the systemic vascular beds are connected by a wide shunt, the ductus arteriosus, it is often difficult to ascertain whether a given circulatory change induced in one bed by a given stimulus is primary or secondary to the circulatory changes occurring in the other bed. To resolve this problem, the effects of various agonists were tested in acute experiments with the ductus open and again with the ductus temporarily closed. The technical details of all these experiments have been published elsewhere (4,5).

## Neonatal and Adult Sheep Studies

The neonatal studies were carried out on newborn lambs delivered spontaneously in our animal facilities at UCLA. In each instance, the age of the neonate was known. Chronic instrumentation was carried out 2-3 days after birth and the same animal was studied daily until it was 7-8 weeks of age. The newborn lambs were housed with their mothers in specially constructed stalls. In this way, normal nursing could continue until the lamb was able to eat and drink on its own in a manner similar to the adult animal.

Assessment of the sympathetic and parasympathetic control of the various cardiovascular functions was made every week and was compared to that of the adult sheep studied simultaneously in the same manner. The tests used to assess sympathetic and parasympathetic tones were the same as described for the fetal studies. Likewise, the responses of the circulatory functions of the neonate to autonomic agonists were tested every week and were compared to those of the adult nonpregnant sheep, as well as to those of the fetus. A dose-response curve for norepinephrine was constructed for the neonate in order to compare its reactivity to that of the fetal lamb. Technical details of the neonatal and adult studies have been reported elsewhere (6).

We shall review the results of all these studies, pointing out the relevance of each finding and its application in perinatal medicine.

## Resting Heart Rate and Its Neural Control

<u>Pattern of fetal, neonatal and adult heart rate</u>. From the clinical and investigative point of view, the heart rate has become one of the most important parameters used to monitor the status of the fetus during intrauterine life and immediatly after birth. In the last quarter of century, the enthusiasm for fetal heart monitoring has increased enormously, mainly because of technological improvement in electronic circuitry, recording devices, and use of computer analysis. Consequently, the medical literature has been deluged by reports concerning patterns of fetal heart rates and their importance as a diagnostic index of fetal well-being.

Yet, despite the prolific literature, very little is known about the pattern of spontaneous changes in the resting heart rate and its neurohumoral control during the fetal and neonatal period.

In Figure I (upper part) are presented comparative data on the resting heart rate in immature, premature, and term fetal lambs, as well as the average weekly heart rate values observed during the first eight weeks of neonatal life. Also listed for comparison are the average resting heart rate values for the adult nonpregnant sheep.

It is clear the during intrauterine life, the resting fetal lamb heart rate remained fairly constant, ranging

Fig. 1. Heart rate levels in the immature, premature, and mature fetal lambs, as well as those of the neonates during the first eight weeks of postnatal age; adult nonpregnant sheep values are plotted for comparison (<u>upper</u> part). Changes in the heart rate observed after cholinergic blockade with atropine (parasympathetic tone) and

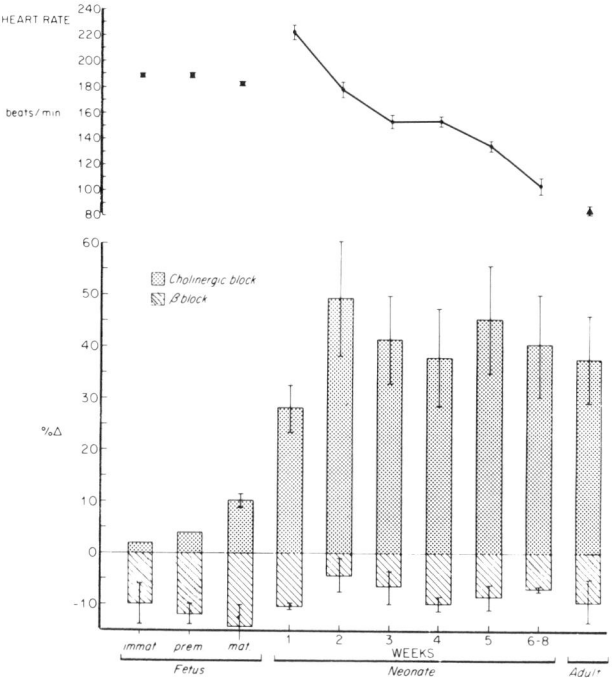

beta adrenergic blockade with propranolol (sympathetic tone) (lower part).

Note that during gestational growth, the resting fetal heart rate remains relatively stable despite the fact that the magnitude of the parasympathetic and sympathetic tones are changing. During the neonatal period, the resting heart rate progressively decreases and by the sixth to eighth neonatal week, it had approached adult levels. During this same period of time, a surge in the parasympathetic tone occurs while the sympathetic tone decreases; both tones remain at adult level. The weekly changes in the neonatal heart rate are, however, independent from the magnitude of the autonomic nervous activities.

between 165 and 185 beats per minute. The heart rate rose appreciably during the first few days after birth. Thereafter, the resting heart rate exhibited a spontaneous and progressive weekly decline, and by 6-8 weeks of age, the neonatal heart rate had approached but remained somewhat above the adult level.

The parasympathetic and sympathetic tones of the resting heart rate. In the lower part of Figure 1 (in bars) are presented the changes in the resting heart rate observed after cholinergic and beta adrenergic blockades.

Cholinergic receptor blockade with atropine in the fetal lamb resulted in cardiac acceleration, the magnitude of which depended on fetal maturity. Immature fetuses (300-1500 g) had insignificant increase (+2%) in heart rate; premature fetuses had about 4% increase; whereas term fetuses had 10% increase in heart rate after cholinergic blockade (Fig. 1).

On the other hand, blockade of the beta adrenergic receptors with propranolol had a greater effect on the resting fetal heart than cholinergic blockade. The decrease in heart rate after propranolol averaged 10% in the immature, 12% in the premature, and 14% in the mature lambs (Fig. 1).

After birth, however, radical changes occur in the magnitude of tones exerted on the resting heart rate by the two divisions of the autonomic nervous system. The data presented in Figure 1 show that during the first week after birth, the changes in heart rate produced by cholinergic blockade were close to 30% (as compared to 10% in term fetus), whereas the changes resulting from beta adrenergic blockade amounted to about 10% (14% in term fetus). From the second to the eighth week of neonatal life, the changes in the heart rate produced by cholinergic and beta adrenergic blockade were relatively constant despite the fact that the *resting* heart rate was declining progressively every week throughout the neonatal period (Fig. 1). The average increase in heart rate after cholinergic blockade with atropine from the second to the eighth week of age ranged from 40 to 50% while the decrease after beta adrenergic blockade with propranolol averaged only 5 to 8%; these changes were similar to those observed in the adult nonpregnant sheep (Fig. 1).

These data indicate the following: (1) During intrauterine life, the resting fetal heart rate is maintained at a fairly constant level by mechanisms which are independent of the activities of the autonomic nervous system. (2) In contrast to the fetus, in which there is sympathetic dominance, the parasympathetic system exhibits a definite and marked increase in its tone in the control of the resting heart rate of the neonatal lamb and adult sheep. (3) The progressive decline of the resting heart rate during the neonatal period is not related to changes in the activities of the autonomic nervous tones but rather to alterations in the intrinsic mechanisms of heart rate control.

We also carried out, in the same group of fetal and neonatal lambs and of adult sheep, simultaneous blockade of both cholinergic and beta adrenergic receptors (pharmacologic heart denervation). Figure 2 summarizes the results and compares the heart rate changes observed after simultaneous blockade with those observed after atropine and propranolol separately. It can be seen that, in the fetus, whether immature, premature, or mature, pharmacologic cardiac denervation shifted the resting heart rate to lower levels; the average decrease amounted to about 8%. In contrast, the same procedure performed on the neonatal lambs and adult sheep shifted the resting heart rate consistently to higher levels which were similar in magnitude to those observed after cholinergic blockade alone. The average increase in the resting neonatal heart rate after pharmacologic denervation amounted to 20-40%. These findings not only confirm the predominance of the sympathetic in the fetus and the parasympathetic in the neonate, but also indicate that the intrinsic mechanisms that play a role in setting the resting heart rate change radically after birth and become similar to those of the adult sheep.

The parasympathetic influence on the fetal heart rate and other circulatory functions were also investigated in a group of premature and mature fetal lambs by performing bilateral cervical vagotomy. This procedure did not produce any significant alteration in the heart rate, arterial pressure, or in the aortic, pulmonary, and ductus arteriosus blood flows (4). These negative results are not surprising since bilateral vagotomy in the adult animal also produced no cardiovascular alterations. Although the exact reason is not totally clear, it is thought to be related to the fact that the cervical vagus nerves contain both sympathetic and parasympathetic fibers.

On the other hand, traction or stimulation of the distal end of the vagus nerve elicits a definite bradycardia and very often arrhythmia (Fig. 3). These effects indicate that the vagus nerve exerts a strong action when stimulated.

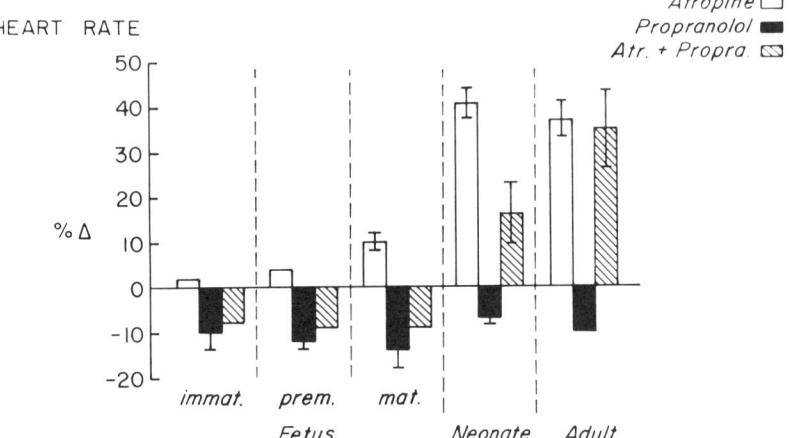

Fig. 2 Comparisons of the changes in heart rate after cholinergic and beta adrenergic blockades performed separately and simultaneously (pharmacologic denervation of the heart) in the fetus, neonate and adult sheep. The ratio of the cholinergic to adrenergic activities in the three groups of animals is evident. Note that pharmacological denervation of the fetal heart shifts the fetal heart rate toward the levels seen after beta blockade. In contrast, the same procedure performed in the neonate and adult sheep shifts the heart rate level toward that seen after cholinergic blockade. These data further emphasize the adrenergic predominance in the fetus and the cholinergic predominance in the neonate and adult animal.

## Resting Systemic and Pulmonary Vascular Pressures and Their Neurohumoral Control

Resting pressures before and after birth. During fetal life, the resting systemic arterial pressure is somewhat lower than the pulmonary artery pressure because of a higher pulmonary than systemic vascular resistance (1-3). Hence, a consistent gradient favoring the pulmonary vascular bed exists and serves to promote the blood flow through the ductus arteriosus.

The resting systemic arterial pressure increases progressively during fetal development. In the immature fetuses, the mean pressure averaged 40mm Hg; it rose to an average of about 50mm Hg in the premature and 58mm Hg in the term fetal lambs (Fig. 4).

After birth, the systemic arterial pressure rises markedly, to an average of about 80mm Hg during the first neonatal week. Thereafter, the arterial pressure remains within these range of values which are comparable to those of the adult nonpregnant levels (Fig. 4).

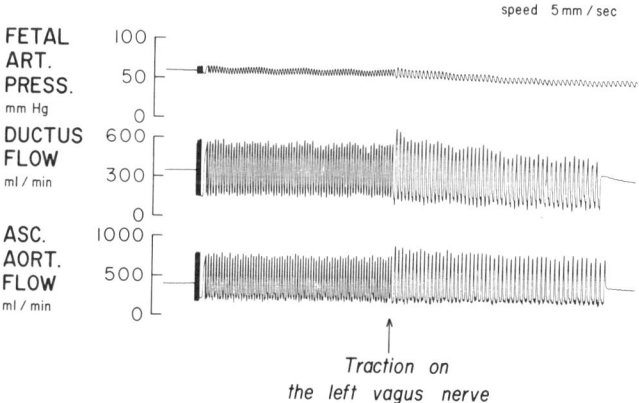

Fig. 3. An illustrative example of the circulatory effects produced by vagal nerve traction. Note the immediate bradycardia and the slight hypotension that follow this procedure.

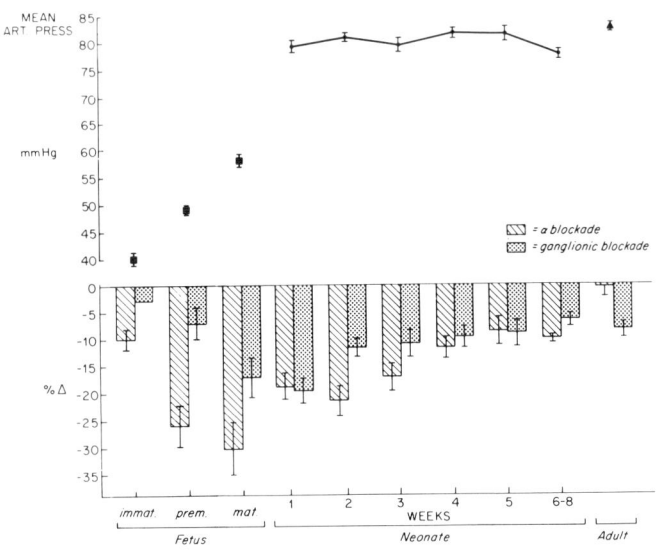

Fig. 4. Comparison of the levels of the resting systemic arterial pressure in the immature, premature, and term fetus to those of the neonate and adult sheep (upper part). Note that the resting arterial pressure increases as the fetal maturity increases. After birth, there is an abrupt rise in the arterial pressure which reaches adult level during the first week and remains at that level thereafter until the adult state.

Lower part: Blood pressure changes observed after alpha adrenergic blockade with phenoxybenzamine and ganglionic blockade with trimethaphan; these changes reflect the magnitude of the adrenergic and total neural tones of the peripheral circulation. The neurohumoral control of the resting arterial pressure increases during the fetal state, reaching a peak at term. After birth, the degree of neurohumoral control decreases weekly and by the sixth to eighth week, it had become negligible and similar to that of the adult sheep (for more information, see text).

The pulmonary artery pressure also shows a progressive increase during fetal development, reaching an average value of about 62mm Hg at term (Fig. 25). After birth, the pulmonary vascular pressure falls dramatically, reaching an average value of about 18 mm Hg during the first neonatal week; thereafter it remains at these levels, which are comparable to those of the adult animal.

Neurohumoral control. Assessment of the neurohumoral control of the systemic arterial and pulmonary circulation was made through (a) alpha adrenergic blockade with doses of phenoxybenzamine which were tested against their agonists, and (b) ganglionic blockade with trimethaphan. The maximum changes in pressures observed within 90 minutes after phenoxybenzamine injections were taken as reflecting the magnitude of the adrenergic tone supporting that pressure while the maximum changes observed after trimethaphan were taken as reflecting the total neural tone (parasympathetic and sympathetic) that contributes to the maintenance of the blood pressure.

Figure 4 shows the changes in the systemic arterial pressure produced by these two forms of autonomic blockades in the immature, premature, and term fetuses, as well as those observed throughout the first eight weeks of the neonatal period; for comparison, the responses of the adult nonpregnant sheep to the same types of autonomic blockade are presented.

Blockade of the alpha adrenergic receptors with phenoxybenzamine produced a 10% decrease in the systemic arterial pressure of the immature, 26% in the premature, and 30% in the term fetal lamb (Fig. 4). The changes in the pulmonary vascular pressure were approximately the same since, in the fetus, the two beds are connected by a wide shunt, the ductus arteriosus (4). The alterations produced by alpha adrenergic blockade in the pulmonary, aortic, and ductus arteriosus blood flows were insignificant (4).

The changes in the arterial pressure observed after phenoxybenzamine indicate a definite pattern of progressively increasing maturation of the alpha adrenergic tone of the peripheral circulation. Whether the same adrenergic control is also exerted over the fetal pulmonary vascular bed cannot be stated from these studies, since pulmonary pressure changes could be secondary to those occurring in the systemic circulation. This problem could only be resolved by performing the same type of blockade with the ductus arteriosus (closed which we did not do).

Blockade of neural transmission at the ganglionic level with constant infusion of trimethaphan produced insignificant changes in the pressures and flows of the immature and premature fetus; only in the term fetuses did the ganglionic blockade produce a significant fall in the systemic arterial pressure, averaging about 18% (Fig. 4); the fall in pulmonary pressure was similar. The changes in heart rate, ascending aortic, and main pulmonary artery blood flows produced by ganglionic blockade were of borderline significance (4).

During the neonatal period, the magnitude of the systemic hypotensive effects produced by alpha adrenergic blockade during the first week was significantly smaller than that observed in the term fetus and was comparable to that elicited by ganglionic blockade (Fig. 4). In the second and third week, the fall in the systemic arterial pressure observed after alpha blockade remained the same and averaged about 20%; the hypotensive effects of ganglionic blockade during these two neonatal weeks also diminished, averaging about 12% (Fig. 4). Thereafter, the blood pressure effects produced by both types of autonomic blockade gradually decreased and, by the eighth week of neonatal life, they had become closely similar to those observed in the adult nonpregnant sheep (Fig. 4).

We may summarize these results as follows: (1) during fetal life, there is a progressive increase in the neurohumoral control of the peripheral circulation reaching a peak at term; the control exerted by the alpha adrenergic system is consistently greater than that mediated through autonomic ganglia; (2) the neurohumoral control of the peripheral circulation decreases progressively in the neonatal period and, by the eighth week of neonatal life, it has become relatively small and similar to that of the adult, nonpregnant sheep.

## Development of Reflexes

The development of the Bezold-Jarisch reflex was studied in fetuses weighing 300-5000 g with veratridine injected intravenously or into the right atrium. In the adult animal, this vagal reflex produces bradycardia, hypotension, and apnea. The effects of this type of afferent vagal stimulation on the fetal heart rate, arterial pressure, and ascending aortic or main pulmonary artery blood flow were recorded continuously. Afferent vagal stimulation was also produced by right atrial stretching with rapid bolus injections of saline or maternal blood; phasic and integrated pressure and flow signals were recorded continuously before and after this procedure. Figure 5 shows an example of the cardiovascular response to veratridine administration in an immature fetal lamb while Figure 6 illustrates an example of the effects in a mature fetus; Figure 7 summarizes the entire data on the Bazold-Jarisch reflex.

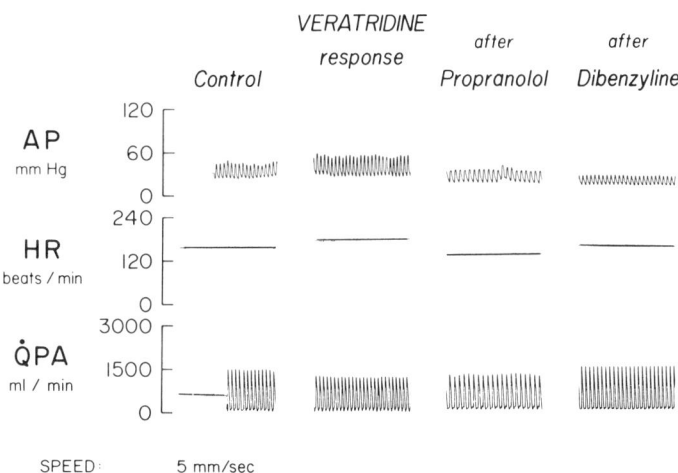

Fig. 5. A representative example of the effects of intra-atrial injection of veratridine in an immature fetal lamb. Note the tachycardia which was blocked by propranolol and the hypertension that was blocked by phenoxybenzamine.

In the immature fetuses, veratridine produced a significant tachycardia and hypertension without an appreciable change in the ascending aortic or pulmonary blood flows (Figs. 5,7). The tachycardia could be blocked by propranolol while the hypertension could be blocked with phenoxybenzamine (Fig. 5). In the premature fetuses (1600-2600 g), veratridine injections in the same dosage produced insignificant changes in the heart rate, arterial pressure, and aortic flow (Fig. 7). This response was construed as a type of transition between those of the immature and term fetuses or adult animal. In the term fetus, however, veratridine given in the same dosage elicited the classical Bazold-Jarisch response, consisting of bradycardia and

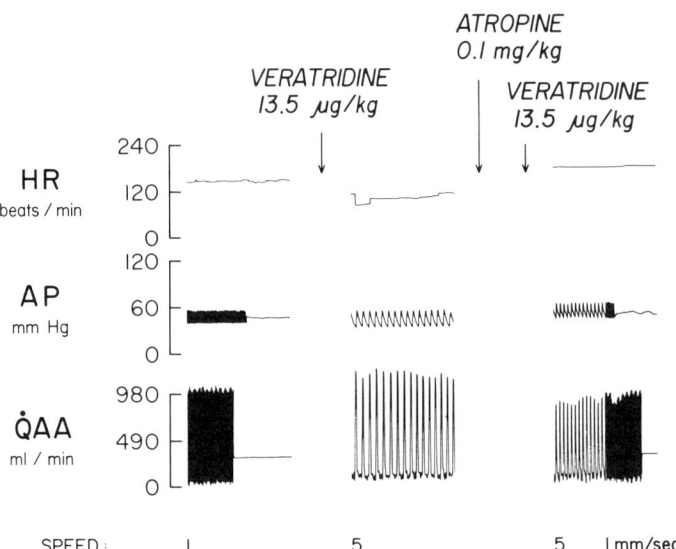

Fig. 6. A representative example of the effects of veratridine in a mature lamb. Note the bradycardia and hypotension which represent the classical response to the Bezold-Jarisch reflex as observed in the adult animal; this response was blocked by atropine, indicating its vagal origin.

hypotension which could be blocked by atropine (Figs. 6,7). The magnitude of the bradycardia and hypotension produced by veratridine in the term fetus was closely similar to that produced in the adult animal.

Right atrial stretching also produced a marked bradycardia and arrhythmia which were much greater in the premature and term fetuses than in the immature ones (Fig. 8).

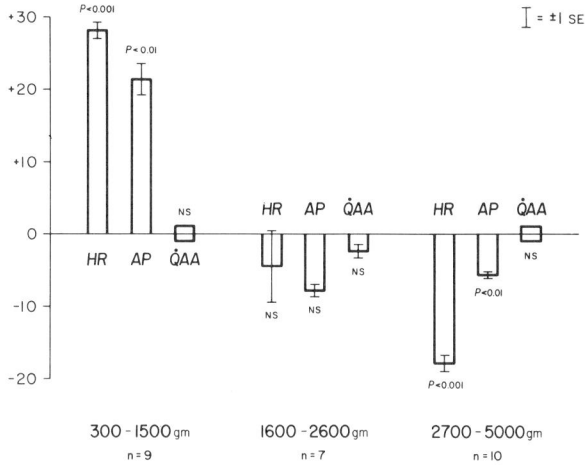

Fig. 7. Average circulatory changes observed after veratridine injections in immature, premature, and mature fetal lambs. Note the tachycardia and hypertension which resulted from adrenergic stimulation produced by veratridine in the immature animal. In the premature lambs, a transitional response occurred as reflected by the insignificant changes in heart rate and arterial pressure, following veratridine administration. In the mature fetuses, veratridine produced the typical circulatory changes of the Bezold-Jarisch reflex.

The arterial pressure and blood flows fell somewhat during the cardiac deceleration; this latter could also be blocked with atropine.

We also studied the effects of stimulation of neural transmission in the autonomic ganglia with 1,1-dimethyl-4-phenylpiperazinium oxide (DMPP) in the immature, premature, and term fetus, as well as in the neonate. This procedure produced initially a transient fall in the arterial pressure and heart rate without much change in the blood flows (Fig. 9); these changes were related to parasympathetic stimulation since they could be blocked by atropine. This initial response was followed by a more prolonged rise in the arterial pressure and heart rate accompanied by a positive inotropic effect as evidenced by the widening of

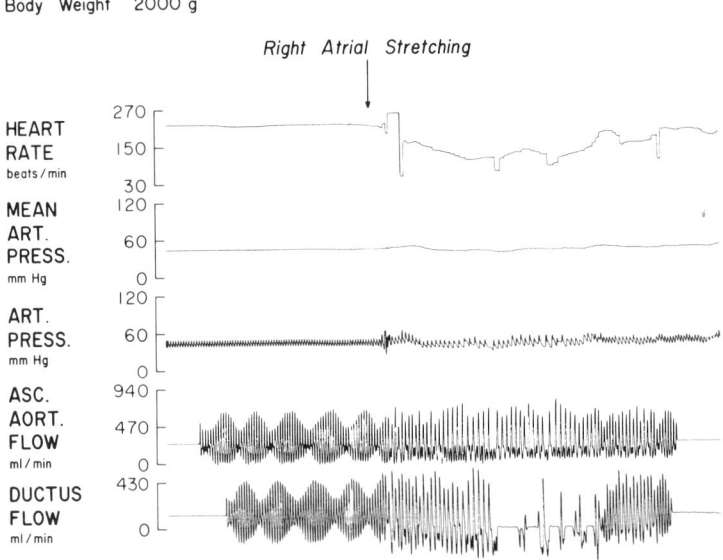

Fig. 8. An illustrative example of afferent vagal stimulation by right atrial stretching in a fetal lamb. Note the marked bradycardia and arrhythmia that follows this procedure.

the phasic flows and the systemic arterial pressure (Fig. 9); these effects were definitely related to sympathetic stimulation since they could be abolished by adrenergic blocking agents.

The conclusions derived from the studies on reflex development and on ganglionic stimulation are: (a) in the immature fetus, veratridine produces alpha and beta adrenergic stimulation instead of the afferent vagal stimulation that is characteristic of this reflex action in the adult animal; (b) this adrenergic stimulation is centrally mediated since it could be reproduced in intracarotid injections of veratridine; (c) the development of the Bezold-Jarisch reflex undergoes a transitional maturation process and

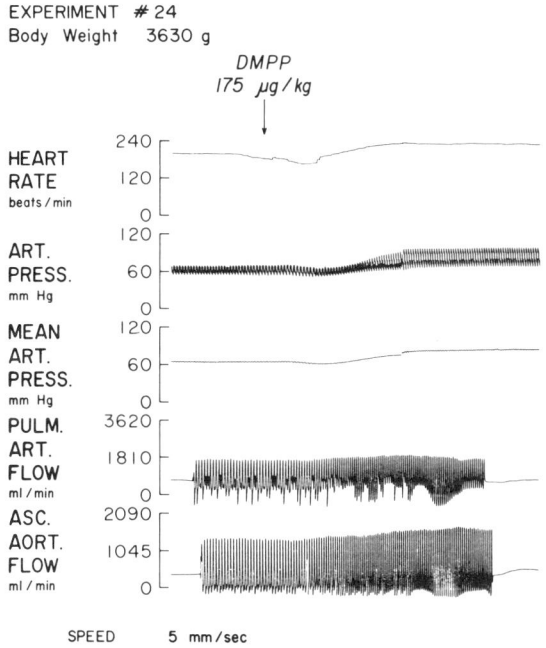

Fig. 9. A representative example of the effects of ganglionic stimulation with DMPP in a term fetal lamb. Note the adrenergic stimulation reflected by the increase in heart rate and arterial pressure that occurred. Widening of the phasic pressure and flow signals reflects positive inotropic effects.

assumes an adult-like pattern at near-term gestation; and (d) the data on the various reflexes provide further evidence that during intrauterine life, the sympathetic system develops and becomes functional earlier than the parasympathetic system and that this latter is capable, when stimulated, of exerting a strong action on the fetal circulatory functions.

## Fetal, Neonatal, and Adult Sheep Responses to Autonomic Agonists

The cardiovascular responses to stimulation of the alpha and beta adrenergic receptors, as well as to stimulation of the cholinergic receptors were studied in immature, premature, and term fetuses. The magnitude of the fetal response to a given dose was compared to that observed in the neonatal lamb studied during the first eight weeks of life. The neonatal and fetal effects were finally compared to those observed in the adult sheep.

All agonists were administered in single bolus injections in doses calculated on the basis of per kg of body weight. In the chronically-instrumented fetus, the body weight was estimated as described in detail elsewhere (3-6).

Fetal effects. In the fetal lambs, norepinephrine, epinephrine, and isoproterenol were administered in doses of 0.1-10 µg/kg. The dosages used in the neonates and adult sheep were considerably smaller (0.02-1 µg/kg) because the neonatal and adult animals proved to be more sensitive to the action of the autonomic agonists than the fetus; the reasons for this difference will be discussed later.

In the fetus, regardless of the gestational age or weight, norepinephrine administered in doses between 0.1 and 1 µg/kg invariably increased the heart rate, systemic arterial, and pulmonary artery pressures and their respective vascular resistances; ascending aortic and main pulmonary artery blood flows also increased; ductus arteriosus blood flow, however, decreased (Fig. 10). The administration of doses larger than 1 µg/kg produced initially tachycardia followed by bradycardia and arrhythmia, particularly in the mature fetuses (Fig. 11).

The magnitude of the circulatory response to norepinephrine increased progressively with fetal maturity, reaching a peak at term (Fig. 10). For instance, the administration of 1 µg/kg to the group of immature fetuses increased the heart rate by 13% and the systemic arterial pressure by 15%; in the premature fetuses, the same dosage produced an average increase in heart rate of 22% and in pressure of 26%; in term fetuses, the heart rate increase averaged 30% and that of the arterial pressure averaged 35% (Fig. 10). The same pattern of changes was observed in relation to the other circulatory parameters (Fig. 10).

The norepinephrine-induced tachycardia, as well as the increase in vertricular outputs produced by norepinephrine, were totally abolished by pretreating the fetus with propranolol (Fig. 12); the systemic and pulmonary hypertension was not affected (Fig. 12). This indicates that the changes produced by norepinephrine in these para-

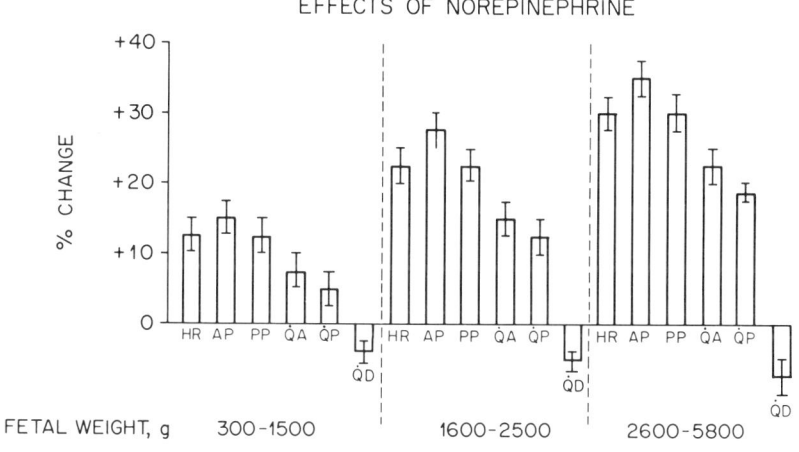

Fig. 10. Average circulatory responses to the I.V. administration of 1 ug/kg of norepinephrine in immature, premature and mature fetal lambs. Note the progressive increase in the magnitude of changes in the various circulatory parameters that occurred with fetal maturation (for more information, see text).

meters were related to stimulation of the beta adrenergic receptors in the fetal heart. On the other hand, the norepinephrine-induced increase in the pulmonary and systemic arterial pressures was abolished by pretreating the fetus with phenoxybenzamine (Fig. 12); this indicates that the changes in these two parameters are related to the stimulation of the alpha adrenergic receptors in the vascular beds.

Since, in the fetus, the pulmonary and systemic vascular beds are connected by the ductus arteriosus, the effects of norepinephrine were tested with the ductus closed in order to assess the exact site of the vasoconstriction. Figure 13 illustrates a typical example of these experiments. It can be seen that ductus closure did not alter the magnitude of the systemic and pulmonary hypertension induced by a given dose of norepinephrine. This indicates that the hypertensive response of each vascular bed is independent of that occurring in the other bed.

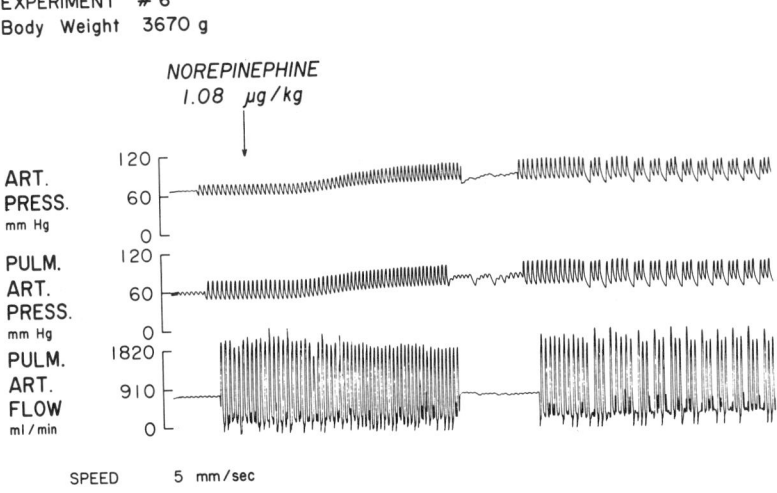

Fig. 11. An illustrative example of the circulatory effects of relatively large doses of norepinephrine in a mature fetal lamb. Note the initial tachycardia followed by bradycardia and arrhythmia.

Figure 14 compares the dose-response relationship curves to norepinephrine, using the changes in the systemic arterial pressure observed in the immature, premature, and mature fetal lambs, as well as those in the neonatal lamb during the first week of life. For any fetal age, the curve relating the pressor response to log norepinephrine dose was typically S-shaped within the dose range of 0.1-1 µg/kg; thereafter the curve became flat (Fig. 14). These data further show that the fetal pressor response to a given dose of norepinephrine increases as the fetus becomes more mature. But despite this increasing responsiveness, the fetus remains considerably less sensitive than the neonate to the action of alpha adrenergic stimulation (Fig. 14).

Assessment of the degree of maturation of the alpha adrenergic receptors was made by calculating the norepinephrine dosage that induced half of the maximal systemic arterial pressure change ($ED_{50}$) in the three groups of fetuses studied. The results showed that the $ED_{50}$ was the same for the immature, premature, and mature lambs (Fig. 14). This suggests that the progressively increasing re-

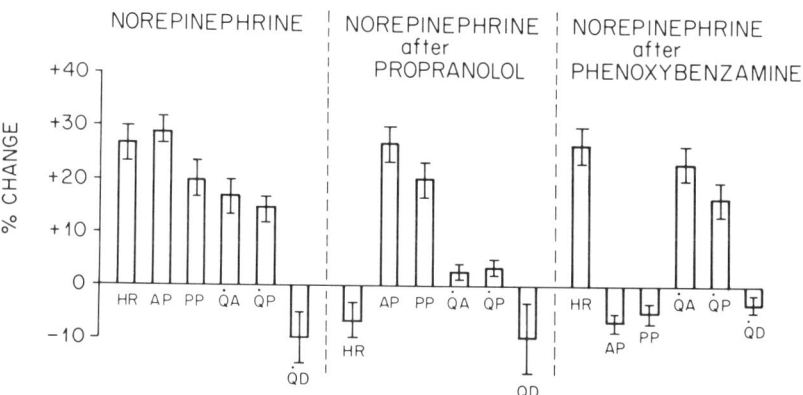

Fig. 12. Average circulatory changes produced by I.V. injections of 1 µg/kg of norepinephrine in immature, premature, and term fetuses before and after pretreatment with propranolol and phenoxybenzamine (for more information, see text).

sponse of the fetus to norepinephrine was not related to increased maturation of the receptors but rather to a maturation of the effector system, namely, the arteriolar walls.

Epinephrine administered in equivalent doses produced circulatory effects which were not significantly different from those of norepinephrine at any fetal age.

Stimulation of the beta adrenergic receptors with isoproterenol produced in the fetus a significant tachycardia accompanied by positive inotropic effects as evi-

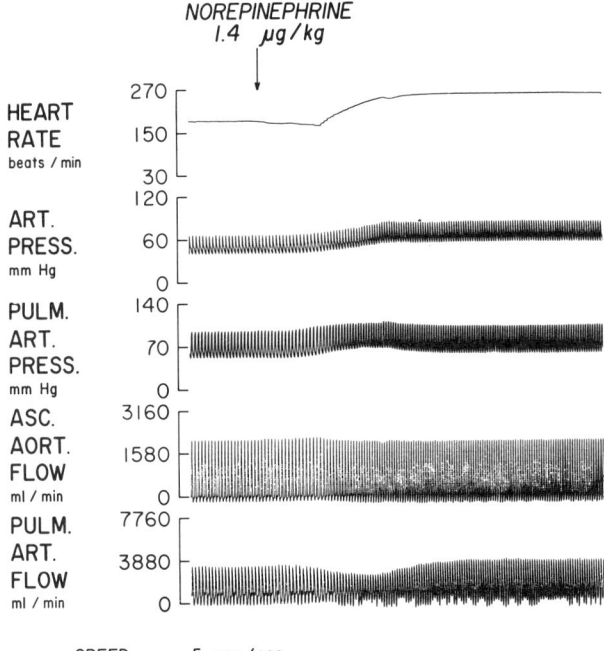

Fig. 13. Illustrative example of the effects of norepinephrine with the ductus closed. Note that despite ductus closure, the rise in pulmonary and systemic vascular pressures produced by norepinephrine persisted; this indicates that both vascular beds respond to alpha adrenergic stimulation independently.

denced by the widening of the pulsatile pressures and flows; the mean pressure did not change significantly (Fig. 15). The magnitude of the tachycardic response to a given dose of isoproterenol increased with fetal age, reaching a peak at term (Fig. 16).

In summary, the effects of alpha and beta adrenergic stimulation in the fetus are: (a) Tachycardia and positive inotropic action resulting in increased right and left ventricular outputs. These effects were observed after stimulation of both of these receptors. (b) In addition, alpha stimulation produces a systemic and pulmonary hypertension which could be blocked by phenoxybenzamine; the effects of beta stimulation on the vascular pressures were insignificant. (c) A persistent decrease in ductus blood flow which occurred after alpha stimulation. (d) All of these changes increased as the gestation approached term.

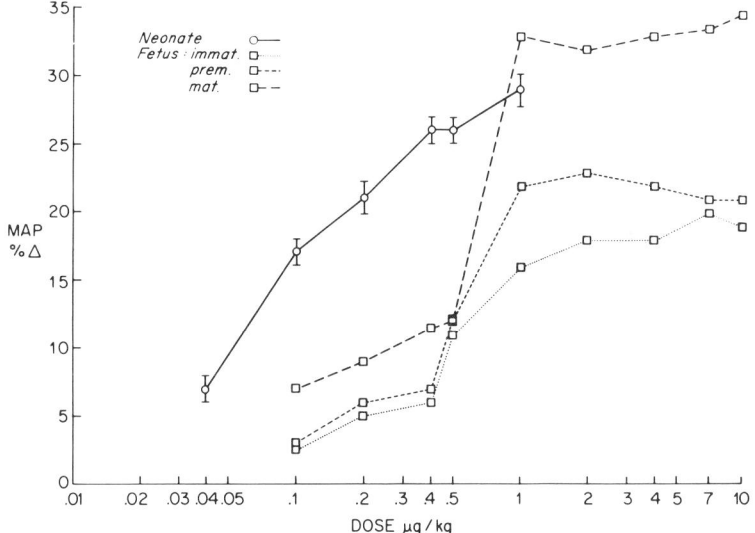

Fig. 14. Comparison of the dose-response curves of norepinephrine in the immature, premature and mature fetuses to those of the neonates. Note the marked shift to the left of the neonatal curve indicating a more marked sensitivity of the neonatal circulation to alpha adrenergic stimulation.

Stimulation of the cholinergic receptors with acetylcholine in the fetal lambs included in our studies regardless of their age produced (a) decreases in heart rate, systemic arterial, and pulmonary artery pressures and ductus arteriosus blood flow and (b) increases in ascending aortic and main pulmonary artery blood flows (Fig. 17). The fetal circulatory response to acetylcholine increased with fetal age, reaching a peak at maturity. For instance, in the immature fetus, the intravenous injections of 1 μg/kg reduced the pulmonary artery pressure by an average of 16%; in the premature fetus, the same dose elicited an average decrease of 25% while in the term fetus, the decrease averaged 32% (Fig. 17). The same progressive changes occurred in the other cardiovascular parameters. Because of the marked increase in the pulmonary blood flow and the decrease in the vascular pressure, the pulmonary vascular resistance decreased strikingly following acetylcholine administration to the fetal lambs.

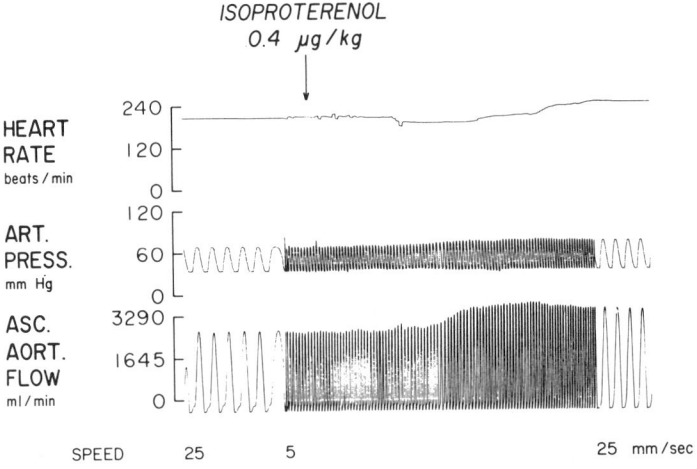

Fig. 15. A representative example of the effects of isoproterenol in a fetal lamb. Note the tachycardia and the positive inotropic effects as evidenced by the widening of the phasic pressure and flow signals.

The dose-response relationship to acetylcholine was studied in the immature, premature, and term fetus, using the changes in the main pulmonary artery pressure as the responding parameter (Fig. 18). These data show that the curves were typically S-shaped and had similar patterns even though the magnitude of pressure changes produced by a given dose was different. These data provide further evidence that the reactivity of the fetal pulmonary vascular bed to acetylcholine increases with fetal maturity.

As in the case of norepinephrine, the acetylcholine dosage that induced half of the maximal pulmonary artery pressure changes ($ED_{50}$) was calculated for the immature, premature, and term fetuses. The values were not significantly different from each other, indicating that the maturation process is probably occurring in the effector system rather than in the cholinergic receptors.

Pretreating the fetus with atropine abolished all of the acetylcholine-induced pulmonary and systemic hemodynamic changes at any fetal age.

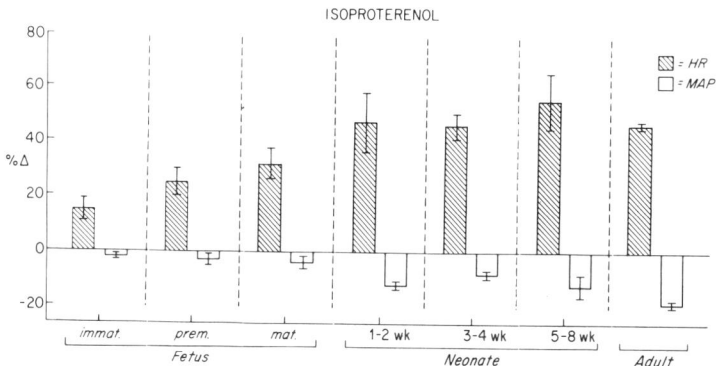

Fig. 16. Comparison of the changes produced by a standard dose of isoproterenol in the fetus, neonate, and adult sheep. Note that the fetal response increased as the fetus approaches maturity. In the neonatal period, the response to a given dose was similar throughout the eight weeks of neonatal age and was comparable to that of the adult sheep.

Figure 19 presents a typical example of the effects of acetylcholine before and after temporary closure of the ductus arteriosus. With ductus open, acetylcholine produced the typical circulatory effects as described above. Ductus closure *per se* produced a significant increase in the pulmonary vascular pressure; the main pulmonary artery blood flow decreased to less than one-third of the values observed prior to ductus closure (Fig. 19). Administration of smaller doses of acetylcholine while the ductus was closed produced a profound fall in the pulmonary artery pressure, the magnitude of which was greater than that seen when the ductus was open; the main pulmonary artery blood flow nearly doubled and pulmonary vascular resistance decreased to below control values. The changes in the sys-

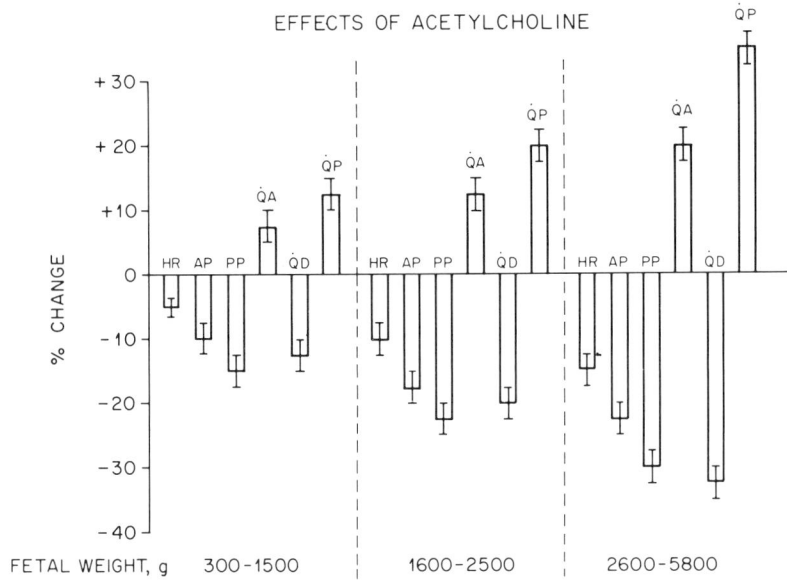

Fig. 17. Cardiovascular responses to 1 µg/kg of acetylcholine of the immature, premature and mature fetuses. Note the fall in the heart rate, as well as in the vascular pressures and ductus flow; pulmonary and ascending aortic flows increased. The response to the same dose increased with fetal age.

temic arterial pressure and in the ascending aortic blood flow produced by acetylcholine during ductus closure were insignificant (Fig. 19). These data provide a definite evidence that the changes in the systemic circulation produced by acetylcholine in the fetus are secondary to those occurring in the pulmonary vascular bed.

<u>Effects in the neonatal lamb and adult sheep</u>. Stimulation of the adrenergic and cholinergic receptors with the same autonomic agonists produced in the neonatal and adult animals circulatory alterations which were qualitatively and quantitatively different from those observed in the fetus.

At any period after birth, the systemic circulation of the neonatal lamb was considerably more sensitive to norepinephrine, isoproterenol, and acetylcholine than that

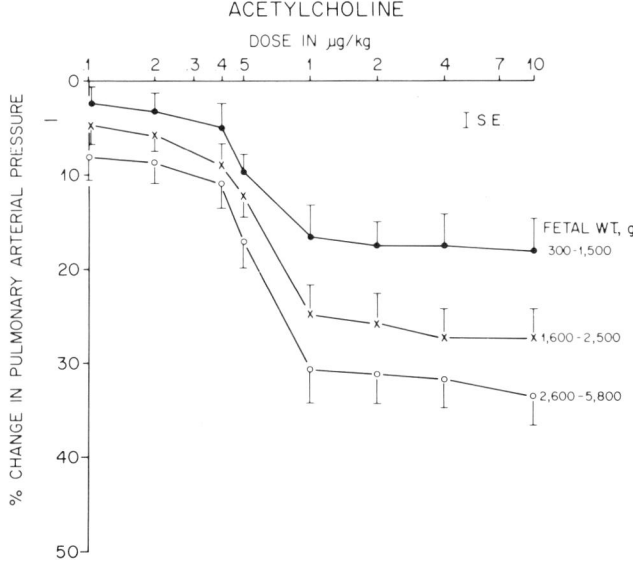

Fig. 18. Dose-response curves to acetylcholine with the pulmonary pressure as the target parameter in the immature, premature and term fetuses. Note the increase in reactivity as the fetus becomes more mature.

of the fetus. In contrast, the pulmonary circulation of
the neonate became almost totally insensitive to these
agents.

Figure 20 illustrates a typical example of the effects
of norepinephrine in a 6-week-old lamb, while Figure 21
compares the responses of the neonatal lambs and adult
sheep to those of the fetuses at different maturity. The
difference in reactivity to a given degree of alpha adren-

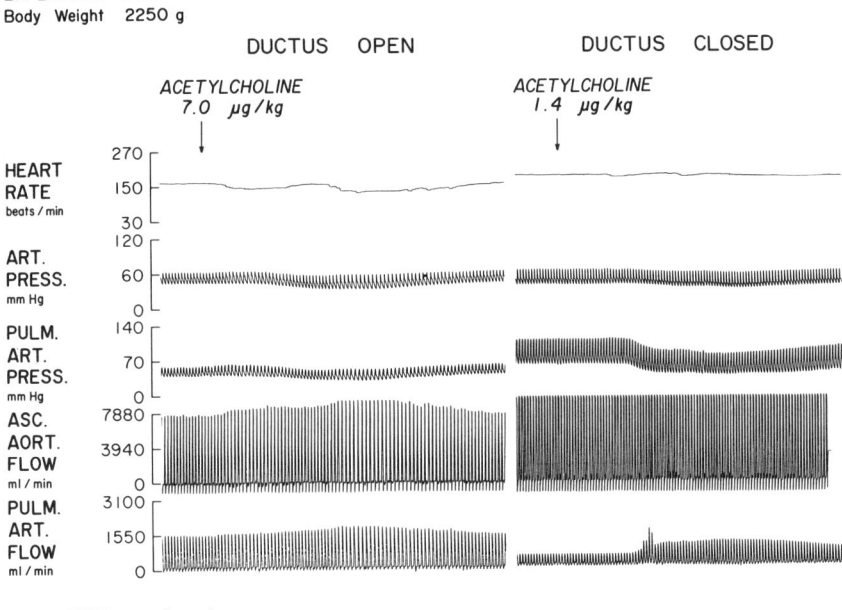

Fig. 19. A representative example of the
effects of acetylcholine in a fetal lamb
with ductus open and closed. Note the parallel changes in pulmonary and systemic arterial
pressures that followed acetylcholine injection when the ductus was open. Ductus closure
per se decreased pulmonary blood flow and
increased pulmonary artery pressure markedly.
Injection of acetylcholine with ductus closed
produced a profound pulmonary vasodilatation
without significant changes in the systemic
pressure; this suggests that the systemic
effects are secondary to the pulmonary effects.

ergic stimulation between the fetus and the neonate can be easily depicted from the dose-response curves displayed in Figure 14.

Administration of norepinephrine to the neonate, regardless of the age or the dose, produced a rise in the systemic arterial pressure accompanied by a consistent bradycardia; the pulmonary vascular pressure and blood flow were not altered (Fig. 20). Such a response was similar to that of the adult sheep (Fig. 21).

The pressor response to a given dose of norepinephrine was four to five times greater in the neonate, even at one week of age, than in the fetus (Fig. 14). In the neonate, doses of norepinephrine larger than 0.5 µg/kg frequently produced arrhythmia along with the bradycardia, whereas cardiac irregularities could be observed in the fetus only after doses larger than 1 µg/kg/.

The norepinephrine-induced bradycardia in the neonate and adult sheep was totally abolished by atrophine; this indicates that it is baroreceptor mediated and secondary to the rise in the systemic arterial pressure.

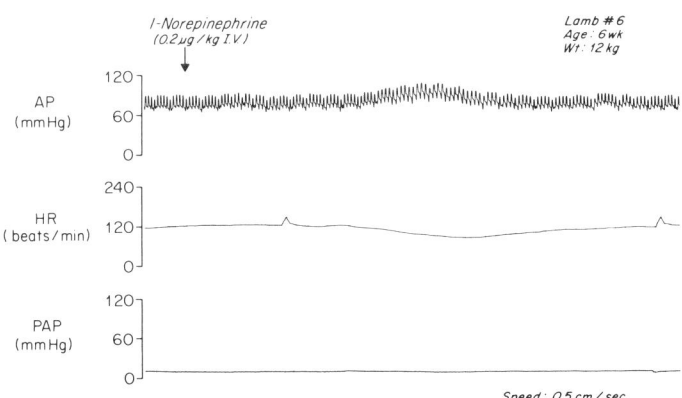

Fig. 20. A representative example of the effects of norepinephrine in a 6-week-old neonatal lamb. Note the bradycardia that occurred concomitantly with the hypertension. Pulmonary artery pressure and flow did not change.

Administration of isoproterenol to the neonate produced a tachycardia and systemic hypotension accompanied by a widening of the pulse pressure. The changes in the pulmonary vascular pressure and flow were negligible (Fig. 22). The magnitude of the changes produced by a given dose of isoproterenol was constant throughout the neonatal period and was similar to that of the adult sheep (Fig. 16); the neonatal circulatory changes were, however, consistently greater than those produced in the fetus (Fig. 16).

Stimulation of the cholinergic receptors with acetylcholine in the neonate produced consistently a systemic hypotension, accompanied by a tachycardia; the pulmonary vascular pressure and blood flow of the neonate were not affected by acetylcholine (Fig. 23). The magnitude of the tachycardia and of the systemic hypotension produced by a given dose of acetylcholine was the same throughout the neonatal period and was similar to that of the adult sheep (Fig. 24). Since in the fetus the changes in the systemic

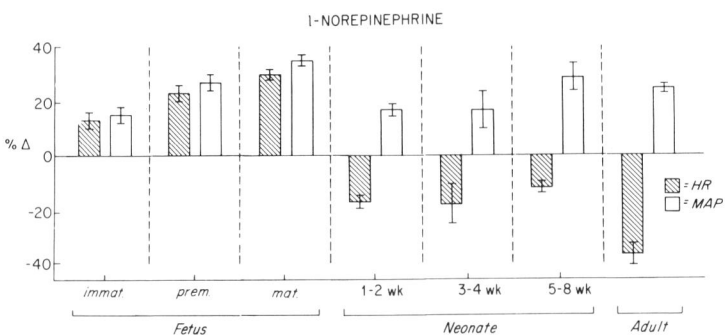

Fig. 21. Comparative data on the effects of norepinephrine in the fetuses and neonates of different ages and in the adult sheep. Note the progressively increasing response to 1 μg/kg of the fetus during the maturation period; the fetus had a consistent tachycardia along with the rise in pressure. In contrast, the neonate and adult sheep had a consistent bradycardia along with the rise in pressure even when the dose administrered was smaller than that used in the fetus (for more information, see text).

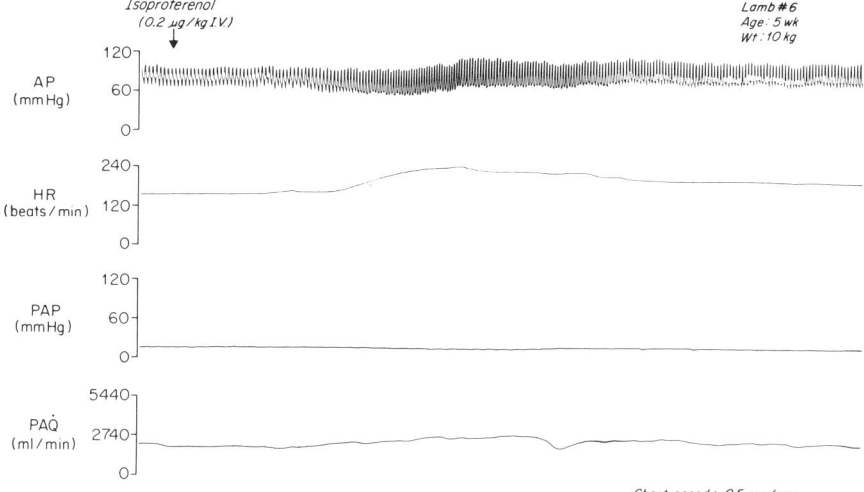

Fig. 22. A representative example of the effects of isoproterenol in a neonatal lamb 6-weeks old. Note the positive chronotropic and inotropic effects on the heart. Pulmonary pressure and flow did not change significantly.

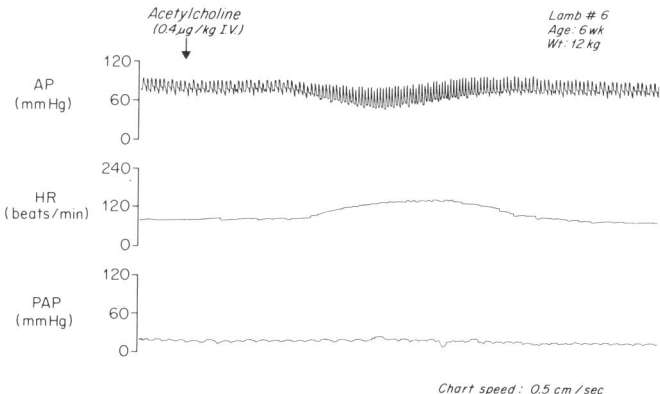

Fig. 23. A representative example of the effects of acetylcholine in a 6-week old lamb. Note the paradoxical tachycardia that occurred concomitantly with the fall in the systemic arterial pressure. Acetylcholine did not alter the pressure and flow in the pulmonary circulation.

arterial pressure were secondary to those occurring in the pulmonary vascular bed (see experiments with ductus closed, Fig. 19), comparison of the systemic effects of acetylcholine in the fetus and neonate may be misleading. Nevertheless, it could be stated that the systemic peripheral circulation begins to show a significant response to cholinergic stimulation only after birth.

In Figure 25, data are presented which depict the response of the pulmonary vascular bed to cholinergic receptor stimulation as a function of the level of the pulmonary artery pressure in the resting state. It can be seen that during intrauterine life, the fetal pulmonary vascular pressure was relatively high and increased progressively as the fetus became more mature; during this period of fetal development, acetylcholine produced a profound pulmonary hypotension, the magnitude of which increased with fetal age (Fig. 25; see also Fig. 17 and 18).

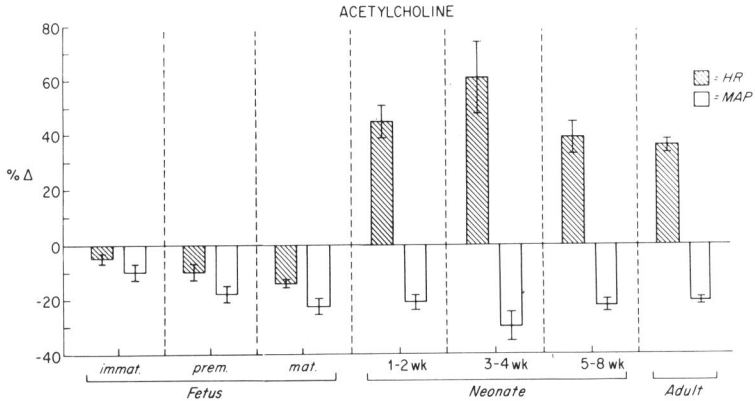

Fig. 24. Data comparing the changes produced by acetylcholine on the heart rate and arterial pressure of fetuses and neonates of different ages and those produced in the adult sheep. Note the progressively increasing bradycardia and hypotension produced in the fetuses. In contrast, the neonate had a marked tachycardia along with the hypotension which were similar to those of the adult sheep.

During the neonatal period, the pulmonary vascular pressure fell to low levels which were similar to those of the adult sheep; acetylcholine administration to both the neonate and adult sheep had no effect on the pulmonary artery pressure (Fig. 25). When, however, the pulmonary vascular pressure of the adult animal was raised by hypoxia, acetylcholine produced pulmonary hypotensive effects closely similar to those observed in the fetus.

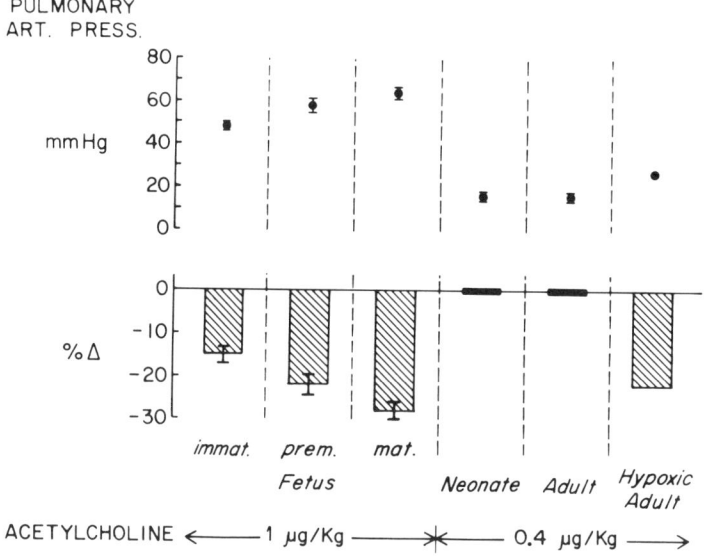

Fig. 25. Response of the pulmonary artery pressure to acetylcholine as a function of the resting status of the pulmonary vascular resistance. During the fetal state, the resting resistance and pressure were relatively high; consequently, acetylcholine elicited a marked hypotensive response. In the neonate and adult sheep, the pulmonary vascular resistance and pressure were very low; the response to acetylcholine was negligible. When the pulmonary resistance and pressure were raised by hypoxia, acetylcholine re-exerted a hypotensive response.

These observations demonstrate clearly the relationship of the response of the pulmonary vascular bed to vasoactive agents and the status of its resistance at the time of drug administration.

## DISCUSSION

The present review provides information on several questions related to the development of the autonomic control of the various circulatory functions during intrauterine life and the changes that occur after birth and up to the adult state.

### Sympathetic and Parasympathetic Control

It is clear from these series of observations that the parasympathetic system exerts very little, if any, control on the resting fetal cardiovascular functions prior to term gestation. This view is supported by the lack of any appreciable effects on these functions produced by the various forms of cholinergic blockade; it also receives support from the absence of a clear-cut manifestation of the Bezold-Jarisch reflex in the immature and premature lambs.

When the fetal lamb approaches maturity, however, the cholinergic tone becomes clearly evident. If one takes the fetal heart rate as an example, it can be seen that at term the magnitude of the parasympathetic tone (reflected by the changes in heart rate after cholinergic blockade) has become closely similar to that of the adrenergic tone. Also, the parasympathetic mediated reflexes begin to assume an adult-like pattern at term gestation.

In contrast to the negligible parasympathetic tone exerted on the fetal circulatory functions prior to maturity, the sympathetic control begins as early as the 60th day of gestation and increases significantly with the progress of gestation. If we again take the heart rate as an example, we can see clearly the predominance of the tone exerted by the sympathetic limb of the autonomic nervous system; the magnitude of the bradycardia observed after propranolol (beta blockade) and after pharmacologic denervation of the heart supplemented by the pattern of responses to veratridine injections all attest to the earlier development of the adrenergic system in the fetal lamb.

But the predominance of the sympathetic over the parasympathetic becomes even more striking when one examines the influence of the two systems on the peripheral circulation of the fetal lamb. The present data show that not only the total neural support of the blood pressure (reflected by the results of ganglionic blockade) reaches its peak near term but also that a marked surge in the magnitude of the humoral control of the peripheral circulation occurs prior to birth. This can be clearly seen by the striking hypotension (30% of control) that follows alpha adrenergic blockade with phenoxybenzamine in the term fetus. Since this alpha adrenergic antagonist inhibits more effectively the total circulating catecholamines than those released at the nerve ending (8), the greater response of the term fetus is probably related to an overall increase in the activity of the fetal adrenal glands. This hypothesis receives support from several reports which indicate that a surge in the adrenal functions occurs as the fetus approaches maturity and that such a *surge* is involved in lung maturation, as well as in the initiation of labor, particularly in the sheep (9,10).

Although the present series of observations show a definite predominance of the sympathetic tone in the peripheral circulation and the heart rate, the parasympathetic system is overwhelmingly predominant in the control of the fetal pulmonary circulation and the ductus arteriosus blood flow. This is illustrated more clearly by the response of these vascular structures to acetylcholine.

After birth, a marked increase occurs in the parasympathetic tone of the circulation, particularly that pertaining to the resting heart rate; such a parasympathetic dominance becomes evident during the first neonatal week and remains at about the same level until the adult condition. Such a parasympathetic dominance of heart rate control has been observed in a variety of animal species by different authors (18,19).

During this same period, the adrenergic tone of the heart as reflected by the response to propranolol becomes considerably smaller that that seen in the fetus. The cholinergic dominance in heart rate control is further supported by the pattern in the heart rate changes observed after pharmacologic heart denervation in the neonatal period.

But despite these continuous changes in the degree of control exerted by the two divisions of the autonomic nervous system from the early fetal and up to the adult state, the heart rate seems to set its own resting level spontaneously at any period during intrauterine and extrauterine life. In other words, the heart does not seem to sense the constantly changing ratio of the adrenergic and cholinergic activities and appears to set its resting rate without any relation to any given autonomic activity; such spontaneous and physiological changes seem to be mediated through modifications in the intrinsic conducting mechanisms of the heart and through changes in the action potential of the myocardial cells (12,13). The autonomic nervous system appears to exert its controlling action during stress or when activated by specific reflexes.

On the other hand, the peripheral circulation seems to continue to be under neurohumoral control during the early part of neonatal life.

Judging by the responses to alpha adrenergic and ganglionic blockades, we are led to believe that the neurohumoral control of the systemic arterial pressure reaches a peak before birth and falls significantly during the first week of neonatal life. Thereafter the neurohumoral support of the peripheral circulation progressively decreases until it reaches insignificant levels in the normal adult state.

In view of these negligible circulatory effects resulting from autonomic blockade in the old neonate and adult sheep, one is faced with the following question: how it the systemic arterial pressure maintained in the resting state? This question has puzzled investigators in the past since the same negligible blood pressure changes are seen in nonpregnant sheep, as well as in normotensive nonpregnant human subjects in the recumbent position (14,15,16).

Although the exact answer is not yet at hand, it is probable that a number of homeostatic factors contribute to maintain a stable normal blood pressure in the absence of neurohumoral tone. The most important of these is probably the intrinsic tone of the arteriolar walls. A number of reports (for review see 15) have shown that the arteriolar system regains rapidly its normal tone after total pharmacological or surgical denervation so that no significant blood pressure alterations are seen; only when the subject assumes the upright position does the blood pressure fall because of venous pooling.

As to the pulmonary vascular beds, it is difficult to state whether, in the fetus, these vessels possess any neurohumoral tone despite the marked fall in pulmonary artery pressure that occurred during alpha adrenergic blockade, since such a fall could have been secondary to the systemic hypotension. This question can only be decided with experiments in which the ductus arteriosus is closed; such studies remain to be performed. In the neonate and adult sheep in which the two vascular beds are separated, the pulmonary circulation does not seem to possess any neurohumoral tone as evidenced by the absence of any response to the various modes of neurohumoral blockades.

### Response to Autonomic Agonists

It is obvious from the studies reviewed in this chapter that the fetal circulation, regardless of the age, is less reactive to the action of autonomic agonists than that of the neonate or adult animal.

Two main factors could be responsible for this difference. The first is related to maturation of the neuroeffector system and the second is concerned with closure of the vascular shunts, including elimination of the umbilico-placental circulation. The first hypothesis seems unlikely since the curves pertaining to the dose-response relationships both both acetylcholine and norepinephrine suggests that both types of receptors had reached full maturation at term (5).

We believe that the second hypothesis is more likely. The striking increase in responsiveness to autonomic agonists that occurs after birth is related to closure of the vascular shunts and to the elimination of the low resistance system of the umbilico-placental vascular bed. The presence of these structures, particularly the umbilico-placental vascular bed, tend to dampen the cardiovascular response of the fetus to any stimulus.

The consistent decrease in heart rate that occurs simultaneously with the rise in pressure following norepinephrine administration to the neonatal lamb and adult sheep was not observed in the fetus.

We believe that the bradycardia is related to baroreceptor stimulation by the rise in pressure mediated through the parasympathetic system since it could be blocked by atropine.

The fact that the fetus responded with tachycardia to norepinephrine, in contrast to the bradycardia of the neonate and adult, adds further evidence to the predominance of the adrenergic system during intrauterine life and the surge in the cholinergic system after birth.

The contrast in the reactivity of the pulmonary vascular bed to acetylcholine, between fetus and neonate, is striking. In the fetus, regardless of the age, a profound pulmonary hypotension occurred following acetylcholine; in the neonate and adult sheep, no changes were observed. Only when the adult pulmonary vascular resistance was increased by hypoxia did an acetylcholine action reappear.

These findings add further evidence to the concept of the relationship between the "resting tone" of a given vascular bed and its reactivity to vasoactive agents (17).

## CONCLUSIONS

The following conclusions can be drawn from the data reviewed:

(1) Both sympathetic and parasympathetic functions mature at varying rates during fetal life, with receptor and effector maturations preceding peripheral neural development.

(2) The sympathetic nervous system develops and becomes functional earlier in fetal life than the parasympathetic system; at 60 days' gestation, the adrenergic tone exerted on the fetal heart rate is two to three times greater than cholinergic tone.

(3) At term gestation, the cholinergic and adrenergic tones exerted on the resting heart rate are nearly balanced.

TABLE 1

Effects of Bilateral Vagal Section on Fetal Circulation

| Parameter | N | Control | After Section |
|---|---|---|---|
| Arterial pressure, mm Hg | 10 | 60 ± 2 | 61 ± 2 |
| Heart rate, beats/min | 10 | 195 ± 8 | 196 ± 7 |
| Pulmonary art. press., mm Hg | 4 | 64 ± 3 | 64 ± 4 |
| Ascend. aort. flow, ml/kg/min | 5 | 95 ± 15 | 96 ± 14 |
| Main pulm. art. flow, ml/kg/min | 4 | 120 ± 16 | 124 ± 12 |
| Ductus art. flow, ml/kg/min | 9 | 86 ± 14 | 82 ± 10 |

¶ Figures represent averages of several readings taken before and after vagal sections. N = number of fetuses in which a given parameter was measured; ± 1 S.E.

After birth, a surge in the parasympathetic tone of the resting heart rate occurs while the adrenergic tone becomes less than that prevailing in the fetus. After the second neonatal week, the parasympathetic system becomes dominant in the tone exerted on the heart rate in a manner similar to that of the adult animal.

(4) Throughout fetal and neonatal life, the resting heart rate is set at a given level largely through changes in the intrinsic control of the heart and not through alterations in the ratio of the activities of the two divisions of the autonomic nervous system.

(5) In the fetus, the peripheral circulation is under a neurohumoral tone, the magnitude of which increases markedly as pregnancy approaches term. After birth, a progressive decline in the neurohumoral tone of the systemic circulation takes place, reaching low levels comparable to those of the adult nonpregnant sheep at 6-8 weeks of age.

(6) In the fetus, the various cardiovascular functions are responsive to the stimulating action of neurotransmitters. The response increases with fetal age until term gestation. The pulmonary vascular bed is primarily responsive to the cholinergic neurotransmitter, acetylcholine, whereas the systemic circulation is primarily the target of norepinephrine. The increased responsiveness with fetal age is related to maturation of the effector system.

(7) After birth, the neonatal cardiovascular system becomes 4-5 times more sensitive to the action of neurotransmitters than the fetal system. Such increased reactivity is related mainly to closure of the vascular shunts and elimination of the umbilico-placental circulation.

(8) In the neonatal lamb and adult sheep, the pulmonary vascular bed loses its responsiveness to neurotransmitters.

## ACKNOWLEDGMENTS

Supported by grants from the National Heart and Lung Institutes and grants from the National Institutes of Health, HL-01755 and HL-13634.

Charles R. Brinkman, III is the recipient of Career Development Award HL-70237. James R. Woods, Jr. is U.S. Army sponsored Postdoctoral Fellow in Fetal-Maternal Medicine and Adrien Dandavino is Canadian Medical Research Council sponsored Fellow in Fetal-Maternal Medicine.

## REFERENCES

1. Assali, N. S.; Bekey, G. A.; and Morrison, L. W. 1968. Fetal and neonatal circulation. In *Biology of Gestation,* vol. 2, ed., N. A. Assali, New York: Academic Press.

2. Kaplan, S. A., and Assali, N. S. 1972. Disorders of circulation. In *Pathophysiology of gestation,* vol. 3, ed., N. A. Assali, New York: Academic Press.

3. Dawes, G. S. 1968. *Fetal and neonatal physiology.* Chicago; Yearbook.

4. Nuwayhid, B.; Brinkman, C. R., III; Su, C.; Bevan, J. A.; and Assali, N. S. 1975. Development of autonomic control of fetal circulation. *Am. J. Physiol.* 228:337-344.

5. ----. 1975. Systemic and pulmonary hemodynamic responses to adrenergic and cholinergic agonists during fetal development. *Biol. Neonate.* 26:301-317.

6. Woods, J. R., Jr.; Dandavino, A.; Murayama, K.; Brinkman, C. R., III; and Assali, N. S. 1977. Autonomic control of cardiovascular functions during neonatal development and in adult sheep. *Circulation Res.* 40:401.

7. Assali, N. S.; Brinkman, C. R., III; and Nuwayhid, B. 1974. Comparison of maternal and fetal cardiovascular functions in acute and chronic experiments in the sheep. *Am. J. Obstet. Gynecol.* 120:411-425.

8. Nickerson, M. 1965. In *The Pharmacological basis of Therapeutics*, eds., L. S. Goodman and A. Gilman, 3rd ed., New York: MacMillan.

9. Liggins, G. C. and Howie, R. N. 1973. Prevention of respiratory distress syndrome by antepartum corticosteroid therapy. In *Proceedings of the Sir Joseph Barcroft Symposium*, Cambridge University Press.

10. Liggins, G. C. 1969. Premature delivery of foetal lambs infused with glucocorticoids, *J. Endocrinol.* 45:515-522.

11. Ballard, P. L., and Ballard, R. A. 1972. Glucocorticoid receptors and the role of glucocorticoids in fetal lung development. *Proc. Nat. Acad. Sc.* 69: 2668-2672.

12. Friedman, W. F. 1973. The intrinsic physiologic properties of the developing heart. In *Neonatal Heart Disease*, ed. W. F. Friedman, M. Leach, and E. H. Sonnenblick. New York: Grune and Stratton.

13. Hopkins, S. F. Jr.; McCutcheon, E. P.; and Weskstein, D. R. 1973. Postnatal changes in ventricular function, *Circulat. Res.* 32:685-691.

14. Assali, N. S. and Prystowsky, H. 1950. Studies on autonomic blockade. I. Comparison between the effects of tetraethylammonium chloride (TEAC) and high selective spinal anesthesia on blood pressure of normal and toxemic pregnancies, *J. Clin. Invest.* 29:1354-1366.

15. Assali, N. S. and Brinkman, C. R., III. 1972. Disorders of maternal circulatory and respiratory adjustments. In *Pathophysiology of gestation*, vol. 1, ed., N. S. Assali, New York: Academic Press.

16. Volle, R. L. and Koelle, G. B. 1965. In *The Pharmacological basis of therapeutics*, 3rd ed., eds., L. S. Goodman, and A. Gilman, New York: MacMillan.

17. Fishman, A. P. 1976. Hypoxia on the pulmonary circulation: how and where it acts. *Circulat. Res.* 38: 221-231.

18. Warner, H. R. and Russell, R. O. Jr. 1969. Effects of combined sympathetic and vagal stimulation on heart rate in the dog. *Circulat. Res.* 24:567-574.

19. Higgins, C. B.; Vatner, S. F. and Braunwald, E. 1973. Parasympathetic control of the heart. *Pharmacol. Rev.* 25:119-155.

# 4

# Postnatal Maturation of the Central Neural Cardiovascular Regulatory System

## P.M. Gootman
Department of Physiology/Downstate Medical Center
State University of New York/Brooklyn, New York

## N.M. Buckley
Department of Physiology
Albert Einstein College of Medicine/Bronx, New York

## N. Gootman
Division of Pediatric Cardiology
Long Island Jewish-Hillside Medical Center and Health Science Center
State University of New York at Stony Brook/New Hyde Park, New York

### INTRODUCTION

The investigations communicated in this chapter were begun because of the finding that stressed human neonates have relatively stable heart rates rather than the marked fluctuations in heart rate seen in adults (20, 87; N. Gootman, unpublished observations), a disparity that has continued to be reported (69, 102). Study of the literature at the onset of our work revealed that little was known about the neonatal cardiovascular controlling system and its postnatal maturation, particularly with reference to asynchronous development of autonomic control. The following review sets our studies of the past ten years, on the capability of the neonatal central nervous system to regulate cardiovascular function and the time course of postnatal maturation of the controlling system, in the context of others in this growing field of interest.

It is well-known that the regulatory role of the controlling system is of prime importance in the organism's responses to changes in the external as well as internal environment. In the adult, vasoactive sites as defined by blood pressure, heart rate, and in some cases, blood flow responses have been located in the central nervous system (e.g., 18, 22, 34, 36, 50, 56, 64, 65, 74, 76, 77, 78, 93). In 1972 (53), we published a preliminary survey in the neonatal pig of some vasoactive sites, particularly medullary, as defined by the occurrence of blood pressure and heart rate changes to stimulation, and discussed their locations relative to such sites in adults of other species. Our results as to medullary pressor sites were later verified by Marshall and Breazile (79.80); however, they reported difficulty in obtaining responses from classical depressor sites. Our work has been extended since then to include peripheral blood flow responses to central stimulation, permitting even more exact localization.

In addition to our investigation of effects of central stimulation, we have been evaluating the maturation of the cardiovascular controlling system also through examination of reflex responses to somatic and afferent stimulation (1, 53, 86), interactions between afferent and central stimulation, and by study of the postnatal maturation of peripheral and cardiac adrenergic receptors (16). The effects of hypercapnia or hemorrhage on afferent (86) and central stimulation is currently under study.

## GENERAL METHODOLOGY

The piglet was chosen as our experimental model because of the accumulating evidence which indicates that the neonatal pig is physiologically quite similar to the neonatal human (1,2,12, 47, 48, 55). All general methods briefly described below were the same in all our studies (16,17,24,25,53,86). The various experimental protocols were carried out on piglets ranging in age from 2 hours to 25 days, and on sexually mature miniature swine. The animals were anesthetized either intraperitoneally with Na-pentobarbital (16), 10-15 mg/kg in piglets and 25 mg/kg in swine, or with 0.25-1.0% halothane in nitrous oxide (17, 25, 86). They were paralyzed with decamethonium-Br, which has only transient effects on the cardiovascular system of piglets (24). The right jugular vein was catheterized to permit continuous i.v. drip of 5% dextrose (25 ml/kg/6 hrs). Rate and volume of ventilation were adjusted to maintain normal piglet arterial pH and $pCO_2$ as determined by analyzing blood samples at intervals on a radiometer electrode assembly. Body temperature was controlled according to the age of the animal.

In all animals, the abdominal aorta was catheterized via the left femoral artery. In 24 piglets and 6 mature swine, partial thoracotomy was performed and the right or left ventricle was catheterized by direct puncture. Right femoral, renal, and carotid flows were obtained by non-cannulating electromagnetic transducers (Biotronix, Narco-Bio, Carolina Medical, or Statham), at least two flows being studied in each animal. These three arterial circulations are known to participate to differing degrees during generalized increases in autonomic activity (60,68). Zero flows were determined by both distal occlusion and induced asystole techniques; transducers were calibrated *in vitro* with piglet vessels and blood for each experiment.

Pressures registered by calibrated Statham P23Db transducers, flows, ECG, and marker pulses for procedures were recorded simultaneously on a Beckman dynagraph or on Electronics for Medicine oscilloscope assemblies. Occasionally, data were also stored on a Sangamo 7-channel tape recorder. Mean aortic pressure (AoP) and mean flows (FF, RF, CAF) were determined, and peripheral resistances (Fem R, Ren R, Car R) were calculated as the ratio of mean pressure to mean flow.

Animals were grouped according to age: less than 1 week ($\leq$ 1 day, 2-4 days), 1 week (6-10 days), 2 weeks (12-17 days), and mature (5-6 months). Each animal was its own control for all observations on effects of experimental interventions.

Mean values ($\bar{X}$) of initial and terminal control data were determined, together with their standard errors (SE), and compared to ascertain the effect of the passage of time alone. Mean values of a maximum response ($\bar{X}_R$) or maximum change ($\bar{X}_\Delta$) in a given cardiovascular function in each animal during or after a given experimental intervention were expressed as absolute or percent change; compared to zero change to establish the statistical significance of the observed change; and cross-compared among age groups to determine age-dependency of observed changes after a given experimental intervention. Statistical analyses were carried out on a programmable Textronix calculator (Model 4661). The 2-tailed Student t-test was used and the null hypothesis was rejected at p values $\leq$ .05.

Specific methodology and the results of our studies are presented below under four major headings.

I. Studies of Reflex Responses to Afferent Stimulation

One major question has been the postnatal development of cardiovascular responses to peripheral stimuli affecting the autonomic nervous system. Alteration of afferent input from the carotid sinus baroreceptors is a well-known test of the capacity for central integration of heart rate and arterial pressure and flow responses in adult mammals (e.g., 11, 13, 28, 60, 68, 99). Bilateral vagotomy would reveal the degree of tonic vagal action on the heart and the contribution of the aortic arch receptors to the maintenance of basal arterial pressure. Alteration of complex afferent inputs from a mixed somatic nerve such as the sciatic would also test the integrative capability of various substations of the cardiovascular regulatory system (e.g., 27, 35, 53, 61, 70, 89). Among the studies of the aortic and carotid baroreceptor reflexes in the neonate (10, 30, 38, 41, 88, 92, 96, 98, 106), most have been designed to determine presence or absence of reflex activity by birth rather than to evaluate its postnatal development. To our knowledge, stimulation of mixed somatic afferents has been studied in neonates only in our laboratories (17, 53, 86).

## Specific Procedures

The carotid arteries were exposed and the vagus nerves isolated for later transection. The carotid baroreceptor reflex was inhibited by bilateral common carotid occlusion (BCCO) for 20 sec. The left carotid sinus was stimulated by sudden high-pressure infusion (CSI) of a small aliquot of saline through a catheter whose tip was located at the level of the sinus (17). The right sciatic nerve (SN) or the median nerve of the brachial plexus (BN) was exposed, tied, and cut; and the central end of the nerve was stimulated through a pair of silver wire electrodes, using 10-sec trains of biphasic pulses of constant width (0.5-2.0 msec). Frequency or intensity series of stimulations were carried out (for further details, see Refs. 17 and 86).

In 18 of the 60 piglets used for sciatic nerve stimulation (SNS) experiments, the stress of hypercapnia was produced by artificial ventilation with a gas mixture containing 10% $CO_2$, leading to arterial $pCO_2$ values which approached 70 torr (86); and then subjected to acute non-shocking hemorrhage of 20-30 ml/kg. During hypercapnia alone, and after hemorrhage was superimposed on hypercapnia, sciatic nerve stimulation was repeated as in control periods.

## Results

*Carotid sinus inhibition* by bilateral common carotid occlusion (BCCO) in 66 piglets was accompanied by similar pressor responses at all ages, which were significantly smaller than those in 8 mature swine, regardless of anesthetic agent employed (17, 53). (Typical aortic pressure and femoral flow responses in a young piglet can be seen in Figs. 12 or 13, top traces, right.) There was no immediate heart rate response in most of the piglets, and no bradycardia at the peak of the pressor effect except in 3 older piglets and in the mature swine. Absence of heart rate changes during inhibition of the carotid sinus baroreceptor reflex has also been noted in young rabbits (30). Only in piglets at least one week old and under pentobarbital anesthesia did femoral and renal resistances increase consistently during bilateral common carotid occlusion as in the mature swine. Intrarenal redistribution of blood flow, shown to occur in puppies but not in adult dogs subjected to this procedure (71), may account for variable renal resistance changes in younger animals.

*Carotid sinus stimulation* by unilateral carotid sinus infusion in 23 piglets under halothane anesthesia led to depressor responses of similar magnitude at all ages (17). (Typical aortic pressure and femoral flow responses in one piglet can be seen in Fig. 11, top trace, right.)

We did not find the age-dependent difference in aortic pressure responses to carotid sinus inhibition or stimulation that other investigators have reported for newborn rabbits (30, 91) and lambs (92). In our piglets, there was no heart rate response in any animal less than one week old; bradycardia occurred in the oldest piglets and in mature swine. Stimulation of baroreceptors in young rabbits under pentobarbital anesthesia, by electrical stimulation of the central end of the aortic depressor nerve or by perfusion of the carotid sinus at different mean pressures, elicits a decrease in systemic arterial pressure, but tachycardia only with the former technique (30). In human infants within the first week after birth, passive tilting head-down leads to tachycardia and a fall in systemic blood pressure and forearm blood flow (83, 101). Fetal and neonatal lambs also exhibit an age-dependent incidence of tachycardia when inflation of a balloon in the descending aorta produces hypertension in the brachial and hypotension in the femoral circulations (88). Unfortunately, the afferent discharge in the aortic depressor and carotid sinus nerves has been recorded only in rabbits during postnatal development (10,30). Discharge was found to be synchronous with the systolic rise in arterial pressure and to vary in frequency as changes in arterial pressure were produced experimentally. A similar relationship between peak discharge frequency and mean arterial pressure was found in newborn, young and adult rabbits (10).

*Somatic nerve stimulation*, whether sciatic (SNS) or the median nerve of the brachial plexus (BNS), led to several patterns of response in piglets of all ages, depending upon the combination of stimulus parameters. Figure 1 illustrates a typical response to median nerve of the brachial plexus stimulation with 1.6 mA at 100 Hz for 10 sec in one piglet of the 38 tested. Table 1 summarizes the results of sciatic nerve stimulation in the 30 frequency series experiments carried out in other piglets under halothane anesthesia (17). The effects of somatic nerve stimulation were first evaluated from graphs of percent changes in pressure, flows, and heart rate plotted as a function of increasing stimulation frequency; then the changes accompanying the maximum mean aortic pressure responses to low and high ranges of stimulation frequency were summarized.

Although heart rate decreased 5 to 13% in mature swine, as aortic pressure decreased 6 to 18% during low-frequency sciatic nerve stimulation, there was no heart rate change in many piglets of different ages. During high-frequency sciatic nerve stimulation, the pressor response was accompanied by an increased heart rate in most piglets, and in the mature swine exhibiting 17 to 47% elevation of mean aortic pressure. The age-dependent magnitude of the maximum pressure effect of high-frequency or high-intensity stimulation of the somatic afferents also indicates that the younger animals are incapable of the total response patterns reported for adult mammals (27, 35, 61, 70, 89).

Hypercapnia at normal $pO_2$ levels led to increased mean aortic pressure and heart rate; and renal carotid resistances increased only when marked acidosis accompanied the hypercapnia (86). The intensity threshold for sciatic nerve stimulation was decreased significantly under hypercapnia, and increased significantly after hemorrhage was superimposed. In the presence of the combined stresses, the total pattern of cardiovascular responses to graded increases in stimulus intensity was markedly depressed (See Fig. 2 in Ref. 86). These observations suggest that the newborn piglet is incapable of responding with the degree of sympathetic activation required to completely compensate for such stresses.

*Bilateral vagotomy* at the end of experiments was followed by elevation of mean aortic pressure and heart rate in all animals regardless of age or type of anesthesia. Table 2 summarizes the statistically significant effects in animals under pentobarbital anesthesia, and emphasizes that the absolute changes in mean aortic pres-

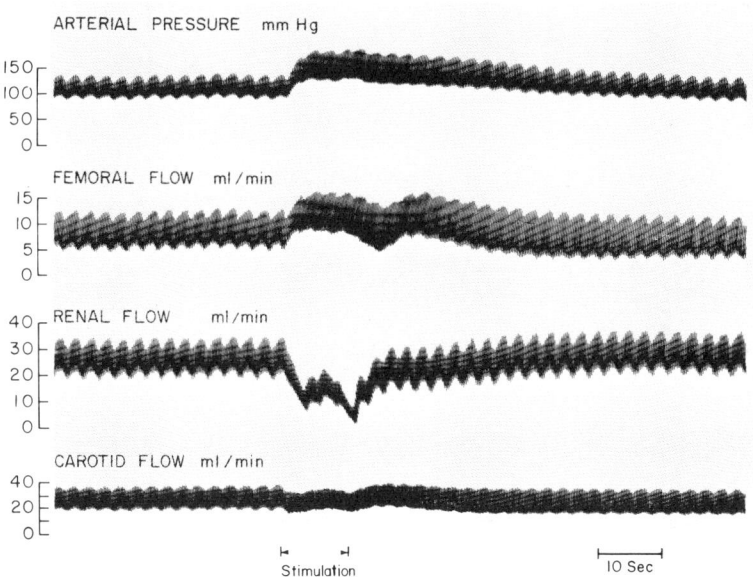

Fig. 1. Reflex effects of stimulation of the central end of the cut right median nerve of the brachial plexus on aortic pressure, renal, femoral, and carotid flows in a 16-day-old piglet with vagi intact (1.6 mA at 100 Hz for 10 sec).

Table 1

Significant Effects of Sciatic Nerve Stimulation (SNS) in Newborn Piglets Under Halothane-$N_2O$ Anesthesia[a]

| Percent Change[b] | ≤2 Days Old | | | 1 Week Old | | | 2-3 Weeks Old | | |
|---|---|---|---|---|---|---|---|---|---|
| | N | $\overline{X}_R$ | SE | N | $\overline{X}_R$ | SE | N | $\overline{X}_R$ | SE |
| Low-Frequency SNS (5-20 Hz) | | | | | | | | | |
| $\overline{AoP}$ | 7 | -14.0 | ±3.03 (.002) | 8 | -15.5 | ±3.64 (.002) | 7 | -11.0 | ±1.83 (<.001) |
| $\overline{Fem\ F}$ | 5 | -4.5 | ±2.27 | 7 | -13.0 | ±5.21 (.03) | 6 | -6.0 | ±2.93 (.05) |
| $\overline{Ren\ F}$ | 4 | -4.5 | ±1.55 (.05) | 5 | -5.0 | ±9.57 | 6 | -10.0 | ±4.27 (.05) |
| $\overline{Car\ F}$ | 6 | -6.0 | ±4.68 | 5 | -12.5 | ±6.25 | 7 | -4.0 | ±1.93 (.04) |
| High-Frequency SNS (50-80 Hz) | | | | | | | | | |
| $\overline{AoP}$ | 7 | +14.0 | ±3.89 (.008) | 8 | +29.5 | ±5.92[c] (.001) | 6 | +20.5 | ±6.73 (.025) |

| | | | | |
|---|---|---|---|---|
| Fem F̄ | 4 | +3.5 ±2.96 | 6 | +25.5 ±9.35[c] (.03) | 6 | +6.0 ±5.37 |
| Fem R | 4 | +9.5 ±4.44 | 6 | +11.0 ±13.1[c] | 6 | +8.0 ±3.47 (.03) |
| Ren F̄ | 6 | +10.5 ±4.32 (.04) | 6 | +20.0 ±10.3 | 6 | +10.5 ±14.4 |
| Car F̄ | 6 | +8.0 ±3.76 (.04) | 6 | +15.5 ±5.20 (.025) | 6 | +8.0 ±2.52 (.02) |

¶ Source: Reference 17, with permission of the publisher, S. Karger AG, Basel.

a Inspired halothane concentration = 0.25% in animals < 1 week old, and 0.5-0.75% in those ≥ 1-week old.

b Percent changes in cardiovascular functions (see test for definitions of abbreviations) given as arithmetic mean ($\bar{X}_R$) of differences between responses and preceding control in individual animals, taken at the peak pressor or depressor response on the frequency or intensity series graph for each animal. SE = standard error; N = number of animals; numbers in parentheses are p values for statistical significance between observed change and zero change.

c $\bar{X}_R$ significantly larger than in piglets ≤ 2 days old (.02 < p < .05).

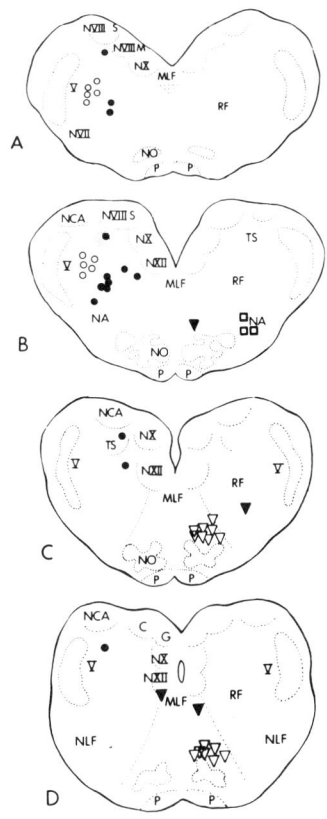

Fig. 2. Diagrammatic coronal sections of piglet brain stem (drawn from histological material) showing regions in medulla from which changes in aortic pressure, regional arterial flows and heart rate were obtained: o = marked increases in pressure and flows; ● = small increases in arterial pressure sometimes accompanied by changes in heart rate; ∇ = marked decreases in pressures and flows; ▼ = small decreases in pressure sometimes accompanied by alterations in heart rate and rhythm; □ = decreased heart rate only. More than one site was occasionally studied in one animal. *Anatomical designations* (see Ref. 105): *C*, Nucleus cuneatus; *G*, Nucleus gracilis; *MLF*, Fasciculus longitudinalis media-

sure were significantly smaller in piglets than in mature swine. In addition to the changes in renal resistance shown here, animals under halothane anesthesia also demonstrated a consistent increase in femoral resistance after vagotomy (17). Responses to exogenous acetylcholine, bilateral vagotomy, and stimulation of the cardiac end of the vagus nerve have been the basis for the conclusion that parasympathetic activity is fully developed at birth in kittens and puppies (58), rabbits (15,91), humans (4,33), and lambs (31, 82, 98). Therefore, the absence of the efferent limb of the reflex activated during peak pressor responses to altered baroreceptor activity could not account for absence of bradycardia during such responses in young animals.

From the foregoing experimental evidence, it can be concluded that afferent limbs of certain autonomic reflexes active in prenatal and/or early postnatal life in many mammals continue to develop after birth in piglets. Further analysis of the controlling system also requires the localization of vasoactive sites within the central nervous system, and the changes in responses to stimulation of such sites as a function of postnatal age.

## II. Studies on Cardiovascular Responses to Direct Central Nervous System Stimulation

Localization of central nervous system vasoactive sites in the neonate was first reported by us in the piglet (53) and later verified again in the piglet by Marshall and Braezile (79, 80), and a hypothalamic site was located in the weanling puppy (14). Very recently there has been a study on fetal lamb arterial pressure and heart rate responses to hypothalamic stimulation (104). Our own work has been expanded to include regional blood flow responses.

---

lis; *NA*, Nucleus ambiguus; *NCA*, Nucleus cuneatus accessorius; *NLF*, Nucleus fasciculi lateralis; *NO*, Nucleus olivaris; *NVII*, Nucleus nervi facialis; *N VIII S*, Nucleus vestibularis inferior; *N VIII M*, Nucleus vestibularis medialis; *NX*, Nucleus dorsalis nervi vagi; *N XII*, Nucleus nervi hypoglossi; *P*, Pyramis; *RF*, Formatio reticularis; *TS*, Tractus solitarius; *V*, Nucleus tractus spinalis nervi trigemeni.

Table 2

Significant Cardiovascular Effects of
Bilateral Vagotomy in Newborn Piglets and
Mature Swine Under Pentobarbital Anesthesia[a]

| Age | ΔHR (bpm) | ΔAop (mm Hg) | ΔRen R (PRU) |
|---|---|---|---|
| ±1 day (N = 8) | +12.5 ±4.2 | +11.6 ±4.0 | +2.6 ±0.67 |
| 2-4 days (N = 11) | +13.4 ±3.2 | +15.1 ±4.3 | (+0.3 ±0.66) |
| 1 week (N = 16) | +15.9 ±3.6 | +10.5 ±2.7 | +2.9 ±1.3 |
| 2 weeks (N = 10) | +14.8 ±5.5 | +5.4 ±2.1 | +0.8 ±0.30 |
| Mature (N = 6) | +18.0 ±4.8 | +39.5 ±9.8[b] | +1.7 ±0.50 |

[a] Values are arithmetic means and standard errors; all means are significantly different from zero change (.001 < p < .04) except values in parentheses.

[b] Significantly different from values in piglets (p = .025).

## Specific Procedures

Animals were placed in a piglet stereotaxic apparatus (53), the cortex and floor of the fourth ventricle exposed and protected with a warmed mixture of vaseline and mineral oil. Exploration of diencephalic, mesencephalic, and medullary regions was successfully carried out in 80 piglets, using bipolar electrodes of nichrome wire insulated with enamel except at the tip (49, 50, 76) (tip diameter = 0.3 mm). Stimulus parameters were: biphasic pulse (width 0.1 msec) 0.01-1.6 mA, 5-100 Hz in 10-sec trains; intervals between stimulus trains, 3-4 min. Grids of points were stimulated along dorsoventral tracks at 0.5-1.0 mm intervals; the bottom of each track was marked by passing direct current; in some experiments the sites of intensity and frequency studies were also marked. All electrode positions were histologically located after the formalin-fixed brain was imbedded in either paraffin (lower brain stem) or celloidin (diencephalon and mesencephalon); sections of 15 and 30 μ thickness, respectively, were made and stained with cresyl violet and luxol fast blue (53).

Aortic pressure and arterial flow responses to stimulation of diencephalon and brain stem were analyzed in terms of observations made as follows: vasoactive sites were located usually by exploring a track with 0.8 mA stimuli at 40 Hz. Thresholds for blood pressure response was found for all vasoactive sites, then frequency was systematically increased while holding current intensity constant; finally, current intensity was systematically increased while holding stimulus frequency constant. To assure that the responses were not affected by the order of stimulus presentation, random changes in intensity and/or frequency were also employed.

## Results

All the observed responses were very well localized, for movement of the stimulating electrode from an active site in the dorsolateral plane usually resulted in marked diminution in magnitude of response, which usually then had a longer latency. Movement of the electrode equal to or greater than 1.0 mm distant from an active locus usually resulted in loss of response.

Medullary Stimulation. Of 138 sites stimulated in the medulla of 18 piglets, 44 were found to be vasoactive. Pressor sites were found in the dorsolateral reticular formation of the medulla, about 1.7 to 3.0 mm rostral to the obex, while depressor sites were found in the ventromedial reticular formation between 0.5 rostral to and 1.0 mm caudal to the obex. Figure 2 is a composite diagram showing stimulation sites in the medulla oblongata from which changes in arterial pressure and flow were obtained. Open circles at the left indicate sites from which increases in aortic pressure and femoral, renal and carotid flows were obtained (ranging from +3 to +100%) with latencies ranging from 0.6 to 2.0 sec, but without heart rate changes. Open triangles at the right indicate sites from which short latency decreases in aortic pressure and flows were obtained (ranging from -45 to -5%). At some loci (solid circles), increases in pressure and flows with longer latency (1.5 to 4.0 sec) were obtained; at these sites cardiac arrhythmias were observed in 4 animals. Solid triangles indicate sites from which longer latency decreases in aortic pressure were obtained. Stimulation of sites near nucleus ambiguus (open squares) produced decreases in heart rate. Changes in stimulus parameters did not usually alter the direction of responses obtained from any of these medullary sites. These locations are similar to those reported for control of cardiovascular function in the adult monkey (74) and cat (18, 49-51, 77). Characteristic responses to selective sites in Figure 2 are shown in Figure 3 and are summarized for two age groups in Table 3. Short latency pressor responses without heart rate changes were generally accompanied by increased flow in femoral, renal and cephalic regions only in the older animals. Short latency depressor responses without heart rate changes were significantly greater in the older age group.

Pressure and flow responses, as a function of stimulus intensity, continuously increased in the oldest animals with increasing stimulus intensity to 1.6 mA. There was an intermediate age group in which a plateau of the pressor responses occurred above stimulus intensities of 0.5 to 0.7 mA. In the youngest animals, responses were smaller at intensities above 0.8 mA. It was also observed that with increasing stimulus frequency the responses tended to diminish in all animals tested (regardless of age), unlike the responses seen in adult cats (49, 103), and in some cases flows reversed direction. Huttenlocker (59) has suggested that myelination increases the ability of central

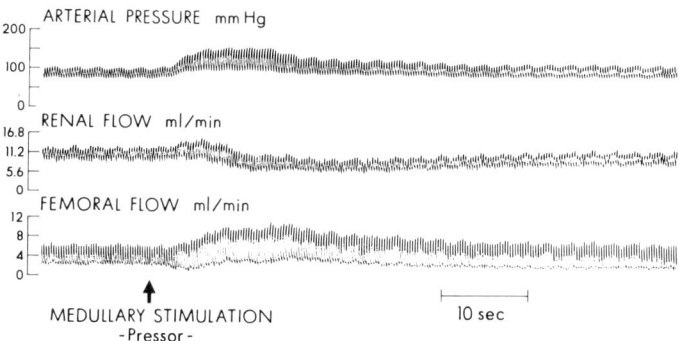

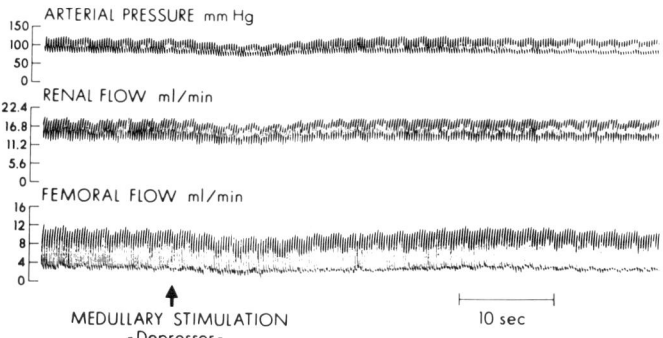

Fig. 3. A. Regional flows and pressor responses to medullary stimulation in a 3-day-old piglet (0.3 mA at 50 Hz for 10 sec at a "o" site in section A of Fig. 2). B. Regional flows and depressor responses to medullary stimulation in a 3-day-old piglet (0.5 mA at 50 Hz for 10 sec at a "∇" site in Section C of Fig. 2.

Table 3

Cardiovascular Responses to High-Frequency Stimulation
of Selected Medullary Sites in Newborn Piglets
Under Halothane Anesthesia[a]

|  | $\overline{\Delta AoP}$ (%) | $\overline{\Delta FF}$ (%) |
|---|---|---|
| Dorsolateral reticular formation [b] (open circles, Fig. 2) | | |
| ≤1day-3 days (N = 7) | +28.2 ±4.0 | (+16.7 ±24.8) |
| 1 week (N = 4) | +42.3 ±11.1 | +13.3 ±2.63 |
| Ventromedial reticular formation [b] (open triangles, Fig. 2) | | |
| ≤1day-3 days (N = 9) | -9.7 ±1.34 | -10.4 ±3.28 |
| 1 week (N = 4) | -18.3 ±3.10[b] | (-2.0 ±5.0) |

[a] Values are arithmetic means and standard errors; all means are significantly different from zero (.001 < p < 0.03) except values in parentheses.

[b] No heart rate changes were elicited by stimulation at these points.

[c] Significantly greater in older group of piglets (p < 0.4).

axons to respond repetitively. The well-known postnatal myelination may explain why responses tend to plateau more markedly in the 3-hour- to 3-day-old piglets than in the older animals; axons within the vasomotor system may become refractory with increasing frequency of electrical stimulation. However, with increasing stimulus intensity there is also a decrease in response magnitude, which may reflect spread of excitation to adjacent inhibitory regions by spatial summation.

Diencephalic Stimulation. Six hundred seventy-nine sites were stimulated in the diencephalon and mesencephalon of 62 piglets and, of these, 65 sites evoked vasomotor responses. Vasoactive sites found in the diencephalic region from the level of the anterior commissure to the mamillary bodies are indicated in the composite diagrams of Figure 4 (open diamonds in Sec. A-D). The locations of these vasoactive sites are similar to those reported for adult mammals of other species (34, 64, 74, 77, 78, 93). Pressure, heart rate, and femoral flow responses to high frequency stimulation of sites in lateral hypothalamus and zona incerta are given in Table 4, upper sections, for two age groups. In can be seen that the blood pressure responses for both sites and the femoral flow responses for sites in zona incerta are of significantly greater magnitude in the older age group.

High frequency stimulation of vasoactive hypothalamic sites, despite a number of different locations, led to consistent pressor responses (ranging from +8 to +69%) generally accompanied by increased femoral and carotid flows (ranging from +7 to +90%) and heart rate (ranging from +4 to +67%). Low frequency stimulation (5-20 Hz) at many of the same hypothalamic sites in the same animal, usually led to consistently depressor responses (ranging from -21 to -7%) accompanied by decreased femoral and renal flows (ranging from -50 to -4%). The two response patterns obtained from the same site in anterolateral hypothalamus in one piglet is illustrated in Figure 5. It was noted that for posterior and lateral hypothalamic sites, the reversal frequency was higher in the older animals (older than 10 days) compared with the younger (up to 4 days). Peak pressure and depressor responses differed in magnitude; the fall in aortic pressure was never more than 20 mm Hg, while increases were as great as 48 mm Hg. Such differences in magnitude of response have been reported to occur to hypothalamic stimulation in adult cats (65). The reversal of

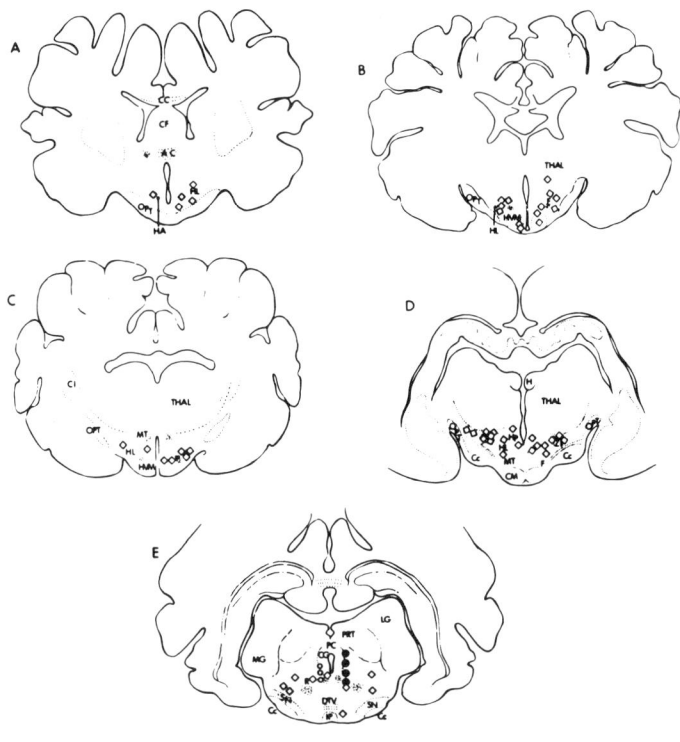

Fig. 4. *Sections A-D*: Diagrammatic coronal sections of piglet diencephalon (drawn from histological material) showing sites from which marked alterations in aortic pressure, heart rate and arterial flows were obtained with stimulation (◇). Results of explorations in planes 0.5 mm anterior and 0.5 mm posterior to each level of section are included. Sites shown are usually from more than one piglet. *Section E*: diagram of coronal section of mesencephalon through the posterior commissure indicating locations of sites from which responses were obtained o and ● sites from which responses were obtained in mapping experiments in 2 different animals; ◇ = sites from which responses were obtained in 9 other piglets. *Anatomical designations* (see Ref. 105): *AC*, Commissura anterior; *CC*, Corpus callosum truncus; *CF*, Corpus fornicus; *CG*, Substantia grisea centralis; *CI*, Capsula Interna;

response with increasing frequency of stimulation of the diencephalon was seen in piglets of all ages. Scherrer (90) reported a similar phenomenon in the adult rat. A possible mechanism to explain the patterns of responses to increasing frequency of stimulation might be the type of synaptic networks stimulated at different frequencies: e.g., at low frequency, large inhibitory postsynaptic potentials due to recurrent inhibition may result in a net reduction of output; high frequency may lead to failure to sustain inhibitory postsynaptic potentials while sustaining excitatory postsynaptic potentials, and thus yield a net increase in output from the region.

When the stimulus intensity was increased beyond a certain level in the high frequency range (60 to 100 Hz), we observed a diminution in response magnitude. This may reflect the immaturity of the integrating and/or conducting systems in the newborn pig, since a decrease in response magnitude might indicate a weakening of temporal summation. Spatial summation seems to be more effective in these newborn animals because increasing stimulus intensity usually resulted in a plateau of response magnitude rather than an actual decrease. The inability of the neonatal responses to follow high-frequency stimulation has also been reported for the efferent sympathetic nerves of the newborn puppy hindlimb circulation (9, 45), newborn rabbit heart (91), and for pyramidal tract neurons to antidromic stimulation in young kittens (50). These investigators reported that the maximum frequency of following increased with increasing age from 1 day to 5 weeks. We noted in the course of our present experiments that the responses of piglets between

---

$CM$, Corpus mamillare; $Cc$, Crus cerebri, $DTV$, Decussatio tegmenti ventralis; $F$, Columna fornicis; $H$, Habenula; $HA$, Nucleus hypothalamicus anterior; $HL$, Nucleus hypothalamicus lateralis; $HP$, Nucleus hypothalamicus posterior; $HVM$, Nucleus hypothalamicus ventromedialis; $IP$, Nucleus interpeduncularis; $LG$, Corpus geniculatum laterale; $MG$, Corpus geniculatum mediale; $MT$, Fasciculus mamillothalamicus; $OPT$, Tractus opticus; $PC$, Commissura posterior; $PRT$, Pretectum; $R$, Nucleus ruber; $SN$, Substantia nigra; $THAL$, Thalamus, $ZI$, Zona incerta.

Table 4

Cardiovascular Responses to High-Frequency Stimulation of Selected
Diencephalic and Mesencephalic Sites in Newborn Piglets Under Halothane Anesthesia [a]

|  | $\overline{\Delta AoP}$ (%) | $\Delta HR$ (%) | $\overline{\Delta FF}$ (%) |
|---|---|---|---|
| Lateral hypothalamus (Sec. B,C,D of Fig. 4) | | | |
| ≤1 day–3 days (N=4) | +25.6 ±3.9 | (+19.6 ±9.7) | (3.5 ±2.0) |
| 1 week old (N=5) | +50.0 ±9.6[b] | (+10.6 ±5.9) | +64.8 ±12.6 |
| Zona incerta (Sec. D of Fig. 4) | | | |
| 1 week old (N=5) | +26.6 ±5.4 | (+6.0 ±3.7) | +9.5 ±1.5 |
| 2 weeks old (N=3) | +68.3 ±11.3[b] | +25.5 ±2.5 | +54.0 ±2.1[b] |
| Mesencephalon (Sec. E of Fig. 4) | | | |
| ≤1 day–3 days (N=5) | +33.3 ±6.1 | (+2.4 ±1.6) | +36.6 ±3.6 |
| 1 week old (N=4) | +31.1 ±5.8 | +19.6 ±5.7 | +20.8 ±4.8 |

a Values are arithmetic means and standard errors; all means are significantly different from zero ($p < 0.4$) except values in parentheses.

b Significantly greater in older group of piglets ($p < 0.4$).

3 and 24 hours tended to diminish more rapidly than did those of 2-week-old pigs.

Hypothalamic stimulation in adult monkeys (36) resulted in flow responses and changes in calculated resistances in skeletal muscle and renal vascular beds similar to our observations, although the microsphere technique was used to measure flow in the monkey. However, with measurement of venous effluent flow by drop counter technique in adult cats, decreases in muscle and kidney flows were reported to accompany the increase in blood pressure and heart rate during hypothalamic stimulation (34). Electrical stimulation of the hypothalamus in fetal lambs (104) elicited increases in blood pressure and heart rate at sites similar to those reported earlier by us (53) and in this communication.

Vasoactive sites were found in the zona incerta of 11 piglets (Sec. D of Fig. 4) in locations similar to those reported for adult mammals of other species (74, 78). A graph of responses to increasing stimulus intensity from one of the piglets is shown in Figure 6. It can be seen that the pattern of responses was a function of intensity

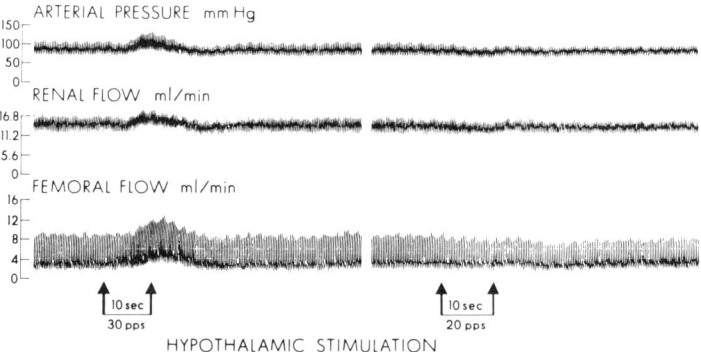

Fig. 5. Effects of hypothalamic stimulation on regional flows and aortic pressure in a 3-day piglet: pressor response (left record) and depressor response (right record) obtained by 0.5 mA stimuli at two different frequencies of stimulation at site shown at left in section A of Fig. 4.

of stimulation, with decreases in blood pressure occurring at low intensity of stimulation and increases at higher intensities. The increases in blood pressure ranged from +13 to +90% and were accompanied by changes in heart rate (+1 to +25%) and femoral, renal and carotid flows (ranging from +6 to +100%).

Mesencephalic Stimulation. In 12 piglets, the stimulation sites were located at the border of the central grey and in the tegmentum (Sec. E of Fig. 4). Stimulation of these vasoactive sites produced increases in mean aortic pressure (ranging from +16 to +166%) accompanied by increases in femoral flows (ranging from +14 to +44%) as seen for the two age groups of Table 4, lower section. Lindgren (76) also reported marked increases in blood pressure and femoral flow with stimulation of sites within the mesencephalon. Renal flow decreased markedly in some piglets, as has also been reported for adult dogs (56). The latency for onset of pressure responses ranged from 1.0 to 2.2 sec; with

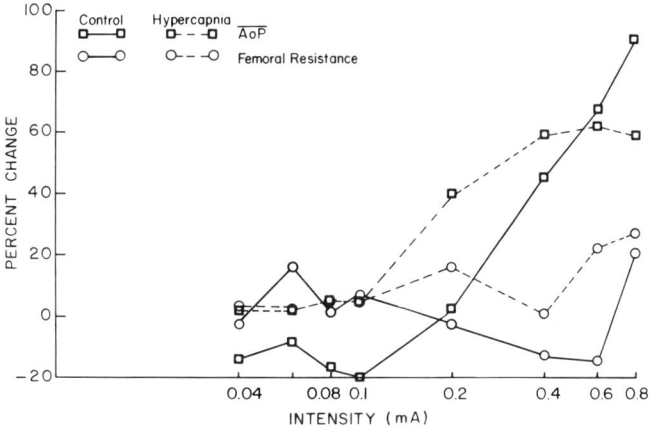

Fig. 6. Graphs relating effect of hypercapnia (pH 7.29, $pCO_2$ 58 torr) on percent change in mean aortic pressure (□), and femoral flow (o) responses to increasing intensity of stimulation at a site in zona incerta in a 13-day-old piglet (each intensity applied at 50 Hz for 10 sec).

sequential changes in stimulus parameters, only increases in pressure and carotid and femoral flows occurred, i.e., there was no difference between the effects of high and low frequency stimulation on the direction of the responses. Kuo, et al. (74) also reported the absence of depressor responses to midbrain stimulation in the adult monkey.

The effects of stresses such as hypercapnia or hemorrhage on the responses obtained to stimulation of vasoactive sites in the diencephalon are presently under investigation. In the majority of the 18 piglets in which hypercapnia was induced by having the animals breath gas mixtures containing 10% $CO_2$, the responses to diencephalic stimulation were either diminished or lost. Hypercapnia

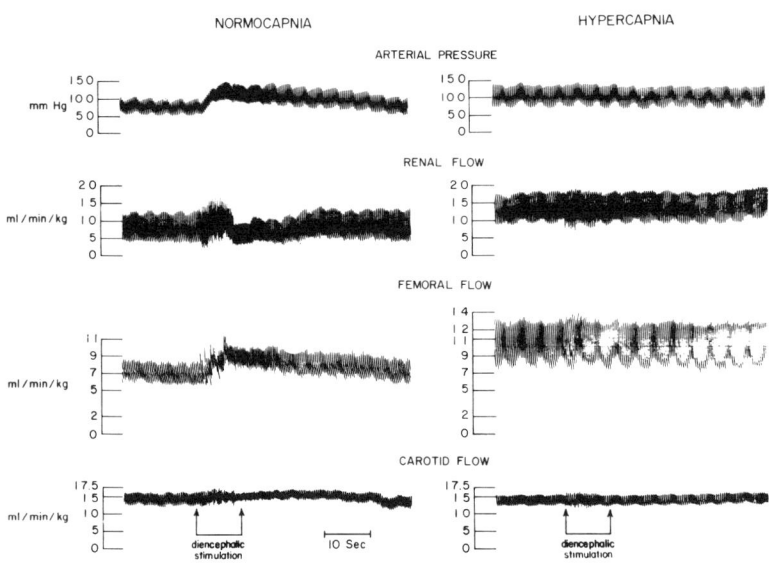

Fig. 7. Effect of hypercapnia (pH 7.27, $pCO_2$ 53 torr) on mean aortic pressure, heart rate and mean regional flow responses to stimulation of a site in ventromedial nucleus of the hypothalamus in a 10-day-old piglet (0.9 mA, 50 Hz, for 10 sec). Decreases in responses under hypercapnia: mean aortic pressure +51% to 7%; heart rate +31% to -3%; $\overline{RF}$ +47% to 8%; $\overline{FF}$ +25% to +9%; $\overline{CF}$ +11% to +4%.

is known in adult mammals to increase the excitability of
many neuronal systems as indicated by, e.g., increased discharge in both single neuron and population recordings
(19, 21, 43, 44, 46). A graph of the effects of hypercapnia on the responses to a variety of stimulus intensities applied to a site in zona incerta is shown in Figure
6; the depression of the responses under hypercapnia can
be seen. The loss of depressor responses, which were obtained to the lower intensities (0.04-0.1 mA), is unexpected since, with sympathetic activation (as indicated by
the increased pressure and flows under hypercapnia (86)),
one would expect an inhibitory intervention to result in
greater responses (42). An example of a loss of pressor
responses to hypothalamic stimulation under hypercapnia is
shown in Figure 7. The decreased magnitude of responses
observed in the animals either with vagi intact or after
vagotomy, would indicate that, while increased sympathetic
discharge is possible, as exemplified by the increased
blood pressure and flows obtained under hypercapnia, the
rostral stations of the cardiovascular controlling system
are either less responsive to the electrical stimulation
or else unable to increase sympathetic activity any further in the hypercapnic state.

In adult cats moderate hemorrhage resulted in increased outflow from the cardiovascular controlling centers
as indicated by increased sympathetic discharge (44, 49).
Gellhorn (42) reported that moderate hemorrhage resulted
in increased responses to hypothalamic stimulation in adult
cats. On the other hand, hemorrhage in the neonatal pig
has a marked depressive effect on the pressor responses to
diencephalic stimulation. The responses were either diminished or lost in the 11 piglets studied to date. A test
of peripheral receptors (0.5 µg NE/kg) showed that the
alpha adrenergic receptors were still functional. In animals 1-day or less of age, responses tended to be lost after moderate hemorrhage (16-20 mg/kg); in animals 3 days
of age or more, responses were still present. A graph of
responses in a 1-day old with a loss of 6.7 ml/kg whole
blood is shown in Figure 8; there were no responses to stimulation with the stress of a hemorrhage of 20 ml/kg. The
results of diencephalic stimulation under the stress of
either hypercapnia or hemorrhage suggest that moderate
stress severely effects the vasoactive sites in the diencephalon.

If one is discussing the effectiveness of activation or inhibition of the sympathetic nervous system, it becomes necessary to know the degree of maturation of the peripheral and cardiac adrenergic receptors present at the various ages.

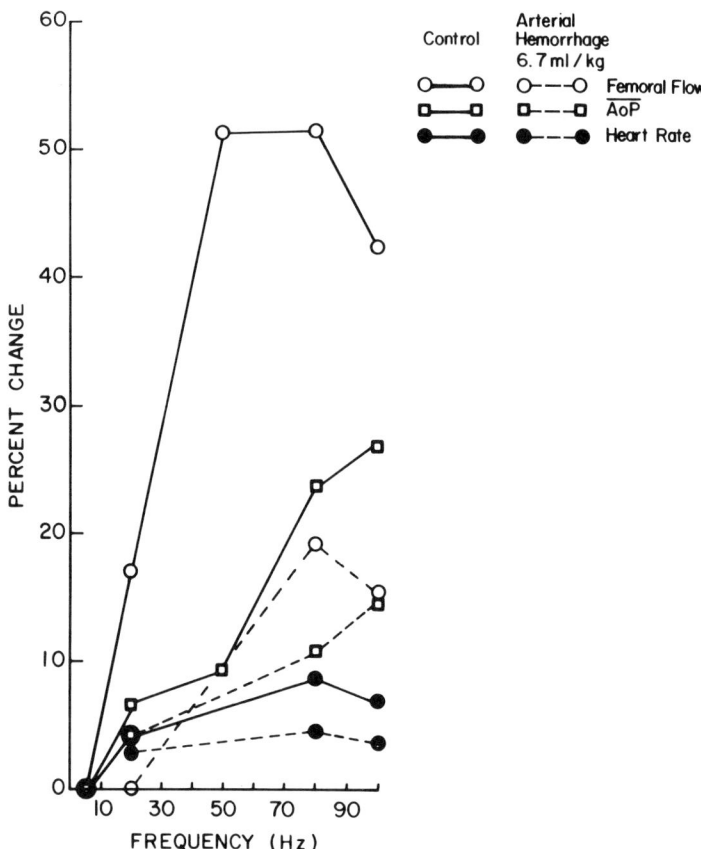

Fig. 8. Graphs relating effect of hemorrhage (6.7 ml/kg) on percent change in mean aortic pressure, (□); heart rate (●) and mean femoral flow (o) responses to increasing frequency of stimulation at a diencephalic site in a one-day-old piglet (0.8 mA for 10 sec at each frequency).

### III. Studies of Peripheral Cardiovascular Adrenergic Mechanisms

In our investigation of the postnatal development of cardiovascular effects of afferent and central stimulation, the question of responsiveness of the peripheral adrenergic mechanisms prompted us to examine the effects of exogenous catecholamines on cardiovascular function in piglets. The maturation of positive inotropic and chronotropic responses to adrenergic stimuli had been studied in fetal and neonatal mammals by testing effects of exogenous catecholamines on heart *in situ* (1, 2, 3, 5, 7, 84, 88) or on isolated myocardium (4, 15, 37); and by electrical stimulation of cardiac sympathetic efferents (8, 30, 31, 38, 41, 58, 88, 98) or field stimulation of isolated myocardium (4, 100). Cardiac β-adrenergic receptors have been demonstrated to be present in fetal and newborn lambs by the use of adrenergic blocking agents (6, 81, 82, 97, 98). The development of adrenergic responses in the peripheral circulation has mainly centered on the renal (55, 62, 63, 71) and the femoral regions (9, 26, 45, 54, 66). We evaluated the effects of graded doses of norepinephrine, epinephrine, and isoproterenol on arterial and ventricular pressures, heart rate, and carotid, renal, and femoral flows in piglets of differing postnatal age, before and after adrenergic blockade (16).

#### Specific Procedures

After establishing control conditions of cardiovascular function in 87 piglets under pentobarbital anesthesia, norepinephrine (NE), epinephrine (E) or isoproterenol (ISP) were administered in a randomized series within the range 0.03-1.0 µg/kg. Single dose intra-atrial injections were made through the catheter introduced via the right jugular vein pushed to the level of the right atrium. Enough time was allowed (at least 5 min) between each injection to permit return to control levels of function. Maximum intraventricular pressure change during isovolumetric contraction (max dP/dt), obtained via calibrated electronic differentiation of the pressure pulse, was used as the index of contractile function. Changes in max dP/dt at known end-diastolic pressure were used to evaluate inotropic responses.

Table 5

Cardiac Effects of Single 0.1 µg/kg Doses of Catecholamines
in Piglets and Miniature Swine Under Pentobarbital Anesthesia [a]

| Age | Inotropic Effect ($\Delta dP/dt$, mm Hg/msec) | | | | Chronotropic Effect ($\Delta HR$, bpm) | | | |
|---|---|---|---|---|---|---|---|---|
| | NE | | E | | NE | | E | |
| | $\bar{X}_\Delta$ | SE | $\bar{X}_\Delta$ | SE | $\bar{X}_\Delta$ | SE | $\bar{X}_\Delta$ | SE |
| ≤ 1 Day | all 0 (N=6) | | +0.03 ±0.12 (N=6) | | +1.4 ±0.81 (N=11) | | +0.6 ±1.17 (N=9) | |
| 2-4 days | all 0 (N=5) | | +0.15 ±0.15 (N=6) | | +4.2 ±1.91 (N=9) | | +8.6 ±3.41[b] (N=7) | |
| 7-10 days | +0.18 ±0.07[b] (N=6) | | +0.06 ±0.04 (N=6) | | +1.7 ±3.15 (N=9) | | +11.2 ±3.48[b] (N=11) | |

| | | | |
|---|---|---|---|
| 12-16 days | +0.05 ±0.03 (N=5) | +0.11 ±0.07 (N=4) | +8.3 ±3.06 (N=8) | +6.0 ±1.67[b] (N=9) |
| 5-6 months | +0.13 ±0.05[b] (N=6) | +0.15 ±0.04 (N=6) | -5.7 ±1.39[b] (N=6) | +6.3 ±1.83[b] (N=6) |

[a] See text. N = number of animals; $\bar{X}_\Delta$ = means of absolute change from pre-injection control value in each animal; SE = standard error of the mean.

[b] Observed change significantly greater than zero change ($p < 0.05$).

## Results

Table 5 summarizes two of the cardiac effects of randomized single (0.1 µg/kg) doses of norepinephrine and epinephrine for the 5 groups of animals. Significant positive chronotropic effects were not observed in the youngest animals until 0.5 µg/kg of either catecholamine was given; in the next oldest, 0.1 µg epinephrine/kg was sufficient to elicit a significant increase in heart rate. These heart rate changes were greater in the 2-4-day old than in the 1-day old piglets, and even greater in those 1-week old, as previously reported for kittens and puppies by Hutchinson (58) and in isolated rabbit hearts by Brus and Jacobowitz (15). After 1 week of age, some piglets exhibited bradycardia after norepinephrine administration, as did the mature minipigs.

Significant positive inotropic effects were not observed in either group of younger piglets until a dose of 0.5 µg/kg of either catecholamine was given. As with heart rate, effects on max dP/dt were greater in the 2-4-day old piglets than in the younger ones, and even greater in the 1-week old animals. The cardiac effects of graded doses of isoproterenol included differences in threshold dose and magnitude of responses, with increasing postnatal age, that are qualitatively similar to those results with norepinephrine and epinephrine, as shown in Figure 9. This contrasts with supersensitivity to norepinephrine but not isoproterenol in papillary muscle from young neonatal lambs (37).

Ventricular heart rate and contractile mechanisms of response to exogenous catecholamines appear to mature relatively slowly within the first 2-3 weeks after birth in pigs. The results of the experiments just described are consonant with findings in other species placing the piglet between the less mature rabbit (30,91) or puppy (8, 38, 41, 58, 84) and the more mature lamb (31, 32, 37, 81, 88, 97).

Table 6 summarizes the most significant peripheral vascular effects of norepinephrine or epinephrine in the 5 age groups of animals. Significant increases in diastolic blood pressure (DBP) were observed in 2 to 4-day old piglets and even in younger ones following administration of 0.1 µg norepinephrine/kg. This represents a significantly lower threshold for arterial pressure effect than for cardiac

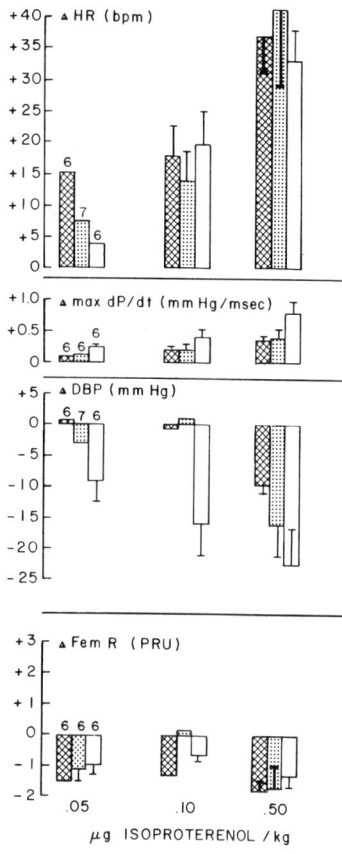

Fig. 9. Dose-response relationship for isoproterenol (randomized intra-atrial doses) in piglets 2 to 4 days old (▨) and one week old (▦) and in mature miniature swine (□): means (and standard errors) of effect on heart rate (top panel), maximum RV dP/dt (second panel), diastolic blood pressure (third panel), and femoral resistance (bottom panel). Numerals above bars for mean responses to .05 µg ISP/kg indicate numbers of animals in each group. Standard errors are shown as perpendiculars for those mean values that are significantly different from zero change ($p \leq .05$).

Table 6

Peripheral Vascular Effects of Single 0.1 µg/kg Doses of Catecholamines in Piglets and Miniature Swine under Pentobarbital Anesthesia[a]

| Age | Arterial Pressure (ΔDBP, mm Hg) | | | | Renal Resistance (ΔRen R, PRU) | | | | Femoral Resistance (ΔFem R, PRU) | | | |
|---|---|---|---|---|---|---|---|---|---|---|---|---|
| | NE | | E | | NE | | E | | NE | | E | |
| | $\bar{X}_\Delta$ | SE | $\bar{X}_\Delta$ | SE | $\bar{X}_\Delta$ | SE | $\bar{X}_\Delta$ | SE | $\bar{X}_\Delta$ | SE | $\bar{X}_\Delta$ | SE |
| ± 1 day | +5.0 ±0.66[b,c] (N=11) | | +2.8 ±1.18 (N=8) | | −0.1 ±0.37 (N=5) | | +1.1 ±0.31[b] (N=5) | | −0.2 ±0.19 (N=8) | | +0.4 ±0.30 (N=9) | |
| 2–4 days | +4.9 ±2.19 (N=9) | | +5.4 ±1.99 (N=7) | | 0.0 ±0.63 (N=6) | | +1.3 ±0.37[b] (N=4) | | +0.3 ±0.71 (N=6) | | −0.1 ±0.74 (N=8) | |
| 7–10 days | +8.3 ±2.14[b,c] (N=8) | | +0.7 ±1.84 (N=11) | | +1.9 ±0.95 (N=6) | | −0.6 ±1.27 (N=5) | | +0.2 ±1.27 (N=6) | | +0.4 ±0.51 (N=9) | |

| | | | | | |
|---|---|---|---|---|---|
| 12-16 days | +4.6 ±1.90[b,c] (N=7) | +4.0 ±4.17 (N=10) | +0.5 ±2.79 (N=5) | +0.3 ±1.11 (N=4) | +0.1 ±0.52 (N=7) | -0.5 ±0.62 (N=9) |
| 5-6 months | +22.0 ±3.97[b] (N=6) | -9.4 ±2.12[b] (N=5) | +0.6 ±0.12[b] (N=6) | +0.6 ±0.34 (N=5) | +0.7 ±0.17[b] (N=6) | 0.8 ±0.19[b] (N=5) |

[a] See text. N = number of animals; $\bar{X}_\Delta$ = mean of absolute change from pre-injection control value in each animal; SE = standard error of the mean.

[b] Observed change significantly greater than zero change ($p < 0.05$).

[c] $\bar{X}_\Delta$ significantly $< \bar{X}_\Delta$ for 5-6 month old animals ($p < 0.01$).

contractile effect of norepinephrine in these young animals. A pressor response to epinephrine was observed with all doses tested in animals up to 4 days of age; however, a dose of 0.5 µg epinephrine/kg was required to evoke a consistently pressor response in animals 1-week old or older. Pressor effects of either catecholamine were accompanied by increased femoral flow and by biphasic renal flow changes, except in the youngest piglets. Larger doses of catecholamines were required to increase renal resistance significantly, as in the puppies studed by Jose et al. (62).

One of the most interesting findings with respect to the peripheral vascular effects of norepinephrine and epinephrine was that low doses of epinephrine had pressor effects in the younger piglets, in contrast to the vasodilator effect in older animals. The vasodilator effect of low doses of epinephrine has been attributed to its action on β-adrenergic vascular receptors. We tested this mechanism by evaluating cardiovascular response to graded test doses of isoproterenol, its more specific agonist. The positive inotropic effect was observed in piglets of all ages, but a vasodilator response was not elicited until they were at least one week old, as shown in Figure 9. There have been few studies on the vascular effects of the β-adrenergic agonist isoproterenol in young animals, although its cardiac effects generally increase during fetal development in lambs (6, 37, 81). In the unanesthetized fetus of that species, differing test methods lead to differing responses: a single dose of 1 µg isoproterenol/kg decreases blood pressure without changing heart rate (97), while 50-100 γ isoproterenol/kg/min increases heart rate without changing blood pressure (5).

The use of α-adrenergic blockade alone as an index of α-adrenergic receptor activity in lambs revealed an age-dependent decrease in arterial pressure after administration of 0.1-5.0 mg phenoxybenzamine/kg (82, 98) or 0.1 mg phentolamine/kg (98) and suggested that α-adrenergic receptors were present at an early age in that species. More definitive evidence on this question was supplied by the demonstration of age-dependence of blockade of the effects of test doses of an α-adrenergic agonist. In fetal lambs, the pressor effect of 1.0 µg norepinephrine/kg was blocked by 0.1-0.2 mg phenoxybenzamine/kg without alteration of the effects on heart rate and pulmonary and aortic flows (81). In piglets of all ages blockade with 0.25 mg phentolamine/kg was accompanied by depression or inhibition of pressor and renal vascular effects of 0.5 µg norepinephrine/kg, the normal positive inotropic response being abolished in many piglets but not in mature swine (16).

The presence of β-adrenergic mechanisms in the cardiovascular system of fetal and neonatal mammals has not only been demonstrated by the cardiac responses to isoproterenol, but also by slowing of the heart rate after administration of β-adrenergic blocking agents to fetal and neonatal lambs (6, 82, 97, 98), and by loss of cardiac but not pressor effects of norepinephrine after blockade with propranolol in fetal lambs (81) and neonatal piglets (16). In piglets older than 2 days of age, 0.1 mg propranolol/kg blocked the depressor and femoral vascular effects as well as the cardiac chronotropic and inotropic effects of 0.1 µg isoproterenol/kg. Although cardiac β-receptor mechanisms are apparent at birth also in this species, vascular β-receptor mechanisms appear to develop more slowly than do α-receptor mechanisms.

## IV. Studies of the Reflex Cardiovascular Responses to Interactions Between the Peripheral and Central Nervous Systems

Since the regulatory system must integrate information from more than one input at a time, we also chose to analyze interactions between peripheral somatic afferents and between somatic and baroreceptor afferents, and interactions between substations of the cardiovascular controlling system and baroreceptor reflexes. Interactions between baroreceptor reflexes and somatic nerve stimulation have been studied in the adult animal by a few investigators (72, 85, 94, 95). There have been a number of studies of interactions between the hypothalamus and baroreceptor reflexes in adult animals (e.g., 23, 29, 40, 57, 67, 73, 75). To our knowledge, our laboratory is the only one currently engaged in a study of either of these patterns of interaction in the neonate.

### Specific Procedures

Reproducible pressor or depressor responses were first obtained by experimental selection of a combination of appropriate stimulus parameters for somatic afferent nerve and central vasoactive sites. The carotid sinus reflex was either inhibited or stimulated as described above. The following sequence pattern of stimulation was carried out after obtaining control responses to individual stimulation procedures: stimulus 1 alone, stimulus 2 alone, 1 followed by 2, 2 followed by 1, and simultaneous stimulation.

### Results

<u>Interactions Between Somatic Afferents</u>. In piglets of all ages, high-frequency sciatic nerve stimulation or stimulation of the median nerve of the brachial plexus resulted in pressor responses accompanied by increases in heart rate and regional flows. Combinations of somatic stimulation produced pressor responses which were greater than the effect of either stimulation alone (upper section of Table 7). Low-frequency stimulation of either nerve resulted in depressor responses (Fig. 10, top traces). When high-fre-

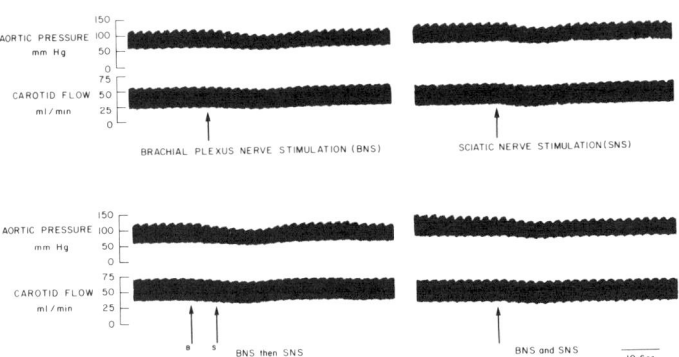

Fig. 10. Responses obtained to interactions between low-frequency stimulation (5 Hz for 10 sec, 0.7 mA) of the median nerve of the brachial plexus (BNS) and sciatic nerve (SNS). *Top set of traces*: left: BNS alone; right: SNS alone. *Bottom set of traces*: left: first, median nerve of the brachial plexus stimulation (B), then sciatic nerve stimulation (S); right: simultaneous median nerve of the brachial plexus stimulation and sciatic nerve stimulation.

quency pressor stimulation was interacted with low frequency depressor stimulation in 14 piglets, depressor responses were lost and pressor responses dominated, as shown for effects of pressor sciatic nerve stimulation and depressor stimulation of the median nerve of the brachial plexus (Table 7, lower section). When low-frequency depressor stimulations were interacted in 11 piglets (Figure 10, lower traces), the decreased blood pressure was accompanied by decreases in femoral and carotid flows without change in heart rate. The magnitude of the responses to any of the combinations of stimulation were not significantly different from the responses to stimulation of either nerve alone.

In the adult, simultaneous stimulation of two somatic afferents with stimulus parameters that evoked falls in blood pressure (depressor responses) resulted in responses that exceeded that of separate stimulation but usually remained below the sum of the latter (94). However, in the piglets the magnitude of the responses to combinations of such stimulation were not significantly different from the depressor responses obtained to each nerve alone. Facilitation between pressor reflexes to afferent somatic nerve stimulation in adult cats has been reported to be more easily obtained (94). Although we have obtained responses which were significantly greater in combination than to each nerve alone, we have never observed facilitation (Table 8).

We have concluded that the piglets were capable of some central nervous system augmentation and inhibition of combined afferent somatic stimulation.

<u>Interactions Between Somatic and Baroreceptor Afferents</u>. Responses to interaction between high frequency sciatic nerve stimulation and carotid sinus inhibition by bilateral common carotid occlusion, each of which had a pressor effect, is summarized in Table 8, upper section. Combined stimulation resulted in greater pressor responses, with larger changes in heart rate and femoral and renal resistances, than to bilateral common carotid occlusion alone, as in the 3 mature swine tested. Responses to interaction between high frequency sciatic nerve stimulation and carotid sinus stimulation by infusion are summarized in the lower section of Table 8. The depressor and other responses to carotid sinus stimulation by infusion were lost during the combinations of stimulations. A typical response to the inter-

Table 7

Significant Changes in Cardiovascular Parameters During Somatic-Visceral Afferent Interactions in Newborn Piglets Under Halothane Anesthesia[a]

|  | $\overline{\Delta AoP}$ (%) | ΔHR (%) | ΔFem R (%) | ΔRen R (%) |
|---|---|---|---|---|
| Pressor SNS: Carotid Sinus Inhibition | | | | |
| SNS alone | +26.6 ±4.7 | +13.0 ±2.9 | +16.2 ±8.0 | (+8.6 ±4.3) |
| BCCO alone | +10.3 ±1.4 | (+1.0 ±2.0) | (+1.8 ±3.4) | (+10.1 ±9.8) |
| S then O | +30.9 ±5.5[c] | (+1.2 ±3.5) | +23.0 ±3.3 | +15.3 ±5.3[b] |
| O then S | +31.4 ±5.5[c] | +4.2 ±2.1 | +16.9 ±6.7 | +29.2 ±9.3[b] |
| S plus O | +38.1 ±6.2[c] | +6.9 ±3.0 | +14.0 ±5.6 | +25.4 ±8.7[b] |

Pressor SNS: Carotid Sinus Stimulation (7 piglets)

| | | | | |
|---|---|---|---|---|
| SNS alone | +27.8 ±6.9 | +8.2 ±2.0 | +18.3 ±5.9 | (+17.3 ±21.5) |
| CSI | -13.5 ±2.9 | (-4.2 ±2.4) | -8.2 ±3.2 | (-4.8 ±4.6) |
| S then I | +28.1 ±8.2[d] | (+6.2 ±4.1) | +16.2 ±3.4[d] | (+17.3 ±5.8) |
| I then S | (+15.2 ±9.5) | (+2.6 ±3.0) | (+15.7 ±14.2) | (+11.2 ±12.0) |
| S plus I | (+13.6 ±7.5) | (+7.5 ±4.8) | (-1.0 ±3.0) | (-6.0 ±7.9) |

[a] Values are arithmetic means and standard errors; all means are significantly different from zero change ($.001 < p < .05$) except values in parentheses.

[b] Significantly different from SNS alone ($p < .04$).

[c] Significantly different from BCCO alone ($p < .04$).

[d] Significantly different from CSI alone ($p < .04$).

Table 8

Significant Changes in Cardiovascular Parameters During Somatic-Somatic Afferent
Interactions in Newborn Piglets Under Halothane Anesthesia[a]

| | $\overline{\Delta AoP}$ (%) | $\Delta HR$ (%) | $\overline{\Delta FF}$ (%) |
|---|---|---|---|
| Pressor SNS: Pressor BNS (N = 13) | | | |
| SNS alone | +27.7 ±4.0 | +13.0 ±2.9 | +43.5 ±11.7 |
| BNS alone | +28.9 ±3.9 | +12.6 ±2.7 | +43.2 ±11.3 |
| S then B | +41.6 ±6.4[b,c] | +19.4 ±3.1 | +81.8 ±21.3[b,c] |
| B then S | +38.6 ±6.6 | +17.6 ±3.0 | +50.6 ±16.9 |
| S plus B | +39.7 ±4.6[b,c] | +18.7 ±3.0 | +57.2 ±10.9 |

Pressor SNS: Depressor BNS (N = 8)

| | | |
|---|---|---|
| SNS alone | +26.2 ±6.1 | +13.1 ±5.1 | +25.3 ±8.2 |
| BNS alone | -13.9 ±1.7[b] | (+2.7 ±3.4) | -13.6 ±3.9[b] |
| S then B | +29.6 ±9.5[c] | +14.1 ±5.2 | +48.9 ±17.7[c] |
| B then S | +21.6 ±9.1[c] | +14.9 ±5.7 | +43.3 ±20.3[c] |
| S plus B | +15.3 ±6.2[c] | +14.3 ±4.9 | (+28.2 ±20.0) |

[a] Values are arithmetic means and standard errors; all means are significantly different from zero change (.001 < p < .03)

[b] Significantly different from SNS alone (.02 < p < .05)

[c] Significantly different from BNS alone (.001 < p < .01)

actions (in 8 piglets) between low frequency sciatic nerve stimulation and carotid sinus stimulation by infusion, each of which has a depressor effect, is shown in Figure 11. The magnitude of the responses to any of the combinations of stimulation were not significantly different from responses to either alone. Interactions between low frequency sciatic nerve stimulation and bilateral common carotid occlusion were also carried out in 12 piglets; the depressor response to low frequency sciatic nerve stimulation was lost during all combinations of stimulation.

In the adult, the combination of bilateral common carotid occlusion with afferent somatic nerve stimulation that leads to depressor responses results in a greater fall in blood pressure and greater muscle vasodilation than

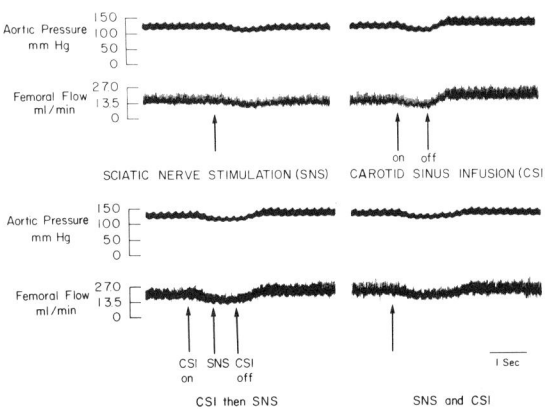

Fig. 11. Responses obtained to interactions between low frequency sciatic nerve stimulation (SNS) (stimulation parameters: 5 Hz for 10 sec, 0.8 mA) and carotid sinus stimulation (CSI) by high pressure (distension). *Top set of traces*: left: sciatic nerve stimulation alone; right: carotid sinus stimulation alone. *Bottom set of traces*: left: first carotid sinus stimulation (on), then sciatic nerve stimulation, for 10 sec, carotid sinus stimulation off; right: simultaneous sciatic nerve stimulation and carotid sinus stimulation.

to just bilateral common carotid occlusion alone (61). This was not observed in the neonate, in which the depressor responses were converted to pressor responses during combinations of stimulation. However, we did observe interaction responses, similar to those reported in adult cats (61), between pressor responses to afferent stimulation and bilateral common carotid occlusion. High frequency and/or intensity of stimulation of somatic afferents leads to increased blood pressure and decreased renal flow, which was markedly diminished when somatic afferent stimulation in adult cats was combined with bilateral common carotid occlusion (61). Significantly increased renal resistance was also observed with all combinations of stimulation between pressor sciatic nerve stimulation and bilateral common carotid occlusion in neonatal piglets.

On the other hand, high-frequency sciatic nerve stimulation has been reported to decrease the open-loop gain of the carotid sinus reflex in adult cats (72). This result is in disagreement with that of Ulmer (95) who did not find a significant change in the slope of the intrasinus pressure-arterial pressure curve. Alterations in the cardiac component of the baroreceptor reflex have been reported by Quest and Gebber (85). They found that the bradycardia evoked by carotid sinus nerve stimulation was blocked by high-frequency (pressor) SNS and augmented by low-frequency (depressor) sciatic nerve stimulation. In newborn piglets, heart rate changes, if present at all, were not significantly greater in combination than to any one stimulus alone. Responses to carotid sinus stimulation in piglets were lost when combined with high-frequency somatic nerve stimulation and were not significantly different when combined with low-frequency somatic nerve stimulation.

We have concluded from these observations that newborn piglets are capable of some central nervous system integration of afferent inputs from somatic nerves and the carotid baroreceptors.

Interactions Between Baroreceptor Afferents and Central Vasoactive Site Stimulation. A wide variety of interactions between the baroreceptor reflex and stimulation of vasoactive sites in the hypothalamus in adult mammals, ranging from no interaction to suppression of the responses to hypothalamic stimulation with carotid sinus stimulation (23), to modification of the chronotropic responses to the carotid sinus reflex by hypothalamic stimulation (29, 76), have been observed. Recently, Kumada et al. (73) reflex reported an increase in gain of the carotid sinus reflex with hypothalamic stimulation. Gebber and Klevans (39) have reported inhibition of baroreceptor-induced bradycardia with stimulation of certain medullary areas in adult cats. They did not study the effect of medullary stimulation on the reflex responses to baroreceptor inhibition.

Our studies of interactions between medullary or hypothalamic vasoactive sites and carotid baroreceptors in the neonate are still in progress. An example of the interactions between medullary pressor site (open circle, level B of Fig. 2) stimulation and carotid sinus inhibition (bilateral common carotid occlusion) from a 3-day old piglet is shown in Figure 12. In 5 piglets in which this sequence

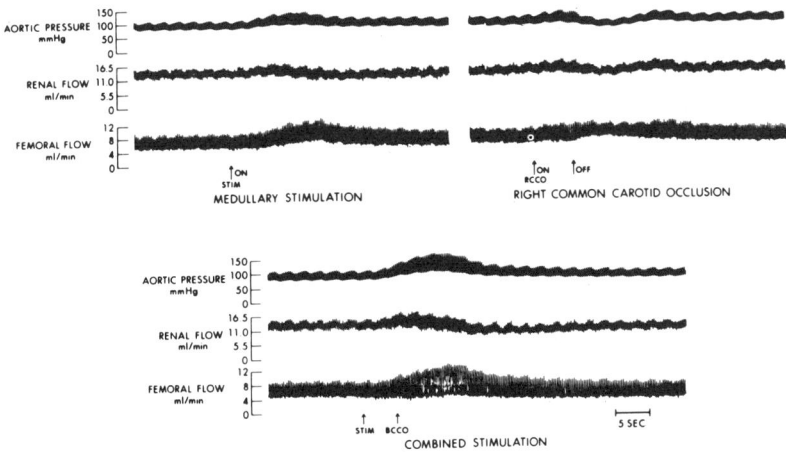

Fig. 12. Aortic pressure and renal and femoral arterial flow responses to stimulation of a site in the medullary pressor area (open circle, Section B in Fig. 2, stimulation parameters: 50 Hz for 10 sec, 0.6 mA) and to right common carotid artery occlusion (left carotid artery cannulated). (top set of traces). Bottom traces: combined procedures: medullary stimulation followed, 5 sec later, by right common carotid artery occlusion.

of interactions was carried out, pressure and flow responses were greater to combination of stimulation than to each stimulus alone. Figure 13 shows a response pattern obtained from a site in lateral hypothalamus in a 2-day old piglet. Arterial pressure responses at this site in piglets were similar to those reported by Gebber and Snyder (40) in adult cats--enhancement of the responses when hypothalamic stimulation was combined with bilateral common carotid occlusion. They, however, did not record regional flow responses.

The results of our experiments on combined peripheral and central interactions suggest that the neonate's cardiovascular controlling system is capable of complex functions.

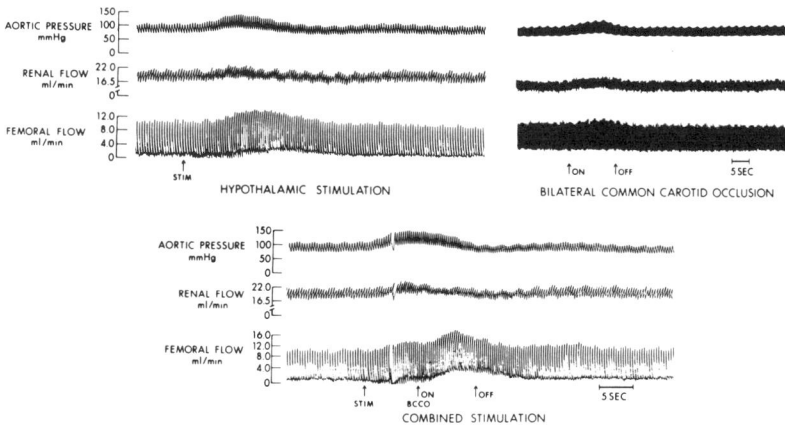

Fig. 13. Aortic pressure and renal and femoral arterial flow responses to hypothalamic stimulation (lateral hypothalamic area, Section c of Fig. 4, 50 Hz, for 10 sec, 0.7 mA) and bilateral common carotid artery occlusion (*top set of traces*). *Bottom traces*: combined procedures: hypothalamic stimulation followed by bilateral common carotid occlusion.

## SUMMARY DISCUSSION

It is becoming apparent that postnatal maturation of the cardiovascular controlling system is asynchronous with respect to control of heart rate, blood pressure, and regional blood flows. This is manifested by a number of findings in investigations of the reflex responses to afferent stimulation, the responses to central stimulation, and the functional presence of peripheral adrenergic receptors. A significant finding is the absence of the total integrated response involving both cardiac and peripheral vascular components.

There is considerable evidence to show that the mammalian heart is innervated at birth. The efferent parasympathetic pathways are functional at birth; for example, in piglets, stimulation of nucleus ambiguus in the medulla resulted in a definite bradycardia, with vagi intact, while vagotomy resulted in a statistically significant increase in heart rate (Table 2). There have been a number of reports which indicate that efferent sympathetic innervation of the heart develops at various rates in different mammalian species (8, 32, 38, 41, 58, 91, 100). Our results (16) have shown that in piglets there is an age-dependent chronotropic response to exogenous catecholamines (Table 5). In addition, we have been able to elicit tachycardia by stimulation of sites within the brain stem and diencephalon. Tachycardia to hypothalamic stimulation in fetal lambs of late gestational age has also been reported (104).

Nevertheless, considerable postnatal maturation of heart rate control can be concluded from the observations of absence or sporadic occurrence of cardiac responses to various perturbations of the cardiovascular system. Immaturity is manifested by the absence of the cardiac component of the reflex responses to baroreceptor stimulation or inhibition and to stimulation of somatic afferents. The pathways of somatosympathetic reflexes from the spinal cord to the medulla are different from those of the baroreceptor afferents (89). Therefore the use of both baroreceptor reflexes and somatosympathetic reflexes permits independent testing of the cardiovascular regulatory system. We found that, in the youngest animals, the peripheral vascular components of the responses to carotid sinus manipulations were present while the heart rate component of the responses was absent. Stimulation of somatic afferents led to changes in pressure and flows (Table 1 and Ref. 17); but again, the cardiac component of the responses to low-frequency stimulation, which was seen in mature swine as in adult mammals of other species (61, 72, 85), was not seen in the younger animals. Furthermore, post-stimulus bradycardia following high frequency stimulation was never observed. The characteristic bradycardia seen at the peak of the pressor response to exogenous norepinephrine in adults was also absent in younger piglets and newborn humans (2, 7, 66). Age-dependent changes in the inotropic response to norepinephrine, epinephrine and isoproterenol have also been observed in mammals, including piglets (Table 5).

Mammalian vascular smooth muscle develops its sympathetic innervation early but there is considerable postnatal maturation of blood pressure control. Pressor responses to baroreceptor inhibition, high frequency or intensity stimulation of somatic afferents and exogenous norepinephrine are detectable at birth. Vasopressor areas in the brain stem and diencephalon can be located at birth in piglets and in fetal lambs. Inhibition of sympathetic regulation of blood pressure is evidenced by depressor responses to baroreceptor stimulation, low-frequency stimulation of somatic afferents, and direct stimulation of sites in the medulla. Depressor responses in response to somatic afferent or medullary stimulation in piglets were generally smaller than are pressor responses, perhaps reflecting the degree of basal sympathetic tone present in the neonate.

The progressive increase in blood pressure with age in neonatal mammals, and the significantly greater responses observed in the older piglets to various procedures such as somatic afferent or central stimulation, also constitute evidence for postnatal maturation of sympathetic control of blood pressure and flows. There were significant differences in the magnitude of the responses to medullary depressor area stimulation between two age groups (Table 3), indicating postnatal maturation of the vasodepressor components of the regulatory system. Furthermore, the magnitudes of the depressor responses, that we observed, may reflect the level of basal sympathetic tone, since, with less sympathetic activity, the responses to depressor area stimulation would be of smaller magnitude. There was no significant difference in the pressor responses between the less-than-1-week and 1-week groups of animals (Table 3); this may indicate little difference in maturation between these two age groups, which were between the 1-day and 1-month old piglets reported by Marshall and Breazile (80) to exhibit a significant difference in the magnitudes of pressor responses.

Immaturity of blood pressure control is also manifested by significantly greater pressor responses in older than in younger piglets to the various procedures and by the absence of depressor effects of exogenous epinephrine or isoproterenol in the youngest piglets. Maturation of vascular β-adrenergic receptors is relatively slow in this mammalian species. Although pressor and depressor responses to various perturbations are present at birth, effects on femoral and renal flows and resistances were not clear cut until at least one week after birth in piglets. There is pharmacologic evidence for α-adrenergic receptors in the renal circulation and β-adrenergic receptors in the femoral circulation by that time.

Immaturity of the central controlling system, *per se*, is suggested by the results of direct stimulation of sites in the medulla, diencephalon and mesencephalon. One significant finding is the absence of the total integrated response involving both blood pressure and vascular resistance changes in the femoral and renal circulations. It will be important to evaluate mesenteric flow changes during central stimulation, because of the role of that circulation in readjustments of total peripheral resistance. Another significant finding is the inability of the young piglets to maintain a pressor response to increasing frequency or intensity of stimulation of the medullary pressor area. It must be kept in mind that with electrical stimulation of the neuraxis, whether cells, fibers or both are being stimulated is unknown. All that can be said is that there are sites, which are localized, from which cardiovascular responses could be obtained with such stimulation. The relatively small number of such sites, particularly in the diencephalon, found in these young animals, compared with adults of other species in the same area (34, 64, 74) may also reflect immaturity or a species difference. This is presently under investigation by more extensive mapping experiments.

Immaturity of the cardiovascular controlling system is particularly manifest when one examines the responses obtained to more complex patterns of afferent stimulation, afferent and central stimulation, and the effects of moderate stresses on the cardiovascular responses to peripheral and central stimulation. The variety of complex interactions reported in the adult mammal to combinations of baroreceptor and somatic afferent stimulation or hypothalamic stimulation were not observed in the piglet, and, in particular, the alterations of the cardiac component of the baroreceptor reflex. Moreover, high-frequency sciatic nerve stimulation, which has been reported to either increase the open-loop gain of the baroreceptor reflex (72) or else not to affect the gain at all (95) in the adult, completely eliminated any response to carotid sinus stimulation in the piglet. Moderate stress such as hypercapnia and/or hemorrhage markedly affected the cardiovascular responses to afferent and central stimulation; these responses were considerably diminished in the presence of such stresses. These results suggest that the neonate's cardiovascular controlling system has a "low safety factor."

There appears, therefore, to be postnatal maturation of the integrating components of cardiovascular reflexes which have differing rates of development. The significantly increased magnitudes of the pressure and flow responses with increasing age to central stimulation reflect one aspect of this asynchrony; the lack of depressor responses to low doses of epinephrine and to isoproterenol in the younger animals indicates another; the delayed appearance of the cardiac components of various reflexes and the diminution of responses under moderate stresses points to still others.

## ACKNOWLEDGMENTS

These investigations were supported in part by Public Health Service Grants NS-12031 and HL-15444 and by Nassau Heart Association Grant No. 443.

The authors would like to thank Drs. G. D. Reddy, L. C. Weaver, and M. E. Salinas-Zeballos, who collaborated on different phases of the work. We would also like to acknowledge the technical assistance of Ms. Linda Crane, Ms. Barbara Buckley, Mrs. Marie Elbert, and Mr. Isaac Frasier.

The miniature swine were purchased from the Thompson Research Foundation, Illinois.

## NOTES

Much of this work has been reported at yearly meetings of the American Heart Association (*Circulation* 38: Suppl. 6: 85, 1968; 44:II-224, 1971; 46:II-5, 1972; 52:II-191, 1975), American Physiology Society (*Fed. Proc.* 32:254, 1973; 33: 444, 1974; 34:406, 1975) and at the Society for Pediatric Research (*Pediatric Res.* 6:343, 1972; 7:703, 1973; 10:312, 1976). Summaries were presented at the 1970 World Congress of Cardiology (*Cardiovascular Res.* 4:153) and 1974 International Congress of Physiological Sciences (*Transactions* 11:330).

## REFERENCES

1. Adams, F. H.; Hirvonen, L.; Lind, J.; and Peltonen, T. 1958. Physiologic studies on the cardiovascular status of newborn pigs. Effect of adrenaline, noradrenaline, acetylcholine and serotonin. *Études Néonatales* 7:53-61.

2. Adams, F. H.; Lind, J.; and Rauramo, L. 1958. Physiologic studies on the cardiovascular status of normal newborn infants. Effect of adrenaline, noradrenaline, 10% oxygen and 100% oxygen. *Études Néonatales* 7:62-70.

3. Alexander, G.; Bell, A. W.; and Setchell, B. P. 1972. Regional distribution of cardiac output in young lambs: effect of cold exposure and treatment with catecholamines. *J. Physiol.* 220:511-528. London.

4. Anderson, K. E.; Gensser, G.; and Nilsson, E. 1970. Contractility of isolated human fetal hearts: influence of contraction rate, acid-base parameters and a local anesthetic. *Acta Physiol. Scand.*: suppl. 353.

5. Assali, N. S.; Brinkman, C. R., III; and Nuwayhid, B. 1974. Comparison of maternal and fetal cardiovascular functions in acute and chronic experiments in sheep. *Amer. J. Obst. Gynec.* 120:411-425.

6. Barrett, C. T.; Heymann, M. A.; and Rudolph, A. M. 1972. $\alpha$- and $\beta$-adrenergic receptor activity in fetal sheep. *Amer. J. Obst. Gynec.* 112:1114-1121.

7. Beard, R. W. 1962. Response of the human fetal heart and maternal circulation to adrenaline and noradrenaline. *Brit. Med. J.* 1:443-446.

8. Boatman, D. L. and Brody, M. J. 1967. Cardiac responses to adrenergic stimulation in the newborn dog. *Arch. Int. Pharmacodyn.* 170:1-11.

9. Boatman, D. L.; Shaffer, R. A.; Dixon, R. L.; and Brody, M. J. 1965. Function of vascular smooth muscle and its sympathetic innervation in the newborn dog. *J. Clin. Invest.* 44:241-246.

10. Bloor, C. M. 1964. Aortic baroreceptor threshold and sensitivity in rabbits at different ages. *J. Physiol.* 174:163-171. London.

11. Bond, R. F. and Green, H. D. 1969. Cardiac output redistribution during bilateral common carotid occlusion. *Am. J. Physiol.* 216:393-403.

12. Book, S. A. and Bustad, L. K. 1974. The fetal and neonatal pig in biomedical research. *J. Animal Science* 38:997-1002.

13. Booth, N. H.; Bredeck, H. E.; and Herin, R. A. 1966. Baroreceptor and chemoreceptor reflex mechanisms in swine. In *Swine in Biomedical Research*, eds. L. F. Bustad, and R. O. McClellan, pp. 331-345. Seattle: Frayn Printing Co.

14. Broadie, T. A.; Davedas, M.; Rysavy, J.; Delaney, J. P.; and Leonars, A. S. 1973. The effect of hypoxia and posterior hypothalamic stimulation on colonic blood flow in the weanling puppy. *J. Ped. Surg.* 8:747-756.

15. Brus, R. and Jacobowitz, D. 1972. Influence of norepinephrine, tyramine and acetylcholine upon isolated perfused hearts of immature and adult rabbits. *Arch. Int. Pharmacodyn.* 200:266-272.

16. Buckley, N. M.; Yellin, E. L.; Miness, S. B.; and Frasier, I. D. 1974. Postnatal development of cardiovascular responses to exogenous catecholamines in piglets. *Transact. 26th Intl. Congr. Physiol. Sci.* 11:329. New Delhi.

17. Buckley, N. M.; Gootman, P. M.; Gootman, N.; Reddy, G. D.; Weaver, L. C.; and Crane, L. A. 1976. Age-dependent cardiovascular effects of afferent stimulation in neonatal pigs. *Biol. Neonate* 30:268-279.

18. Chai, C. Y. and Wang, S. C. 1962. Localization of central cardiovascular control mechanism in the lower brain stem of the cat. *Am. J. Physiol.* 202:25-30.

19. Chalazonitis, N. 1963. Effects of changes in $P_{CO_2}$ and $PO_2$ on rhythmic potentials from giant neurons. *Ann. N.Y. Acad. Sci.* 109:451-479.

20. Cheek, D. B. and Rowe, R. D. 1966. Aspects of sympathetic activity in the newborn, including the respiratory distress syndrome. *Ped. Clin. N.A.* 13:863-877.

21. Cohen, M. I. 1968. Discharge patterns of brain-stem respiratory neurons in relation to carbon dioxide tension. *J. Neurophysiol.* 31:142-165.

22. Coote, J. H.; Hilton, S. M.; and Zbrozyna, A. W. 1973. The ponto-medullary area integrating the defense reaction in the cat and its influence on muscle blood flow. *J. Physiol.* 229:257-274. London.

23. Coote, J. H. and Perez-Gonzales, J. F. 1972. The baroreceptor reflex during stimulation of the hypothalamic defence region. *J. Physiol.* 224: 74-75P. London.

24. Crane, L. A.; Gootman, N.; and Gootman, P. M. 1974. The effects of decamethonium (C-10) on blood pressure and heart rate in newborn piglets. *Arch. Int. Pharmacodyn.* 208:52-60.

25. ----. 1975. Age-dependent cardiovascular effects of halothane anesthesia in neonatal pigs. *Arch. Int. Pharmacodyn.* 214:180-187.

26. Dawes, G. S.; Lewis, B. V.; Milligan, J. E.; Roach, M. R.; and Talner, N. S. 1968. Vasomotor responses in the hindlimbs of foetal and new-born lambs to asphyxia and aortic chemoreceptor stimulation. *J. Physiol.* 195:55-81. London.

27. De Molina, A. F.; Achard, O.; and Wyss, O. A. M. 1953. Respiratory and vasomotor responses to stimulation of afferent fibers in somatic nerves. *Helv. Physiol. et Pharmacol. Acta* 11:1-19.

28. DiSalvo, J.; Parker, P. E.; Scott, J. B.; and Haddy, F. J. 1971. Carotid baroreceptor influence on total and segmental resistances in skin and muscle vasculatures. *Am. J. Physiol.* 220:1970-1978.

29. Djojosugito, A. M.; Folkow, B.; Kylstra, P. H.; Lisander, B.; and Tuttle, R. S. 1970. Differentiated interaction between the hypothalamic defence reaction and baroreceptor reflexes. I. Effects on heart rate and regional flow resistance. *Acta Physiol. Scand.* 78:376-385.

30. Downing, S. E. 1960. Baroreceptor reflexes in newborn rabbits. *J. Physiol.* 150:201-213. London.

31. Downing, S. E.; Milgram, E. A.; and Halloran, K. H. 1971. Cardiac responses to autonomic stimulation during acidosis and hypoxia in the lamb. *Am. J. Physiol.* 220:1956-1963.

32. Downing, S. E.; Talner, N. S.; Campbell, A. G. M.; Halloran, K. H.; and Wax, H. B. 1969. Influence of sympathetic nerve stimulation on ventricular function in the newborn lamb. *Circulation Res.* 25:417-428.

33. Dupuis, C.; Dupuis, B.; Adams, F. H.; Lind, J.; and Peltonen, T. 1958. Further studies on the cardiovascular status of normal newborn infants. IV. Effect of adrenaline, acetylcholine, 10% oxygen and 100% oxygen on the electrocardiogram. *J. Pediat.* 52:649-661.

34. Feigl, E. O. 1964. Vasoconstriction resulting from diencephalic stimulation. *Acta Physiol. Scand.* 60:372-380.

35. Fell, C. 1968. Changes in blood flow distribution produced by central sciatic nerve stimulation. *Am. J. Physiol.* 214:561-565.

36. Forsyth, R. P. 1970. Hypothalamic control of the distribution of cardiac output in the unanesthetized rhesus monkey. *Circulation Res.* 26:783-794.

37. Friedman, W. F. 1972. The intrinsic physiologic properties of the developing heart. *Progr. Cardiovasc. Dis.* 15:87-111.

38. Gauthier, P.; Nadeau, R. A.; and de Champlain, J. 1975. The development of sympathetic innervation and the functional state of the cardiovascular system in newborn dogs. *Can. J. Physiol. Pharmacol.* 53:763-776.

39. Gebber, G. L. and Klevans, L. R. 1972. Central nervous system modulation of cardiovascular reflexes. *Fed. Proc.* 31:1245-1252.

40. Gebber, G. L. and Snyder, D. W. 1970. Hypothalamic control of baroreceptor reflexes. *Am. J. Physiol.* 218:124-131.

41. Geis, W. P.; Tatooles, C. J.; Priola, D. V.; and Friedman, W. F. 1975. Factors influencing neurohumoral control of the heart in the newborn dog. *Am. J. Physiol.* 228:1685-1689.

42. Gellhorn, E. 1962. Effect of hemorrhage, reinjection blood and dextran on the reactivity of the sympathetic and parasympathetic systems. *Acta Neuroveg.* 22:291-299.

43. Gerard, R. W. 1936. Factors controlling brain potentials. *Cold Spring Harbor Symp. Quant. Biol.* 4:292-298.

44. Gernandt, B.; Liljestrand, G.; and Zotterman, Y. 1946. Efferent impulses in the splanchnic nerve. *Acta Physiol. Scand.* 11:230-247.

45. Gerová, M.; Gero, J.; Doležel, S.; and Konečny, M. 1974. Postnatal development of sympathetic control in canine femoral artery. *Physiol. Bohemoslov.* 23:289-296.

46. Gibbs, F. A.; Williams, D.; and Gibbs, E. L. 1940. Modification of the cortical frequency spectrum by changes in $CO_2$, blood sugar, and $O_2$. *J. Neurophysiol.* 3:49-58.

47. Gibson, E. A.; Blackmore, R. J. J.; Wijeratne, W. V. S.; and Wrathal, A. E. 1976. The "barker" (neonatal respiratory distress) syndrome in the pig: its occurrence in the field. *Vet. Rec.* 98:476-478.

48. Glauser, E. M. 1966. Advantages of piglets as experimental animals in pediatric research. *Exptl. Med. Surg.* 24:181-190.

49. Gootman, P. M. and Cohen, M. I. 1970. Efferent splanchnic activity and systemic arterial pressure. *Am. J. Physiol.* 219:897-903.

50. ----. 1971. Evoked potentials produced by electrical stimulation of medullary vasomotor regions. *Exptl. Brain Res.* 13:1-14.

51. ----. 1973. Periodic modulation (cardiac and respiratory) of spontaneous and evoked sympathetic discharge. *Acta Physiol. Polon.* 24:99-109.

52. ----. 1974. The interrelationship between sympathetic discharge and central respiratory drive. In *Central-rhythmic and regulation*, eds. W. Umbach, and H. P. Koepchen, pp. 195-209. Stuttgart:Hippokrates-Verlag.

53. Gootman, N.; Gootman, P. M.; Buckley, N. M.; Cohen, M. I.; Levine, M.; and Spielberg, R. 1972. Central vasomotor regulation in the newborn piglet (*Sus scrofa*). *Am. J. Physiol.* 222:994-999.

54. Gray, S. D. 1974. Catecholamine effects on neonatal arterial smooth muscle. *Proc. 26th Intl. Cong. Physiol. Sci.* 11:43. New Delhi.

55. Grusken, A. B.; Edelmann, A. B., Jr.; and Yuan, S. 1970. Maturational changes in renal blood flow in piglets. *Pediatric Res.* 4:7-13.

56. Haas, E.; Goldblatt, H.; Rowland, V.; and Vrtunski, P. 1974. Neurogenic pressor response due to mesencephalic electrical stimulation-effect of blockade on renin-angiotensin system. *Am. J. Physiol.* 226: 771-775.

57. Humphreys, P. W.; Joels, N.; and McAllen, R. M. 1971. Modification of the reflex response to stimulation of carotid sinus baroreceptors during and following stimulation of the hypothalamic defence area in the cat. *J. Physiol.* 216:461-482. London.

58. Hutchinson, E. A.; Percival, C. J.; and Young, I. M. 1962. Development of cardiovascular responses in the kitten. *Quart. J. Exp. Physiol.* 47:201-210.

59. Huttenlocker, P. R. 1970. Myelination and the development of function in immature pyramidal tract. *Exptl. Neurol.* 29:405-415.

60. Iriuchijima, J.; Koike, H.; and Kurihara, M. 1971. Vascular area which most contributes to the carotid occlusion reflex. *Pflügers Arch. Ges. Physiol.* 325: 279-286.

61. Johansson, B. 1962. Circulatory responses to stimulation of somatic afferents. *Acta Physiol. Scand.* 57: suppl. 198.

62. Jose, P. A.; Slotkoff, L. M.; Lilienfeld, L. S.; Calcagno, P. C.; and Eisner, G. M. 1974. Sensitivity of neonatal renal vasculature to epinephrine. *Am. J. Physiol.* 226:796-799.

63. Jose, P. A.; Slotkoff, L. M.; Montgomery, J.; Calcagno, P. C.; and Eisner, G. 1975. Autoregulation of renal blood flow in the puppy. *Am. J. Physiol.* 229:983-988.

64. Jurf, A. N., and Blake, W. D. 1972. Renal response to electrical stimulation in the septum and diencephalon of rabbit. *Circulation Res.* 30:322-331.

65. Kabat, H.; Magoun, H. W.; and Ranson, S. W. 1935. Electrical stimulation of points in the forebrain and midbrain. *Arch. Neurol. Psych.* 34:933-955.

66. Karlberg, P.; Moore, R. E.; and Oliver, T. K., Jr. 1965. Thermogenic and cardiovascular responses of the newborn baby to adrenaline. *Acta Paed. Scand.* 54:225-238.

67. Keith, I. C.; Kidd, C.; and Penna, P. E. 1973. Modification of sympathetic chronotropic carotid sinus reflex responses by hypothalamic stimulation. *J. Physiol.* 232:77-78P. London.

68. Kendrick, E.; Öberg, B.; and Wennergren, G. 1972. Vasoconstrictor fiber discharge to skeletal muscle, kidney, intestine and skin at various levels of arterial baroreceptor activity in the cat. *Acta Physiol. Scand.* 85:464-476.

69. Kero, P. 1974. Heart rate variation in infants with the respiratory distress syndrome. *Acta Paed. Scand.* 64: suppl. 250:5-70.

70. Khayutin, V. M. 1966. Specific and nonspecific responses of the vasomotor center to impulses of spinal afferent fibers. *Acta Physiol. Acad. Sci. Hung.* 29: 131-144.

71. Kleinman, L. I. and Lubbe, R. J. 1972. Factors affecting maturation of glomerular filtration rate and renal plasma flow in the new-born dog. *J. Physiol.* 223:395-410. London.

72. Kumada, M.; Nogami, K.; and Sagawa, K. 1975. Modulation of carotid sinus baroreceptor reflex by sciatic nerve stimulation. *Am. J. Physiol.* 228:1535-1541.

73. Kumada, M.; Schramm, L. P.; Altmansberger, R. A.; and Sagawa, K. 1975. Modulation of carotid sinus baroreceptor reflex by hypothalamic defence response. *Am. J. Physiol.* 228:34-35.

74. Kuo, J. S.; Chai, C. Y.; Lee, T. M., Liu, C. N.; and Lim, R. K. S. 1970. Localization of central cardiovascular control mechanism in the brain stem of the monkey. *Exptl. Neurol.* 29:131-141.

75. Kylstra, P. H., and Lisander, B. 1970. Differentiated interaction between the hypothalamic defence area and baroreceptor reflexes. II. Effects on aortic blood flow as related to work load on the left ventricle. *Acta Physiol. Scand.* 78:386-392.

76. Lindgren, P. 1955. Mesencephalon and the vasomotor system. *Acta Physiol. Scand.* 35: suppl. 121.

77. Löfving, B. 1961. Cardiovascular adjustments induced from the rostral cingulate gyrus. *Acta Physiol. Scand.* 53: suppl. 184.

78. Manning, J. W., Jr. and Peiss, C. N. 1960. Cardiovascular responses to electrical stimulation in the diencephalon. *Am. J. Physiol.* 198:366-370.

79. Marshall, A. E. and Braezile, J. E. 1974. Localization of cardiovascular centers in myelencephalon of newborn and older pigs. *Am. J. Vet. Res.* 35:223-229.

80. ----. 1974. Evidence for maturation of myencephalic cardiovascular control in the postnatal pig. *Am. J. Vet. Res.* 35:231-236.

81. Nuwayhid, B.; Brinkman, C. R., III; Su, C.; Bevan, J. A.; and Assali, N. S. 1975. Systemic and pulmonary hemodynamic responses to adrenergic and cholinergic agonists during fetal development. *Biol. Neonate* 26:301-317.

82. ----. 1975. Development of autonomic control of the fetal circulation. *Am. J. Physiol.* 228:337-344.

83. Picton-Warlow, G. G. and Mayer, F. E. 1970. Cardiovascular responses to postural changes in the neonate. *Arch. Dis. Child.* 45:354-359.

84. Privitera, P. J.; Loggie, J. M. H.; and Gaffney, T. E. 1969. A comparison of the cardiovascular effects of biogenic amines and their precursors in newborn and adult dogs. *J. Pharm. Exp. Ther.* 166:293-298.

85. Quest, J. A. and Gebber, G. L. 1972. Modulation of baroreceptor reflexes by somatic afferent nerve stimulation. *Am. J. Physiol.* 222:1252-1259.

86. Reddy, G. D.; Gootman, N.; Buckley, N. M.; Gootman, P. M.; and Crane, L. 1974. Regional blood flow changes in neonatal pigs in response to hypercapnia, hemorrhage and sciatic nerve stimulation. *Biol. Neonate* 25:249-262.

87. Rudolph, A. J.; Vallbona, C.; and Desmond, M. N. 1965. Cardiovascular studies in the newborn. III. Heart rate patterns in infants with idiopathic respiratory distress syndrome. *Pediatrics* 36:554-559.

88. Rudolph, A. M. and Heymann, M. A. 1974. Fetal and neonatal circulation and respiration. *Ann. Rev. Physiol.* 36:187-207.

89. Sato, A. and Schmidt, R. F. 1973. Somatosympathetic reflexes: afferent fibers, central pathways, discharge characteristics. *Physiol. Rev.* 53:916-947.

90. Scherrer, H. 1959. Hypothalamic stimulation and blood pressure homeostasis. *Acta Neuroveg.* 920: 205-218.

91. Schwieler, G. H.; Douglas, J. S.; and Bouhuys, A. 1970. Postnatal development of autonomic innervation in the rabbit. *Am. J. Physiol.* 219:391-397.

92. Shinebourne, E. A.; Vapaavuori, E. K.; Williams, R. L.; Heymann, M. A.; and Rudolph, A. M. 1972. Development of baroreflex activity in unanesthetized fetal and neonatal lambs. *Circulation Res.* 31:710-718.

93. Smith, O. A., Jr.; Stelhenson, R. B.; and Randall, D. C. 1974. Range of control of cardiovascular variables by the hypothalamus. In *Recent Studies of Hypothalamic Function*, eds., K. Lederis and K. E. Cooper, pp. 294-305. Basel: Karger.

94. Such, G. 1970. Spatial summations of vasomotor reflexes. *Acta Physiol. Acad. Sci. Hung.* 37:215-232.

95. Ulmer, H. V. 1969. Der Einfluss elektrischer Reizung des N. ischiadicus auf den Verlauf der Carotissinus-Aortendruck-Kennlinie des Hundes. *Z. Biol.* 116: 235-240.

96. Vallbona, C.; Desmond, M. M.; Rudolph, A. J.; Pap, L. F.; Hill, R. M.; Franklin, R. R.; and Rush, J. B. 1963. Cardiodynamic studies in the newborn. II. Regulation of the heart rate. *Biol. Neonate* 5: 159-199.

97. Van Petten, G. R., and Willes, R. F. 1970. β-adrenoceptive responses in the unanesthetized ovine foetus. *Brit. J. Pharmacol.* 38:572-582.

98. Vapaavouri, E. K.; Shinebourne, E. A.; Williams, R. L.; Heymann, M. A.; and Rudolph, A. M. 1973. Development of cardiovascular responses to autonomic blockade in intact fetal and neonatal lambs. *Biol. Neonate* 22:177-188.

99. Vatner, S. F.; Boettcher, D. H.; Heyndrickx, G. R.; and McRitchie, R. J. 1975. Reduced baroreflex sensitivity with volume loading in conscious dogs. *Circulation Res.* 37:236-242.

100. Walker, D. 1975. Functional development of the autonomic innervation of the human fetal heart. *Biol. Neonate* 25:31-43.

101. Walsh, S. Z.; Meyer, W. W.; and Lind, J. 1974. *Human Fetal and Neonatal Circulation*, pp. 167-172. Springfield, Ill.: Charles C. Thomas.

102. Watanabe, K.; Iwase, K.; and Hara, K. 1973. Heart rate variability during sleep and wakefulness in low-birth weight infants. *Biol. Neonate* 22:87-98.

103. Wilkus, R. J. and Peiss, C. N. 1963. Stimulation parameters for cardiovascular responses from medulla and hypothalamus. *Am. J. Physiol.* 205:601-605.

104. Williams, R. L.; Hof, R. P.; Heymann, M. A.; and Rudolph, A. M. 1976. Cardiovascular effects of electrical stimulation of the forebrain in the fetal lamb. *Pediatric Res.* 10:40-45.

105. Yoshikawa, T. 1968. *Atlas of the Brains of Domestic Animals*, Univ. of Tokyo Press, Tokyo and the Penn. State Univ. Press. University Park, Pennsylvania.

106. Young, M. 1965. The patterns of arterial pressure responses to a reduction in pulse pressure in the kitten. *J. Physiol.* 178:4-6P. London.

# 5

## Development of Fetal Cardiovascular Responses to Alpha-Adrenergic Agonists

G.R. Van Petten, W.H. Harris, and G.J. Mears
Faculty of Pharmacy and Pharmaceutical Sciences
The University of Alberta
Edmonton, Alberta, Canada

INTRODUCTION

The effects of a variety of autonomic drugs on fetal cardiovascular functions have been studied in attempts to obtain further information on development of autonomic control of the fetal circulation. In several instances these experiments have also suggested that the cardiovascular actions of the specific drugs used undergo developmental changes with advancing gestation. Thus, in the fetal lamb Vapaavouri and associates (10) found a maximal response to parasympathetic and alpha-adrenergic blocking drugs by 120 days of gestation but a continuing increase in the response to a beta-adrenergic blocking agent with advancing gestation. In contrast, the threshold dose of the beta-adrenergic agonist, isoproterenol, remained the same and others (9) have observed that the maximal degree of tachycardia produced by isoproterenol did not change during the last third of gestation. Moreover, Joelsson and co-workers (4) found that both alpha- and beta-adrenergic blocking drugs produced only minor changes in fetal heart rate and blood pressure during the last third of gestation, suggesting minimal basal autonomic tone. With respect to alpha-adrenergic agonists, the magnitude of the systemic and pulmonary pressor response to norepinephrine (5) and the systemic response to tyramine (10) have been shown to increase with advancing gestation in the fetal lamb.

A number of major factors known to be important determinants of the observed response to drugs may be involved in such developmental changes in drug action. First, the *pharmacokinetics* of both drug distribution to sites of action and the mechanisms involved in termination of drug action, including metabolism and elimination, may change during fetal life; many of these processes undergo marked change after birth. Second, the degree of *presynaptic nerve development* might be expected to affect the action of certain drugs, for example, the action of indirectly acting pressor amines such as tyramine or ephedrine which depend upon available pools of endogenous norepinephrine for their action. Third, the character of the *drug receptor interaction* or the number of available receptors may alter drug response. Fourth, the development of the *receptor-effector system* may well play an important role in developmental changes in drug action.

In view of the large range of drugs such as tricyclic antidepressives, phenothiazine tranquilizers and pressor amines which may be used during pregnancy and which have cardiovascular effects mediated through the alpha-adrenergic receptor-effector system, it is important to fully understand how fetal responses to adrenergic drugs may change with advancing gestation. Moreover, a better understanding of the development of drug-responsiveness may help to further our appreciation of the development of autonomic cardiovascular control mechanisms. In an attempt to provide further information in this area, this report summarizes some of our recent data on the developmental changes in (a) the pressor response of the unanesthetized fetal lambs to indirectly acting (e.g. ephedrine) and directly acting pressor amines (e.g. phenylephrine) and (b) the *in vitro* responses of the ear artery of the sheep to electrical and chemical stimulation.

## METHODS

*In vivo* experiments were carried out in the chronically cannulated ovine fetus of known gestational age, the newborn and the non-pregnant adult sheep. Briefly, surgical procedures carried out under halothane anesthesia, as previously described (2, 11), included implantation of silicone rubber cannulas in the femoral artery and saphenous vein of the fetus, implantation of non-occluding polyvinyl cannulas in the carotid artery and jugular vein of the adult and newborn and implantation of subcutaneous ECG leads on the chest of the fetus. Experiments were not done until at least 48 h after surgery and 2 to 5 days were allowed between experiments in the same animal. All drugs were given intravenously via the above cannulas and experiments carried out only in animals where the heart rate and variation were within the normal range for their age; where possible, the fetal blood gases were measured and the femoral arterial $pO_2$ found to range from 18-23 mm Hg.

*In vitro* experiments were carried out using helical strips of central ear artery obtained from the fetal lamb, newborn lamb or adult sheep. The methods used have been described in detail elsewhere (12, 13). Briefly, the helical strips of central ear artery were suspended between two parallel platinum wire electrodes in Krebs solution maintained at 36 ±1$^0$C and continuously bubbled with 5% $CO_2$ in oxygen. Using force displacement transducers, contractions were recorded isometrically in response to transmural electrical stimulation or addition of agonists including norepinephrine, 5-hydroxytryptamine and lysine vasopressin.

## RESULTS

### *In Vivo* Experiments

The typical changes in mean arterial blood pressure (MABP) and heart rate produced by administration of ephedrine (1.0 mg/kg) to the 120-129 day fetus are shown in Figure 1. While blood pressure rose to a peak at 1 minute and then declined to pre-drug values by 10 minutes, heart rate first slowed (maximum decrease at 2 min), then rose to pre-drug levels at 10 minutes and thereafter increased and then remained above control values for 30 to 40 minutes.

The cardiac slowing following ephedrine administration was blocked by preadministration of atropine (1.0 mg/kg) to the fetus and the later cardiac acceleration was blocked by preadministration of metoprolol, a cardioselective beta-adrenoceptor blocking agent. The time course of these actions of ephedrine were similar in the fetus from 112 days to term and in newborn lambs (96h) and adult non-pregnant sheep (2, 3).

In contrast to ephedrine, the time course of the pressor effect of phenylephrine was markedly different in the fetus at different ages and in the newborn lamb (Fig. 2). Thus, a dose of 10 µg/kg phenylephrine produced a pressor response with a duration in excess of 15 minutes in the fetus at 100-109 and 120-129 days of gestation, of about

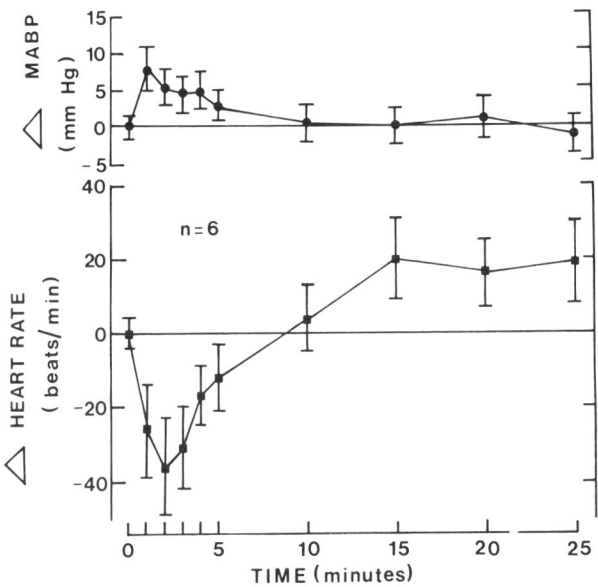

Fig. 1. The change in mean arterial blood pressure (MABP) and change in heart rate are shown at different times after administration of ephedrine (1.0 mg/kg) at time zero to fetuses (n=6) of 120-129 days gestational age. Vertical bars represent S.E.M.

15 minutes at 130-145 days, but of less than 5 minutes in the newborn lamb (96 h). The time course of the response in the newborn lamb was nearly identical to the time course of the pressor response produced in the adult sheep by an equi-effective dose of phenylephrine. In all age groups, the time course of the rise in blood pressure produced by phenylephrine was mirrored almost exactly by a decrease in heart rate (3).

It is also apparent from the peak of the pressor response produced by phenylephrine (Fig. 2) that the fetal pressor response to this drug increased with advancing gestation with a sharp increase in the response of the newborn shortly after birth. Similar increases in the pressor res-

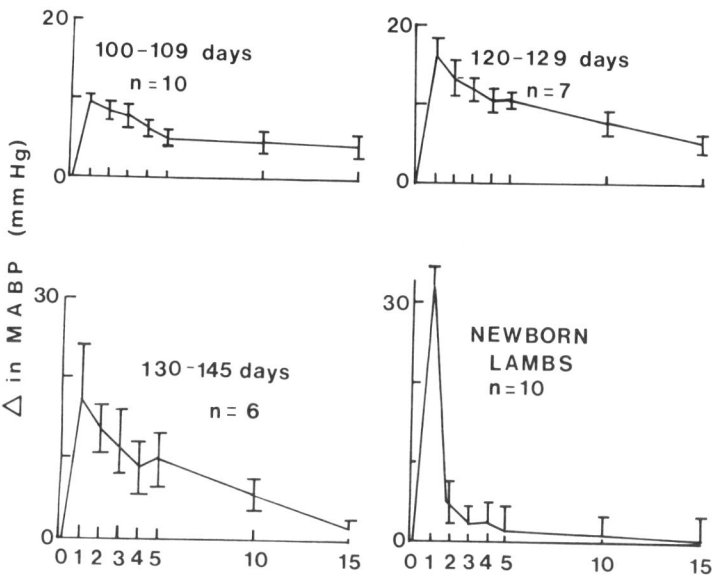

Fig. 2. The time-course of the change in mean arterial blood pressure (MABP) produced by administration of phenylephrine (10 µg/kg) to fetuses at different gestational ages (100-109; 120-129; 130-145 days) and in newborn lambs is shown in each of the four panels. The time in minutes is shown in the horizontal axis of each panel and vertical bars represent S.E.M.

ponses to ephedrine were also seen. Figure 3 shows the regression of the pressor responses to ephedrine and phenylephrine on fetal gestational age. Also, it should be noted that the rise in blood pressure produced by ephedrine in the newborn lamb tended to be larger, although not significantly so, than in the late term fetus (3). On the other hand, compared with the late term fetus, the response to phenylephrine increased significantly after birth but at 4 days of age was still well below adult levels (3). In contrast to the increase in the pressor response produced by phenylephrine with advancing development, the same dose of the alpha-adrenoceptor blocking agent, namely, 0.15 mg/kg of phentolamine was required to produce approximately

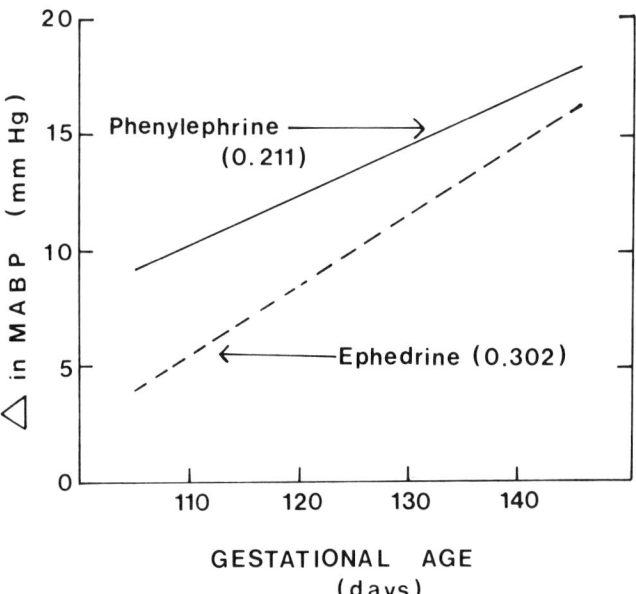

Fig. 3. The regression of the change in mean arterial blood pressure (MABP) produced by phenylephrine or ephedrine on gestational age. The maximum increase in MABP produced by administration of each drug to fetuses at different gestational ages was used and the slopes obtained are indicated in brackets.

50% inhibition of the pressor response to 10 µg/kg phenylephrine in all age groups studied (112-119, 120-129, and 130-145 day fetuses and adult sheep (3).

### In Vitro Experiments

The isolated helical strips of central ear artery from the adult sheep responded in a typical dose-related fashion to norepinephrine and in a frequency-related manner to transmural electrical stimulation (Fig. 4). The maximum contraction which could be achieved by electrical stimulation was about 43 percent of the maximum contraction that was produced by norepinephrine. When tissue from fetuses at 110-115 days and 133-137 days, or newborn lambs at 3-5 days and 2-3 weeks were used, the norepinephrine dose-response curves expressed as a percentage of the maximum contraction at each age were not significantly different (13). Moreover, there was no significant difference between the $ED_{50}$ for norepinephrine in tissue of any of the groups and the $pA_2$* for phentolamine blockade of the norepinephrine responses was virtually identical in all groups. In addition, the dose-response curves for 5-hydroxytryptamine and lysine vasopressin did not differ in tissues obtained from animals at these different ages.

In contrast, the maximum absolute response which could be obtained with norepinephrine or electrical stimulation did change markedly with development (Table 1). Thus, in tissue from 110-115 day fetuses, only 4 of 8 tissues responded to norepinephrine and the maximum contraction was only 3.1% of the adult response. The ability of the ear artery to contract in response to norepinephrine increased considerably by late term (133-137 days) but did not change between late term and the newborn period (3-5 days). There was a further increase in the maximum contraction produced by norepinephrine in tissue from 2-3 week old lambs but the response was still only 30.8 percent of that of the adult. The response to electrical stimulation of the tissue from the 110-115 day fetus was essentially nonexistent. By late term (133-137 days) and the early newborn period (3-5 days) the response to transmural stimulation was still small and even at 2-3 weeks of age was still very small compared with the adult.

*$pA_2$ is the negative logarithm of the molar concentration of an antagonist which reduces a double dose of an agonist to that of a single dose.

## DISCUSSION

These data clearly confirm and extend previous suggestions (10) that the systemic pressor actions of sympathomimetic amines increase with advancing gestation in the fetal lamb. Thus, the maximum increase in blood pressure produced by several doses of ephedrine and phenylephrine progressively increased throughout the last third of gestation in the fetal lamb. In both cases the effectiveness

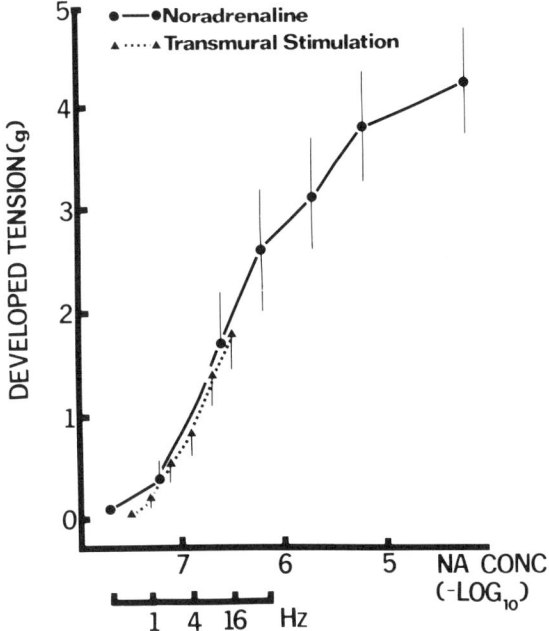

Fig. 4. Norepinephrine (NA) cumulative dose-response curve and frequency response curve of helical strips of central ear artery of adult sheep. Responses are expressed in terms of developed tension. The abscissa for the NA curve is the negative $\log_{10}$ of the norepinephrine concentration and abscissa for the frequency response curve is frequency of a 30-sec train of electrical stimulation. Each point is the mean from tissue from 8 animals and the vertical lines are the S.E.M.

Table 1

Maximal Responses of Isolated Central Ear Artery
to Norepinephrine and Transmural Stimulation

|  | N | Maximal Responses as Percent of Adult Norepinephrine Response | |
|---|---|---|---|
|  |  | Norepinephrine | 16 Hz |
| 110-115-day fetus | 4[a] | 3.1 ± 1.2 | 1.2 |
| 133-137-day fetus | 4[b] | 18.5 ± 3.8 | 4.0 ± 2.1 |
| 3-5-day newborn | 7 | 18.0 ± 4.3 | 2.6 ± 0.7 |
| 2-3-week newborn | 8 | 30.8 ± 8.5 | 9.0 ± 4.5 |
| Adult | 8 | 100 | 42.9 ± 9.0 |

Source: After Wyse, Van Petten, Harris (13)

[a] Of eight experiments, only 4 gave recordable responses; the responses to 16-Hz-electrical stimulation were near the limits of detection.

[b] Of five experiments, only 4 gave recordable responses.

of the drugs in increasing blood pressure was well below
adult levels both at term and in the newborn lamb (2, 3).
In the case of ephedrine, there was only a small, statistically insignificant, increase in the pressor response
between the late term fetus and the newborn lamb. However, with phenylephrine the response to a given dose in
the newborn lamb was significantly above the response in
the late term fetus (130-145 days). The latter observation might be expected in view of Dawes' (1) observations
that a rise in fetal blood pressure was accompanied by an
increase in the percentage of cardiac output perfusing
the placenta and that placental vessels are probably relatively unresponsive to sympathomimetic amines; that is,
the placental circulation might be expected to have a dampening effect on the pressor action of drugs. The smaller
increase in the response to ephedrine in the newborn lamb
compared with the late term fetus may be related to the
somewhat lower potency, compared with phenylephrine, of
this drug in the fetus which in turn may be related to inadequate stores of norepinephrine available for release.

The observed developmental increases in the pressor
responses to ephedrine and phenylephrine may be due to
maturation of several processes including the alpha-adrenergic receptor, presynaptic nerve development and the receptor-effector system including biochemical, physiological
and anatomical components. The present data are in agreement with other results (5, 10) which suggest that maturation of the alpha-adrenergic receptor is complete prior to
the last third of gestation. Thus, *in vivo*, the dose of
phentolamine required to produce approximately 50 percent
inhibition of the phenylephrine pressor response remained
the same from 100 days of gestation through to adulthood.
Similarly, *in vitro*, the norepinephrine, 5-hydroxytryptamine and lysine vasopressin dose-response curves were parallel in the central ear artery of the 110-115 and 133-137
day fetus, the newborn lamb at 3-5 days or 2-3 weeks and
the adult. Moreover, the $pA_2$ of phentolamine against norepinephrine induced contractions was the same in this tissue
at all ages studied. Thus, no evidence was obtained for
a change in the character of the alpha-adrenergic drug-
receptor interation after 100 days of fetal life in the
sheep.

In contrast, these experiments do indicate that,
functionally, pre-synaptic nerve development is incomplete
in the fetus and undergoes development for a considerable

period of time after birth. For example, the slightly greater slope of increase in pressor response with gestational age observed with ephedrine compared with phenylephrine could be due in part to development of adequate stores of norephinephrine available for release. Such a hypothesis is supported by the *in vitro* data with the isolated central ear artery. Thus, the maximal contraction which could be induced by transmural electrical stimulation was essentially non-existent in the 110-115 days fetus, was only 5-10% of the adult response in the 133-137 day fetus and 3-5 day newborn, and had risen only to 21% of the adult value by 2-3 weeks of age. Since the ability of this technique to depolarize post-ganglionic adrenergic nerve endings without direct depolarization of smooth muscle cells has been well documented (6, 7, 8), these results indicate that the development of functional sympathetic innervation in this vascular tissue is far from complete at birth and lags well behind the ability of the vessel to contract in response to stimulation with drugs acting directly on smooth muscle receptors.

The present data also suggest that the receptor-effector system of vascular smooth muscle of the lamb undergoes considerable development throughout the last third of gestation and well beyond birth. For example, since the character of the drug-receptor interaction appears not to change, the increase in the response to the directly acting drug, phenylephrine, over the last third of gestation might be explained by maturation of the receptor-effector system. Similarly, the increase in the maximal response of the isolated ear artery to norepinephrine observed between 110 days and birth and between the 3-5 day newborn and the 2-3 week newborn could be due to maturation of the receptor-effector system. Whether such maturation involves the biochemical and physiological events coupling receptor activation with smooth muscle contraction, structural changes, or changes in the type of smooth muscle cells in the vasculature or a combination of these factors remains to be elucidated.

It is of interest that the *in vivo* experiments with phenylephrine indicated a small decrease in the duration of pressor action between 110 days and term in the fetus and a marked decrease in the duration of action in the newborn lamb compared with the late term fetus. Similar changes were seen with other directly acting amines including norepinephrine. These observations suggest a very rapid development shortly after birth of the processes, for example, uptake of norepinephrine, responsible for terminating the action of these drugs. The very rapid transition from the relatively long duration of action in the late-term fetus to a duration of effect virtually identical to that seen in the adult contrasts with the apparent progressive development of vascular post-ganglionic adrenergic innervation and receptor-effector mechanisms.

Indeed, in summary, these and other observations strongly suggest that the major factors involved in the action of adrenergic drugs in the developing fetus and newborn achieve maturity at different rates and ages. Thus, vascular alpha-adrenergic receptors appear to be well developed by 0.7 gestation in the lamb. Functional post-ganglionic adrenergic innervation of blood vessels may be poor even at birth and require a considerable period of time after birth for maturation. The receptor-effector system in vascular smooth muscle seems to develop in a more progressive manner throughout the last part of gestation and well after birth while other factors such as those responsible for terminating the action of drugs such as norepinephrine undergo only slow maturation during fetal life with rapid change occurring shortly after birth.

## ACKNOWLEDGMENTS

Special thanks are due to Mrs. Beverly Graham (R.N.) and Ms. Yvette Brideau (B.A.) for their skilled assistance in the surgical procedures. The expert technical assistance of Mrs. Heather Mathison (B.Sc.), Mr. R. Royer and Ms. Gail Jones during the course of these experiments was greatly appreciated. This work was supported by research grants from the MRC (Canada) and the Alberta Heart Foundation. W.H.H. was an MRC Fellow whose present address is Department of Biomedical Sciences, University of Guelph, Guelph, Ontario N1G 2W1, Canada.

## REFERENCES

1. Dawes, G. S.; Mott, J. C.; and Rennick, B. R. 1956. Some effects of adrenaline, noradrenaline and acetylcholine on the foetal circulation in the lamb. J. Physiol. 134:139-148.

2. Harris, W. H. and Van Petten, G. R. 1975. Development of the pressor response to sympathomimetic amines in the fetal lamb. *Proc. Can. Fed. Biol. Soc.* 18:75.

3. ----. 1978. Development of cardiovascular responses to sympathomimetic amines and autonomic blockade in the unanesthetized fetus. *Can. J. Physiol. Pharmacol.* 56:400-409.

4. Joelsson, I.; Barton, M. D.; Daniel, S.; James, S.; and Adamsons, K. 1972. The response of the unanesthetized sheep fetus to sympathomimetic amines and adrenergic blocking agents. *Am. J. Obstet. Gynec.* 114:43-50.

5. Nuwayhid, B.; Brinkman, C. R.; Su, C.; Bevan, J. A.; and Assali, N. S. 1975. Systemic and pulmonary hemodynamic responses to adrenergic and cholinergic agonists during fetal development. *Biol. Neonate* 26:301-317.

6. Patterson, G. 1965. The response to transmural stimulation of isolated arterial strips and its modification by drugs. *J. Pharm. Pharmacol.* 17:341-349.

7. Su, C. and Bevan, J. A. 1970. The release of $H^3$-norepinephrine in arterial strips studied by the technique of superfusion and transmural stimulation. *J. Pharmacol. Exp. Ther.* 172:62-68.

8. Van Houtte, P.; Clement, D., and Leusen, I. 1967. The reactivity of isolated veins to electrical stimulation. *Arch. Int. Physiol. Biochem.* 15:641-657.

9. Van Petten, G. R. and Willes, R. F. 1970. β-adrenoceptive responses in the unanesthetized ovine fetus. *Br. J. Pharmacol.* 38:572-582.

10. Vapaavouri, E. K.; Shinebourne, E. A.; Williams, R. L.; Heymann, M. A.; and Rudolph, A. M. 1973. Development of cardiovascular responses to autonomic blockade in intact fetal and neonatal lambs. *Biol. Neonate* 22:177-188.

11. Willes, R. F.; Van Petten, G. R.; and Truelove, J. F. 1970. Chronic exteriorization of vascular cannulas and ECG electrodes from the ovine fetus. *J. Appl. Physiol.* 28:248-250.

12. Wyse, D. G. 1973. Inactivation of exogenous and neural noradrenaline by elastic and muscular arteries. *Can. J. Physiol Pharmacol.* 51:164-168.

13. Wyse, D. G.; Van Petten, G. R.; and Harris, W. H. 1977. Responses to electrical stimulation, noradrenaline, serotonin and vasopressin in the isolated ear artery of the developing lamb and ewe. *Can. J. Physiol. Pharmacol.* 55:1001-1006.

# 6

## Reactivity of Fetal Vascular Smooth Muscle to Sympathetic Nerve Stimulation and Vasoactive Agents

C. Su, B.L. Pegram, J.A. Bevan, N.S. Assali, and C.R. Brinkman, III

Departments of Pharmacology and Obstetrics and Gynecology
University of California
School of Medicine
Los Angeles, California

The vascular smooth muscle tone in the mammalian adult is under the influence of numerous factors. In most vascular beds under normal conditions, the sympathetic nervous system plays the dominant role in maintaining the muscular tone. Exceptionally, the sympathetic nerves exert vasodilation in certain vascular beds (1). Certain regions of the vascular tree, including some peripheral veins (2), are devoid of sympathetic innervation or are under minimal influence of this innervation. Another factor consists of the humoral vasoactive agents, for example, angiotensin. The plasma level of this polypeptide is believed to be associated with several pathophysiological states. The third major factor is metabolic in nature. Alterations of the pH, $PO_2$ or $P_{CO_2}$ of the local environment, or liberation of a vasoactive material, e.g., adenosine, due to metabolic activities, can affect the vascular muscle tone. The role of humoral and metabolic influences must not be underesti-

mated especially during the fetal and neonatal life, before
the sympathetic nervous system takes full control of the
circulation.

The time course of development of the autonomic nervous system varies with the species. In the rat, a contractile response of the isolated portal vein to transmural electrical stimulation, indicative of functional adrenergic innervation, appears one week after birth (3). Sympathetic fibers develop later and reach a full maturity two months after birth in most organs in the dog (4). Interestingly enough, nervous development may not be synchronous throughout the vascular tree in view of data of the puppy. Thus, 1-4 weeks after birth the femoral artery contains dense adrenergic nerves according to histological evidence, and it constricts powerfully to stimulation of the lumbar sympathetic chain. Both these parameters regress 2-4 months after birth (5).

Development of the autonomic influence on the circulatory system of the lamb has been investigated *in vivo* by a number of workers (6-9). The sympathetic control of the peripheral circulation exists in early fetal life prior to parasympathetic influence on the heart. The peripheral sympathetic fibers in the lamb are believed to complete development during the first few days after birth (10). However, it is difficult to retrace the sympathetic nervous development to the time of inception by the *in vivo* approach, and many details of the local mechanisms are unknown.

The present chapter primarily deals with the sympathetic adrenergic innervation of the lamb fetal blood vessels in relation to development and regional variation. When considering functional innervation, one is in fact concerned with a neuroeffector system, which consists of three structural units: the neuron, synaptic cleft, and effector cell. This work is directed at the typical vascular adrenergic neuroeffector system, consisting of the adrenergic neuron and vascular smooth muscle cell which contracts in response to the adrenergic transmitter, l-norepinephrine.

For a successful neuroeffector transmission, the adrenergic neuron must be capable of synthesizing, storing, and releasing norepinephrine. Further, the neuron should be prepared to recover much of the released norepinephrine, since the neuronal membrane transport of norepinephrine ("uptake$_1$") is believed to be the major mode of inactivation of the adrenergic transmitter. The smooth muscle cell must, of course, contract in response to the transmitter, through the adrenergic alpha-receptor system as well as coupling and contractile mechanisms. The muscle cell also bears partial responsibility for inactivating the transmitter. This requires muscle cell membrane transport of norepinephrine ("uptake$_2$") and metabolic degradation of norepinephrine by two enzyme systems, monoamine oxidase (MAO) and catechol-0-methyltransferase (COMT), contained in the muscle cell. Finally, the neuronal membrane has to be drawn near the smooth muscle cell membrane, to form a functional synapse in order for the transmitter to act on the muscle cell in sufficient concentration.

The neuron and smooth muscle cell each contains a number of component mechanisms essential for its performance. Our objective was to determine when and in what order these components develop, and whether and how the components, individually or as integrated functional whole, vary among different blood vessels in the fetal lamb.

## MATERIALS AND METHODS

Ewes of dated pregnancies were used. Following spinal anesthesia to the ewe with pontocaine, the fetus was rapidly delivered by cesarean section. The fetal head was covered with a saline-filled rubber glove to prevent breathing. Under local anesthesia with lidocaine, the carotid arteries and jugular veins were sectioned near their rostral ends to exsanguinate the fetus. The thoracic aorta, ductus arteriosus, pulmonary artery, common carotid artery, anterior mesenteric artery, renal artery, and saphenous vein were rapidly excised. They were kept in Krebs bicarbonate solution equilibrated with 95% oxygen and 5% $CO_2$ at room temperature until the experiment.

Fetuses of estimated 53-90 days' gestation (body weight 0.05 to 0.76 kg) were grouped as the immature; those of 115-130 days' gestation (1.45 to 2.4 kg), the premature; and those over 140 days in gestation (2.78 to 3.9 kg) were referred to as the mature fetuses. These gestational periods roughly correspond to those adopted previously (8). The body weights are lighter in the present work since twin fetuses were used. Of each pair of twins, one was used for studies of contractile responses and the other for histofluorescence and biochemical assays.

The contractile response was measured on a 3-mm long segment cut from each vessel. Two fine stainless steel wires were inserted through the lumen; one was anchored to a stationary rod and the other connected to a Statham strain gauge for recording developed forces on a Grass Polygraph. The vascular ring was bathed in Krebs bicarbonate solution equilibrated with 95% oxygen and 5% $CO_2$ at $38^\circ C$. A passive stretch of 1 g was applied. Norepinephrine was applied in graded concentrations cumulatively ($3 \times 10^{-8}$ to $3 \times 10^{-4}$ M) to each preparation. After washing three times, the vessel ring was allowed to recover from contraction completely and graded concentrations of serotonin (5HT) ($2.5 \times 10^{-7}$ to $2.5 \times 10^{-4}$ M) were applied. This was followed similarly by concentrations of acetylcholine (ACh) ($5 \times 10^{-8}$ to $10^{-3}$ M). From the dose-response curves the maximal effect and median effective concentration ($ED_{50}$, the concentration for half-maximal response were estimated. For the purpose of transmural nerve stimulation, a platinum wire electrode was placed on either side of the vascular ring segment. Rectangular, short duration (0.3 ms) pulses of supramaximal voltage were applied. Previous studies of the rabbit pulmonary artery showed that selective nerve stimulation was achieved by these parameters (11). The stimulation was applied at a frequency of 25 Hz until the contraction attained a plateau. This frequency was chosen as the maximal effective based on frequency-response studies of the fetal carotid artery.

At the end of these experiments, the vessel ring length (L) and width between the two steel wires were measured with a caliper, before the ring was dismounted and blotted on filter paper and weighed. The vascular wall cross-sectional area perpendicular to the contractile developed force was estimated by dividing the weight by the width (W), assuming that the tissue specific gravity was approximately unity. The circumferential area of vascular wall per unit weight was derived by weight/thickness of the wall and thickness was calculated from weight/2WL. Under the experimental conditions, all other factors being equal, the contractile developed force should be proportional to the cross-sectional area of the contractile elements. In order to facilitate comparisons between blood vessels of varying wall thicknesses, the developed forces were expressed per unit cross-sectional area of the vascular wall. It is assumed that smooth muscle cells occupy a constant proportion of the wall regardless of its thickness. In most larger arteries the adrenergic neurons form a two-dimensional plexus, and it would be more relevant to relate the neuronal norepinephrine uptake to the area of such plexus than to tissue mass. The area of the plexus was in some cases assumed to be approximately equal to the circumferential area calculated as described above, and the neuronal uptake of $^3$H-norepinephrine was calculated per unit circumferential area and used as an index of neuronal density in this study.

In selected vessel preparations transverse sections were made and specific catecholamine histofluorescence was examined by the method of Falck et al. (12). The monoamine oxidase activity was assayed radiochemically using $^{14}$C-tryptamine (13) and the catechol-O-methyltransferase with $^{14}$C-adenosylmethionine and 1-norepinephrine (14). For the uptake studies each tissue was divided into halves. One half was preincubated in Krebs bicarbonate solution at 38°C for 30 min. Then $^3$H-norepinephrine ($10^{-8}$ M) and $^{14}$C-sucrose ($2 \times 10^{-6}$ M) were added and after 60 min the tissue was rinsed for 1 sec in Krebs solution containing no $^3$H or $^{14}$C, blotted on filter paper, and weighed. It was then digested with Solyene (Packard Corp., Downers Grove, Ill.) and assayed for $^3$H and $^{14}$C activities by liquid scintillation spectrometry using appropriate settings for double label counting. The other half of the tissue was treated similarly except that cocaine ($10^{-4}$ M) was added to the Krebs solution throughout the preincubation and 60-min incubation periods. An aliquot of the incubation medium was also assayed for $^3$H and $^{14}$C. The uptake was expressed as the moles $^3$H-norepinephrine was calculated per unit circumferential vascular wall or per weight, or as the tissue-to-medium ratio in radioactivity (i.e., the ml incubation medium cleared per g tissue weight). The difference between the cocaine treated and untreated $^3$H-norepinephrine uptake was taken as the neuronal uptake. The $^{14}$C-sucrose uptake was used as the extracellular space; the extraneuronal uptake of $^3$H-norepinephrine was derived from the $^3$H-norepinephrine uptake in the presence of cocaine, minus the $^{14}$C-sucrose uptake.

Drugs used included 1-norepinephrine bitartrate, serotonin creatinine sulfate, acetylcholine chloride (Calbiochem, La Jolla, Calif.), bretylium tosylate (Burroughs Wellcome, Tuckahoe, N.Y.) and cocaine hydrochloride (Mallincrodt, St. Louis, Mo.). 1-Norepinephrine-7-$^3$H, 7.9-9.2 Ci/mmol, was obtained from New England Nuclear (Boston, Mass.) and $^{14}$C-sucrose, uniformly labeled, 600 mCi/mmol, from Amersham/Searle (Arlington Heights, Ill.). Krebs bicarbonate solution contained (mM): $Na^+$ 144.2, $K^+$ 4.9,

$Ca^{++}$ 1.6, $Mg^{++}$ 1.2, $Cl^-$ 126.7, $HCO_3^-$ 25.0, $SO_4^{--}$ 1.2, glucose 1.1, ethylenediamine tetraacetic acid 0.024 and ascorbic acid 1.1. Analysis of data was performed by the analysis of variance test for single variable of classification, and multiple comparisons among the means based on the F statistics, according to the format of Dixon and Massey (15). Where necessary, differences between paired observations were also tested by the Student t test. Statistical inferences were made at 5% significance level unless otherwise stated.

## RESULTS

A. Gestational Changes of Adrenergic Mechanisms

The common carotid artery of the lamb fetus was examined in some detail in regard to adrenergic neuroeffector mechanisms from early gestation to near term. The characteristics of the coronary arteries from the mature fetuses, including the near term, will first be described as a basis of gestation-related comparisons. These arteries were relatively thick-walled (0.41 ± 0.03 mm, mean and s.e.m., 6 fetuses from 6 ewes) for the small diameter (2.1 ± 0.1 mm). In a recent study (16), electrical transmural stimulation elicited a contraction in all preparations. When the stimulation was applied at the maximal frequency until the contraction attained a plateau, the developed force was 1.47 ± 0.33 $kdyn/mm^2$ cross sectional area (n = 6). The contractile responses of all vessel preparations from this and other groups of fetuses were invariably abolished by the administration of an adrenergic neuron blocking agent, bretylium (25 μM). This indicated that the response to

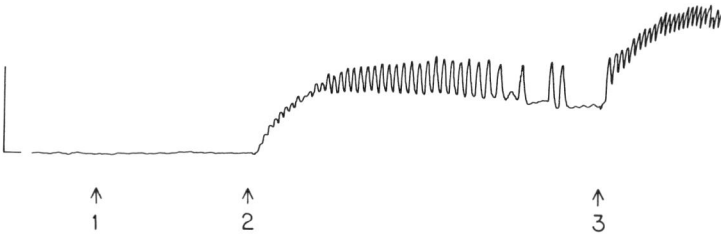

Fig. 1. Tracing of contractile responses of fetal lamb carotid artery. At arrows 1, 2, and 3, norepinephrine (0.25, 2.5, and 25μM) was added. Calibration: 500 dyn, 1 min.

transmural stimulation was mediated by excitation of adrenergic neurons within the vascular wall, rather than direct stimulation of vascular smooth muscle.

Norepinephrine and serotonin also elicited a contraction in the carotid arteries of all six mature fetuses. Figure 1 illustrates the contractile response of a carotid artery preparation to serotonin. In this artery rhythmic phasic contractions were frequently induced, superimposed upon a tonic sustained contraction. The maximal responses to both serotonin and norepinephrine were slightly greater than that to transmural nerve stimulation (Fig. 2).

A possible index of the adrenergic nerve density, the neuronal uptake of $^3$H-norepinephrine in the carotid arteries was quite variable but distinctly demonstrable (17.2 ± 2.8 ml cleared per g tissue per hr, or 172 pmol per g per hr). The extraneuronal uptake of $^3$H-norepinephrine (12.3 ± 1.6 pmol/g/hr) and monoamine oxidase (0.62 ± 0.08 nmol/mg/hr) and catechol-O-methyltransferase (12 ± 2 pmol/mg/hr) activities, which primarily represent the extraneuronal capacity for inactivation of the adrenergic transmitter, were all present.

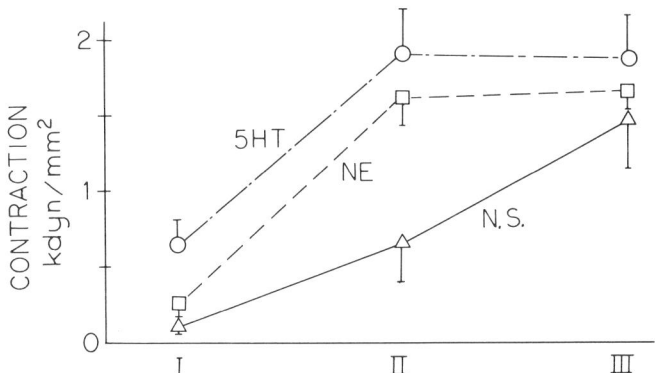

Fig. 2. Maximal responses of the fetal lamb carotid artery to serotonin, norepinephrine, and transmural adrenergic nerve stimulation (N.S.). Vertical bars represent S.E.M. Numbers of observations of the immature (I), premature (II), and mature (III) fetal groups are 9, 6, and 6, respectively.

In the younger, premature fetus group, all these inactivation activities were detectable in the carotid arteries. The neuronal and extraneuronal norepinephrine uptake and monoamine oxidase and catechol-O-methyltransferase activities were 160 ±86 (7) and 18.0 ±2.5 (8) pmol/g/hr and 0.63 ±0.11 (6) nmol/mg/hr and 14.3 ±3.8 (6) pmol/mg/hr. They were not significantly different from those in the mature fetuses. The carotid arterial wall thickness was somewhat smaller (0.34 ± 0.05 mm, n = 8) but, as illustrated in Figure 2, the maximal responses to norepinephrine and serotonin corrected for the wall thickness were nearly equal to those of the mature carotid arteries. However, the maximal response to transmural adrenergic nerve stimulation was less than half. Although the variation was large, the results are suggestive that the response to nerve stimulation gradually increases with gestation through term.

It was not until the gestation was retraced beyond the midpoint, to the immature fetal period, that markedly smaller arterial reactivity could be revealed. The carotid artery of this group had a thin wall (0.16 ± 0.04 mm, n = 9) compared to the older ones. Norepinephrine and serotonin elicited much smaller responses even corrected for the wall thickness, on average about 16% and 34%, respectively, of those in the mature fetuses. Moreover, the response to nerve stimulation was almost negligible in comparison to that in older fetuses (Fig. 2). Of the 9 immature fetuses tested, only the two largest, weighing 0.67 and 0.76 kg (88 and 90 days gestation), were those whose carotid artery gave a contractile response to nerve stimulation at all. A response, though of limited magnitude, was obtained with norepinephrine in all carotid arteries from 6 fetuses weighing 0.28 kg or more (76 days gestation or older). Serotonin effectively caused a constriction in all cases except the youngest one with a 0.05 kg body weight (53 days gestation). From the small neurogenic response, one might expect sparse adrenergic nerve supply and, therefore, low neuronal norepinephrine uptake. The carotid arterial neuronal uptake of $^3$H-norepinephrine in the immature fetal group was rather high (296 ± 54 pmol/g/hr) compared to that in the premature and mature fetuses on tissue weight basis. On the other hand, when the neuronal uptake was corrected for the outer surface of the arterial segments, the trend was reversed: 46 ± 10 (8), 54 ± 9 (9), and 77 ± 17 (5) pmol/mm$^2$/hr for the immature, premature, and mature fetuses, respectively. The extraneuronal norepinephrine uptake, expressed on the tissue weight basis, in the immature group (27.1 ± 4.0 pmol/g/hr, n = 8) was slightly higher than in the premature group.

### B. Variation of Blood Vessels in Reactivity

In order to evaluate the variation of the lamb fetal blood vessels, comparisons were made primarily on the basis of the maximal contractile responses to several vasoactive agents. The responses to norepinephrine and serotonin are summarized in Table 1.

The analysis of variance indicates that the mean maximal responses of different vessels to norepinephrine or serotonin were unequal at a 5% significance level within each of the three gestational periods. Multiple comparisons of these responses were performed to identify the individual differences. During the mature or near-term period, the aorta, pulmonary artery, and ductus arteriosus were significantly smaller in responses than the renal artery, saphenous veins, and carotid artery; the carotid artery was smaller than the renal artery, and the mesenteric artery smaller than the renal artery or saphenous vein. Thus, the vessels can be categorized into three groups, consisting of the three largest arteries, the carotid and mesenteric arteries, and the renal artery and saphenous vein. A similar pattern was seen among vessels of the premature fetuses with respect to norepinephrine and serotonin. There is suggestive evidence that the vessels from the immature fetuses are unequal in their maximal responses to these agents, but no significant differences could be established due to large variances (Table 1).

Transmural nerve stimulation elicited a distinct constrictor response in all vessels of the mature fetuses. This is in contrast to the vessels of the immature fetuses which gave a response no greater than 0.1 kdyn/mm$^2$ on average, or no response at all. The mean response varied widely among different vessels (Table 2). The saphenous vein, renal artery, and carotid artery were the most reactive; the anterior mesenteric artery was intermediate and the aorta, pulmonary artery, and ductus arteriosus were the least, just as they were with respect to norepinephrine or serotonin. Similar differences can be found in the neuronal uptake of $^3$H-norepinephrine in various vessels (Table 2) which, however, did not closely parallel the neurogenic constrictor response.

Table 1

Maximal Constrictor Responses of Fetal Blood Vessels to Norepinephrine and Serotonin, $kdyn/mm^2$

| Blood Vessel | Immature Fetuses | | | | | |
|---|---|---|---|---|---|---|
| | Norepinephrine | | | Serotonin | | |
| | Mean | SE | n | Mean | SE | n |
| Aorta | 0.48 | 0.11 | 9 | 0.05 | 0.02 | 8 |
| Ductus Arteriosus | 0.12 | 0.03 | 9 | 0.01 | 0.05 | 9 |
| Pulmonary artery | 0.13 | 0.02 | 9 | 0.01 | 0.01 | 8 |
| Carotid artery | 0.27 | 0.09 | 9 | 0.64 | 0.18 | 9 |
| Mesenteric artery | 0.62 | 0.35 | 4 | 0.15 | 0.10 | 4 |
| Renal artery | 0.88 | 0.12 | 3 | 0.40 | 0.13 | 3 |
| Saphenous vein | 0.53 | 0.17 | 5 | 0.42 | 0.12 | 5 |

Table 1 (continued)

| Premature Fetuses | | | | | | Mature Fetuses | | | | | |
|---|---|---|---|---|---|---|---|---|---|---|---|
| Norepinephrine | | | Serotonin | | | Norepinephrine | | | Serotonin | | |
| Mean | SE | n | Mean | SE | n | Mean | SE | n | Mean | SE | n |
| 0.67 | 0.16 | 8 | 0.13 | 0.06 | 4 | 0.53 | 0.11 | 6 | 0.04 | 0.02 | 6 |
| 0.21 | 0.06 | 8 | 0.07 | 0.03 | 4 | 0.16 | 0.05 | 6 | 0.02 | 0.03 | 6 |
| 0.18 | 0.06 | 8 | 0.04 | 0.03 | 4 | 0.15 | 0.05 | 6 | 0.02 | 0.01 | 6 |
| 1.78 | 0.30 | 5 | 2.00 | 0.34 | 3 | 1.67 | 0.17 | 6 | 1.88 | 0.28 | 6 |
| 2.72 | 0.65 | 7 | 1.25 | 0.49 | 3 | 1.23 | 0.25 | 5 | 0.62 | 0.17 | 5 |
| 2.33 | 0.42 | 7 | 1.73 | 0.16 | 4 | 2.81 | 0.21 | 6 | 2.11 | 0.42 | 6 |
| 1.67 | 0.35 | 4 | 1.53 | 0.50 | 4 | 2.65 | 0.45 | 6 | 2.66 | 0.51 | 6 |

Table 2

Neurogenic Constriction and Neuronal
$^3$H-Norepinephrine Uptake of Mature Fetal Blood Vessels

|  | Constriction (kdyn/mm$^2$) | | | $^3$H-Norepinephrine Uptake (ml/g/hr) | | |
|---|---|---|---|---|---|---|
| Blood Vessel | Mean | SE | n | Mean | SE | n |
| Aorta | 0.03 | 0.02 | 6 | 7.4 | 1.3 | 6 |
| Ductus arteriosus | 0.03 | 0.01 | 6 | 5.7 | 2.2 | 6 |
| Pulmonary artery | 0.03 | 0.01 | 6 | 9.5 | 1.9 | 6 |
| Carotid artery | 1.47 | 0.33 | 6 | 17.2 | 2.8 | 6 |
| Mesenteric artery | 0.26 | 0.17 | 5 | 18.1 | 3.1 | 5 |
| Renal artery | 1.14 | 0.53 | 6 | 18.1 | 2.7 | 6 |
| Saphenous vein | 1.91 | 0.35 | 6 | 26.1 | 6.1 | 6 |

Histofluorescence of the transverse sections of selected vessels show that the catecholamine-containing structures, presumably representing adrenergic nerve plexus, vary markedly in localization among different vessels. Fairly intense specific fluorescence was associated both with the anterior mesenteric artery (Fig. 3) and saphenous vein (Fig. 4) in keeping with their high neuronal $^3$H-norepinephrine uptake. The fluorescence in the mesenteric artery was entirely localized to the adventitio-medial junction in all three premature fetuses tested. In contrast, in the saphenous vein fluorescent bands were found in the outer 2/3 of tunica media. This medial distribution of nerve terminals can probably account for the far larger neurogenic constriction of this vein than that of the mesenteric artery. It may be noted that the response of the mesenteric artery to exogenous norepinephrine was, by comparison, quite considerable (Table 1). Similarly, the carotid artery showed specific fluorescence in the media and, somewhat to a lesser degree, at the adventitio-medial junction. This artery, too, gave a much greater neurogenic response than the mesenteric, even though the neuronal $^3$H-norepinephrine uptake did not exceed that in the mesenteric artery. The renal artery, on the other hand, had fluorescence primarily localized to the junction with occasional intrusion into the outer medial layers. As may be predicted, this artery gave a rather limited neurogenic response considering its exceptionally large response to exogenous norepinephrine. In the thoracic aorta, ductus arteriosus, and pulmonary artery, large intensely fluorescent bodies were found scattered within tunica media (Fig. 5). Exposure to hydrochloric acid did not affect the appearance of these bodies.

It can be seen in Table 1 that the renal artery and saphenous vein, as well as the carotid artery, increased in their maximal response to norepinephrine with advancing gestation, even after the responses were corrected for the increasing thickness of vascular wall. The procedure of multiple comparisons showed that the mean maximal response of any of these three vessels from the mature fetuses was significantly greater than that from the immature fetuses. No significant changes with gestation was associated with the aorta, pulmonary artery, and ductus arteriosus. The mesenteric artery appeared to reach a peak at the premature fetal period and then hold at the level thereafter. The gestation-related changes in the maximal response to serotonin were similar to those to norepinephrine.

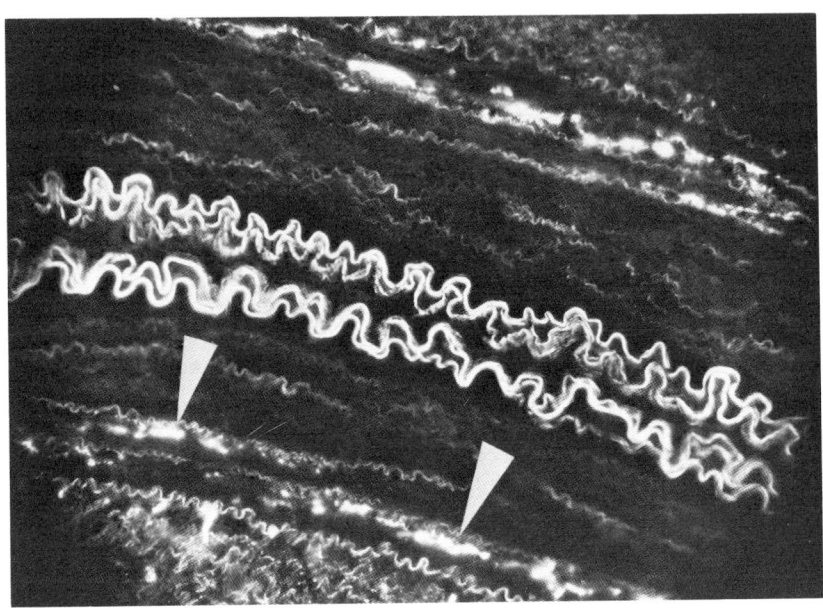

Figure 3

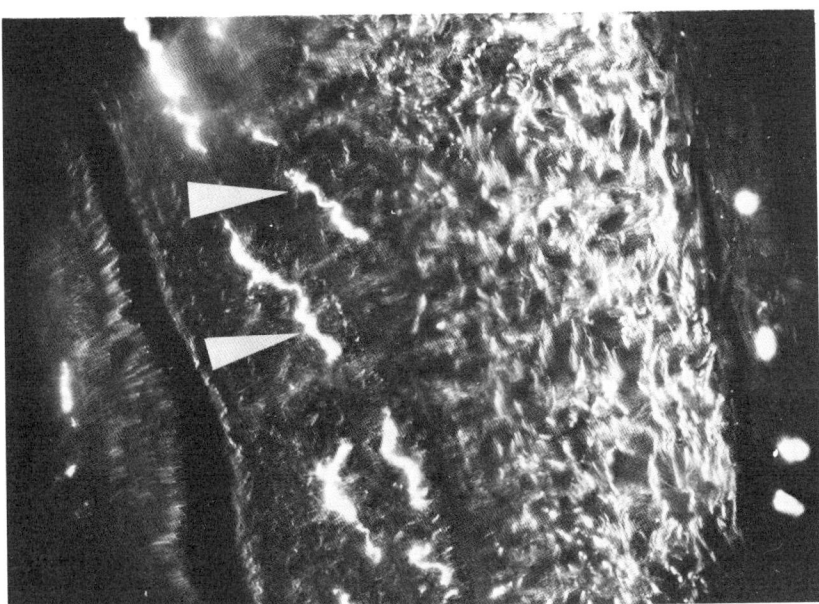

Figure 4

Another vasoactive agent, acetylcholine, induced a small contractile response in most vessel preparations from mature fetuses, but less consistently in those from younger animals. The vessels tended to improve in the response to acetylcholine with gestational growth, although not statistically significantly due to large variances. The only exception noted was the ductus arteriosus, which invariably constricted in response to acetylcholine regardless of the gestational age. The maximal response to acetylcholine per vessel wall sectional area appeared to increase initially with fetal growth, until the fetuses weighed 1-2 kg and were up to 130 days in gestational age. Further growth through term appeared to be accompanied by a decline in the maximal response to acetylcholine.

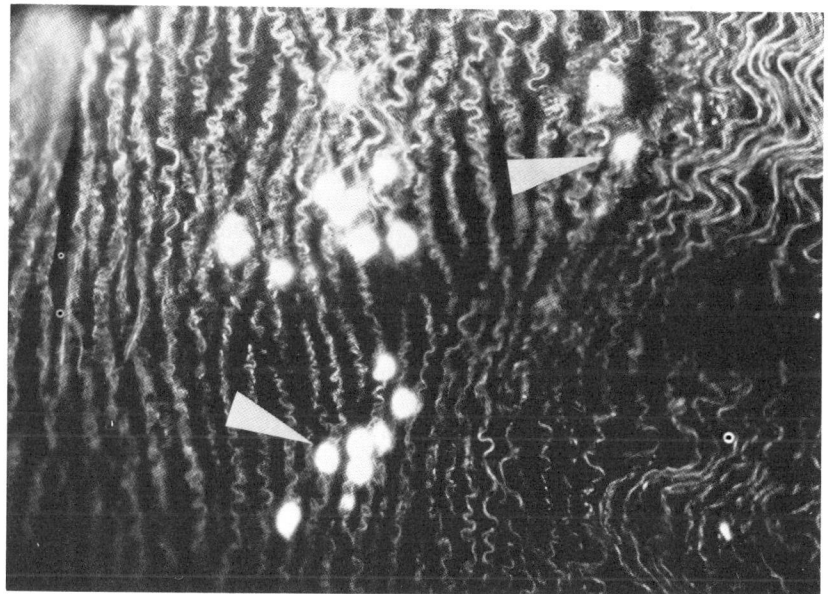

Figure 5

Figs. 3, 4 and 5. Histofluorescence photographs of the transverse section of fetal lamb anterior mesenteric artery. Saphenous vein, and thoracic aorta, respectively. Triangular pointers indicate catecholamine fluorescence (Figs. 3 and 4) and fluorescent bodies (Fig. 5).

## DISCUSSION

The *in vivo* and *in vitro* methods both have their merits and limitations. The latter approach permits detailed histomorphological, biochemical, and pharmacological studies of the individual fetal blood vessels, from an early gestational period when the fetuses are too small for *in vivo* circulatory measurements. The present results indicate that the adrenergic neuroeffector synapse in the common carotid artery is operational in very young fetuses. This finding supports the *in vivo* observations that the autonomic nervous system functions in the lamb fetus before term (6-10). Examination of the synaptic component mechanisms during an early gestational period reveals that these mechanisms do not develop in unison.

Vascular smooth muscle of fetuses 0.3 kg or larger was capable of constriction upon application of the adrenergic transmitter substance, norepinephrine, as well as a nonadrenergic vasoconstrictor agent, serotonin. It had previously been demonstrated that the effects of these agents were prevented by specific receptor blocking agents, phentolamine and methysergide. Thus, specific receptors were already present and the sensitivity of the alpha-adrenergic receptors to norepinephrine was no lower in the immature fetuses than in older fetuses (16). Nevertheless, electrical currents sufficient to excite any intramural neurons elicited a constriction only in carotid arteries from fetuses 0.7 kg and over. Since the effector smooth muscle cells in younger fetuses were reactive to the transmitter, it must be concluded that the transmitter was not delivered to the effector cells, at least not in sufficient quantity to reach the threshold concentration of the receptors. In view of the neuronal uptake of $^3$H-norepinephrine which was demonstrable in all fetuses, some neuronal elements capable of transmitter storage had obviously been formed. It is not clear whether the neurons had not attained a critical density within the vascular wall, or they were yet unprepared to synthesize and/or deliver it. If any catecholamines circulated in the fetal blood they can conceivably be neuronally accumulated, but it is probably the transmitter newly synthesized intraneuronally that is essential for the release during nerve stimulation. Another possibility cannot be ruled out that, even though the neurons had sufficiently matured, the neuronal and muscle cell membranes only belatedly formed an appropriate synaptic relationship. This is in view of the report that norepinephrine was present in the postganglionic sympathetic neurons in the 8-week human fetuses, but mature-looking synapses appeared in fetuses several weeks later (17).

Subsequent to this early period, the smooth muscle increased in responses to norepinephrine and serotonin and appeared to mature fully during the premature fetal period, in so far as the maximal responses did not further increase towards term. The neurogenic response, on the other hand, increased more slowly with gestational growth. It was on average about 1/3 of the responses to norepinephrine and serotonin among the premature fetuses, although it approximated the latter responses near term. Interestingly, the capacity for neuronal $^3$H-norepinephrine uptake did not materially vary between the immature and premature fetuses, and the small difference between the premature fetuses would seem insufficient to account for the doubling of neurogenic response. The explanation for the small neurogenic response especially in the premature fetuses might therefore be sought in the effector, smooth muscle cells. Recently, Ljung and Stage (3) showed that spontaneous rhythmic contractile activity appeared in the rat portal vein abruptly during the third postnatal week. This was accompanied by an increase in the sensitivity to norepinephrine and to transmural adrenergic nerve stimulation. The authors suggested that the myogenic propagation as revealed by the rhythmic contractions was important to the neuronal control of this vein. The carotid artery of the lamb fetus also displayed some spontaneous contractions (Fig. 1) suggestive of myogenic propagation. Although the onset of such activity during the course of gestations has not been determined, this factor may conceivably be associated with the neurogenic contraction. On the other hand, the sensitivity of the carotid artery to norepinephrine during the premature period was not significantly lower than during the last gestational period (16). Therefore, the lagging neurogenic response probably cannot be attributed to the lack of myogenic propagation but more likely to delayed maturation of the neurons and/or synapses.

The extraneuronal uptake of $^3$H-norepinephrine in younger fetuses was, if any, higher than in later stages. The monoamine oxidase and catechol-O-methyltransferase in the premature fetuses were as high as in the mature ones. Thus, the extraneuronal means of transmitter disposition seem to be well prepared prior to the genesis and full deployment of adrenergic neuroeffector synapses in the lamb carotid artery. These mechanisms may perhaps be taken as a safeguard measure in anticipation of the adrenergic transmitter, designed to terminate the action of the latter. It is noteworthy that the same mechanisms would also be useful against several other biogenic amines, including serotonin and epinephrine, that may be blood-borne before or after the adrenergic nervous system is established. The possible role of such vasoactive humoral agents in the regulation of fetal circulation yet remains to be elucidated.

Based on the above results, it may be speculated that the extraneuronal, muscle cellular mechanisms for the inactivation of the adrenergic transmitter becomes operational very early. Closely following them are the contractile reactivity of the muscle cells to the transmitter and other vasoactive agents through specific receptor sites. Subsequently the adrenergic neurons capable of norepinephrine uptake and storage begin to ramify within the vascular wall. The transmitter release from these neurons may be achieved later and sufficiently dense neuronal plexus or formation of functional synaptic nerve-muscle relationship may only then be established. These all seem to take place by the midpoint of gestation at least in the carotid artery. Thereafter, the neurons and/or synapses continue to mature through and possibly beyond term. The time course may well vary with vessels and in some vessels the effector cells, too, may continue to improve in reactivity at least through term, in view of the comparisons of fetal blood vessels (see below).

Regional variation is an outstanding feature of blood vessels of the adult animals. This pertains not only to their reactivity to vasoactive agents but to the nature and density of motor innervation (18-19). The fetal vascular regional variation has not been systematically investigated. The present study demonstrates that the lamb fetal vessels are highly heterogeneous in a number of parameters.

The maximal responses to norepinephrine and serotonin vary by as much as two orders of magnitude among different vessels. In general, the renal artery and saphenous vein were the most reactive, followed by the carotid and mesenteric arteries, while the three large arteries including the thoracic aorta, pulmonary artery, and ductus arteriosus were the least reactive. Such a pattern was common to both the mature and premature fetuses and there was suggestive evidence that the vessels of the immature fetuses were also dissimilar in these responses. Our preliminary studies indicate that the aorta, pulmonary artery, and ductus arteriosus differ in sensitivity to norepinephrine, expressed as the median effective concentration ($ED_{50}$), either in the immature, premature or mature period, although the sensitivity of a given vessel does not significantly change in the course of gestation.

Wide variation was also noted in the reactivity to adrenergic nerve stimulation at least among the mature fetal vessels. Again, the saphenous vein and renal and carotid arteries led all other vessels and the mesenteric artery was only slightly more reactive than the three largest arteries. This ranking was somewhat reflected in the relative capacities of these vessels for neuronal $^3$H-norepinephrine uptake, which is a function of adrenergic nerve density. However, histofluorescence studies indicate that the intramural distribution of the adrenergic nerve plexus markedly varied among blood vessels. Those containing the plexus within the tunica media, in particular the saphenous vein, gave a large neurogenic constriction compared to those in which the plexus was confined to the adventitio-medial junction, e.g., the anterior mesenteric artery. This relationship is analogous to that in some adult rabbit vessels [20]. It can probably be explained at least in part by a reduced diffusional loss of the transmitter where the nerve plexus is inbedded deep in the media. It has been shown that the transmitter diffuses six times faster through the adventitia than the media in the central ear artery of the rabbit, in which the nerve terminals are confined to the adventitio-medial border [21].

"Brightly fluorescent cells" have been described to be diffusely scattered throughout the myocardium of fetal lambs 75-85 days in gestational age. Based on the resistance of the fluorescence to hydrochloric acid, they were considered to contain dopamine rather than norepinephrine [22]. "Small intensely fluorescent cells" have been shown in the heart of rabbit fetuses and neonates and they have been suggested to be secretary cells [23]. The present work demonstrates the presence of similar fluorescent bodies in the large blood vessels, including the pulmonary artery, thoracic aorta, and ductus arteriosus. Since the fluorescence was not altered by exposure to hydrochloric acid, dopamine may be contained in these bodies but their functional significance is unknown.

The fetal vessels were dissimilar even in the temporal pattern of their gestation-related changes. In broad outlines, the reactivity to serotonin and norepinephrine of the aorta, ductus arteriosus, and pulmonary artery remained low and virtually unchanged throughout gestation on the basis of vessel wall cross-sectional area. The reactivity of the carotid and mesenteric arteries rose from a low level to a plateau during the premature period, whereas that of the renal artery and saphenous vein continued to improve through term. As an exception, the reactivity of the ductus arteriosus to acetylcholine diminished towards term. This is of interest particularly because other vessels tended to increase in this response with progress in gestation.

It is evident that the lamb fetal blood vessels differentiate early in gestation. It appears that the vessel-related differences and gestation-related changes in the vascular neuroeffector mechanisms occur independently of the maturation of adrenergic innervation and there is as yet no identifiable extrinsic factor that dictates these differences and changes.

## SUMMARY

Blood vessels isolated from the lamb fetuses of varying gestation were studied in regard to the reactivity to sympathetic, adrenergic nerve stimulation, and several vasoactive agents. The developmental pattern of component mechanisms of the adrenergic neuroeffector synapse was also investigated. In the common carotid artery, the extraneuronal mechanisms for inactivation of the adrenergic transmitter, norepinephrine (NE), and neuronal uptake of norepinephrine were present since early gestation. Most of the carotid arteries from the youngest fetuses (53-90 days gestation) were reactive to norepinephrine and serotonin (5HT) but only those from fetuses of 88 days gestation or older contracted in response to adrenergic nerve stimulation. In the older fetuses, the neurogenic response of the carotid artery increased later than the responses to norepinephrine and serotonin with advance in gestation. Of seven blood vessels from the young or mature fetuses, the renal artery and saphenous vein produced the largest maximal contractile responses to norepinephrine and serotonin; these vessels were followed by the carotid and anterior mesenteric arteries; the thoracic aorta, pulmonary artery, and ductus arteriosus were the least reactive. Adrenergic nerve stimulation elicited the greatest contraction in the first three vessels. This was roughly paralleled by the relatively high adrenergic nerve density as assessed by the neuronal uptake of tritiated norepinephrine. It was probably also attributable, at least in part, to the presence of adrenergic nerves within the medial coat, demonstrated by the catecholamine histofluorescence, in the saphenous vein and carotid artery. The renal and mesenteric arteries and saphenous vein, as well as the carotid artery, significantly increased in the response to norepinephrine and serotonin with advance in gestation, whereas other vessels showed little change. The ductus arteriosus reached a peak reactivity to acetylcholine early in gestation but then declined in this parameter towards term.

These results suggest that many quantitative and qualitative differences exist among blood vessels from early fetal life on. They probably occur independently of the adrenergic innervation or at leart prior to maturation of this innervation. Within a given vascular segment, the mechanisms for the inactivation of the adrenergic transmitter develop early. They are followed by the reactivity of the smooth muscle cell to the transmitter. Only then does the adrenergic neuron acquire the capacity to deliver the transmitter and the neuronal and muscle cell membranes form a functional synaptic relationship, to complete a neuroeffector system.

## ACKNOWLEDGMENT

This work was supported by USPHS grants HL 01755 and HL 08359 and The American Heart Association-Greater Los Angeles Affiliate grant 408 IG.

## REFERENCES

1. Su, C. 1977. Adrenergic and nonadrenergic vasodilator innervation. In *Factors influencing vascular reactivity,* eds. O. Carrier, Jr., and S. Shibata, pp. 156-168. Tokyo: Igaku Shoin.

2. Bevan, J. A.; Hosmer, D. W.; Ljung, B.; Pegram, B.; and Su, C. 1974. Norepinephrine uptake, smooth muscle sensitivity and metabolizing enzyme activity in rabbit veins. *Circ. Res.* 34:541-548.

3. Ljung, B. and Stage (McMurphy), D. 1975. Postnatal ontogenic development of neurogenic and myogenic control in the rat portal vein. *Acta Physiol. Scand.* 94:112-127.

4. Gauthier, P.; Nadeau, R. A.; and de Champlain, J. 1975. The development of sympathetic innervation and the functional state of the cardiovascular system in newborn dogs. *Canad. J. Physiol. Pharmacol.* 53: 763-776.

5. Gerova, M.; Gero, J.; Dolezel, S.; and Konecny, M. 1974. Postnatal development of sympathetic control in canine femoral artery. *Physiol. Bohemoslov.* 23: 289-295.

6. Milligan, J. E. 1972. Control of circulation in the fetus with special references to the autonomic nervous system and chemoreceptor activity. In *Physiological biochemistry of the fetus,* eds. Hodari and Mariona, pp. 204-212. Springfield, Ill.: Thomas.

7. Vapaavouri, E. K.; Shinebourne, E. A.; Williams, R. L.; Heyman, M. A.; and Rudolph, A. M. 1973. Development of cardiovascular responses to autonomic blockade in intact fetal and neonatal lamb. *Biol. Neonate* 22:177-188.

8. Nuwayhid, E.; Brinkman, C. R., III; Su, C.; Bevan, J. A.; and Assali, N. S. 1975. Development of autonomic control of fetal circulation. *Am. J. Physiol.* 228: 337-344.

9. ----. 1975. Systemic and pulmonary hemodynamic responses to adrenergic and cholinergic agonists during fetal development. *Biol. Neonate* 26:301-317.

10. Rudolph, A. M. and Heymann, M. A. 1974. Fetal and neonatal circulation and respiration. *Ann. Rev. Physiol.* 36:187-207.

11. Su, C. and Bevan, J. A. 1970. The release of $H^3$-norepinephrine in arterial strips studied by the technique of superfusion and transmural stimulation. *J. Pharmacol. Expt. Ther.* 172:62-68.

12. Falck, B.; Hillarp, N. A.; Thieme, G.; and Torp, A. 1962. Fluorescence of catecholamines and related compounds condensed with formaldehyde. *J. Histochem. Cytochem.* 10:348-354.

13. Wurtman, R. and Axelrod, J. 1963. Sensitive and specific assay for the estimation of monoamine oxidase. *Biochem. Pharmacol.* 12:439-441.

14. Krakoff, L. R.; Buccino, R. A.; Spann, J. F., Jr., and de Champlain, J. 1968. Cardiac catechol-O-methyl-transferase and monoamine oxidase activity in congestive heart failure. *Am. J. Physiol.* 215:549-552.

15. Dixson, W. J. and Massey, F. J., Jr. 1969. *Introduction to statistical analysis.* New York: McGraw-Hill.

16. Su, C.; Bevan, J. A.; Assali, N. S.; and Brinkman, C. R., III. 1977. Development of neuroeffector mechanisms in the carotid artery of the fetal lamb. *Blood Vessels* 14:12-24.

17. Kaverna, L.; Hervonen, A.; and Hervonen, H. 1974. Morphological characteristics of the ontogenesis of the mammalian peripheral adrenergic nervous system with special remarks on the human fetus. *Med. Biol.* 52:144-153.

18. Bevan, J. A., and Su, C. 1973. Sympathetic mechanisms in blood vessels: nerve and muscle relationships. *Ann. Rev. Pharmacol.* 13:269-285.

19. Su, C., and Lee, T. F-J. 1976. Regional variation of adrenergic and nonadrenergic nerves in blood vessels. *Proc. 2nd Intl. Symp. on Vascular Neuroeffector Mechanisms,* pp. 35-42. Basel: Karger.

20. Bevan, J. A. 1977. Some functional consequences of variation in adrenergic synaptic cleft with and in nerve density and distribution. *Fed. Proc.* 36:2439-2443.

21. Bevan, J. A. and Su, C. 1974. Variation of intra- and perisynaptic adrenergic transmitter concentrations with width of synaptic cleft in vascular tissues. *J. Pharmacol. Expt. Ther.* 190:30-38.

22. Lebowitz, E. A.; Novick, J. S.; and Rudolph, A. M. 1972. Development of myocardial sympathetic innervation in the fetal lamb. *Pediat. Res.* 6:887-893.

23. Papka, R. E. 1974. A study of catecholamine-containing cells in the hearts of fetal and postnatal rabbits by fluorescence and electron microscopy. *Cell Tiss. Res.* 154:471-484.

# 7

## The Electrical Properties of Embryonic Chick Cardiac Cells

Nick Sperelakis and Michael J. McLean
Department of Physiology
University of Virginia
School of Medicine
Charlottesville, Virginia

INTRODUCTION

Important changes occur in myocardial cells during embryonic development, including ultrastructural, metabolic, pharmacologic, and electrophysiologic changes. For example, striking changes occur in the electrical properties of ventricular myocardial cells during embryonic development of chick heart. The electrical properties at each stage of development determine many of the functional properties of the heart at that stage. Strangely enough, studies on the electrophysiological properties of embryonic heart cells have been comparatively neglected until recently, even though such studies would be useful, not only for elucidating the changes during differentiation, but also for understanding the complex electrophysiology of adult hearts. Important clues for analyzing the function of adult hearts have been obtained by such developmental studies.

The discussion and findings to be reviewed here will be based heavily on work from the authors' laboratory for the sake of convenience. However, some of the important contributions made by other laboratories will be mentioned whenever appropriate. In addition, most of the data presented is for the chick, although we have reason to believe that similar changes occur in the developing rat heart (1, 4). Besides considering data from the intact developing hearts, we will also present some findings on organ-cultured hearts and cultured heart cells for purposes of comparison and illustration of related phenomena. Finally, we shall briefly discuss the effect of positive inotropic agents on membrane cation channels, and the control exerted by the myocardial cell on its calcium ion influx.

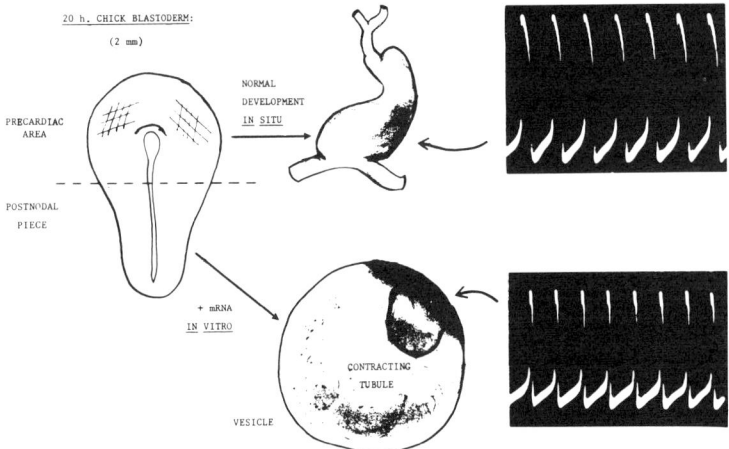

Fig. 1. Sketch of 20-hr chick blastoderm showing the precardiac areas (cross-hatched regions) and the postnodal piece. The precardiac areas give rise to the heart (upper middle sketch); spontaneous action potentials recorded from 3-day old tubular heart shown in upper right. The isolated postnodal piece (posterior part of the blastoderm transected 0.6 mm behind Hensen's node) when placed in culture in the presence of adult heart RNA extracts, rounds up to form a vesicle containing a contracting tubule within several days (lower middle sketch). Action potentials recorded from this tubule (lower right) are similar to those recorded from the intact 3-day heart *in situ*.

## CHANGE IN PROPERTIES OF INTACT HEARTS DEVELOPING *IN SITU*

### Precardiac Areas of the Blastoderm

The early embryo (up to 1 somite) possesses cells which are destined to form the heart. These cells migrate through the early embryo and congregate bilaterally into the so-called precardiac areas (mesoderm) of the anterior half of the flattened 16-17 hr (head process) blastoderm (42). Explants of these areas develop spontaneous electrical activity after several days in culture, which include spontaneous action potentials of about 50 mV amplitude (20). In culture, a tubular heart develops within a vesicle and it beats spontaneously for several days, but further differentiation does not proceed *in vitro*. The precardiac area can be treated with trypsin to facilitate mechanical separation of the three germ layers, and culture of the precardiac mesoderm gives rise to a solid mass of cells which fire spontaneous action potentials and contract (36). If the postnodal piece (posterior third of the blastoderm) is dissected from the 19-hr chick blastoderm and placed into culture, it does not normally give rise to heart tissue. If, however, the postnodal piece is cultured in the presence of an RNA-enriched fraction obtained from adult chicken hearts, a typical spontaneously-beating tubular heart forms within a vesicle (Fig. 1), as in the case of the precardiac areas described above (31). The beating cells in the induced hearts were shown to possess some myofilaments. In addition, the induced tubular hearts can be shown to exhibit spontaneous action potentials (Fig. 1) (27). Thus, it appears that either RNA or some other material within the extract obtained from the adult heart can induce cells in the postnodal piece, normally not destined to form the heart, to take on many of the properties of cardiac myoblasts. Subsequently in development, the twin tubular primordia formed bilaterally from the precardiac mesoderm fuse to form a single tubular heart.[1] The tubu-

---

[1] This fusion begins from the head end and proceeds posteriorly by a zipper-like process, such that the first region fused (ventricle) begins contracting first; the atria are added posteriorly later. The latest portion added has the highest intrinsic pacemaker rate, and thereby drives the entire heart (34). Cutting the 2-day heart into bulbus, ventricle, and sinoatrium regions shows that each region has its characteristic automaticity, the sinoatrium being the fastest.

lar heart begins contracting spontaneously at 30-40 hr (9-19 somite stage), and it begins to propel blood shortly thereafter. The blood pressure is very low at this stage (1-2 mm Hg), and it increases progressively during embryonic development (approximately 30 mm Hg by day 18) and increases further during postembryonic life (9, 41). The velocity of propagation of the peristaltic contraction wave in 3-day hearts is approximately 1 cm/sec (41).

The heart rate of the chick embryo increases from about 50 beats/min at day 1.5 to the maximal value of about 220 beats/min by day 8, and it has been suggested that the increase in beating rate results from the concomitant rise in blood pressure (41). The influence of temperature on heart rate decreases markedly during development (41). We found that the $Q_{10}$ decreases from about 3.6 on day 3 to about 2.0 on day 18 for the same temperature range (unpublished observations). Breaks in the Arrhenius plots also occur at different temperatures in young versus old hearts (41).

## Resting Potentials, $K^+$ Permeability

The transmembrane resting potential ($E_m$), measured by intracellular microelectrodes, of the ventricular portion of the heart increases during embryonic development, as shown in Figure 2 and Table 1. In a 2-day old heart, the mean resting potential is about -40 mV, and this increases to about -51 on day 3. As can be seen in Figure 2, the greatest changes occur between days 2 and 7, and thereafter the increase is smaller. The resting potential is close to -80 mV by day 12, nearly the final adult value. As will be discussed below, the large increase in resting potential during the first few days may be due mainly to an increase in potassium ion permeability ($P_K$) and not in potassium equilibrium potential ($E_K$) (or $[K^+]_i$). Some other investigators (20) have reported larger values of resting potential for the young hearts, i.e., less of a change during development. It is possible that low recorded potentials in young hearts can be caused by improper sealing of the microelectrode; such current leakage around the electrode tip would be most prominent in cells having a high input resistance. However, substantially similar results to ours were found by others (21, 32, 50, 65).

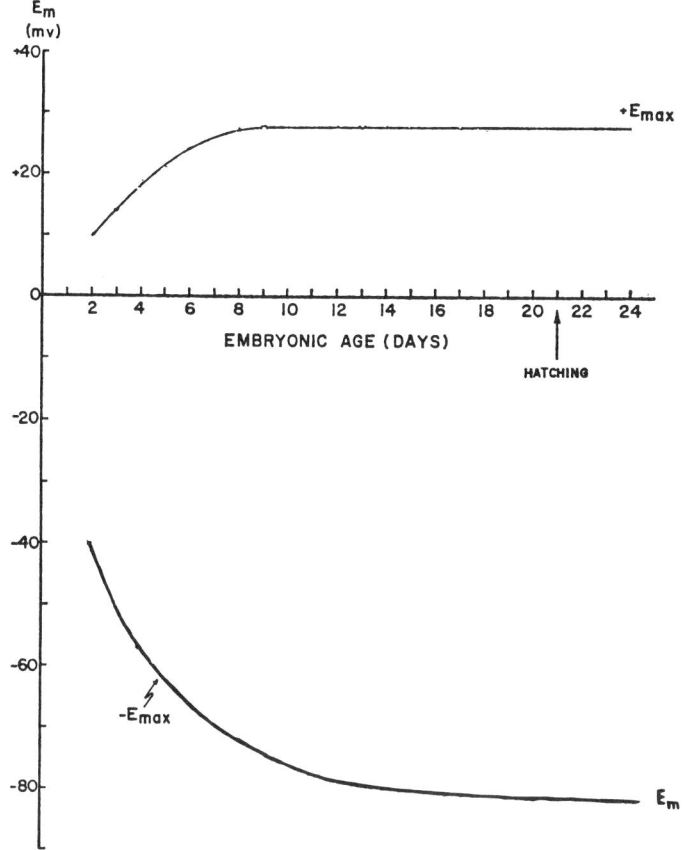

Fig. 2. Graphic representation of the resting potential ($E_m$), action potential peak overshoot potential ($+E_{max}$), and action potential amplitude (difference between the two curves) of intact embryonic chick hearts (ventricle) as a function of developmental age. These potentials increase markedly during development, the greatest changes occurring between days 2 and 8. Hyperpolarizing afterpotentials were present until about day 8, and the maximum diastolic potential ($-E_{max}$) was plotted instead of resting potential for these ages. The estimated $K^+$ equilibrium potential ($E_K$) is already large in young hearts and does not increase very much during development. (Modified from Sperelakis and Shigenobu, 1972.)

Table 1

Summary of Data Obtained from $E_m$ vs. $\log [K^+]_o$ Curves for Chick Embryonic Hearts (Ventricular Cells) at Various Stages of Development and from Input Resistance ($r_{in}$) Measurements

| Embryonic Age (days) | $E_m$ (mV) | Slope (mV/decade) | Extrapolated $[K^+]_i$ (mM) | $E_K$ (mV) | $P_{Na}/P_K$ ratio | $r_{in}$ (MΩ) |
|---|---|---|---|---|---|---|
| 2 | −40 | 30 | 125 | −100 | 0.21 | 13.0 |
| 3 | −51 | 40 | 130 | −101 | 0.17 | 8.5 |
| 4 | −57 | 46 | 140 | −103 | 0.08 | 6.5 |
| 5−6 | −58 | 50 | 130 | −101 | 0.08 | 5.5 |
| 7−9 | −71 | 51 | 145 | −104 | 0.07 | 5.5 |
| 11−13 | −80 | 53 | 145 | −104 | 0.07 | 4.7 |
| 14−20 | −78 | 52 | 155 | −106 | 0.05 | 4.5 |

Note: All values given are approximate. The resting potential ($E$) values are given for a $[K^+]_o$ of 2.7 mM. The slope is the average at $[K^+]_i$ linear of 10 to 100 mM. $[K^+]_i$ was estimated from the extrapolation of fitted levels curves to zero potential. The $P_{NA}/P_K$ ratios were calculated from the Goldman constant-field equation at every $[K^+]_o$ level for which $E_m$ was measured, and an average value was calculated for each heart; some individual values for hearts in the 14–20-day age group were as low as 0.02.

Source: Data were taken from Sperelakis and Shigenobu, 1972 and from Sperelakis et al., 1975.

Similar values for slope and $[K^+]_i$ were obtained by Pappano (1972) for embryonic chick atrium at 4 days, 6 days, and 12 days; for 18-day values were −59 mV/decade and 125 mM $[K^+]_i$, respectively.

## Resting Potential vs. log $[K^+]_o$ Curves

The relationship between resting potential and external $K^+$ concentration ($[K^+]_o$) is illustrated in Figure 3 for three hearts of different ages (data points). Also plotted are theoretical curves calculated from the Goldman constant-field equation, given in the inset of the figure, for five different ratios of $P_{Na}/P_K$: 0.001, 0.01, 0.05, 0.1, and 0.2. For these calculations it was assumed that $[Na^+]_i$ was 30 mM, as indicated in the figure. $[K^+]_i$ was estimated to be 150 mM by extrapolation of the curves connecting the data points to zero potential. It is seen that the data points for the 3-day heart most closely fit the theoretical curve for a $P_{Na}/P_K$ ratio of 0.2; that for the 5-day heart fits the curve for a $P_{Na}/P_K$ of 0.1, and that for the 15-day heart most closely fits between the 0.05 and 0.01 curves. These data suggest that the $P_{Na}/P_K$ ratio is very high in young hearts, and that this accounts for the low measured resting potential. That is, the low resting potential is not due to a low $[K^+]_i$ and potassium equilibrium potential ($E_K$). As shown in Table 1, $[K^+]_i$ is already quite high, 125 mM, in 2-day hearts, and the subsequent increase is only small; a maximum value of 155 mM, about the adult value, is attained by day 14. Table 1 also shows that only a very small increase in the calculated potassium equilibrium potential ($E_K$) occurs during development: potassium equilibrium potential ($E_K$) is about -100mV on day 2 and -106 mV on days 14-20. Thus, in the young hearts, the resting potential is far from potassium equilibrium potential ($E_K$) due to the high $P_{Na}/P_K$ ratio. A similar finding has been obtained in adult sino-atrial nodal cells (14).

In old embryonic or adult hearts, the resting potential vs log $[K^+]_o$ curve is nearly linear above 10 mM$[K^+]_o$ with a slope approaching the theoretical 60 mV/decade (from the Nernst equation), indicating a virtually completely $K^+$-selective membrane in high $[K^+]_o$. If the slope were exactly 60 mV/decade, then the resting potential would be equal to potassium equilibrium potential ($E_K$). The data in Figure 2 and Table 1 show that the slope for hearts 7 to 20-days old is 51-53 mV/decade, whereas that for 4-day, 3-day, and 2-day hearts is 46, 40, and 30 mV/decade, respectively. As can be seen, the curves for young hearts and high $P_{Na}/P_K$ ratios are continually bending, and so the values given above are for the average slope. Similar values

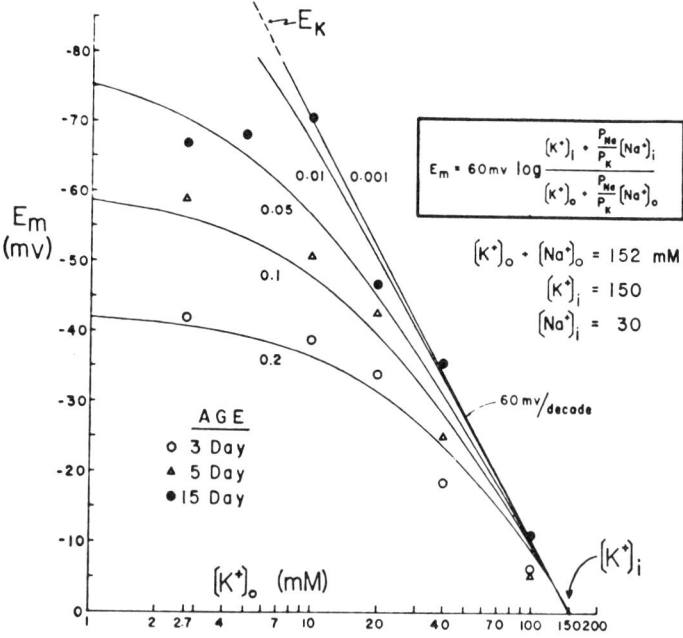

Fig. 3. Resting potential ($E_m$) plotted as a function of $[K^+]_o$ on a logarithmic scale for three representative hearts of different ages. $[K^+]_o$ was elevated by substitution of $K^+$ for equimolar amounts of $Na^+$. Continuous lines give theoretical calculations from the constant-field equation (inset) for $P_{Na}/P_K$ ratios of 0.001, 0.01, 0.05, 0.1, and 0.2. Calculations were made assuming $[K^+]_i$ and $[Na^+]_i$ values shown. For a $P_{Na}/P_K$ ratio of 0.001, the curve is linear over the entire range with a slope of 60 mV/decade, i.e., it closely follows $E_K$. Symbols give representative data obtained from embryonic chick hearts at days 3 (o), 5 (△) and 15 (●). The data for the 3-day heart follow the curve for a $P_{Na}/P_K$ ratio of 0.2, those for the 5-day heart follow the curve for 0.1, and those for the 15-day heart fall between the curves for 0.01 to 0.05. The estimated intracellular $K^+$ activities ($[K^+]_i$) obtained by extrapolation to zero potential are nearly the same for all ages. (Modified from Sperelakis and Shigenobu, 1972.)

for $[K^+]_i$ and slope were found for embryonic chick atrial cells at various stages of development (32).

## Membrane Resistance

In order to help determine whether the $P_{Na}/P_K$ ratio is high in young hearts because of a high sodium permeability or because of a low potassium permeability, input resistance was determined from steady-state voltage/current curves. The input resistance ($r_{in}$) of the ventricular cells is high (13 M$\Omega$) in young 2-day-old hearts, and rapidly declines over the next few days, reaching the final adult value (4.5 M$\Omega$) by day 14 (Table 1). If the average cell size and the electrical arrangement of the cells remains unchanged during this period, the high $r_{in}$ of the young hearts would suggest that membrane resistivity ($R_m$) is very high. The latter would be consistent with a low potassium conductance and potassium permeability in the young hearts. These results would suggest that the $P_{Na}/P_K$ ratio is high in young hearts because potassium permeability is low, and not because sodium permeability is high.

Consistent with our interpretation that potassium permeability is low in young hearts, is the finding that the chronaxie of young hearts (2-day old) is about four fold higher than that of 9-16 day hearts (50). This indicates that the membrane time constant is about fourfold higher in young hearts, and if membrane capacitance remains constant, membrane resistivity must be about four-fold higher.

Carmeliet and coworkers (2) have reported, on the basis of $^{42}K$ flux measurements, that potassium permeability is about 2-3 fold lower in 6-8 day hearts than in 18-20 day hearts, consistent with the conclusions from the electrical studies described above. (Although they reported that the calculated potassium permeability for 3-5 day hearts was nearly as high as that for the 18-20 day hearts, which disagrees with the electrical measurements, they did not control for spontaneous beating and action potentials, hence increased $K^+$ efflux, which occurs in the 3-5 day hearts; this would cause them to calculate an erroneously high potassium permeability.) The potassium permeability coefficient ($P_K$) was $13.2 \times 10^{-8}$ cm/sec for 7-day hearts and about $27.5 \times 10^{-8}$ cm/sec for 19-day hearts (at a $[K]_o$ of 2.5 - 5 mM). ($P_K$ was greatly reduced in 0 mM external $K^+$.) The $P_{Na}/P_K$ ratios calculated from the constant-field equation were 0.018 for the 19-day hearts and 0.037 for the 7-day hearts. The sodium permeability, as calculated from $P_K$ and the $P_{Na}/P_K$ ratio, suggested that this parameter did not change during development (constant at about $0.50 \times 10^{-8}$ cm/sec).

## Sensitivity to $[K^+]_o$

Since flattening of the resting potential vs log $[K^+]_o$ curve at lower $[K^+]_o$ levels is much more prominent for young hearts, i.e.. they are depolarized less by a given increment in $[K^+]_o$ (see Fig. 3), young hearts should be less affected by elevation of $[K^+]_o$. As illustrated in Figure 4, young hearts are indeed less affected by elevation of $[K^+]_o$ than are older hearts. This is true for both

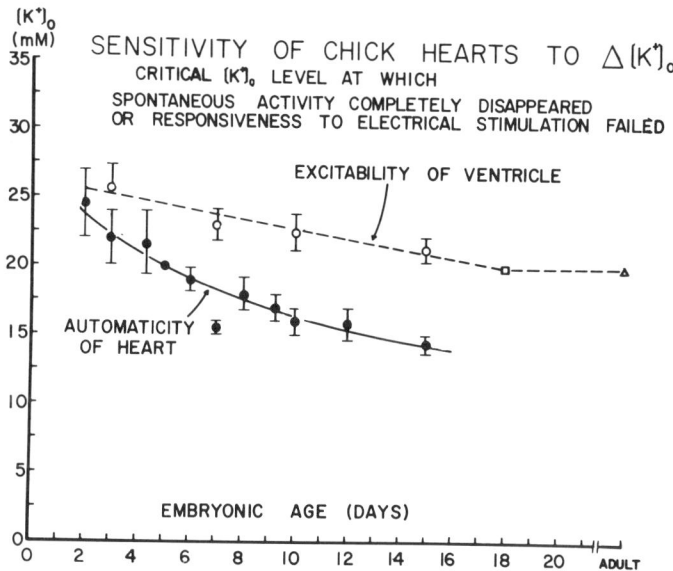

Fig. 4. Data summarizing the sensitivity to $[K^+]_o$ of chick hearts at various stages of development. The ordinate gives the critical $[K^+]_o$ level at which (a) membrane excitability and contractions of the ventricular myocardium in response to electrical stimulation are abolished, or (b) automaticity of the heart completely disappears. (Modified from Sperelakis and Shigenobu, 1972.)

inhibition of automaticity of the whole heart as well as for loss of excitability of the ventricle to electrical stimulation. Automaticity and excitability of 2-3 day hearts fail at about 25 mM $K^+$, whereas failure occurs at about 15 mM (automaticity) and 20 mM (excitability) for the 14-20 day hearts. Depression of automaticity was evident by 12 mM in hearts of all ages.

## Lack of effect of acetylcholine on potassium permeability

The young ventricular cells are completely insensitive to acetylcholine, even though a large hyperpolarization is theoretically possible because the resting potential is much below the potassium equilibrium potential. Therefore, it is likely that acetylcholine does not significantly increase potassium permeability in ventricular cells. In old ventricular muscle also, acetylcholine has no effect on shortening the action potential, whereas the old embryonic chick atrial action potential is markedly shortened. The atrial cells of young hearts are slightly depolarized by acetylcholine in normal medium, and slightly hyperpolarized in Na-free medium, suggesting that acetylcholine increases both sodium conductance and potassium conductance in young hearts (22, 32, 33).[2]

## $(Na^+, K^+)$-ATPase Activity

The specific activity of the $(Na^+, K^+)$-ATPase is low in young embryonic chick hearts and rises during development (51). As indicated in Table 2, the average value on day 4 is about 35% of that on day 16, that on day 6 is about 42%, that on day 9 is about 57%, and that on day 13 is about 77%. The ATPase activity is highest on day 20 (about 144% of that on day 16). The adult level is about equal to that on embryonic day 16.

Thus, while potassium permeability is increasing during development, and hence the outward passive leak of

---

[2] Pappano interprets the small hyperpolarization produced by acetylcholine in Na-free solution to be consistent with a low potassium permeability--namely, that few $K^+$ channels are available to be opened by acetylcholine (32).

Table 2

$(Na^+, K^+)$-ATPase Specific Activity of Embryonic Chick Hearts at Different Stages of Development

| Embryonic Age (days) | $(Na^+, K^+)$-ATPase Activity | | | |
|---|---|---|---|---|
| | Series I | | Series II | |
| | Absolute (μmoles $P_i$/hr/mg protein) | Rel. to Day 16 (%) | Rel. to Day 16 (%) | |
| 4 | 1.3 (1) | 18 | 43 | (2) |
| 6 | 3.0 ± 0.9 (8) | 41 | 43 ± 7 | (10) |
| 9 | 4.2 ± 0.8 (12) | 57 | 56 ± 5 | (12) |
| 13 | 5.9 ± 1.0 (15) | 80 | 73 ± 4 | (17) |
| 16 | 7.4 ± 0.7 (14)[a] | 100 | 100 | (28) |
| 20 | 11.0 ± 1.3 (4) | 149 | 140 ± 7 | (8) |
| 23 | 8.9 ± 0.9 (6) | 120 | 115 ± 11 | (7) |
| 30 | 9.3 ± 1.1 (5) | 126 | 126 ± 4 | (5) |
| Adult | --- | --- | 96 ± 10 | (8) |

Source: Data from Sperelakis, 1972.

[a] In a third series of separate experiments, the activity of the enzyme for 16-day hearts was generally running between 30 and 40 μmoles/hr/mg protein, and that for cultured heart cells prepared from 16-day hearts was 3 to 10-fold lower (Sperelakis and Lee, 1971).

$K^+$ and inward leak of $Na^+$ (due to the increased electrochemical driving force), the capability of the $Na^+$-$K^+$ pump is increasing. That is, the increased cation pump capability tends to compensate for the increased demand on the pump due to increased cation leak. However, the pumping capacity of the very young hearts must be sufficient to maintain the high $[K^+]_i$ and low $[Na^+]_i$ already present in the young cells.

When the ventricular myocardial cells from 16-day hearts are placed into monolayer cell culture, the specific activity of the ($Na^+$, $K^+$)-ATPase decreases by more than threefold (56). The lower $Na^+$-$K^+$ pumping capability of the cultured cells is consistent with the lower potassium permeability and somewhat lower $[K^+]_i$ generally observed in these cells.

### Automaticity

Large changes in automaticity of the ventricular cells also occur during development, as would be predicted from the changes in potassium permeability. The incidence of hyperpolarizing afterpotentials and pacemaker potentials is very high (80-100%) in the young hearts, and this incidence decreases to 0% in the old embryonic hearts (Table 3) (59). If a portion of the ventricle is cut and isolated to remove drive from the nodal cells, the incidence of pacemaker potentials observed in the impaled cells is 100% for embryos up through day 10 (Fig. 5), whereas the incidence is 0% in embryos day 12 or older (Table 3; Fig. 6). These results indicate that the ventricular myocardial cells possess automaticity capability when they are young, but that this capability diminishes as the cells become older.

However, the results discussed below indicate that the old ventricular cells again become automatic when trypsin-dispersed and placed into cell culture. The major requirements for automaticity appear to be: (1) a low chloride conductance, as is generally true for myocardial cells, and (2) a low potassium conductance. The low potassium conductance enhances membrane inductance in series with the negative slope conductance of one type of $K^+$ channel, and tends to cause oscillations in membrane potential. The low potassium conductance also produces some depolarization (moves the resting potential farther from the potassium equilibrium potential) and places the membrane potential in the region than can support pacemaker oscillations.

Table 3

Incidence of Pacemaker Potentials and Hyperpolarizing Afterpotentials Found in Ventricular Myocardial Cells During Embryonic Development of Chick Hearts

| Embryonic Age (days) | Hyperpolarizing Afterpotentials[a] (Intact Hearts) (%) | Pacemaker Potentials[b] Intact Hearts (%) | Cut Ventricles (%) |
|---|---|---|---|
| 2-3 | 81 | 80-100 | 100 |
| 4 | 63 | 60-100 | 100 |
| 7 | 38 | 20-40 | 100 |
| 10 | 0 | 0 | 100 |
| 12 | 0 | 0 | 0 |
| 17 | 0 | 0 | 0 |
| 20 | 0 | 0 | 0 |
| 27 | 0 | 0 | 0 |

[a] Usually, cells which exhibit pacemaker potentials also possess hyperpolarizing afterpotentials, but the converse is not always true. Source: Data from Sperelakis and Shigenobu, 1972.

[b] Source: Data from Sperelakis and McLean, unpublished observations.

## Ion Content

### Sodium

Tissue electrolyte analyses of chick embryonic hearts (ventricles) indicate that the total tissue content of $Na^+$ in young hearts is very high, and that it decreases gradually until about day 13, after which the level remains constant (17, 65) Sodium ion exchangeability is low in young hearts and rises gradually during development to about 70% near hatching. Thus, there may be a great amount of bound $Na^+$ in young hearts. Much of the $Na^+$ may be bound in the nucleus (18) and in the extracellular (subendocardial) mucopolysaccharide cardiac jelly (which serves a valvelike function) in the young heart (62). The data from our electrophysiological studies (59) indicated that the thermodynamically-active free intracellular $Na^+$ must not be too high because the $Na^+$-dependent action potentials overshoot to +11 mV in day 2 hearts; the overshoot increases rapidly, reaching +28 mV by day 7. The latter value is the same as the adult value. Carmeliet et al. (2) measured $[Na^+]_i$ values respectively, and Harsch and Green (11) reported that the $[Na^+]_i$ levels remained relatively constant (23-38 mM) between days 8 and 18). Thus, there seems little doubt that the free $[Na^+]_i$ is already low in young hearts.

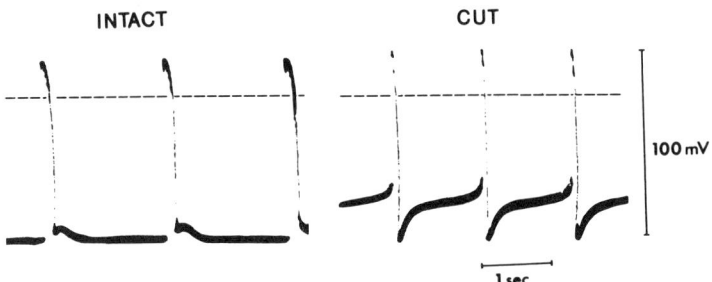

Fig. 5. Electrophysiological recordings from an intact 6-day old embryonic chick heart (ventricle) (left panel) and from a ventricular fragment cut from the apex of the same heart (right panel). *Left panel:* Ventricular action potentials (maximal rate of rise 70 V/sec) recorded in the spontaneously-contracting heart in response to conduction of the impulse from the sinoatrial pacemaker; note stable resting potential, i.e., absence of pacemaker potentials. *Right panel:* Pacemaker potentials preceded the action potentials in the cells of the cut ventricular fragment, revealing intrinsic automaticity. The cut fragment contracted spontaneously.

**Embryonic Chick Ventricular Fragments**

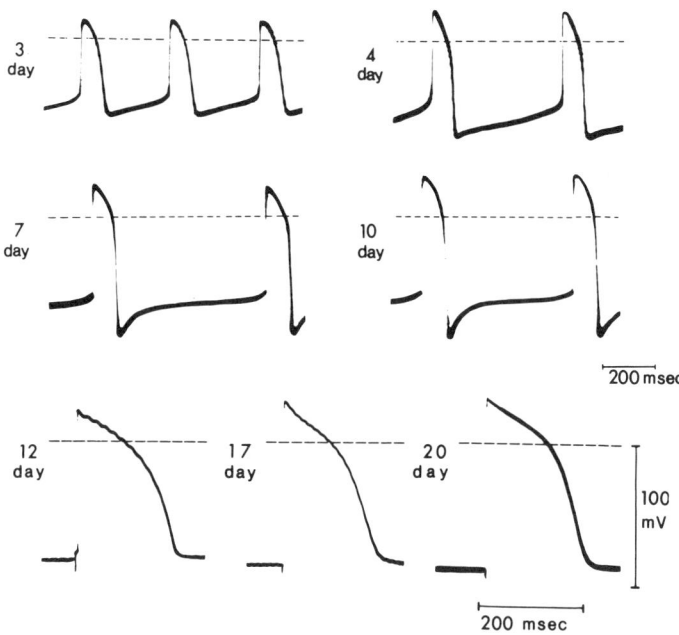

Fig. 6. Electrophysiological recordings from ventricular fragments cut from chick embryonic hearts of various ages. *Upper two rows:* Pacemaker potentials preceded the action potentials in impaled cells within fragments from 3,4,7, and 10-day old hearts. Maximal rate of rise and maximal diastolic potentials in the fragments were the same as in the intact heart. The fragments contracted spontaneously.
*Bottom row:* Cut ventricular fragments from hearts 12 days old and older did not exhibit automaticity. Action potentials with fast rates of rise (equal to those recorded from cells in the intact hearts of the respective ages) were elicited by field stimulation (note shock artifacts) from fragments of 12,17, and 20-day ventricles. Time calibration in the middle row applies to top two rows.

There is a decrease in extracellular space during development. The inulin space was reported to decrease from 39% at day 8 to 19% at day 18 (11). The extracellular space determined by radioactive chromium-EDTA decreased from 34% on days 6-8 to 27% on days 18-20(2). The estimates of intracellular $Cl^-$ of 36-45 mM for days 8-18 are too high to be consistent with the measured resting potentials (of -70 to -80 mV), assuming $Cl^-$ to be passively distributed (11).

## Potassium

It was reported that the $K^+$ content of chick hearts gradually increases during development, from about 68 mmoles/kg on days 2-3 to a plateau level of about 86 mmoles/kg on day 13 (17). Others have reported that the calculated $[K^+]_i$ levels of chick ventricles are 145 mM on day 8 and 91 mM on day 18 (11). Carmeliet, et al. (2) calculated $[K^+]_i$ values of 151 mM and 122 mM for chick hearts on days 6-8 and days 18-20, respectively. The electrophysiological data from resting potential vs log $[K^+]_o$ curves (59) indicate that $[K^+]_i$ is about 125 mM on day 2, and that it increases gradually to about 155 mM on day 2, and that 1). Pappano (32) has also reported high values of $[K^+]_i$ 145 mM) on day 4 for chick atrial cells. Thus, there is no question that $[K^+]_i$ is already high in young hearts. Hence, the cardiac cells must actively transport cations before day 2, and even during the precardiac blastoderm stage.

## Changes in Voltage-Dependent Membrane Channels

The action potentials of the cells of intact chick hearts undergo sequential changes during development *in situ* (48, 59, 61), as shown in Figure 7. *Young* (2-4 days *in ovo*) myocardial cells possess slowly rising (10-30 V/sec) (Fig. 8) action potentials preceded by pacemaker potentials (Fig. 7A). Tetrodotoxin (TTX), a specific blocker of fast $Na^+$ channels, has no effect on the action potential rate of rise or overshoot (Fig. 7B) (Table 4). Hyperpolarizing current pulses do not greatly increase the rate of rise of the action potential (Fig. 9), and excitability is not lost until the membrane is depolarized to less than -25 mV (Fig. 9). The action potential upstroke in young hearts is generated by $Na^+$ influx through tetrodotoxin-insensitive slow $Na^+$ channels, as indicated by the dependence of the action potential

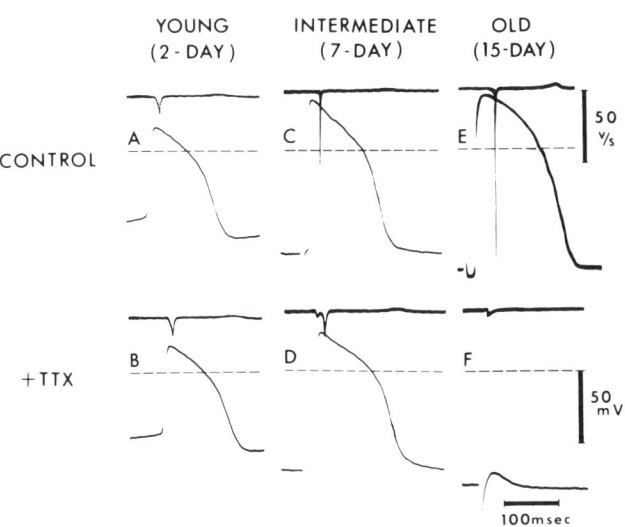

Fig. 7. Development of sensitivity to tetrodotoxin (TTX) of intact embryonic chick hearts with increasing embryonic age. *A-B*: Intracellular recordings from a 2-day-old heart before (*A*) and 20 min after (*B*) the addition of tetrodotoxin (20 µg/ml). *C-D*: Recordings from a 7-day-old heart before (*C*) and 2 min after (*D*) the addition of tetrodotoxin (2 µg/ml). Note depression of the rate of rise in *D*. *E-F*: From a 15-day-old heart prior to (*E*) and 2 min after (*F*) the addition of tetrodotoxin (1 µg/ml). The action potentials were abolished and excitability was not restored by strong field stimulation in *F*. The upper traces give dV/dt; this trace has been shifted relative to the V-t trace to prevent obscuring dV/dt. The horizontal broken line in each panel represents zero potential. dV/dt calibration (in *E*) and voltage and time calibrations (in *F*) pertain to all panels. (Modified from Sperelakis and Shigenobu, 1972.)

overshoot (Fig. 10) and rate of rise (Fig. 11) on the external Na$^+$ concentration. The slope of overshoot as a function of $[Na^+]_o$ approaches the theoretical 60 mV/decade from the Nernst equation at the lower $[Na^+]_o$ concentrations (Fig. 10).

Kinetically fast Na$^+$ channels, which are sensitive to tetrodotoxin, make their initial appearance on about day 5, which is about the time that innervation reaches the heart and circulation is established to the chorioallantoic membrane for gas exchange. At this time, the maximal rate of rise of the action potential ($+\dot{V}_{max}$) suddenly jumps to about 50-70 V/sec (Fig. 8). During this *intermediate* stage of development (from about day 5 to day 8), both slow and fast Na$^+$ channels coexist in the membrane; tetrodotoxin causes a reduction in the maximal rate of rise of the action potential to about the rate observed in young

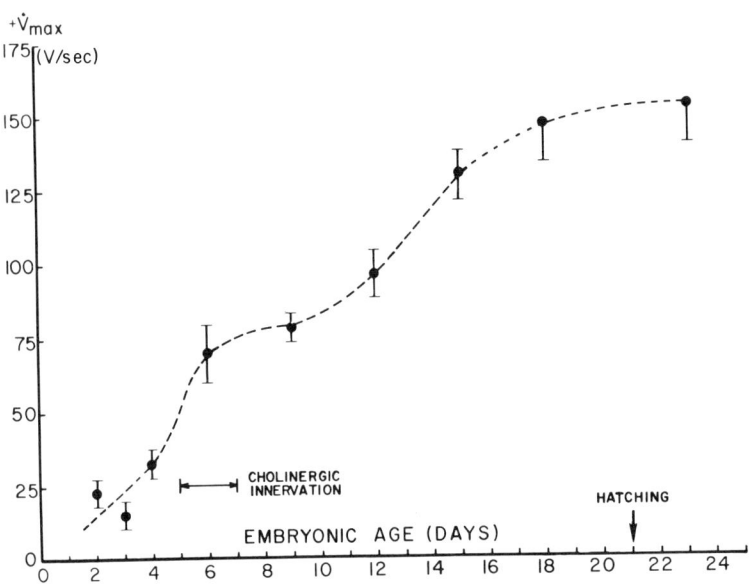

Fig. 8. Changes in the maximum rate of rise of the action potential ($+\dot{V}_{max}$) during embryonic development. Most of the data were grouped into 3-day periods. There may be a jump in $+\dot{V}_{max}$ on about day 5. The approximate period of cholinergic innervation of the heart is indicated. Vertical bars = ±1 SE. (Modified from Sperelakis and Shigenobu, 1972.)

### Table 4
### Effect of Tetrodotoxin (TTX) on the Action Potential Maximal Rate of Rise ($+\dot{V}_{max}$) of Chick Embryonic Hearts (Ventricular Muscle) as a Function of Developmental Age

| Embryonic Age (days) | $+\dot{V}_{max}$ (V/sec) Control | +TTX | TTX Sensitivity |
|---|---|---|---|
| 2-3 | 10-20 | 10-20 | None |
| 5-6 | 50-60 | 10-20 | Partial |
| 8-10 | 70-85 | 0 | Complete |
| 12-16 | 90-140 | 0 | Complete |
| 17-21 | 140-160 | 0 | Complete |

The $+\dot{V}_{max}$ values given are the approximate ranges.

Source: Data from Sperelakis and Shigenobu, 1972.

cells, i.e., 10-20 V/sec, but the action potentials and accompanying contractions persist (Fig. 7C and D) (Table 4).

After about day 8, the action potentials are completely abolished by tetrodotoxin despite increased stimulus intensity (Fig. 7E and F) (Table 4), and depolarization to less than -50 mV now abolishes excitability (Fig. 9). This indicates that the action potential-generating $Na^+$ channels consist predominantly of fast $Na^+$ channels. The dependence of the inward current on $[Na^+]_o$ during the action potential of *old* embryonic hearts is also shown in Figures 10 and 11. The density of fast $Na^+$ channels continues to increase until about day 18, when the adult maximal rate of rise of about 150 V/sec is achieved (Fig. 8). Most of the slow $Na^+$ channels appear to have been inactivated (either by removal from the membrane or by masking), and insufficient numbers remain to support regenerative excitation, particularly at the high level of resting potential. However, it is not clear to what extent the density of slow channels is reduced in old hearts, because the

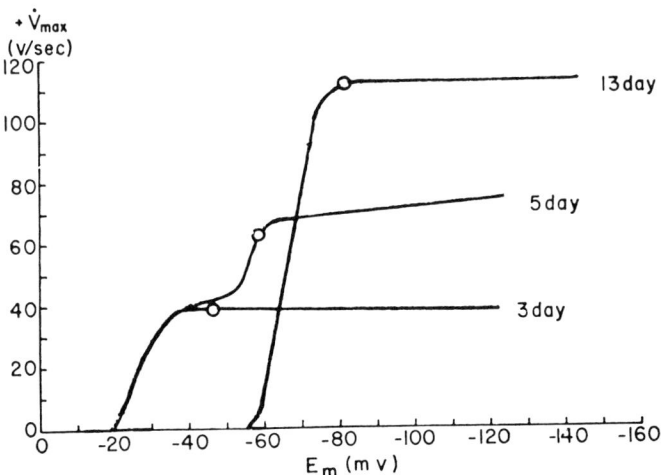

Fig. 9. Evidence for two types of Na$^+$ channels. Data illustrating changes in the maximal rate of rise of the action potential (+$\dot{V}_{max}$) as a function of membrane potential ($E_m$) for three representative hearts of different embryonic ages (3, 5, and 13 days). The transmembrane potential was changed by applying rectangular polarizing current pulses of long duration (several seconds). The circles give the mean resting potential (polarizing pulses not applied). Note that complete inactivation for the 13-day heart occurs at about -55 mV, whereas complete inactivation does not occur until about -25 mV for the 3-day and 5-day hearts. For the 3-day and 13-day hearts, there appears to be only one type of channel: slow and fast, respectively. The data for the 5-day heart (transition period) suggest that there are two sets of channels, one set inactivating at about -55 mV and the other set at -25mV. (Modified from Sperelakis and Shigenobu, 1972.)

simultaneous increase in resting potential should render propagation more difficult for any given density of slow channels. Addition of some positive inotropic agents increases the number of slow $Ca^{++}$-$Na^+$ channels in the membrane, and leads to the regaining of excitability in cells whose fast Na$^+$ channels have been inactivated, as discussed below.

Verapamil and D-600 block the action potentials of the young embryonic hearts, indicating that these agents block slow $Na^+$ channels as well as slow $Ca^{++}$ and slow $Ca^{++}$-$Na^+$ channels (47). Thus, these agents are not specific for blockade of $Ca^{++}$ current. In contrast, $Mn^{++}$ at 1 mM does not depress the action potentials of young hearts, although it does block the contractions, indicating a greater specificity for slow $Ca^{++}$ and $Ca^{++}$-$Na^+$ channels.

## Changes in Other Membrane Properties

Changes in other membrane properties also occur during development. For example, glucose uptake is very high in young hearts, and decreases during development. A car-

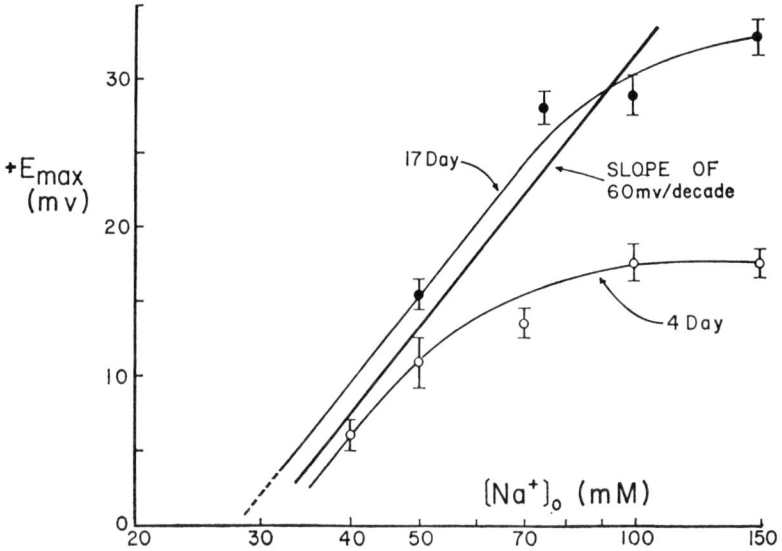

Fig. 10. Evidence that $Na^+$ is the predominant carrier of inward current throughout development. The mean overshoot values ($+E_{max}$) for one representative 4-day heart and one representative 17-day-old heart plotted as a function of external $Na^+$ concentration on a logarithmic scale. A line having a slope of 60 mV/decade has been arbitrarily superimposed. The curves are nearly linear with a slope approaching 60 mV/decade at lower $[Na^+]_o$, but flatten at higher $[Na^+]_o$ levels. (Modified from Sperelakis and Shigenobu, 1972.)

rier-mediated saturable glucose transport system appears on about day 7, and an enhancement of glucose uptake by insulin can be first demonstrated shortly thereafter (10). Amino acid uptake also decreases with development (6).

There are also changes in membrane fluidity (microviscosity) during development of chick hearts (19). The

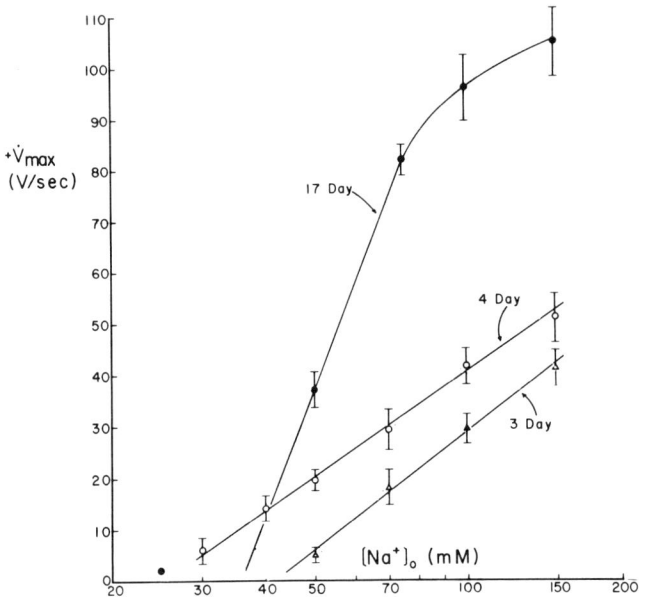

Fig. 11. Evidence suggesting that the major inward current is carried by $Na^+$ in both young and old hearts. The effect of variation in $[Na^+]_o$ on maximum rate of rise of the action potential ($+\dot{V}_{max}$) is illustrated. $[Na^+]_o$ (plotted on a logarithmic scale) was lowered by replacing the NaCl with equimolar amounts of choline-Cl. Plotted are the mean $+\dot{V}_{max}$ ($\pm 1$ SE) values of multiple penetrations into two 3-day old hearts, three 4-day-old hearts, and one 17-day old heart. The curve for the old heart is linear at lower $[Na^+]_o$ levels, and flattens at higher levels; for the young hearts, the curves are linear over the entire range. (Modified from Sperelakis and Shigenobu, 1972.)

cholesterol/phospholipid ratio of the sarcolemma increases during development, concomitant with an increase in the number of unsaturated fatty acid residues. In general, the changes seem to be too complex to correlate with changes in the electrical properties, as, for example, with changes in potassium permeability. It is of interest that the membrane fluidity of cultured chick skeletal myoblasts increases concomitant with fusion and myotube formation (7).

Table 5

Cyclic AMP Content of Embryonic Chick Hearts at Different Stages of Development

| Embryonic Age (days) | Cyclic AMP (pmoles/mg protein) |
|---|---|
| 4 | 116 ± 21  (8) |
| 5 | 41 ± 7  (3) |
| 7-8 | 35 ± 2  (31)[a] |
| 11-12 | 24 ± 2  (18) |
| 15-17 | 9.4 ± 0.5  (24) |
| 18-20 | 11.4 ± 1.0  (20) |
| Cultured heart cells[b] | 16 ± 2  (24) |

Source: Data from McLean et al., 1975.

[a] Isoproterenol ($10^{-6}$ M) raised the cyclic AMP level of 7-8 day hearts to 80 ±15 (12) at 3 min (time of peak effect).

[b] The cultured heart cells were prepared from 15-20 day old hearts.

## Cyclic AMP Levels

The cyclic AMP level is high in young hearts and it decreases during development (23, 39). In one set of experiments, the cyclic AMP level was 116 pmoles/mg protein on day 4, and this decreased sharply by day 5 to about 41 (Table 5). There was a gradual further decline until the adult level was reached by about day 16. The relationship, if any, between changes in membrane properties and changes in cyclic AMP levels during development of heart remains to be clarified. However, in cultured skeletal muscle, the cyclic AMP level decreases sharply as the myoblasts fuse into myotubes, i.e., when the cells further differentiate (40). In addition, as will be discussed below, increase in cyclic AMP level is associated with increase in the number of available slow channels.

Cultured heart cells isolated from 16-day hearts had a slightly higher cyclic AMP level in one set of experiments (Table 5), but had a significantly lower level in another set of experiments (39) (Table 6). The cyclic AMP level of young 4-day hearts placed into organ culture for 2 weeks declines to about the adult value, even though the cells do not further differentiate electrically or ultrastructurally (39). Isoproterenol was capable of markedly elevating the cyclic AMP level in all cases: young or old, cultured or noncultured (Table 6), thus demonstrating the presence of functional beta-adrenergic receptors.

## Ultrastructure and Metabolism

Thin myofilaments are clearly present at 30 hr (they may actually appear by 18 hr--the head fold stage (20), and they begin to collect into groups at about 36 hr (12). Thin filaments are found without thick filaments, but thick filaments usually occur in association with thin filaments. The myofibrils in the 2-3 day hearts are relatively sparse and in various stages of formation(Fig. 12A). The myofibrils are not aligned, and they run in all directions, including perpendicular to one another. Bundles of myofilaments attached to one Z line often radiate in several directions (Fig. 12A). Free cytoplasmic polyribosomes are abundant and rough ER tubules lined with ribosomes are prominent in young hearts; the ribosomes are frequently observed in close association with the developing myofibrils. The sarcomeres

Table 6

Cyclic AMP Response to Isoproterenol of Young and Old Embryonic Chick Hearts Before and After Organ Culture and Cell Culture

|  | 4-day old | | 16-day old | |
| --- | --- | --- | --- | --- |
|  | Control | Isoproterenol | Control | Isoproterenol |
| Fresh noncultured | 33.6 ± 2.2 (13) | 118.9 ± 2.3 (7) | 11.7 ± 1.5 (17) | 22.0 ± 1.5 (8) |
| Organ cultured | 6.0 ± 2.4 (20) | 37.7 ± 4.7 (13) | 5.3 ± 1.2 (24) | 17.9 ± 2.0 (6) |
| Cell cultured | 5.6 ± 1.2 (4) | 28.8 ± 2.0 (4) | 4.5 ± 0.4 (49) | 23.6 ± 3.0 (14) |

Source: Data from Renaud et al., 1977. Data given as means ± 1 SE (N) and as pmoles/mg protein. Isoproterenol concentration used was $5 \times 10^{-5}$ M.

### A. INTACT 3-DAY

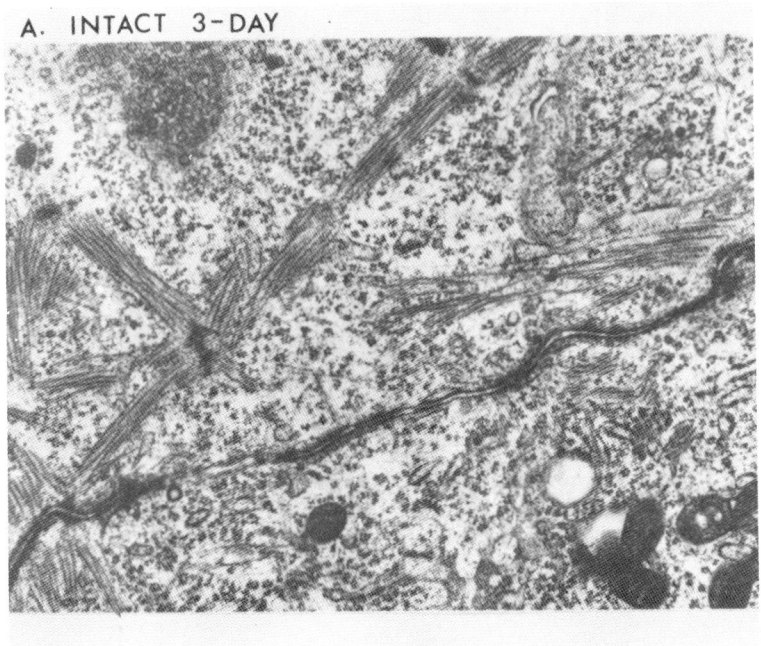

### B. INTACT 19-DAY

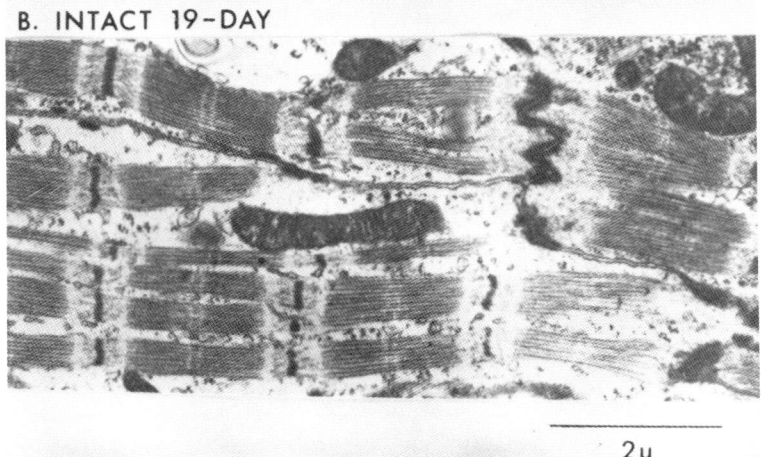

2μ

Fig. 12. Cell ultrastructure of young (3 days *in ovo*) and old (19 days *in ovo*) intact embryonic chick hearts *in situ*. A: 3-day ventricular cell demonstrating paucity and non-alignment of myofibrils. Ribosomes are abundant in the cytoplasm. The contiguous cells are held in close apposition by desmosomes. B:

are usually incomplete early in their formation, and the myofibrils are short. H zones first become obvious at 8 days, and M lines do not appear until about day 18. By day 18, the embryonic myocardial cell closely resembles the adult cell, and has a close packing of completed myofibrils with rows of mitochondria in between (Fig. 12B). The myofibrils have the typical hexagonal array of six thin actin filaments surrounding each thick myosin filament. In embryonic chick skeletal muscle, the adult level of the tropomyosin-troponin complex is reached by day 16 (13).

Sarcoplasmic reticulum is found in the young hearts, and it increases with development, forming a network of sarcoplasmic reticulum around each myofibril. Subsarcolemmal cisterns ("diads," "junctional couplings," or "junctional complexes"), regions in which elements of the sarcoplasmic reticulum come in close apposition to the surface sarcolemma, are observed in young hearts; the junctional cistern becomes flatter and denser with age (54). The junctional sarcoplasmic reticulum is continuous with the network sarcoplasmic reticulum. A transverse (T) tubular system is not found in chicken hearts.

The young hearts have large pools of glycogen, and their metabolism appears to be mainly anaerobic glycolysis. The circulation to the chorioallantoic membrane under the eggshell for gas exchange is not established until about day 5. Hearts of young chick embryos utilize the phosphogluconate pathway to a greater extent, relative to the tricarboxylic acid cycle pathway, than do hearts of older embryos or adults (3), and proliferating cells in general are characterized by high activity of pentose cycle enzymes (35). Enzymes of the pentose shunt pathway, such as glucose-6-P dehydrogenase and 6-P-gluconic dehydrogenase, decrease from day 4 to day 20, whereas enzymes of the Krebs cycle, such as isocitric dehydrogenase and α-ketoglutaric dehydrogenase, increase during development (45). In addition, hexokinase activity increases several fold (45). In general, the capacity to metabolize long-chain fatty acids appears late in development (after day 12) (64).

---

19-day ventricular cells with abundant and aligned myofibrils. A convoluted intercalated disc appears between contiguous cells. (Taken from Sperelakis and McLean, 1976.)

## ORGAN-CULTURED EMBRYONIC HEARTS

Cultivation of embryonic hearts *in vitro* provides a powerful means of analyzing the changes which occur during normal development *in situ*. When young (3-day-old) embryonic chick hearts, which have not yet become innervated by either cholinergic or adrenergic fibers, are placed into organ culture, they fail to gain tetrodotoxin-sensitive fast $Na^+$ channels (55,60). Instead, the action potentials continue to be generated by tetrodotoxin-insensitive slow $Na^+$ channels, the rates of rise remain slow, and pacemaker potentials precede the action potentials (Fig. 13). These action potentials resemble those recorded from 3-day-old hearts *in situ*. Similar findings were obtained when the young hearts were grafted on to the chorioallantoic membrane of host chicks for blood perfusion (38). Thus, organ-cultured young hearts do not differentiate further *in vitro*, but appear to be arrested in the young embryonic state.

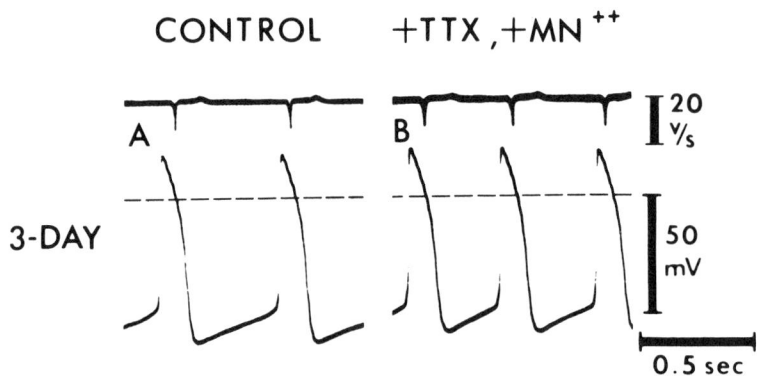

Fig. 13. Failure of appearance of fast $Na^+$ channels in a 3-day heart, organ-cultured for 6 days. The action potentials recorded from one cell prior to (A) and another cell after (B) the addition of tetrodotoxin are almost identical; dV/dt values are similar for the two cells. Subsequent addition of $Mn^{++}$ (1mM; panel B) in the presence of tetrodotoxin had no further effect on the maximal rate of the rise of the action potential. The upper traces give dV/dt. (Modified from Sperelakis and Shigenobu, 1972.)

Results differing from these have been reported by DeHaan and co-workers (5).

Recordings of action potentials from organ-cultured intermediate (5- to 7-day-old) hearts reveal the presence of both fast and slow $Na^+$ channels, just as *in situ*. The action potentials have moderate rates of rise (40-70 V/sec), and tetrodotoxin addition reduces the maximal rate of rise of the action potential to about 10 V/sec. Cultured 17-day hearts tend to retain their fast-rising action potentials and complete sensitivity to tetrodotoxin, but survival is limited to a few days. Thus, in organ-culture, the embryonic hearts tend to retain the state of differentiation achieved at the time of placement into culture.

When young hearts which have been arrested in the early developmental state are treated with messenger RNA-enriched fractions obtained from adult chicken hearts, they gain fast $Na^+$ channels and become completely sensitive to tetrodotoxin (Fig. 14E and H) (26). That is, young hearts which have been arrested in development by placement *in vitro* can be induced to undergo further membrane differentiation. Cultured young hearts may thus provide a useful model for studying the genetic regulation of membrane differentiation during embryonic life.

## CULTURED HEART CELLS

Various stages of cardiac electrical differentiation can be simulated *in vitro* using cell culture techniques. The electrophysiological properties observed depend in large part on the age of the hearts from which the cells are isolated, and on the method of cell culture.

When cells are dispersed from *old* embryonic hearts using trypsin and standard monolayer cultures are prepared, the cells are found to possess slowly-rising, tetrodotoxin-insensitive action potentials with pacemaker potentials (Fig. 15C and D), similar to those recorded from young (2- to 3-day-old) hearts, rather than the old hearts from which the cells were taken (Fig. 15A and B). It thus appears that cell separation results in a rapid reversion toward the young embryonic state (24, 37, 52, 57). This reversion can be partially prevented (or reversed) by separating the cells in trypsinizing solutions containing elevated $K^+$ concentrations (12-60 mM; achieved by isosmolar substitution

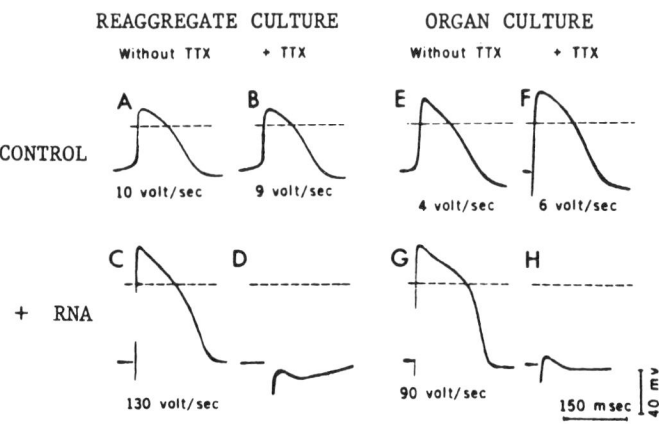

Fig. 14. Arrest of development of young (3 d.) embryonic myocardial cells in cultured reaggregates (A-B) and in organ-cultured intact hearts (E-F), and induction of further differentiation by culturing with RNA-enriched extracts from adult hearts (C-D, G-H). A-B: Records from one cell in a spherical reaggregate before (A) and after (B) tetrodotoxin (TTX) ( 1 µg/ml), illustrating the failure of membrane differentiation, including lack of fast $Na^+$ channels. C-D: Records from one cell in a spherical reaggregate culture treated with RNA showing a fast rate of rise (130 V/sec) and a high resting potential (-75 mV) (C); tetrodotoxin (0.1 µg/ml) rapidly abolished the action potential (D). E-F: Records from one cell in a 3-day-old intact heart organ-cultured for 10 days before (E) and after (F) the addition of tetrodotoxin (1 µg/ml), and showing the arrest of development. G-H: Records from one cell in an organ-cultured 3-day-old heart treated with RNA showing a fast rate of rise (90 V/sec) (G) and complete blockade by tetrodotoxin (H). Electrical field stimulation applied in C, D, F, G, and H. (Taken from Sperelakis and McLean, 1976.)

of $K^+$ for $Na^+$) (24). Action potentials recorded from cells prepared in this manner fire from moderately high stable resting potentials of about -60 mV, and they are completely sensitive to tetrodotoxin; however, the maximal rate of rise of the action potential is still rather slow, averaging about 30 V/sec (Fig. 15E and F). The mechanism whereby elevated $K^+$ helps to prevent reversion of the cultured cells is not known, although $[K^+]_i$ is reduced in the reverted cells to about 90-100 mM and this does not occur in highly differentiated cultured cells. Cells with highly differentiated electrophysiological properties can be obtained following trypsinization in solutions containing elevated $K^+$ and ATP (5 mM) and subsequent reaggregation to form small spheres (0.1 to 0.5 mm in diameter) *in vitro* (25). Action potentials recorded from cells in such reaggregates possess rapid rates of rise (up to 200 V/sec) and fire from high stable resting potentials (about -80 mV), and tetrodotoxin completely abolishes all excitability (Fig. 15G and H). The intracellular records are indistinguishable from those made from the intact 16-day ventricle (Fig. 15A and B). As in the case of the intact old embryonic hearts, positive inotropes induce slowly-rising electrical responses in highly differentiated cultured cells following blockade of the fast $Na^+$ channels with tetrodotoxin (8, 16, 25, 63). As expected, acetylcholine was without effect on cultured ventricular cells, both reverted (21) and highly differentiated. The production of cultured cells resembling various stages of differentiation may facilitate elucidation of the mechanisms operating during normal cardiogenesis.

When *young* (2- to 3-day-old) embryonic hearts are trypsin-dispersed and the cells are allowed to reaggregate *in vitro*, the cells retain electrical activity characteristic of the intact young heart, i.e., the cells have low resting potentials, and they fire spontaneous slowly-rising action potentials which are unaffected by addition of tetrodotoxin. However, as observed in the case of the young organ-cultured intact hearts, the addition of RNA-enriched extracts from adult chicken hearts to the reaggregated young myocardial cells induces the appearance of rapidly-rising (100 V/sec) action potentials which fire from high stable resting potentials (-70 to -80 mV) and are completely sensitive to tetrodotoxin (Fig. 14A and D) (26). This further differentiation results in the achievement of adult-like electrophysiological properties, as occurs *in situ*, and thus appears to faithfully simulate the course of events

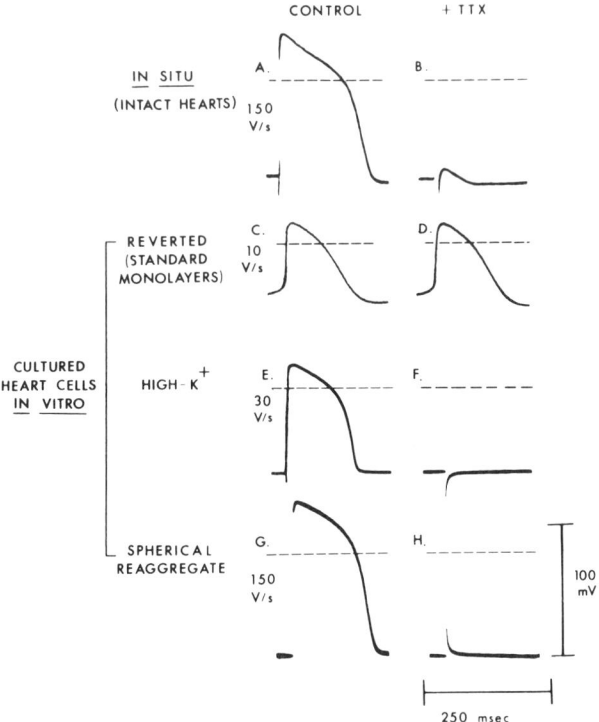

Fig. 15. Comparison of electrophysiological properties of the intact old (16-day) embryonic chick heart *in situ* (A-B) with those of trypsin-dispersed old ventricular myocardial cells in cultures prepared by three different methods (C-H). A-B: Intact heart; control action potential (A) was rapidly-rising (150 V/sec), had a high stable resting potential (about -80 mV), and was completely abolished by tetrodotoxin (TTX; 0.1 µg/ml) (B). C-D: Standard reverted monolayers; control action potential was slowly-rising (10 V/sec), was preceded by a pacemaker potential, the resting potential was low (about -50 mV) (C), and tetrodotoxin did not alter the action potential (D). E-F: Partially reverted cells cultured as monolayers in media containing elevated $K^+$ concentration (25 mM); control action potential had a rate of rise of 30 V/sec (E), lacked a pacemaker potential, had a moderately high resting potential (-60 mV), and was completely abolished

occurring in normal development. The mechanism of action of the active principal in the RNA extract remains to be elucidated.

In addition to the factors discussed above, Jones and co-workers (15) have recently found that multilayer-cultured ventricular cells can be made to exhibit highly differentiated morphology (e.g., densely packed and aligned myofibrils with mitochondria between the fibrils), as well as advanced electrical properties, if the serum concentration is lowered from 10% to 0.1 % after the first few days, and the cells are subsequently allowed to age *in vitro* for several weeks.

## INDUCTION OF SLOW $Ca^{++}$ CHANNELS
## BY POSITIVE INOTROPIC AGENTS

Acetylcholine has a negative chronotropic action on the early tubular heart (2-3 day) due to direct action on the atrial cells (22, 32). Ventricular cells are not affected by acetylcholine at any time during development (29,59). Catecholamines have a positive chronotropic effect on the early heart; furthermore, catecholamines elevate cellular cyclic AMP content as early as day 4 (23, 49). However, the positive inotropic effect of catecholamines does not become apparent until day 6 or 7 (see Ref. 53; also, for arguments that the positive inotropic effect may occur earlier in development, see Ref. 46).

Many positive inotropic agents exert their effect by increasing $Ca^{++}$ influx during the action potential, probably by increasing the number of available slow $Ca^{++}$ or slow $Ca^{++}$-$Na^+$ channels. Inward $Ca^{++}$ current through slow channels has been shown to affect cardiac excitation-contraction coupling (28, 30). Following the blockade of the fast $Na^+$ channels by tetrodotoxin, or their inactivation in

---

by tetrodotoxin (*F*). *G-H*: Highly differentiated cells in spherical reaggregate culture; control action potentials were rapidly-rising (150 V/sec), the resting potentials were high (-80 mV) (*G*), and tetrodotoxin abolished the action potentials (*H*). (Taken from Sperelakis and McLean, 1976.)

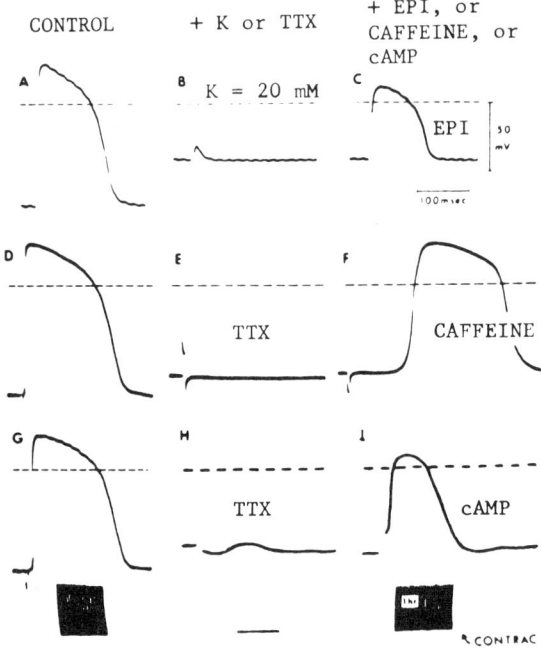

Fig. 16. Induction of slow responses by catecholamines and cAMP in chick embryonic hearts (15- to 17-days-old) whose fast $Na^+$ channels have been inactivated by elevated $[K^+]_o$ (A-C) or tetrodotoxin (TTX) (D-F, G-I). A-C: $K^+$ inactivation of the fast $Na^+$ channels. A: Control action potential. B: Abolition of excitability by elevation of $[K^+]_o$ to 20 mM, despite a 10-fold increase in stimulus intensity. C: Slowly-rising electrical response produced rapidly (1 min) by addition of epinephrine (Epi, $1 \times 10^{-5}$ M); contractions (not shown) accompanied the electrical responses. Addition of $Mn^{++}$ or $La^{3+}$ (1 mM) blocked the slow responses within 5 min (not shown). D-F: Tetrodotoxin inactivation of the fast $Na^+$ channels. D: Control action potential. E: Abolition of action potentials 2 min after exposure to tetrodotoxin (4 μg/ml). F: Restoration of electrical (and contractile) activity by addition of epinephrine. G-I: Induction of the slow channels by cAMP. G: Control action

elevated $K^+$ (raising $[K^+]_o$ to 27 mM lowers the resting potential to about -40 mV, well above the fast $Na^+$ channel inactivation potential), catecholamines produce a slowly-rising, overshooting, electrical response which resembles the plateau component of the action potential (43, 49). This slow response is blocked by $Mn^{++}$, $La^{+++}$, verapamil or its congener, D-600, and by zero $Ca^{++}$, indicating dependence on $[Ca^{++}]_o$. $Ca^{++}$ and $Na^+$ appear to compete for admittance to the same channel. The slow response is also dependent on cellular energy, as evidenced by abolition of the response when ATP content is reduced by anoxia, ischemia, or metabolic poisons (44). Many agents which elicit the slow response increase cyclic AMP concentration concomitantly. For example, catecholamines and histamine stimulate adenylate cyclase activity by binding to the beta-adrenergic receptor and histamine type-2 receptor, respectively (16, 49). Methylxanthines inhibit phosphodiesterase, preventing hydrolysis and leading to a buildup of cyclic AMP. Cyclic AMP and its dibutyryl derivative also evoke the slow response, although more slowly, presumably due to slow entry into the cells. Thus, it appears that cyclic AMP may somehow control the number of channels available for $Ca^{++}$ influx during the plateau. Schneider and Sperelakis (43, 58) hypothesized that the increase in cyclic AMP due to positive inotropes might activate a protein kinase which phosphorylates a protein constituent of the slow channels, thereby making the channels available for voltage activation.

Some positive inotropes, which also act by inducing slow channels, do not increase cyclic AMP. Angiotensin may act by directly stimulating the protein kinase (see 63), bypassing the need for a cyclic AMP increase and leading to enhanced slow channel phosphorylation. Fluoride ions also elicit the slow response without increasing cyclic AMP (63); fluoride ions may act to prevent channel de-

---

potential. *H*: Abolition of excitability in tetrodotoxin. *I*: Restoration of slowly-rising electrical responses 20-60 min after addition of cyclic AMP (3 mM). The insets in *G-I* give the contractions recorded on a penwriter at slow speed. Note that the contractile force in *I* is about equal to that in *G*. (Modified from Shigenobu and Sperelakis, 1972.)

phosphorylation by means of inhibition of phosphatases, thereby maintaining augmented numbers of phosphorylated channels in the voltage-activatable state.

## SUMMARY

Striking changes occur in the electrical properties of the myocardial cell membrane during embryonic development of chick hearts. The young tubular hearts (2-4 days old) have a low resting potential of about -40 mV, even though $[K^+]_i$ is nearly as high as the adult value. Analysis shows that the low resting potential is caused by a low potassium ion permeability. The low potassium permeability can account for the high degree of automaticity observed in the ventricular cells of the young embryonic heart. Potassium permeability increases rapidly during development, nearly attaining the final adult value by day 12; the resting potential increases to about -80 mV, and automaticity of the ventricular cells is suppressed. The young heart has a low $(Na^+,K^+)$-ATPase activity, and this enzyme activity increases during development along with potassium permeability.

The young hearts have slow-rising action potentials (10-20 V/sec) which are dependent mainly on $[Na^+]_o$ and which are not affected by tetrodotoxin; hyperpolarization does not greatly increase the action potential rate of rise. Thus, it appears that fast $Na^+$ channels are absent in young hearts, the inward current during the action potential being carried predominantly through tetrodotoxin-insensitive slow $Na^+$ channels. The slow $Na^+$ channels are blocked by verapamil, but not by 1 mM $Mn^{++}$. $Mn^{++}$ does block the contractions, presumably by blockade of the $Ca^{++}$ influx during the action potential. In many respects, the young tubular hearts resemble pulsating blood vessels, and their electrical properties are similar to vascular smooth muscle. On about day 5, which is about the time of arrival of the innervation at the heart, there is a large jump in the maximal rate of rise of the action potential $(+\dot{V}_{max})$ to between 50-80 V/sec, and tetrodotoxin now reduces the maximal rate of rise of the action potential to 10-20 V/sec, i.e., to the value observed in 2-4 day old hearts. Thus, tetrodotoxin-sensitive fast $Na^+$ channels make their first appearance on day 5, and they progressively increase in number until they attain the final adult level by day 18. Between day 5 and day 8, fast $Na^+$ channels and a high density of slow $Na^+$ channels coexist. After day 8, tetrodotoxin usually completely abolishes excitability, suggestive that the slow $Na^+$ channels have decreased in number sufficiently so as to not support regenerative excitation.

Young hearts (3 days old), removed prior to innervation and placed into organ culture for 2 weeks or grafted on the chorioallantoic membrane of a host chick for blood perfusion, do not gain fast $Na^+$ channels or otherwise differentiate. This suggests that something in the *in situ* condition, such as neurotrophic factors, is required for triggering further differentiation. If, however, these young organ-cultured hearts are exposed for several days to a messenger-RNA-enriched fraction obtained from adult chicken hearts, they do gain tetrodotoxin-sensitive fast $Na^+$ channels. This induction is blocked by cycloheximide. Thus, the synthesis of specific membrane proteins controls the appearance of fast $Na^+$ channels. Hence, cardiac myoblasts whose development has been arrested *in vitro* can be induced to differentiate electrically.

Trypsin-dispersed myocardial cells obtained from young embryonic chick hearts and placed in cell culture for several weeks also do not proceed with differentiation *in vitro*, unless exposed to the messenger RNA extract from adult heart. Cultured heart cells prepared from old embryonic hearts (ventricles) rapidly revert back to the young embryonic state. That is, they lose their fast $Na^+$ channels, gain slow $Na^+$ channels, and gain automaticity because of a low potassium permeability; they also lose many of their myofibrils. If, however, the cells are allowed to reaggregate into small spheres and to age for several weeks, they sometimes regain their highly differentiated properties, and in some conditions, regain a dense packing of parallel myofibrils. These cells also retain their membrane receptors for catecholamines, histamine, and angiotensin. Thus, highly differentiated myocardial cells can be maintained in cell culture under appropriate conditions.

Positive inotropic agents, such as norepinephrine and histamine, rapidly induce slow $Ca^{++}$-$Na^+$ channels in old embryonic myocardial cells. Following blockade of the fast $Na^+$ channels with tetrodotoxin, these agents rapidly allow the production of slowly rising electrical responses by increasing the number of slow channels available for voltage activation. Cyclic AMP produces the same effect, but much more slowly. Since norepinephrine and histamine rapidly elevate intracellular cyclic AMP levels, these results suggest that cyclic AMP is somehow related to the number of operational slow channels. We postulate that phosphorylation of a membrane protein constituent of the slow channel makes it available for voltage activation. Thus, $Ca^{++}$ influx into the myocardial cells is controlled by both intrinsic and extrinsic factors.

## ACKNOWLEDGMENTS

The work of the authors summarized and reviewed in this chapter was supported by grants from the U.S. Public Health Service (HL-18711 and HL-11155). M. J. M. was a Predoctoral Trainee supported by an NIH Training Grant (HL-05815). Many of the studies on intact hearts were performed in collaboration with my former research associate, Dr. K. Shigenobu.

## REFERENCES

1. Bernard, C. 1976. Establishment of ionic permeabilities of the myocardial membrane during embryonic development of the rat. In *Developmental and physiological correlates of cardiac muscle*, eds. M. Lieberman and T. Sano. pp. 169-184, New York: Raven Press.

2. Carmeliet, E. E.; Horres, C. R.; Lieberman, M.; and Vereecke, J. S. 1976. Developmental aspects of potassium flux and permeability of the embryonic chick heart. *J. Physiol.* 254:673-692. London.

3. Coffey, R.; Chendelin, V.; and Newburgh, R. 1964. Glucose utilization by chick embryo heart homogenates. *J. Gen. Physiol.* 48:105-112.

4. Couch, J. R.; West, T. C.; and Hoff, H. E. 1969. Development of the action potential of the prenatal rat heart. *Circ. Res.* 24:19-31.

5. DeHaan, R. L.; McDonald, T. F.; and Sachs, H. G. 1976. Development of tetrodotoxin sensitivity of embryonic chick hearts *in vitro*. In *Developmental and physiological correlates of cardiac muscle*, ed., M. Lieberman and T. Sano. pp. 155-168, New York: Raven Press.

6. Elsas, L. J.; Wheeler, F. B.; Danner, D. J.; and DeHaan, R. L. 1975. Amino acid transport by aggregates of cultured chicken heart cells. Effect of insulin. *J. Biol. Chem.* 250:9381-9390.

7. Herman, B. A. and Fernandez, B. S. 1976. Developmental changes in membrane fluidity of cultured myogenic cells. *Physiologist* 19:223 (abs.)

8. Freer, R.; Pappano, A. J.; Peach, M. J.; Bing, K.; McLean, M. J.; Vogel, S. M.; and Sperelakis, N. 1976. Mechanism of the positive inotropic effect of angiotensin II on isolated cardiac muscle. *Circ. Res.* 39:172-183.

9. Girard, H. 1974. Arterial pressure in the chick embryo. *Am. J. Physiol.* 224:454-460.

10. Guidotti, G.; Kanemeishi, D.; and Foa, P. P. 1961. Chick embryo heart as a tool for studying cell permeability and insulin action. *Am. J. Physiol.* 201: 863-868.

11. Harsch, M. and Green, J. W. 1963. Electrolyte analyses of chick embryonic fluids and heart tissue. *J. Cell. Comp. Physiol.* 62:319-326.

12. Hibbs, R. G. 1956. Electron microscopy of developing cardiac muscle in chick embryos. *Am. J. Anat.* 99: 17-52.

13. Hitchcock, S. E. 1970. The appearance of a functional contractile apparatus in developing muscle. *Dev. Biol.* 23:399-423.

14. Irisawa, H. and Noma, A., personal communication.

15. Jones, J. K.; Paull, K.; Proskauer, C. C.; Jones, R.; Lepeschkin, E.; and Rush, S. 1975. Ultrastructural changes produced in cultured myocardial cells by electric shock. *Fed. Proc.* 34:972 (abs.), and personal communication.

16. Josephson, I.; Renaud, J.-F.; Vogel, S.; McLean, M.; and Sperelakis, N. 1976. Mechanism of the histamine-induced positive inotropic action in cardiac muscle. *Europ. J. Pharmacol.* 35:393-398.

17. Klein, R. L. 1960. Ontogenesis of K and Na fluxes in embryonic chick heart. *Am. J. Physiol.* 199:613-618.

18. Klein, R. L.; Horton, C. R.; and Thureson-Klein, A. 1970. Studies on nuclear amino acid transport and cation content in embryonic myocardium of the chick. *Am. J. Cardiol.* 25:300-310.

19. Kutchai, H.; King, S. L.; Martin, M.; and Daves, E. D. 1977. Glucose uptake by chicken embryo hearts at various stages of development. *Devel. Biol.* 55:92-102.

20. Le Douarin, G.; Obrecht, G.; and Coraboeuf, E. 1966. Déterminations régionales dans l'air cardiaque presomptive mises en évidence chez l'embryon de poulet par la méthode microelectrophysiologique. *J. Embryol. Exp. Morph.* 15:153-167.

21. Lehmkuhl, D., and Sperelakis, N. 1963. Effect of current on transmembrane potentials in cultured chick heart cells. *J. Gen. Physiol.* 47:895-927.

22. Löffelholz, K. and Pappano, A. J. 1974. Ontogenetic changes in the pacemaker activity in chick heart. *Life Sciences* 14:1755-1763.

23. McLean, M. J.; Lapsley, R. A.; Shigenobu, K.; Murad, F.; and Sperelakis, N. 1975. High cyclic AMP levels in young chick embryonic hearts. *Develop. Biol.* 42:196-201.

24. McLean, M. J. and Sperelakis, N. 1974. Rapid loss of sensitivity to tetrodotoxin by chick ventricular myocardial cells after separation from the heart. *Exp. Cell Res.* 86:351-364.

25. McLean, M. J. and Sperelakis, N. 1976. Retention of fully differentiated electrophysiological properties of chick embryonic heart cells *in vitro*. *Develop. Biol.* 50:134-142.

26. McLean, M. J.; Renaud, J.-F.; Sperelakis, N.; and Niu, M. C. 1976. mRNA induction of fast $Na^+$ channels in cultured cardiac myoblasts. *Science* 191:297-299.

27. McLean, M. J.; Renaud, J.-F.; and Sperelakis, N. 1978. Cardiac-like action potentials in spontaneously-contracting vesicles developed from post-nodal pieces of chick blastoderm exposed to an RNA-enriched fraction from adult heart. Differentiation (in press).

28. Morad, M. and Goldman, Y. 1973. Excitation-contraction coupling in heart muscle: membrane control of development of tension. *Prog. Biophys. Molec. Biol.* 27:259-313.

29. Nakanishi, H. and Takeda, H. 1969. Effect of acetylcholine on the electrical activity of cultured chick embryonic heart. *Jap. J. Pharmacol.* 19:543-550.

30. New, W. and Trautwein, W. 1972. The ionic nature of slow inward current and its relation to contraction. *Pflüg. Archiv.* 334:24-38.

31. Niu, M. C. and Deshpande, A. K. 1973. The development of tubular heart in RNA-treated post-nodal pieces of chick blastoderm. *J. Embryol. Exp. Morph.* 29:485-501.

32. Pappano, A. J. 1972. Sodium-dependent depolarization of non-innervated embryonic chick heart by acetylcholine. *J. Pharmacol. Exp. Therap.* 180:340-350.

33. ----. 1976. Action potentials in chick atria: ontogenic changes in the dependence of tetrodotoxin-resistant action potentials on calcium, strontium, and barium. *Circ. Res.* 39:99-105.

34. Patten, B. M. 1956. The development of the sinoventricular conduction system. *Univ. Mich. Med. Bull.* 22:1-21.

35. Paul, J. 1965. In *Cells and tissues in culture*, vol. 1, ed., E. N. Wilmer, New York: Academic Press. p. 239.

36. Renaud, D. 1973. Etude electrophysiologique de la différentiation cardiaque chez l'embryon de poulet. Thesis. Univ. of Nantes.

37. Renaud, J.-F. 1973. Etude de l'évolution en culture primaire des cardiomyoblastes embryonnaires de poulet. Thesis. Univ. of Nantes.

38. Renaud, J.-F., and Sperelakis, N. 1976. Electrophysiological properties of chick embryonic hearts grafted and organ cultured *in vitro. J. Molec. Cell Cardiol.* 8:889-900.

39. Renaud, J.-F.; Sperelakis, N.; and LeDouarin, G. 1978. Increase of cyclic AMP levels induced by isoproterenol in cultured and non-cultured chick embryonic hearts. *J. Molec. Cell. Cardiol.* 10:281-286.

40. Reporter, M. 1972. An ATP pool with rapid turnover within the cell membrane. *Biochem. Biophys. Res. Comm.* 48:598-604.

41. Romanoff, A. 1960. *The Avian embryo, structure and functional development.* New York: Macmillan.

42. Rosenquist, G., and DeHaan, R. L. 1966. Migration of precardiac cells in the chick embryo. A radio-autographic study. *Contrib. Embryol. Carnegie Inst. Wash.* 263:113-121.

43. Schneider, J. A., and Sperelakis, N. 1975. Slow $Ca^{++}$ and $Na^+$ current channels induced by isoproterenol and methylxanthines on isolated perfused guinea pig hearts whose fast $Na^+$ channels are inactivated in elevated $K^+$. *J. Molec. Cell. Cardiol.* 7:249-273.

44. ----. 1974. The demonstration of energy dependence of the isoproterenol-induced transcellular $Ca^{++}$ current in isolated perfused guinea pig hearts--an explanation for mechanical failure of ischemic myocardium. *J. Surg. Res.* 16:389-403.

45. Seltzer, J. L. and McDougal, D. B. 1975. Enzyme levels in chick embryo heart and brain from 1-21 days of development. *Develop. Biol.* 42:95-105.

46. Shideman, F. E. 1974. Responsiveness of the 4-day-old embryonic chick heart to catecholamines. *Circ. Res.* 34:268-269.

47. Shigenobu, K.; Schneider, J. A.; and Sperelakis, N. 1974. Blockade of slow $Na^+$ and $Ca^{++}$ current in myocardial cells by verapamil. *J. Pharmacol. Exp. Therap.* 190:280-288.

48. Shigenobu, K., and Sperelakis, N. 1971. Development of sensitivity to tetrodotoxin of chick embryonic hearts with age. *J. Molec. Cell. Cardiol.* 3:271-286.

49. ----. 1972. $Ca^{++}$ current channels induced by catecholamines in chick embryonic hearts whose fast $Na^+$ channels are blocked by tetrodotoxin. *Circ. Res.* 31:932-952.

50. Shimizu, Y., and Tasaki, K. 1965. Electrical excitability of developing cardiac muscle in chick embryos. *Tohoku J. Exp. Med.* 88:49-56.

51. Sperelakis, N. 1972. ($Na^+$, $K^+$)-ATPase activity of embryonic chick heart and skeletal muscles as a function of age. *Biochim. Biophys. Acta* 266:230-237.

52. ----. 1972. Electrical properties of embryonic heart cells. In *Electrical phenomena in the heart*, pp. 1-61, ed., W. C. De Mello. New York: Academic Press.

53. ----. 1974. Do young embryonic chick hearts exhibit a positive inotropic response to catecholamines? *Circ. Res.* 34:269-271.

54. Sperelakis, N.; Forbes, M.; and Rubio, R. 1974. The tubular systems of myocardial cells: ultrastructure and possible function. In series, *Recent advances in studies on cardiac structure and metabolism, Myocardial Biol.*, pp. 163-194, vol. 4, eds., N. S. Dhalla, and G. Rona. Baltimore: Univ. Park Press.

55. Sperelakis, N.; Forbes, M. S.; Shigenobu, K.; and Coburn, S. 1974. Organ-cultured chick embryonic heart cells of various ages. Part II. Ultrastructure. *J. Molec. Cell. Cardiol.* 6:473-483.

56. Sperelakis, N., and Lee, E. C. 1971. Characterization of ($Na^+$,$K^+$)-ATPase isolated from embryonic chick hearts and cultured chick heart cells. *Biochim. Biophys. Acta* 233:562-579.

57. Sperelakis, N. and Lehmkuhl, D. 1965. Insensitivity of cultured chick heart cells to autonomic agents and tetrodotoxin. *Am. J. Physiol.* 209:693-698.

58. Sperelakis, N. and Schneider, J. A. 1976. A metabolic control mechanism for calcium ion influx that may protect the ventricular myocardial cell. *Am. J. Cardiol.* 37:1079-1085.

59. Sperelakis, N. and Shigenobu, K. 1972. Changes in membrane properties of chick embryonic hearts during development. *J. Gen. Physiol.* 60:430-453.

60. ----. 1974. Organ-cultured chick embryonic hearts of various ages. Part I. Electrophysiology. *J. Molec. Cell. Cardiol.* 6:449-471.

61. Sperelakis, N.; Shigenobu, K.; and McLean, M. J. 1976. Membrane cation channels--changes in developing hearts, in cell culture, and in organ culture. In *Developmental and physiological correlates of cardiac muscle*, pp. 209-234, eds., M. Lieberman and T. Sano. New York: Raven Press.

62. Thureson-Klein, A., and Klein, R. L. 1971. Cation distribution and cardiac jelly in early embryonic heart: a histochemical and electron microscopic study. *J. Molec. Cell. Cardiol.* 2:31-40.

63. Vogel, S.; Sperelakis, N.; Josephson, I.; and Brooker, G. 1977. Fluoride stimulation of slow $Ca^{++}$ current in cardiac muscle. *J. Molec. Cell. Cardiol.* 9:461-475.

64. Warshaw, J. B. 1972. Cellular energy metabolism during fetal development. IV. Fatty acid activation, acyl transfer and fatty acid oxidation during development of the chick and rat. *Develop. Biol.* 28:537-544.

65. Yeh, B. K. and Hoffman, B. F. 1967. The ionic basis of electrical activity in embryonic cardiac muscle. *J. Gen. Physiol.* 52:666-681.

# 8

# Development of Electrical Activity in Embryonic Myocardial Cells

Melvyn Lieberman, C. Russell Horres, Joyce E. Purdy, Linda R. Halperin

Department of Physiology
Duke University Medical Center
Durham, North Carolina

The evolution of hypotheses associated with experimental and theoretical attempts to describe the generation of the cardiac action potential (for reviews, see Refs. 1-3) has stimulated many cardiac electrophysiologists to search for and interrogate simple preparations of cardiac muscle. Considerable attention has been focused on the use of embryonic heart preparations, both isolated and in tissue culture, to sort out the seemingly complex electrophysiology of adult hearts as well as the sequence of changes of intracellular ion concentrations and membrane permeability which occur during cardiogenesis (e.g., 4-6).

The electrical potential difference developed across the cell membrane of embryonic cardiac muscle, as in other excitable tissues, is determined primarily by the distribution of the inorganic ions, potassium, sodium, and chloride, between the cytoplasmic and interstitial fluid compartments. In evaluating the magnitude of the resting membrane potential, one must consider both the electrochemical potential difference of these ions across the membrane and transport mechanisms, which not only actively maintain the ionic gradients but may generate ionic currents that could contribute to the membrane potential. Indeed, experiments designed to obtain such data are beginning to provide information which relate the changes in membrane properties to their possible role in the regulation of cardiac development and differentiation. Of equal significance, the insights developing as a result of these studies are also applicable to the mechanisms which govern electrical activity of the adult heart. However, even in studies with embryonic heart muscle, the morphology and cellular heterogeneity of the preparations continually challenge the application of methods, such as radioisotopic flux and intracellular recordings, to separate the true properties of the biological mechanisms from the experimentally induced artifacts. Some major problems which confront the cardiac physiologist have been selected to illustrate the difficulties involved in describing the development of electrical activity in embryonic myocardial cells.

## TRANSMEMBRANE POTASSIUM FLUX

Although potassium (K) ions are generally considered to be the major charge carriers in the resting membrane of cardiac muscle and other excitable cells, a close correlation between electrical- and flux-measured conductances has not generally been reported (7). The apparent morphologic simplicity of embryonic heart muscle (8, 9) would seem to offer a significant advantage for radioisotopic flux measurements. However, as shown in Figure 1, the efflux of $^{42}K$ from a preparation of 19-day embryonic chick heart (ventricular strips, 5 mm x 0.5 mm), when plotted on a logarithmic scale, cannot be described by a single rate constant. Our initial concept of similar data (see Ref. 10) described the slow component (dashed lines) as radiotracer exchange from the functional myocardial cells and the rapid component as the exchange from extracellular space. In a subsequent study (11), we became concerned with the possibility that diffusional limitations might be responsible for the deviation from single compartment kinetics shown in Figure 1. Consequently, $^{42}K$ efflux data from the intact embryonic heart in various external potassium concentrations was compared with values obtained by applying the radial diffusion equation of MacDonald et al.

(12), in which case, the solutions were based on the flux values predicted from the constant field equation for the respective potassium concentrations. By trial and error, a radial diffusion time of 75 seconds could closely approximate the experimental data (Fig. 2). Such a diffusional delay is not only reasonable, considering the thickness of the preparation (0.5 mm diameter), but also offers an explanation for the decreased $^{42}K$ efflux rate constant in low external potassium reported for sheep Purkinje fiber (13) and embryonic chick heart (10).

Experimental evidence to support the effects of radial diffusion on $^{42}K$ exchange was obtained from preparations of embryonic chick heart cells that were growth

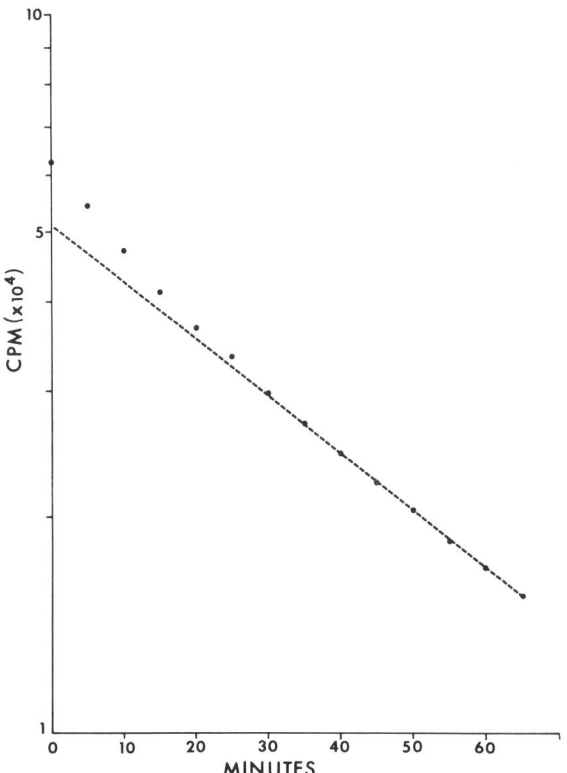

Fig. 1. Potassium-42 efflux kinetics of a 19-day embryonic chick heart (ventricular preparation) in 5.0 mM external potassium at 37°C.

oriented in tissue culture about a continuous winding of
nylon monofilament (for details, see Ref. 14). The preparations, consisting of 70 segments, 5 mm in length, provided a sufficient mass of tissue for isotopic flux studies
while maintaining minimal diffusion distances on the order
of 20 to 100 μm. Since the diameter of the individual segments of the growth-oriented "polystrand" preparation could
be increased by seeding a greater number of cells from a
given suspension in the growth chambers, it was possible to
compare the loss of $^{42}K$ content between preparations of
different weights. In addition, by altering the proportion
of cell types in the seeding population the influence of

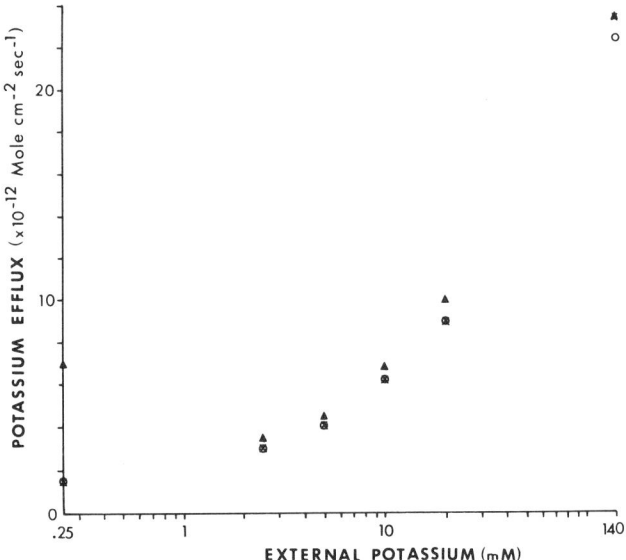

Fig. 2. Potassium efflux as a function of
external potassium in 6-8-day embryonic chick
ventricle. (o) Efflux data obtained from Carmeliet, et al., (10). (▲) Constant field predicted flux from membrane potentials, assuming no effective diffusional limitations at a
potassium concentration of 140 mM. (x) Solution of radial diffusion equation using constant
field predicted flux values. (For details, see
Ref. 11.)

cellular heterogeneity on $^{42}$K efflux kinetics could be tested. This procedure promoted the development of preparations which were either spontaneously active or quiescent, the former containing an inner core of muscle cells and an outer fibroblast-like sheath and the latter being composed solely of fibroblast-like cells (14).

When preparations that have been equilibrated with $^{42}$K are perfused with medium free of potassium, they provide the basis for a valid test to determine the presence of diffusional limitations (i.e., tracer reflux). In such cases, the extracellular space of the preparation should

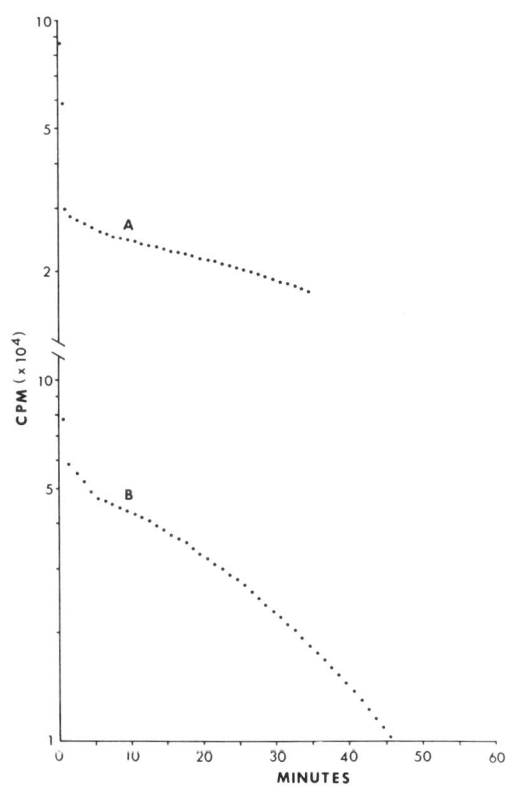

Fig. 3. Potassium-42 efflux kinetics of growth-oriented fibroblast-like cells (derived from primary heart cell cultures) in potassium-free medium. A: large preparation (dry weight = 0.416 mg); B: small preparation (dry weight = 0.054 mg). (For details, see Ref. 11.)

undergo a rapid decline in specific activity as $^{42}K$ is removed from the preparation or, as a consequence of diffusional limitations, recycles into the cell from the extracellular space. Tracer kinetics obtained from quiescent preparations in K-free medium are shown in Figure 3. Following the initial decline in activity (due to the washout both of the carryover isotopic loading solution and extracellular space), the initial low rate of $^{42}K$ loss from the large preparation (A) was followed by a slight increase occurring at 25 minutes in contrast to a rapid increase in tracer efflux from the small preparation (B). Although such data support the hypothesis that extracellular diffusion, in certain experimental conditions, might significantly limit potassium efflux, the diffusional limitations are unable to completely account for the results shown in Figure 1.

To test whether two-compartment efflux kinetics are related to cellular heterogeneity, the efflux kinetics of quiescent preparations of fibroblast-like cells were compared with that obtained from contractile preparations of muscle and fibroblast-like cells. The results shown in Figure 4 clearly demonstrate that single compartment kinetics are obtained only when preparations consist of a single cell type, i.e., fibroblasts. In such cases, the rate constant (0.015 $min^{-1}$) is comparable to that of the slow component (0.021 $min^{-1}$) obtained from intact strips of embryonic heart muscle (e.g., Fig. 1 and Ref. 10). The recent application of compartmental analysis to the $^{42}K$ efflux kinetics of cardiac muscle cells has resulted in a rate constant of 0.067 $min^{-1}$, a value approximately fourfold faster than fibroblast-like cells (15). Consequently, interpretation of the developmental aspects of cardiac membrane permeability previously obtained from measurements of the rate coefficients for $^{42}K$ exchange in the embryonic chick heart (10) should be reevaluated.

## ACTIVE TRANSPORT

The resting membrane potential of embryonic cardiac muscle, either in its intact form or reaggregated in tissue culture, is less negative than the calculated potassium equilibrium potential. To maintain homeostatic conditions, the resultant net passive efflux of potassium must be balanced by the active transport of potassium into the cell. Since the results of an earlier report indicated a relatively low specific activity of membrane Na-K ATPase from embryonic heart muscle in tissue culture (16), the polystrand preparation of growth oriented heart cells was used to measure the percentage of potassium influx linked to active transport. In this study, ouabain ($10^{-4}$ M) produced an almost immediate cessation of contractile activity and reduced the potassium influx by an average of 72% (Table 1).

Subsequent studies with the polystrand preparation determined the extent of coupling between the active transport of potassium influx and sodium efflux. In brief, 10 ml of an appropriate solution was used to bathe preparations for periods up to 60 min after which time they were rinsed for 15 sec in cold isotonic choline chloride (see Fig. 5) and dried overnight at 110°C before weighing. The potassium and sodium content of each preparation was measured by atomic absorption flame spectrophotometry. Table 2 is a summary of the results obtained from preparations

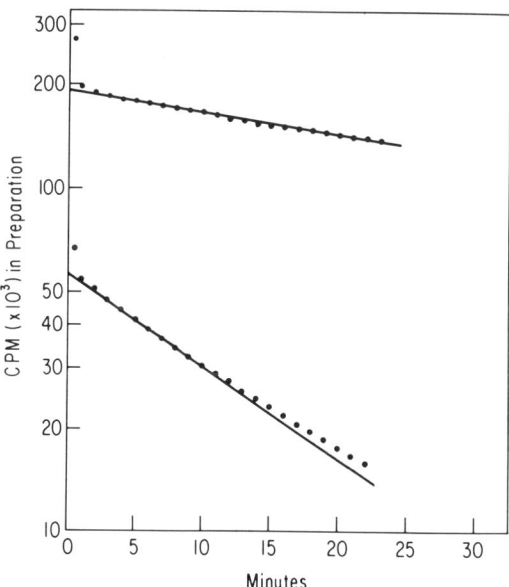

Fig. 4. Kinetics of potassium-42 efflux from polystrand preparations of fibroblast-like cells (*upper curve*) and of a mixed population of cardiac muscle and fibroblast-like cells (*lower curve*).

Table 1

Effects of Ouabain on Potassium-42 Uptake

| Experiment[a] | Normal Uptake[b] (µM/min) | Inhibited Uptake[c] (µM/min) | Decrease (%) |
|---|---|---|---|
| 156[d] | 5.14 | 1.41 | 72.6 |
| 157[d] | 4.29 | 1.34 | 68.8 |
| 158[e] | 6.04 | 1.56 | 74.2 |
| Mean | | | 71.9 |

[a] Five-minute measurements between 90-minute incubation periods in culture medium--preparations not stimulated.

[b] Uptake in 5.4 mM/l external potassium.

[c] Uptake in a salt solution containing 5.4 mM/l external potassium and $1 \times 10^{-4}$ M/l ouabain.

[d] Normal uptake measured first.

[e] Inhibited uptake measured first.

exposed to low-temperature, ouabain- and a potassium-free medium for varying periods of time. The kinetics of net ionic exchange in these studies reflects the combined contribution of muscle and fibroblast-like cells in the preparations. In each case, the loss of cell potassium is reasonably well accounted for by a reciprocal exchange with sodium.

In other experiments, when preparations consisting predominantly of cardiac muscle cells were equilibrated with $^{42}$K and then perfused with a K-free medium at a rate of 3.5 ml per minute, the intracellular potassium content decreased by 55% of the control value within 10 min (11).

Comparable electrophysiological experiments (17) resulted in a significant depolarization of the membrane potential from -66 mV in the presence of 5.4 mM K to a value of -43 mV in the absence of potassium, a finding in all probability due to the diminished intracellular potassium concentration rather than to a decrease in potassium permeability (10, 13). The reintroduction of potassium rapidly (ca. 2 sec) hyperpolarized the cells to -88 mV, a value clearly in excess of the newly established equilibrium potential for potassium ($E_K$ = -69 mV) and thereby provided evidence for an electrogenic transport mechanism. In the future, studies with embryonic heart muscle should therefore take into consideration not only the Na-K coupled active transport system which functions to maintain steady-state conditions, but also the extent to which an electrogenic ion transport mechanism may contribute to the resting potential and membrane current/voltage relationship of developing cardiac muscle.

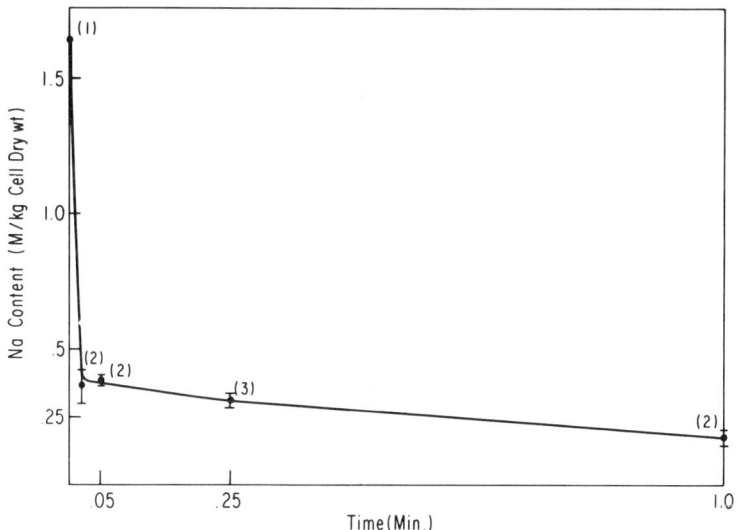

Fig. 5. Sodium content of secondary heart cell cultures grown as polystrand preparations and rinsed in cold isotonic choline chloride. Bar indicates the range of measurements.

Table 2

Intracellular Na and K Content of
Growth-Oriented Cultured Heart Cells

| Time (min) | $n^a$ | K (moles/kg dry wt) | Na (moles/kg dry wt) | K:Na |
|---|---|---|---|---|
| CONTROL (37°C) | | | | |
| 0-120 | 8 | 0.59 ± 0.03 | 0.24 ± 0.02 | 2.5 ± 0.2 |
| LOW TEMPERATURE (1°C) | | | | |
| 0 | 2 | 0.58 | 0.21 | 2.7 |
| 30 | 2 | 0.52 | 0.29 | 1.8 |
| 60 | 2 | 0.46 | 0.33 | 1.4 |
| OUABAIN ($10^{-4}$ M) | | | | |
| 0 | 2 | 0.57 | 0.22 | 2.6 |
| 30 | 2 | 0.47 | 0.32 | 1.5 |
| 60 | 2 | 0.37 | 0.32 | 1.1 |
| K-FREE MEDIA | | | | |
| 0 | 2 | 0.50 | 0.24 | 2.1 |
| 30 | 2 | 0.48 | 0.32 | 1.5 |
| 60 | 3 | 0.36 | 0.32 | 1.1 |

$a$ n represents number of preparations consisting of 70 strands of cultured heart cells.

## ELECTRICAL ACTIVITY

Transmembrane potentials recorded from embryonic chick and rat ventricular muscle between the ages of 2-19 days tend to increase in magnitude (18-20). During the first week of development of the embryonic chick heart, changes in the resting potential and peak action potential amplitude are relatively large and thereafter increase only gradually. Given the reported difficulties in maintaining stable intracellular impalements in the early embryonic heart (10, 18, 19), it is reasonable to consider the possibility that such measurements are unreliable and should not be quantitatively compared with data obtained from older preparations. The difficulty of recording from early embryonic heart muscle is not surprising if one compares the morphologic differences between young and old embryonic hearts (21). The early myocardium is a loosely organized network of immature myocytes, relatively devoid of intercellular junctional specializations (9, 21, 22), whereas in older embryonic hearts, myocardial cells are tightly packed and contain well-oriented myofibrils and prominent junctional specializations in the form of desmosomes, intercalated discs, and nexuses (9, 20). The difficulty of achieving a tight seal between the cell membrane and microelectrode in the early embryonic myocardium would be consistent with these morphological findings and could account for the low resting potentials that have been recorded from these cells. Support for this hypothesis is evident when comparing transmembrane potentials recorded from heart cells in dense and sparse areas of the same culture and noting that the magnitude of action potentials appear to coincide with the morphology of the preparation (Fig. 6).

Input resistance measurements from embryonic myocardial cells in young hearts (2 days) are reported to be considerably higher than the values obtained from hearts older than 10 days (18). Such findings have been interpreted solely in terms of an age dependent decrease in membrane resistance, even though junctional specializations, presumably of low resistance, become more numerous with embryonic age (9, 22). Since input resistance is inextricably associated with the effectiveness of intercellular coupling, age dependent differences in the above described measurements of resting potential (18, 19) and input resistance (18) do not have a unique interpretation. Developmental changes in the ratio of the permeability of sodium

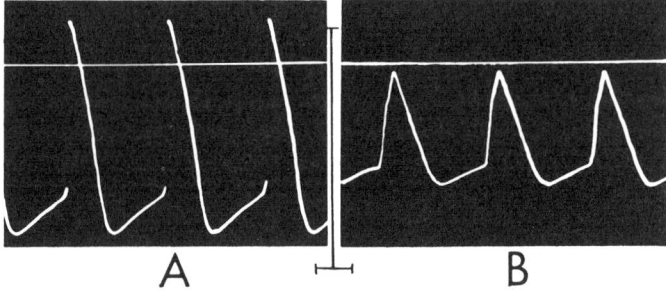

Fig. 6. Transmembrane potentials recorded from dense (A) and sparse (B) areas of a sheet-like culture of heart cells in medium containing 1.4 mM potassium. Vertical bar = 100 mV. Horizontal bar = 200 msec. Modified after (23).

to potassium ($P_{Na}/P_K$) based on these measurements should be clarified before hypotheses are promulgated into theories. Furthermore, the recent finding that potassium efflux is highest in embryonic chick hearts between 3-5 days of age (10) is inconsistent with the derived low membrane permeability reported above.

In recent years, membrane differentiation of embryonic heart cells has also been studied by comparing the effects of tetrodotoxin on membrane potentials at different stages of development (20, 24, 25). Tetrodotoxin, a cationic molecule, is known to specifically block sodium 'channels' in nerve membrane (26), and the presence or absence of an effect of this toxin on embryonic heart cells has been interpreted as reflecting differences in the state of membrane differentiation (20, 24, 25). However, several discussions have recently considered the possibility of a relationship between variations in tetrodotoxin sensitivity and its accessibility to membrane-specific sites of cultured heart cells (27). Since myocardial cells are embedded in a complex extracellular matrix containing fibroblasts and polyanionic macromolecules (8, 9, 21, 28, 29), it is important to consider whether this matrix, the composition of which changes throughout development, complicates the interpretation of physiological responses previously attributed to the state of membrane differentiation of the myocardial cells. The extracellular matrix can be clearly demonstrated in preparations of embryonic heart muscle (Fig. 7) and tissue cultured embryonic chick heart cells (30, 31). Figure 8 illustrates that ruthenium red, a cationic dye which binds to surface acidic glycoproteins (32), stains the extracellular matrix of cultured heart cells from 11-day chick embryos. An additional barrier to membrane-specific sites for cationic molecules might be the fibroblast-like cells which usually surround the muscle core within reaggregate cultures of embryonic heart cells (see Figure 8). Likewise, fibroblast-like cells can be seen overlying cardiac muscle cells in densely grown, multi-layered cultures (Fig. 9). In both instances, the presence of these "encapsulating" (and perhaps secretory) fibroblast-like cells can reduce the effectiveness of a given dose of tetrodotoxin to cardiac cells and raise questions concerning the significance and interpretation of tetrodotoxin binding to embryonic and cultured heart cells (27). Thus, acceptance of the hypothesis that the degree of sensitivity of myocardial cells to ion-specific blocking agents such as tetrodotoxin is an accurate description of differences in the development of ion-specific sites on the cardiac cell membrane should be interpreted with caution until the influence of both the extracellular matrix and fibroblast-like cells is more clearly defined.

In conclusion, to improve our understanding of the development of electrical activity in embryonic myocardial

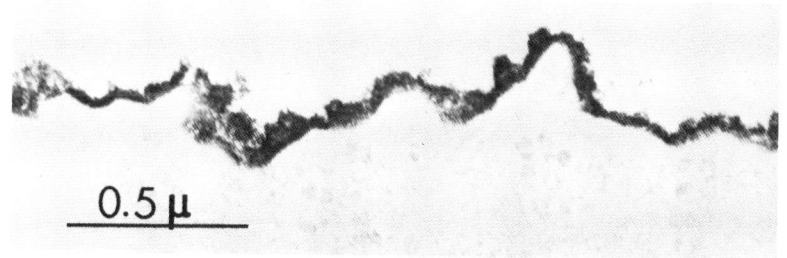

Fig. 7. Cell coat (basal lamina) of an apical myocardial cell from a 12-somite (ca. 40 hrs) embryonic chick heart is stained with ruthenium red, suggesting the presence of anionic groups. Scale, 0.5 µm; x 50000. Modified with permission from Manasek (28).

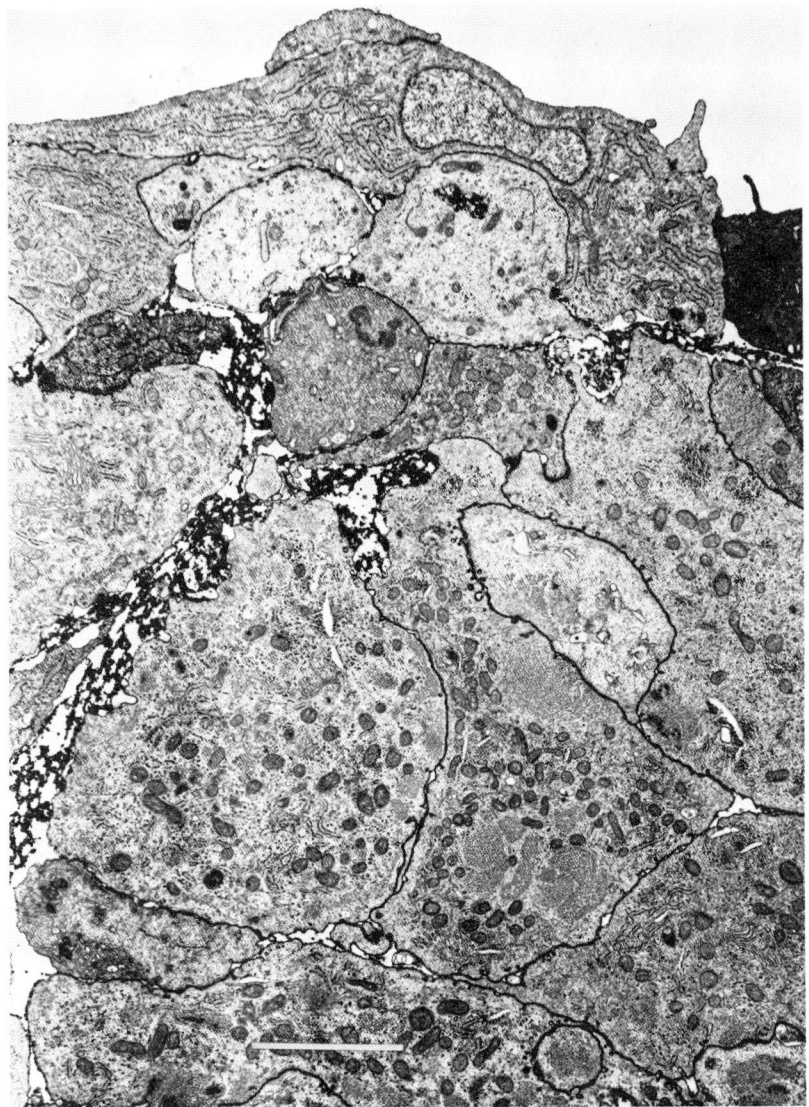

Fig. 8. Cross-section of a synthetic strand (6) of cardiac muscle. The central core of muscle cells is surrounded by a sheath of flattened, fibroblast-like cells. Cell surfaces and extra-cellular material are stained darkly by ruthenium red, suggesting the presence of glycoproteins. Scale, 2.5 μm; x 8000.

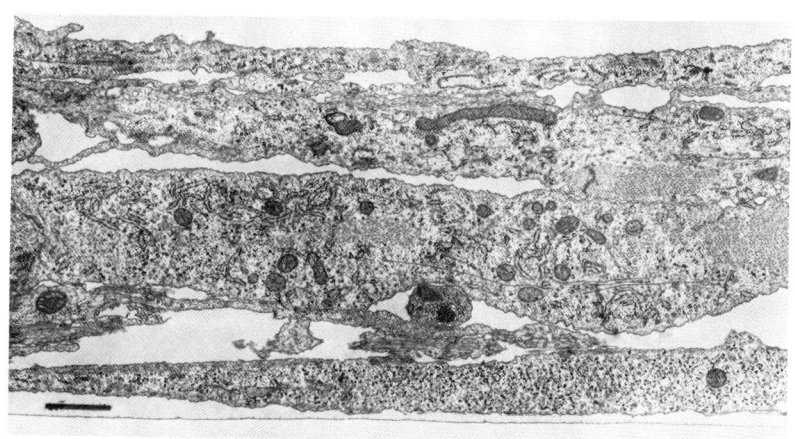

Fig. 9. An electron micrograph of a cross-sectional view of a 4-day primary culture of embryonic heart cells. The bottom cells rest adjacent to the surface of the culture dish. Note the presence of flattened fibroblast-like cells overlying the muscle cells. Scale, 1.0 μm; x 12000.

cells, it is essential to discriminate between the experimental results which are directly attributable to the physiological properties of heart cells and the results induced by the complex anatomy of the tissue. Heart muscle, even in its embryonic form, unfortunately, is sufficiently complex to often preclude our ability to arrive at an unequivocal interpretation of experimental data.

## ACKNOWLEDGMENTS

This research was supported in part by grants from the National Institutes of Health (HL12157), North Carolina Heart Association and an Established Investigatorship Award from the American Heart Association to M. Lieberman.

## REFERENCES

1. Johnson, E. A. and Lieberman, M. 1971. Heart: excitation and contraction. *Ann. Rev. Physiol.* 33:479-532.

2. Fozzard, H.A. and Beeler, G. W., Jr. 1975. The voltage clamp and cardiac electrophysiology. *Circ. Res.* 37:403-413.

3. Coraboeuf, E.; Deroubaix, E.; and Hoerter, J. 1976. Control of ionic permeabilities in normal and ischemic heart. *Circ. Res.* 38 (suppl):92-97.

4. Sperelakis, N. 1972. The electrical properties of embryonic heart cells. In *Electrical phenomena in the Heart*, pp. 1-61, ed., W. C. DeMello. New York: Academic Press.

5. DeHaan, R. L. and Fozzard, H. A. 1975. Membrane response to current pulses in spheroidal aggregates of embryonic heart cells. *J. Gen. Physiol.* 65:207-222.

6. Lieberman, M.; Sawanobori, T.; Kootsey, J. M.; and Johnson, E. A. 1975. A synthetic strand of cardiac muscle: its passive electrical properties. *J. Gen. Physiol.* 65:527-550.

7. Haas, H. G. and Kern, R. 1966. Potassium fluxes in voltage clamped Purkinje fibres. *Pflugers Arch.* 291:69-84.

8. Manasek, F. J. 1968. Embryonic development of the heart. I. Lights and electron microscopic study of myocardial development in the early chick embryo. *J. Morphol.* 125:329-365.

9. Challice, C. E., and Viragh, S. 1973. The embryologic development of the mammalian heart. In *Ultrastructure of the Mammalian Heart*, pp. 91-126, eds., C. E. Challice and S. Viragh. New York: Academic Press.

10. Carmeliet, E. E.; Horres, C. R.; Lieberman, M.; and Vereecke, J. S. 1976. Developmental aspects of potassium flux and permeability of the embryonic chick heart. *J. Physiol.* 254:673-692. London.

11. Horres, C. R. 1975. Potassium tracer kinetics of growth oriented heart cells in tissue culture. Ph.D. dissertation. Duke University. Durham, N.C.

12. MacDonald, R. L.; Mann, J. E., Jr.; and Sperelakis, N. 1974. Derivation of general equations describing tracer diffusion in any two-compartment tissue with application to ionic diffusion in cylindrical muscle bundles. *J. Theor. Biol.* 45:107-130.

13. Carmeliet, E. E. 1961. *Chloride and Potassium Permeability in Cardiac Purkinje Fibres*, pp. 1-152. Brussels: Editions Arscia, S.A.

14. Horres, C. R.; Lieberman, M.; and Purdy, J. E. 1977. Growth orientation of heart cells on nylon monofilament: determination of the volume to surface area ratio and intracellular potassium. *J. Mem. Biol.* 34:313-329.

15. Horres, C. R. and Lieberman, M. 1977. Compartmental analysis of potassium efflux from growth oriented heart cells. *J. Mem. Biol.* 34:331-350.

16. Sperelakis, N. and Lee, E. C. 1971. Characterization of ($Na^+, K^+$)-ATPase isolated from embryonic chick hearts and cultured chick heart cells. *Biochim. Biophys. Acta.* 233:562-579.

17. Chapman, J. B.; Horres, C. R.; Lieberman, M.; and Johnson, E. A. Simulated recovery from sodium loading by an electrogenic pump in cardiac muscle. *Biophys. J.* 16(2):31 (abs.)

18. Sperelakis, N., and Shigenobu, K. 1972. Changes in membrane properties of chick embryonic hearts during development. *J. Gen. Physiol.* 60:430-453.

19. McDonald, T. F., and DeHaan, R. L. 1973. Ion levels and membrane potential in chick heart tissue and cultured cells. *J. Gen. Physiol.* 61:89-109.

20. Bernard, C. 1975. Establishment of ionic permeabilities of the myocardial membrane during embryonic development of the rat. In *Developmental and Physiological Correlates of Cardiac Muscle*, pp. 169-184, eds., M. Lieberman and T. Sano. New York: Raven Press.

21. Manasek, F. J. 1970. Histogenesis of the embryonic myocardium. *Amer. J. Cardiol.* 25:149-168.

22. Spira, A. W. 1971. Cell junctions and their role in transmural diffusion in the embryonic chick heart. Z. Zellforsch. 120:463-487.

23. Lieberman, M. 1967. Effects of cell density and low K on action potentials of cultured chick heart cells. Circ. Res. 21:879-888.

24. DeHaan, R. L.; McDonald, T. F.; and Sachs, H. G. 1975. Development of tetrodotoxin sensitivity of embryonic chick heart cells in vitro. In Developmental and Physiological Correlated of Cardiac Muscle, pp. 155-168, eds., M. Lieberman and T. Sano. New York: Raven Press.

25. Sperelakis, N.; Shigenobu, K.; and McLean, M. J. 1975. Membrane cation channels-changes in developing hearts, in cell culture and in organ culture. In Developmental and Physiological Correlates of Cardiac Muscle, pp. 209-234, eds., M. Lieberman and T. Sano. New York: Raven Press.

26. Narahashi, T.; Moore, J. W.; and Scott, W. R. 1964. Tetrodotoxin blockade of sodium conductance increase in lobster giant axons. J. Gen. Physiol. 47:965-974.

27. Lieberman, M.; Sawanobori, T.; Shigeto, N.; and Johnson, E. A. 1975. Physiologic implications of heart muscle in tissue culture. In Developmental and Physiological Correlates of Cardiac Muscle, pp. 139-154, eds., M. Lieberman and T. Sano. New York: Raven Press.

28. Manasek, F. 1975. The extracellular matrix of the early embryonic heart. In Developmental and Physiological Correlates of Cardiac Muscle, pp. 1-20, eds., M. Lieberman and T. Sano. New York: Raven Press.

29. Gros, D.; Mocquard, J. P.; Challice, C. E.; and Schrével, J. 1975. Évolution de la surface des cellules myocardiques de la souris au cours de l'ontogènese. I. Étude cytochimique des glucides. J. Microscop. Biol. Cell. 23:249-270.

30. Purdy, J. E.; Lieberman, M.; Roggeveen, A. E.; and Kirk, R. G. 1972. Synthetic strand of cardiac muscle:

formation and ultrastructure. *J. Cell Biol.* 55: 563-578.

31. Shimada, Y. and Fischman, D. A. 1975. Cardiac cell aggregation by scanning electron microscopy. In *Developmental and Physiological Correlates of Cardiac Muscle*, pp. 81-101, eds., M. Lieberman and T. Sano. New York: Raven Press.

32. Manasek, F. 1976. Macromolecules of the extracellular compartment of embryonic and mature hearts. *Circ. Res.* 38:331-337.

# 9

# Metabolic Maturation in the Fetal Mouse Heart

## Kern Wildenthal

Pauline and Adolph Weinberger Laboratory for Cardiopulmonary Research
Departments of Physiology and Internal Medicine
University of Texas
Health Science Center at Dallas
Dallas, Texas

Metabolism in the fetal heart differs in many ways from that in adults. The activities of many cardiac enzymes, the ability of the heart to utilize various substrates, and the heart's responsiveness to hormonal stimuli vary in the fetus and neonate as compared to the adult (1-5, 7, 10, 11, 14, 18-20, 22-25, 28-30, 32, 34, 35). Moreover, even within the fetal period, these factors may change considerably and rapidly as term approaches.

To analyze and describe quantitatively the phenomenon of cardiac metabolic maturation one would prefer, ideally, to make measurements using hearts of intact fetuses of various ages *in utero*, under precisely controlled experimental conditions. Unfortunately, the technical problems of such an approach are insurmountable for most purposes, and inferences about patterns of fetal metabolism *in vivo* have had to be drawn primarily from measurements on nonbeating tissue slices and homogenates. In an effort to measure metabolic characteristics in beating hearts while avoiding the inherent complexities that preclude well-controlled studies *in utero*, I have used an organ culture system to maintain and study intact hearts of fetal mice under precisely controlled conditions *in vitro* (26). Organ culture offers the general advantage of providing viable tissue with normal intracellular connections in a system in which functional characteristics remain stable over a period of several days. Fibroblastic "overgrowth" and tissue "de-differentiation" (which pose potential problems in metabolic studies using cell cultures) are avoided, and variables such as the composition of the medium or, in these experiments, the age of the fetus from which the tissue is obtained can be altered easily and selectively while other factors are held constant.

## METHODS

Hearts were removed aseptically from mouse fetuses of 14 to 21 days' gestational age (term = 21-22 days). The hearts were prepared for organ culture by techniques described in detail previously (26). Briefly, the pericardium and vessels were dissected away, and the isolated heart was gently washed to remove blood and body fluids. The hearts were then transferred to culture chambers containing 1.5 ml of medium. The chambers were incubated at 37-38°C in an atmosphere of 95% $O_2$ plus 5% $CO_2$. The media in which the hearts were maintained was "Earle's salt solution" or "medium 199," containing glucose (5-25 mmoles/liter) and/or sodium octanoate (0.5-2.0 mmoles/liter) as the energy-yielding substrate. In other studies the medium also included 35% calf serum (Wellcome Research Labs or Colorado Serum Co.) with or without supplemental oleic or octanoic acid (0.5-2.0 mmoles/liter). Purified insulin, glucagon, or norepinephrine were added in some experiments as described below.

Uptake of glucose and fatty acids by the hearts was calculated from measurements of the differences between their concentrations in the medium before and after a 2-day culture period following initial explantation. Release of lactate by the hearts was calculated similarly. Glucose was analyzed by the glucose oxidase technique, fatty acids by a modified method of Trout (21), and lactate by the procedure of Hohorst et al. (9). Protein content of the hearts was measured by the Lowry technique (13) and glycogen by that of Karlsson (12), after homogenization in a 0.1% Triton solution with a Polytron homogenizer.

Beating rates were monitored visually or with a specially constructed capacitance probe (31), and rate changes in response to hormonal and pharmacological stimuli were measured as described previously (27).

## RESULTS

### Fatty Acids, Glucose, and Lactate Metabolism

At all stages of development during the last third of pregnancy, fetal mouse hearts were able to take up large amounts of octanoate, a medium-chain fatty acid, but only small amounts of long-chain fatty acids (Table 1). Utili-

Table 1

Fatty Acid Utilization in Cultured Fetal Mouse Hearts

| Process | < 16 days | > 20 days |
|---|---|---|
| Uptake of medium-chain fatty acids (nmoles/mg/hour) | 2.6±0.27 | 1.0±0.07 |
| Uptake of long-chain fatty acids (nmoles/mg/hour) | 0.1±0.05 | 0.2±0.05 |

Hearts from young (< 16 days gestational age) and older (> 20 days) fetal mice were cultured for 2 days in medium 199 + 35% calf serum + 50 µg/ml insulin + 0.5 mmole/liter of sodium octanoate (medium-chain fatty acid) or oleate (long-chain fatty acid). The values represent the mean ± 1 S.E.M. of 12 hearts.

zation of long-chain acids increased slightly as term approached, whereas utilization of octanoate actually decreased as the fetus matured (but still remained much higher than that of the long-chain acids). The hearts survived and beat spontaneously for many days in the presence of octanoate as the sole external substrate (30), indicating that this compound not only was taken up but was oxidized effectively to meet the energy requirements of the tissue.

Utilization of glucose by cultured mouse hearts, like that of octanoate, decreased as the fetus matured (Table 2). Simultaneously, release of lactate from the hearts also fell. The ratio of lactate released to glucose consumed was not constant: as the heart matured, a smaller proportion of the glucose taken up could be accounted for by the production of lactate. Such a result might constitute evidence in favor of an improved capacity

Table 2

Glucose Uptake and Lactic Acid Release in Cultured Fetal Mouse Hearts

| Process | < 16 days | > 20 days |
|---|---|---|
| Uptake of glucose (nmole/mg/hour) | 8.4 ± 1.08 | 4.0 ± 0.43 |
| Release of lactate (nmole/mg/hour) | 7.1 ± 1.03 | 2.3 ± 0.17 |
| Ratio of lactate release/glucose uptake | 0.84 ± 0.047 | 0.55 ± 0.047 |

Note: Hearts were cultured for 2 days in medium 199 (glucose = 100 mg/100 ml) + 0.5 mmole/liter Na octanoate. Each value represents the mean ± 1 S.E.M. of 24 hearts.

of the heart to utilize oxidative and/or other pathways for glucose and pyruvate metabolism. Alternatively, the data might be explained by the younger hearts simply taking up much more glucose and shunting the excess to lactate, without there being a significant difference in the capacity of the alternative pathways. If the latter were true, an increase in glucose uptake in the older hearts should also cause increased shunting to lactate and result in a rise in the ratio of lactate released to glucose consumed. When glucose uptake was quadrupled experimentally in the older hearts by supplying excess glucose (20-25 mmoles/liter) in the external medium, however, the lactate:glucose ratio remained less than 0.60 (compare Table 2), suggesting that glycolysis was indeed becoming less dominant and alternative pathways were assuming greater relative importance as term approached.

At the same time that these changes in glycolytic capacity were occurring, glycogen stores were also changing. Freshly isolated hearts from 14 to 16 day fetuses had over 50% more glycogen than those at term. The younger hearts contained 129 ± 19 μmoles glucose units/mg wet weight vs 84 ± 12 at 20-21 days, as compared to adult values of less than 60 (17).

Adult hearts are known to display a preference for utilization of fatty acids over glucose if both are present in high concentrations (15, 16). To test whether such a preference has already developed before maturation, litter-matched fetal mouse hearts were supplied media containing varying quantities of glucose and octanoic acid. As shown in Table 3, fatty acids caused a reduction in glucose uptake and an increase in lactate release. Results were qualitatively similar at 14-17 days gestational age as at 19-21 days. In contrast, elevated glucose levels had no effect on fatty acid utilization (30).

## Responses to Hormones

The susceptibility of cardiac metabolism to control by insulin was tested by maintaining hearts in the presence or absence of high concentrations of the hormone (50 μg/ml). High doses of insulin were anabolic at all ages tested, so that hearts cultured in the presence of the hormone had improved amino acid balance and increased protein mass compared to controls (Table 4). The net anabolic action of insulin was brought about both by increased protein synthesis and by reduced protein degradation (33). Insulin also caused an apparent increase in utilization of glucose

Table 3

Influence of Fatty Acids on Glucose Uptake and
Lactic Acid Release in Cultured Fetal Mouse Hearts

|  | Glucose uptake (nmoles/mg/hour) | Lactate Release (nmoles/mg/hour) |
|---|---|---|
| Control (14-17 days) | 8.0 ± 1.12 | 4.5 ± 0.71 |
| + 2.0 mM octanoate | 6.4 ± 0.93* | 8.7 ± 0.56* |
| Control (19-21 days) | 4.7 ± 0.29 | 1.6 ± 0.14 |
| + 2.0 mM octanoate | 4.0 ± 0.25* | 4.6 ± 0.38* |

Note: Control hearts were cultured in Earle's salt solution (glucose = 100 mg/100 ml) and hearts of matched-litter mates were cultured in the same medium supplemented with 2.0 mM Na octanoate. Values represent the mean ± 1 S.E.M. of 6-8 matched experiments.

\* = $p < 0.05$ compared to matched controls by Student's t test.

and octanoic acid in cultured hearts (Table 4); when uptake was corrected for the alterations in cardiac size and protein content that insulin caused, however, the increased total utilization of substrates in the presence of insulin seemed to be simply a reflection of the increased cardiac mass. No preferential switch to glucose uptake was observed in young or older fetal hearts. Similarly, after 2 days in culture, glycogen concentrations were similar in insulin-deprived hearts as those supplied the hormone.

In contrast to insulin's effects, glucagon (0.5-50.0 µg/ml) failed to alter substrate utilization in any way. In experiments made in collaboration with Clark and Allen (32), this metabolic unresponsiveness to glucagon was found to be associated with a failure of the hormone to stimulate adenylate cyclase and increase cyclic AMP. Such unresponsiveness could have been due to delayed maturation of the receptor, such as has been reported for rat heart (5). However, further studies have indicated that unres-

Table 4

Influence of Insulin on Metabolism of
Cultured Fetal Mouse Hearts

|  | < 16 days | > 20 days |
|---|---|---|
| Total cardiac protein | +18 ± 3% | +13 ± 3% |
| Glucose uptake (per heart) | + 7 ± 3% | +12 ± 4% |
| Octanoate uptake (per heart) | +16 ± 5% | +13 ± 5% |

Note: Hearts were cultured for 2 days in medium 199 (glucose = 100 mg/100 ml) + 0.5 mmoles/liter Na octanoate. Hearts supplied insulin (50 µg/ml) were compared with control hearts from matched litter-mates maintained under similar conditions in the absence of insulin. Values represent the average % change from control for 8-12 matched pairs ± 1 S.E.M. All changes were statistically significant ($p < 0.05$), and no differences were apparent between responses by younger and older hearts.

---

ponsiveness persists in the mouse throughout adult life (6), thus suggesting that in this species, as in guinea pigs (8), that the heart never develops the capacity to respond metabolically to glucagon.

It is of special interest in this regard that fetal mouse hearts did develop the capability of responding to glucagon by increasing heart rate and contractility (29, 32). Maturation of beating responsiveness occurred fairly rapidly between 17 and 20 days (Fig. 1). This apparent disparity between glucagon's effects on beating and on the adenylate cyclase-cyclic AMP system suggests the possibility that the glucagon-induced increases in force and rate of contraction are not mediated by cyclic AMP and not necessarily linked to metabolic alterations (32).

Responses to catecholamines were quite different from those to glucagon (29). Tachycardia in response to norepinephrine occurred at an earlier age, and there was a major increment in the hormone's effect just before birth (Fig. 2). Throughout the latter part of pregnancy catecholamines also stimulated adenylate cyclase and caused major increases in cyclic AMP levels in fetal mouse hearts (32), as well as a 30-40% reduction in tissue glycogen within 30 minutes of exposure.

## DISCUSSION AND CONCLUSIONS

Metabolism in the late-fetal mouse heart is characterized by (1) high glycolytic activity and extensive glycogen stores, and (2) a capacity to utilize large amounts of medium-chain fatty acids (which are not present in the

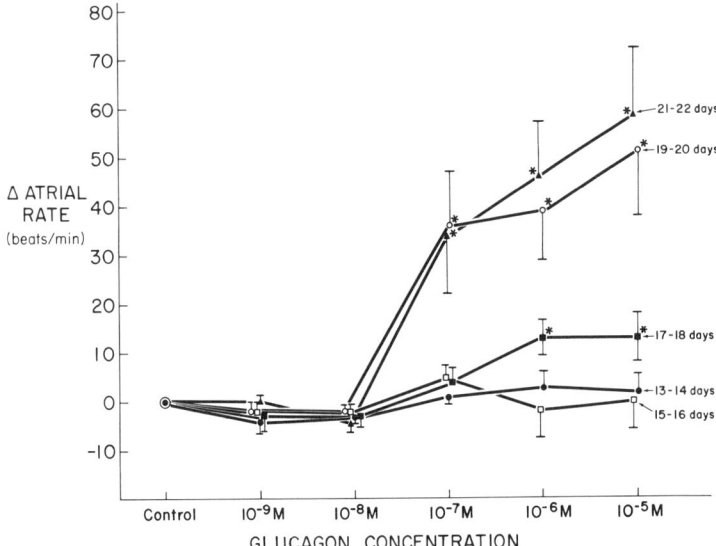

Fig. 1. Influence of glucagon on spontaneous atrial rate at various fetal ages. Each group comprised 10-14 hearts. Rates at each concentration of the drug are expressed as changes from the control rates. The vertical bars represent 1 S.E.M., and asterisks signify statistically significant changes from control rates. Hearts were unresponsive at 13-16 days; they responded minimally at 17-18 days, and well after 19 days. The figure is reproduced from Ref. 29.

animal physiologically) but an inability to utilize long-chain fatty acids to the same degree. During the last week before term (i.e., the last third of pregnancy), major changes occur in the use of all these substrates. Glucose utilization decreases, as has also been reported during the same period in fetal rat hearts (2, 4, 22), and the proportion of glucose that is converted to lactate also falls, suggesting increased diversion of glucose or pyruvate to other pathways, e.g., pyruvate oxidation; nevertheless, glucose uptake and the lactate:glucose ratio remain impressively high.

Simultaneously, the ability to utilize long-chain fatty acids increases somewhat (but still remains suboptimal). Although medium-chain fatty acids freely enter mitochondria where they are broken down by beta-oxidation, long-chain fatty acids cannot do so without the presence of sufficient quantities of carnitine and acylcarnitine transferase (15, 16). The present results suggest, there-

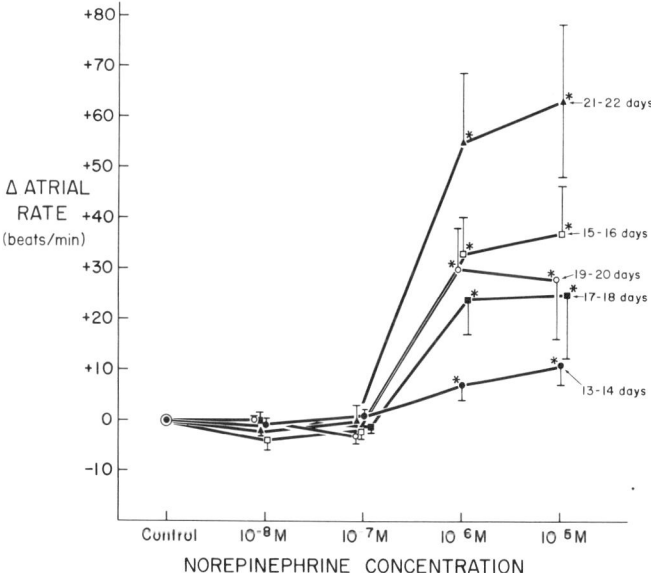

Fig. 2. Influence of l-norepinephrine on spontaneous atrial rate at various fetal ages. Symbols and conditions are identical to those in Figure 1. The figure reproduced is from Ref. 29.

fore, that the capacity to metabolize intramitochondrial fatty acids is already developed by the beginning of the last third of gestation, but that deficiency of the acylcarnitine transferase system persists in the mouse heart and prevents optimal utilization of long-chain fatty acids until after birth. The deficiency may become relatively less marked by term, but it remains present even then. It should be noted that important species differences exist in the exact time at which the heart acquires adequate levels of carnitine and acylcarnitine transferase and the capacity to oxidize long-chain fatty acids (1, 4, 20, 24, 28, 34, 35), and maturation of this system can precede birth in some species.

Fetal mouse hearts, like fetal rat hearts (2-4, 22), have already developed the ability to respond to insulin during the last third of gestation. In mice the response is especially prominent for the hormone's actions on protein and amino acid balance. Insulin appears not to cause a preferential increase in glucose uptake over fatty acids at this age in mouse hearts nor to induce glycogen depletion, in contrast to its effects in adults and in fetuses of other species (2, 4, 22). Unlike rats, which develop glucagon responsiveness at the weanling stage (5), mouse hearts appear never to develop the capacity to respond to glucagon metabolically (6), although the ability to respond to the hormone with increases in beating appears by 17-19 days of gestation. Responsiveness to catecholamines develops even earlier, and both beating and metabolism can be altered markedly by these agents throughout the last third of pregnancy.

In summary, the metabolic changes that accompany maturation of the fetal heart suggest that, as term approaches, the heart is becoming better able to accomodate its metabolism to an imminent milk diet and higher arterial $pO_2$. Nevertheless, glycolytic capacity, although lower than earlier, remains quite high and thus allows the heart to remain partially protected against possible anoxia at birth. Responsiveness to catecholamines is fairly well developed before term, indicating that these agents can exert their beneficial effects in meeting the stress of birth.

Finally, it should be reemphasized that important species differences may occur in the exact timing of the various changes that characterize the maturation of cardiac metabolism in the fetus. Nevertheless, the general pattern of maturation is probably broadly similar in most animals. Isolated hearts *in vitro* have provided useful systems for analyzing changes in the metabolic capabilities of the fetal heart at various stages of development and in its ability to respond to hormones as gestation proceeds. It will be of special interest in the future to make similar investigations of the metabolism of hearts *in utero*.

## ACKNOWLEDGMENTS

Some of the work described in this chapter was done in collaboration with Drs. D. O. Allen, C. M. Clark, Jr., J. S. Ingwall, J. Karlsson, and J. R. Wakeland, to whom I am extremely grateful. The work was supported by grants from the American Heart Association, The Moss Heart Fund, and the National Heart and Lung Institute (HL 14706, HL 06296, and Career Development Award HL 70125). The publishers of the *Journal of Clinical Investigation* and the *Journal of Molecular and Cellular Cardiology* kindly permitted reproduction of some of the figures and tables published in Refs. 29 and 30.

## REFERENCES

1. Breuer, E.; Barta, E.; Zlatos, L.; and Papová, E. 1968. Developmental changes of myocardial metabolism. *Biol. Neonat.* 12:54-64.

2. Clark, C. M., Jr. 1971. Carbohydrate metabolism in the isolated fetal rat heart. *Amer. J. Physiol.* 220: 583-588.

4. ----. 1973. The stimulation by insulin of amino acid uptake and protein synthesis in the isolated fetal rat heart. *Biol. Neonate* 19:379-388.

5. Clark, C. M., Jr.; Beatty, B.; and Allen, D. O. 1973. Evidence for delayed development of the glucagon receptor of adenylate cyclase in the fetal and neonatal rat heart. *J. Clin. Invest.* 52:1018-1025.

6. Clark, C. M., Jr.; Waller, D.; Kohalmi, D.; Gardner, R.; Clark, J.; Levey, G. S.; Wildenthal, K.; and Allen, D. 1976. Evidence that cyclic AMP is not in-

volved in the chronotropic action of glucagon in the adult mouse heart. *Endocrinology* 99:23-29.

7. Cox, S. J. and Gunberg, D. L. 1972. Metabolite utilization by isolated embryonic rat hearts *in vitro*. *J. Embryol. Exp. Morph.* 28:235-245.

8. Henry, P. D.; Dobson, J. G., Jr.; and Sobel, B. E. 1975. Dissociations between changes in myocardial cyclic adenosine monosphosphate and contractility. *Circ. Res.* 36:392-400.

9. Hohorst, H. J.; Kreutz, F. H.; and Bücher, T. 1959. Über Metabolitgehalte und Metabolitkonzentrationen in der Leber der Ratte. *Biochemische Zeitschrift.* 332:18-45.

10. Ingwall, J.S. and Wildenthal, K. 1977. Fetal mouse hearts in organ culture: studies in cardiac metabolism. In *Recent advances in the study of cardiac structure and metabolism*, eds., T. Sano, et al. Baltimore: Univ. Park Press. (In press).

11. Jolley, R. L.; Cheldelin, V. H.; and Newburgh, R. W. 1958. Glucose catabolism in fetal and adult heart. *J. Biol. Chem.* 233:1289-1294.

12. Karlsson, J. 1971. Lactate and phosphagen concentrations in working muscle of man. *Acta Physiol. Scand. Suppl.* 358:1-72.

13. Lowry, O. H.; Rosebrough, N. J.; Farr, A. L.; and Randall, R. J. Protein measurements with the Folin phenol reagent. *J. Biol. Chem.* 193:265-275.

14. Mersmann, H. J. and Phinney, G. 1973. *In vitro* fatty acid oxidation in liver and heart from neonatal swine. (*Sus. domesticus*) *Compar. Biochem. Physiol.* 44B: 219-224.

15. Neely, J. R. and Morgan, H. E. 1974. Relationship between carbohydrate and lipid metabolism and the energy balance of heart muscle. *Ann. Rev. Physiol.* 36:413-459.

16. Opie, L. H. 1969. Metabolism of the heart in health and disease. *Amer. Heart J.* 76:685-698; 77:100-122, 383-410.
17. Shelley, H. J. 1961. Glycogen reserves and their changes at birth and in anoxia. *Brit. Med. Bull.* 17:137-143.
18. Sippel, T. O. 1954. The growth of succinoxidase activity in the hearts of rat and chick embryos. *J. Exper. Zool.* 126:205-221.
19. Skála, J.; Drahota, Z.; and Hahn, P. 1970. Succinate dehydrogenase activity of rat heart muscle mitochondria during development. *Physiol. Bohemoslov.* 19:15-17.
20. Tomec, R. J., and Hoppel, C. L. 1975. Carnitine palmitoyltransferase in bovine fetal heart mitochondria. *Arch. Biochem. Biophys.* 170:716-723.
21. Trout, D. L.; Estes, E. H., Jr.; and Friedberg, S. J. 1960. Titration of free fatty acids of plasma: a study of current methods and a new modification. *J. Lipid Res.* 1:199-202.
22. Vinicor, F.; Clark, J. F.; and Clark, C. M., Jr. 1975. Development of hormone receptors in the isolated fetal rat heart. In *Early diabetes in early life*, pp. 105-114, eds., R. A. Camerini-Davalos and H. S. Cole. New York: Academic Press.
23. Warshaw, J. B. 1970. Cellular energy metabolism during fetal development. III. Deficient acetyl-CoA synthetase, acetylcarnitine transferase and oxidation of acetate in the fetal bovine heart. *Biochim. Biophys. Acta* 223:409-415.
24. ----. 1972. Cellular energy metabolism during fetal development. IV. Fatty acid activation, acyl transfer and fatty acid oxidation during development of the chick and rat. *Developmental Biol.* 28:537-544.
25. Warshaw, J. B., and Terry, M. L. 1970. Cellular energy metabolism during fetal development. II. Fatty acid oxidation by the developing heart. *J. Cell. Biol.* 44:354-360.

26. Wildenthal, K. 1971. Long-term maintenance of spontaneously beating mouse hearts in organ culture. *J. Appl. Physiol.* 30:153-157.

27. ----. 1972. Studies of isolated fetal mouse hearts in organ culture. Evidence for a direct effect of triiodothyronine in enhancing cardiac responsiveness to norepinephrine. *J. Clin. Invest.* 51:2702-2709.

28. ----. 1973. Foetal maturation of cardiac metabolism. In *Proceedings of the Sir Joseph Barcroft symposium: foetal and neonatal physiology*, pp. 181-185, eds., R. M. Comline, et al. Cambridge: Cambridge University Press.

29. ----. 1973. Maturation of responsiveness to cardioactive drugs. Differential effects of acetylcholine, norepinephrine, theophylline, tyramine, glucagon and dibutyryl cyclic AMP on atrial rate in hearts of fetal mice. *J. Clin. Invest.* 52:2250-2258.

30. ----. 1973. Studies of foetal mouse hearts in organ culture: metabolic requirements for prolonged function *in vitro* and the influence of cardiac maturation on substrate utilization. *J. Molec. Cell. Cardiol.* 5:87-99.

31. Wildenthal, K.; Harrison, D. R.; Templeton, G. H.; and Reardon, W. C. 1973. Method for measuring the contractions of small hearts in organ culture. *Cardiovascular Res.* 7:139-144.

32. Wildenthal, K.; Allen, D. O.; Karlsson, J.; Wakeland, J. R.; and Clark, C. M., Jr. 1976. Responsiveness to glucagon in fetal hearts. Species variability and apparent disparities between changes in beating, adenylate cyclase activation and cyclic AMP concentration. *J. Clin. Invest.* 57:551-558.

33. Wildenthal, K.; Griffin, E. E.; and Ingwall, J. S. 1976. Hormonal control of cardiac protein and amino acid balance. *Circ. Res.* (suppl 1): 138-144.

34. Wittles, B. and Bressler, R. 1965. Lipid metabolism in the newborn heart. *J. Clin. Invest.* 44:1639-1646.

35. Wood, J. M. 1975. Carnitine palmityltransferase in neonatal and adult heart and liver mitochondria. Effect of phospholipase C treatment. *J. Biol. Chem.* 250:3062-3066.

# 10

## Glycolytic Control Mechanisms in Cardiac Muscle From Fetal Rhesus Monkeys

Clarissa H. Beatty, Rose Mary Bocek and
Martha K. Young

Division of Perinatal Biology/Oregon Regional Primate Research Center
Beaverton, Oregon, and
Department of Biochemistry
University of Oregon Health Sciences Center/Portland, Oregon

The scarcity of published data on the biochemistry and metabolism of fetal cardiac ventricular muscle contrasts with the abundance of similar information on the adult heart. Studies on cardiac muscle from fetal rhesus monkeys (*Macaca mulatta*), which are biologically similar to human beings, are therefore of particular interest. The significant features of the growth curve of the rhesus monkeys are similar to those of man, and Van Wagenen and Catchpole (47) have concluded that the rhesus is a suitable paradigm of man for investigation of many phases of development.

Many publications are available on the identification of rate-limiting reactions for the glycolytic pathway in adult cardiac muscle (28, 32). Experimental approaches to identifying regulatory enzymes in metabolic pathways include determining mass-action ratios, assaying maximal enzyme activities *in vitro*, measuring the effects of al-

See Note on p. 297.

tered flux rates on the tissue content of pathway intermediates, measuring the rates of product formation from various precursors, and investigating the kinetic and allosteric properties of enzymes. Such studies have provided evidence that the reactions catalyzed by hexokinase, phosphofructokinase, and pyruvate kinase are rate-limiting for glycolysis in adult tissues (28, 32, 34).

As a first step in our investigation we determined the levels of all metabolic intermediates and cofactors in the glycolytic pathway of cardiac muscle from fetal rhesus monkeys, to identify enzymes which catalyze nonequilibrium or possible rate-limiting reactions. The evidence shows that hexokinase, phosphofructokinase, and pyruvate kinase—the same enzymes previously reported to be nonequilibrium and rate-limiting in adult cardiac muscle (32, 34)—are also nonequilibrium and therefore possibly rate-limiting in fetal muscle by midterm. We have also studied the kinetics and the effects of a number of ligands on phosphofructokinase and pyruvate kinase from fetal heart muscle. The sensitivity of glycolytic enzymes to effector molecules is especially important since the values for maximum enzyme capacity seldom mirror their estimated activity *in vivo*. The enzyme capabilities of a given cell are often utilized to a small extent and may vary with intracellular milieu; e.g., erythrocyte phosphofructokinase from adults appears to operate *in vivo* at least 1% of capacity (38).

## MATERIALS AND METHODS

In these experiments we used cardiac ventricular muscle from rhesus fetuses (*Macaca mulatta*), 90-155 days of gestational duration, and from adult rhesus monkeys. The average gestational age in our colony is 165 days. The monkeys were anesthetized with 1% or less Halothane (2-bromo-2-chloro-1,1,1-trifluorethane) in a mixture of 75% oxygen and 25% nitrous oxide or Ketalar (dl-2-[0-chlorophenyl]-2-[methylamino] cyclohexane hydrochloride), 5 mg/kg. The ventricle was rapidly dissected free of adipose and connective tissue, immediately rinsed in ice-cold 150 mM KCl, blotted, and frozen in clamps cooled with liquid nitrogen. The muscle was then wrapped in aluminum foil and stored at liquid nitrogen temperature.

The frozen tissue was homogenized and analyzed for glycolytic intermediates as previously described (2). An extract was also prepared for incubation experiments in the Dubnoff shaker (2).

To prepare homogenates for kinetic studies of phosphofructokinase and pyruvate kinase, we placed 0.5-1.0 g of frozen muscle on 1.0 ml of frozen medium (11 mM $MgCl_2$, 1 mM DTT [dithiothreitol], 82 mM KCl, 55 mM $N$-tris [hydroxymethyl] methyl-2-aminoethanesulfonic acid [TES], pH 7.4, at room temperature) in a glass homogenizing tube packed in dry ice; the preparation was chopped with a scalpel and homogenized as previously described (2). Homogenates were centrifuged for 30 min at 21,000 g (2°). Phosphofructokinase activity in the supernatant was stable at 4°C for 2 days. For the pyruvate kinase determination, the 21,000 g supernatant was quick frozen and stored in a liquid nitrogen refrigerator. The adult enzyme preparation was stable for at least a month at this temperature; the 100- and 150-day fetal preparation was stable for a week.

A Beckman DU spectrophotometer with a Gilford absorbance indicator and automatic cuvette positioner with recorder was used to determine the phosphofructokinase and pyruvate kinase activities by measuring the decrease in the extinction coefficient of NADH at 340 nm with a coupled reaction system.

The final concentrations in the cuvette (3.15 ml) were 50 mM HEPES (n-2-hydroxyethypiperazine-$N^1$-2-ethanesulfonic acid), pH 7.0 or 8.1, 74 mM KCl, 0.15 mM NADH, 5 and 10 mM $MgCl_2$ for phosphofructokinase and pyruvate kinase respectively, and 45 and 5 mg % dialyzed bovine serum albumin (BSA) for phosphofructokinase and pyruvate kinase respectively. Effector molecules were added as indicated in the figures. For phosphofructokinase assay different amounts of ATP and fructose-6-phosphate and excess amounts of aldolase, triosephosphate isomerase, and glycerophosphate dehydrogenase were also added (29); for pyruvate kinase assay various amounts of phosphoenolpyruvate (monocyclohexylamine salt) and ADP and an excess of lactic dehydrogenase were added (6). Supernatant (5 to 25 μl, 4°) was added to start the reaction; assays were run at 28°. The μmoles of fructose diphosphate formed $min^{-1}$ were linear over the range of protein used in these assays. All biochemicals were obtained from Sigma Chemical Company.

Enzyme activities are expressed as μmoles fructose diphosphate formed 10 mg supernatant protein$^{-1}$ min$^{-1}$ for phosphofructokinase and μmoles of NADH oxidized mg supernatant protein$^{-1}$ min$^{-1}$ for pyruvate kinase.

Protein was determined by Lowry's method, with bovine serum albumin as a standard (26).

## RESULTS

### Mass-Action Ratios and Rate-Limiting Reactions

The first step in identifying rate-limiting reactions is the identification of the enzymes that catalyze reactions far removed from equilibrium. Enzymes that catalyze equilibrium reactions cannot control the overall rate of flux in metabolic pathways. Therefore we measured the tissue levels of all metabolic intermediates and cofactors of the glycolytic pathway and calculated the mass-action ratios for each reaction. Similarity between the apparent equilibrium constant and the mass-action ratio indicates an equilibrium reaction. A larger equilibrium constant indicates a nonequilibrium or possible rate-limiting reaction. All regulatory enzymes must be nonequilibrium, but not all nonequilibrium enzymes are necessarily rate-limiting. According to Rolleston (40), if the apparent equilibrium constant is at least 20 times greater than the mass-action ratio, the reaction is probably rate-limiting.

The mass-action ratios for the glycolytic pathway of fetal and adult cardiac muscle and the apparent equilibrium constants at pH 7.4 are shown in Table 1. The fetal and adult values do not differ significantly. The tissue concentrations for the glycolytic intermediates and cofactors from which the mass-action ratios are calculated are shown in Table 2. The apparent equilibrium constants for hexokinase, phosphofructokinase, and pyruvate kinase in these tissues were over 800 times larger than the mass-action ratios at all ages studied. These data suggest that as early as 50% of term these three enzymes are rate-limiting for glycolysis in fetal cardiac muscle. It is difficult, however, to assess the nature of rate-limiting enzymes in a metabolic pathway from calculated apparent equilibrium constants and mass-action ratios since the conditions under which these values are obtained may not reflect the situation *in vivo*. No experimental values for 1, 3-di-P-gly-

Table 1

Comparison of Mass-Action Ratios with Apparent Equilibrium Constants for Glycolytic Reactions of Fetal and Adult Cardiac Muscle

|  | Apparent Equilibrium Constant[a] | Mass-Action Ratios | | |
|---|---|---|---|---|
|  |  | Fetal (90-day) | Fetal (150-day) | Adult |
| Hexokinase[b] | $3.9-5.5 \times 10^3$ | $1 \times 10^{-2}$ | $5.2 \times 10^{-2}$ | $1.2 \times 10^{-1}$ |
| Phosphogluco-isomerase | $0.4-0.5$ | $0.23$ | $0.16$ | $0.27$ |
| Phosphofructo-kinase[b] | $1 \times 10^3$ | $0.94$ | $1.19$ | $1.28$ |
| Aldolase (M) | $7-13 \times 10^{-5}$ | $1.1 \times 10^{-5}$ | $3.2 \times 10^{-5}$ | $1.2 \times 10^{-5}$ |
| Triosephosphate isomerase | $3.6-4.5 \times 10^{-2}$ | $0.9$ | $1.0$ | $2.8$ |
| Glyceraldehyde-3-phosphate dehydrogenase plus phospho-glycerate kinase ($M^{-1}$) | $0.2-1.5 \times 10^3$ | $9.4 \times 10^2$ |  | $2.6 \times 10^2$ |
| Phosphoglycerate mutase | $0.1-0.2$ | $0.2$ | $0.4$ | $0.3$ |
| Enolase | $2.8-4.6$ | $1.1$ | $0.9$ | $0.9$ |
| Pyruvate kinase[b] | $2-20 \times 10^3$ | $5.8$ | $6.1$ | $6.6$ |

Adult and 90-day fetal series, duplicate determinations on 2 monkeys; one monkey in 150-day series. Since determination of the content of 1,3-diphosphoglycerate in tissues is difficult, it is more convenient to combine the reactions producing and using this compound and to calculate the mass-action ratio for the combined glyceraldehyde phosphate dehydrogenase and phosphoglycerate kinase reaction (32, p. 141)

[a] Values from Newsholme and Start (32) and Williamson (48).

[b] Rate-limiting.

Table 2

Levels of Glycolytic Intermediates and Cofactors in Cardiac Ventricular Muscle from Fetal and Adult Rhesus Monkeys

|  | Wet Weight (nmoles/g) | | |
|---|---|---|---|
|  | Fetal | | Adult |
|  | 90-day | 150-day |  |
| Glucose[a] | 4055 | 2654 | 450 |
| Glucose-6-P | 123 | 351 | 228 |
| Fructose-6-P | 27.7 | 57.5 | 61.5 |
| Fructose-1,6-diP | 79.7 | 173 | 334 |
| Glyceraldehyde-P | 30.0 | 75.5 | 37.9 |
| Dihydroxyacetone-P | 28.2 | 73.8 | 108 |
| 3-Phosphoglyceric acid | 53.5 | 32.6 | 45.3 |
| 2-Phosphoglyceric acid | 8.3 | 14.1 | 12.5 |
| Phosphoenolpyruvic acid | 9.5 | 12.1 | 10.5 |
| Pyruvate | 18.2 | 29.4 | 16.8 |
| Lactate | 4763 | 7401 | 9898 |
| Inorganic P | 1997 | 5621 | 7745 |
| ATP | 2100 | 2681 | 3384 |
| ADP | 689 | 1063 | 800 |
| AMP | 7 | neg | 158 |
| Adenylate energy charge[b] | 0.88 | 0.86 | 0.87 |

The 90-day fetal and adult values are averages of duplicate determinations on 2 animals. The 150-day series includes only one animal. The glycolytic intermediates were assayed with appropriate coupling enzymes by the disappearance of added NADH or the formation of NADPH from added NADP, as described by Maitra and Estabrook (27) and Lowry and Passonneau (25) with modifications, $P_i$ was measured by the method of Lowry et al. (23). AMP and ADP were converted to ATP by incubation in HEPES buffer with myokinase and pyruvate kinase (18), and the ATP was determined by the luciferase enzyme in a liquid scintillation spectrometer (out of coincidence) (42).

[a] Intracellular level.

[b] (ATP + 0.5 ADP)/(ATP + ADP + AMP) (1).

cerate levels are available, and the mass-action ratios of glyceraldehyde-P dehydrogenase and P-glycerate kinase cannot be individually calculated. The overall reaction may be nonequilibrium in other tissues such as myometrium and taenia coli (3). However, the apparent equilibrium constants for the combined reaction of the two enzymes, as well as the remaining reactions in the glycolytic pathway for fetal and adult cardiac muscle, appear to be equilibrium reactions since the apparent equilibrium constants are less than 20 times greater than the mass-action ratios (Table 1).

Although the ATP levels (nmoles/g wet wt) are lower in the fetal (Table 2) than in the adult heart, the energy charge, as defined by Atkinson (1) (ATP + ½ ADP ÷ ATP + ADP + AMP), is similar to that in the adult series and greater than 0.8. This value for the adenylate energy charge represents a stable metabolic state and presumably indicates a physiological situation in the cell (28). The ATP levels per mg nitrogen were 146 nmoles for the adult and 145 nmoles for the fetal series.

## Phosphofructokinase Activity

It is difficult to assess the nature of rate-limiting enzymes in a metabolic pathway from calculated apparent equilibrium constants and mass-action ratios since the conditions under which these values were obtained may not reflect the situation *in vivo*. Therefore, the next step was to study some of the numerous effector molecules that might modify nonequilibrium reactions, particularly ATP, citrate, inorganic phosphate, and cyclic AMP. In Figure 1 we have compared the phosphofructokinase activity of fetal and adult cardiac muscle extracts at pH 7.0 and increasing levels of ATP. This extract is similar to a heart extract studied by Mansour (29) and to extracts of several other tissues, such as sperm (14, 31). At pH 7.0, ATP had a marked effect on the phosphofructokinase activity of both fetal and adult muscle, first increasing and then decreasing enzyme activity. This biphasic response of phosphofructokinase from adult tissues to increasing levels of ATP (at low pH values) has been reported by others (13, 28). At the lower ATP levels the phosphofructokinase activity was similar in the fetal and adult muscle extracts; however, the maximum phosphofructokinase activity of the adult and 150-day (term) series was higher than that of

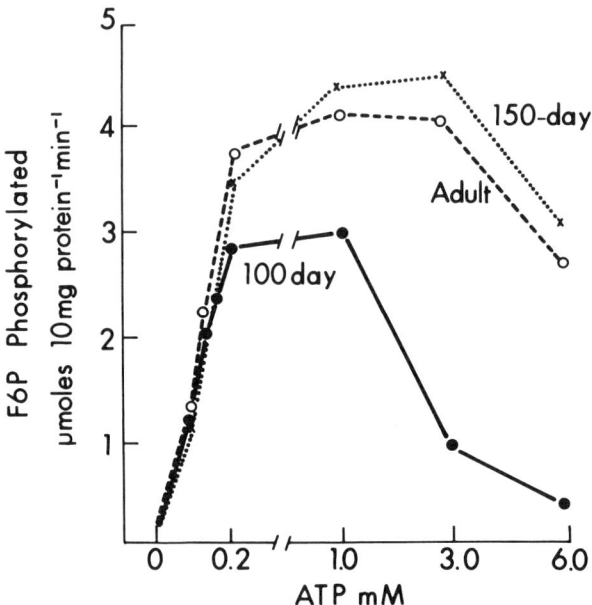

Fig. 1. Effect of ATP on PFK activity (F6P phosphorylated) of extracts (21,000 g supernatant of homogenate) of cardiac muscle from rhesus monkeys at different stages of development. The incubation medium was 1mM F6P, 80 mM TES and KCl, 2 mM NAD, 0.2 mM $MgCl_2$, ATP added as indicated ($MgCl_2$ maintained at equimolar concentrations). One-half milliliter of muscle extract was added to 0.5 ml incubation medium, pH 7.0, and the mixture was incubated in a Dubnoff shaker at $37°C$. The reaction was stopped by the addition of 30 μl 60% $HClO_4$. The value for F6P phosphorylated (PFK activity) was obtained by subtracting the levels of G6P and F6P left after incubation with exogenous F6P from the original values in the incubation medium. Protein was determined by the method of Lowry et al. (26). Endogenous levels of ATP in the medium were 0.02 mM or less, and $P_i$ was 0.1 to 0.2 mM.

the 100-day, and the enzyme activity was less sensitive to ATP inhibition in the older series.

When the concentration of fructose-6-phosphate was doubled, the total amount converted to fructose diphosphate also doubled, the curve shifted to the right, and there was a marked change in the effect of ATP on phosphofructokinase activity of fetal cardiac muscle (Fig. 2). The level of fructose-6-phosphate markedly influences the degree of inhibition with ATP. For instance, the inhibitory effect of 3 mM ATP decreased from almost 100% with 1 mM fructose-6-phosphate to less than 10% with 2 mM fructose-6-phosphate and was eliminated at 3 mM fructose-6-phosphate. Although at substrate levels of 1 mM fructose-6-phosphate, 1 mM ATP

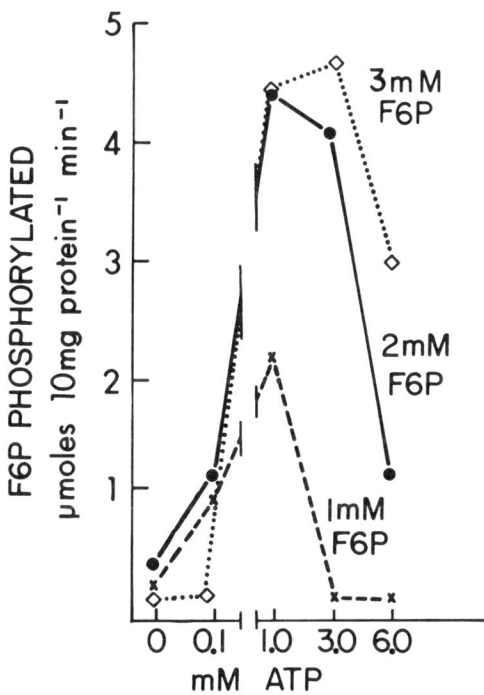

Fig. 2. Effect of increasing levels of F6P on the activity of PFK (F6P phosphorylated) at increasing levels of ATP in extracts of cardiac muscle from a 100-day fetus. Experimental conditions same as Figure 1.

is not inhibitory in this system, at 0.1 mM fructose-6-phosphate, 1 mM ATP decreased phosphofructokinase activity by 80% (data not shown).

The *in situ* intracellular level of fructose-6-phosphate in heart muscle at all ages studied was estimated to be 0.04-0.1 mM (Table 2) and the ATP level was 3 mM or more, conditions assuring a high degree of phosphofructokinase inhibition. Presumably, the operation of the glycolytic pathway in both fetal and adult cardiac muscle *in vivo* depends on the presence of effector molecules capable of relieving ATP inhibition.

One of the activators of phosphofructokinase activity in many adult tissues is fructose diphosphate (30, 32). In our system the level of fructose diphosphate at the end of incubation varied from 0.1 mM to 0.3 mM, levels that are high enough to affect phosphofructokinase activity. Therefore, we also measured phosphofructokinase activity in a coupled reaction system where the fructose diphosphate level was negligible (35). At pH 7.0 the results with the two methods are similar (Figs. 1 and 3). The Vmax for phosphofructokinase is highest in the adult series, both at pH 7.0 and 8.1; increasing the pH from 7.0 to 8.1 almost doubles both the apparent Km and Vmax at all ages (Table 3). With mammalian phosphofructokinase the enzyme generally loses its susceptibility to regulation as the pH increases above 7.6 (4). At pH 8.1 the inhibitory effect of ATP disappears in both the fetal and the adult series (Table 3). Both fetal and adult enzymes show the same kinetic characteristics with increasing levels of ATP (Fig. 4). Staal et al. (41) reported a similar kinetic pattern with adult human erythrocyte phosphofructokinase.

Three mM ATP produced 90% inhibition of 100-day fetal phosphofructokinase activity (Fig. 5). This inhibition was not relieved by 0.4 µM cyclic AMP, but 0.6 mM inorganic phosphate slightly more than doubled the activity and together the effectors caused a 6-fold increase in total activity to about 75% of maximum (Fig. 5). Two µM cyclic AMP alone tripled the phosphofructokinase activity and, in combination with 3 mM inorganic phosphate returned phosphofructokinase activity to the control level. Eight µM cAMP alone increased phosphofructokinase to 75% of the control activity (data not shown). Four mM ATP produced greater inhibition than 3 mM; 2 µM cyclic AMP had little effect on phosphofructokinase activity under these conditions, although it had a synergistic effect in combination with 3mM inorganic phosphate (Fig. 5). These concentrations of cyclic AMP are within the expected range *in vivo*. If one assumes uniform distribution in the intracellular compartment, the concentration of cyclic AMP in adult rhesus heart muscle is about 0.3 µM; this level is increased about 4-fold in the 100-day fetal series (5). Comparable levels of cyclic AMP (0.4 µM) have been reported for adult guinea pig heart (20). Inorganic phosphate levels in cardiac muscle are also high

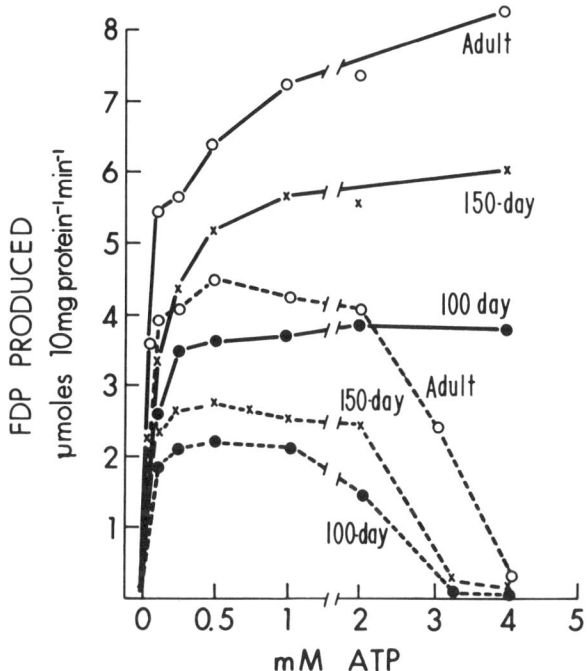

Fig. 3. Effect of ATP on PFK activity (FDP-formed) of extracts of cardiac muscle from rhesus monkeys of different ages. Solid lines, pH 8.1; broken lines, 7.0; both at 28°C. The incubation medium was 5 mM $MgCl_2$, 45 mg% BSA, 74 mM KCl, 50 mM HEPES, 0.15 mM NADH, 0.5 mM F6P, variable amounts of ATP and excess aldolase, triosephosphate isomerase and glycerophosphate dehydrogenase. The PFK was determined by the decrease in the extinction coefficient of NADH at 340 nm. Protein was estimated by the method of Lowry et al. (26).

Table 3

Comparison of Kinetic Data on Cardiac Phosphofructokinase at pH 7.0 and 8.1

|  | Fetal | | Adult |
|---|---|---|---|
| Series | 100-Days | 150-Days | |
| pH 7.0 | | | |
| Km µM ATP | 29 | 30 | 28.5 |
| Vmax[a] | 2.4 | 2.9 | 4.7 |
| pH 8.1 | | | |
| Km µM ATP | 55 | 55 | 48 |
| Vmax[a] | 4.0 | 5.6 | 7.4 |

Values are averages of duplicate determinations on the same heart. The level of endogenous $P_i$ in the assay medium was ~ 5 µM and ATP 1-2 µM. The supernatant was diluted so that the reaction rates at the different ages were similar. Experimental conditions same as Figure 3. The level of F6P was 0.5 mM.

[a] Micromoles FDP produced 10 mg protein$^{-1}$ min$^{-1}$.

enough to relieve inhibition. The sensitivity of fetal and adult phosphofructokinase to cyclic AMP and inorganic phosphate differed somewhat, however, together they returned the enzyme activity to almost the control level (Fig. 6).

Increasing the concentration of citrate from 0.2 mM to 1.3 mM at pH 7.0 progressively increased its inhibitory action of fetal and adult phosphofructokinase (Fig. 7). A concentration of 0.2 mM citrate is within the physiological range. Thus citrate inhibits phosphofructokinase activity as early as midterm at levels that may well exist in the cell. This data agrees with results obtained by others for adult cardiac muscle from rodents (12, 35, 36). Garland et al, (11) reported citrate levels of 0.4 µmoles gm$^{-1}$ wet

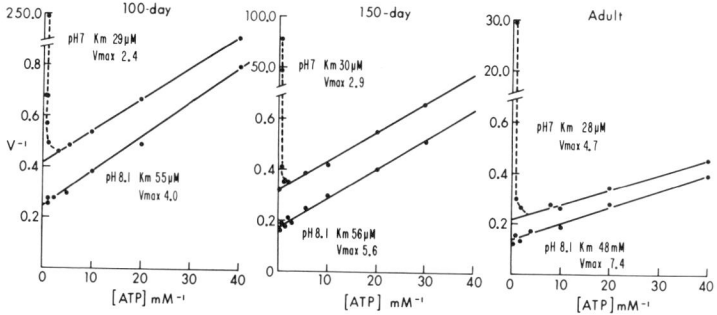

Fig. 4. Kinetic plot of PFK activity in a muscle extract from a 100-day and a 150-day rhesus fetus and an adult rhesus. Experimental conditions as in Figure 3. V = μmoles FDP formed min$^{-1}$ 10 mg protein$^{-1}$.

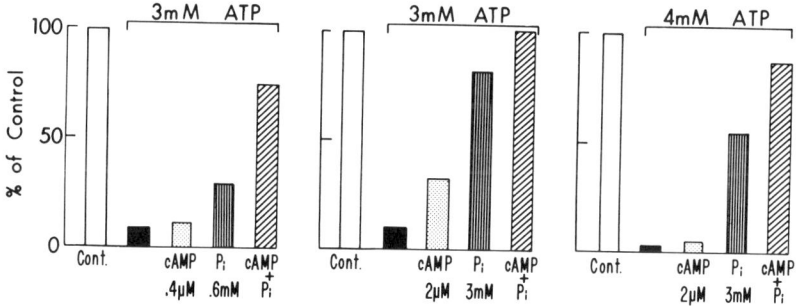

Fig. 5. Relief by cAMP of ATP inhibition of the PFK activity in an extract of a 100-day fetal heart. Experimental conditions same as Figure 3, ATP level in control series 0.2 mM.

weight for cardiac muscle from adult rats and this value increased as much as 5-fold in hearts from starved or diabetic rats.

Although phosphofructokinase activity did not change as the ATP levels were increased from 0.1 to 1.0 mM at pH 7.0 (Fig. 3), the inhibitory effect of citrate was greatest at the higher ATP concentrations, indicating a synergism between ATP and citrate in the fetal as well as in the adult series (Fig. 7). Pogson and Randle (35) also reported a synergism between ATP and citrate in their study of inhibition of phosphofructokinase from adult rat hearts. At pH 8.1, citrate had little inhibitory effect, even in the presence of 1 mM ATP. Citrate inhibition of 100-day fetal phosphofructokinase activity was relieved by cyclic AMP and inorganic phosphate (Fig. 8).

## Pyruvate Kinase Activity

Activities of pyruvate kinase in both fetal and adult cardiac muscle reached a maximum with a substrate level of 0.3 mM phosphoenolpyruvate at pH 7.0 and 3 mM ADP. The activity of the fetal phosphofructokinase at midterm

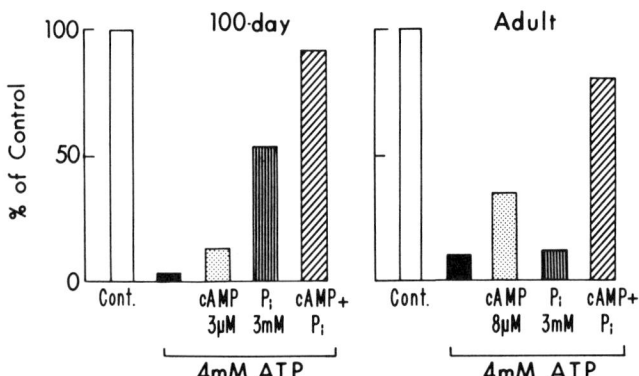

Fig. 6. Relief of PFK inhibition with 4 mM ATP by cAMP and $P_i$ alone and in combination. Experimental conditions same as Figure 3, ATP level in control series 0.2 mM.

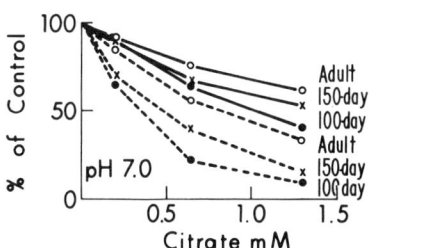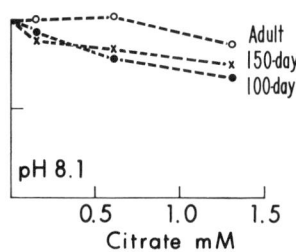

Fig. 7. Effect of citrate on the inhibition of PFK activity at pH 7.0 and 8.1 in extracts of cardiac muscle at different ages. Substrate level of F6P 0.5 mM; solid lines, 0.1 mM ATP; broken lines, 1.0 mM ATP. Experimental conditions same as Figure 3.

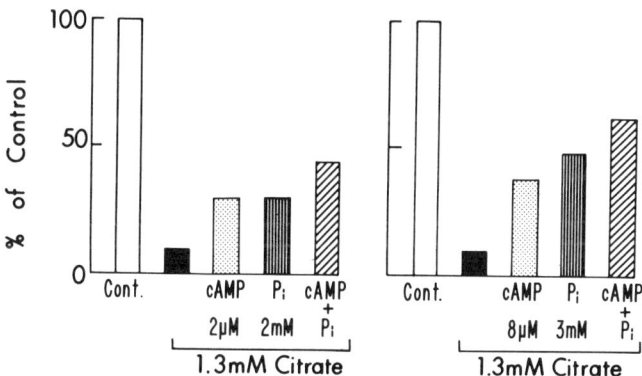

Fig. 8. Relief of PFK inhibition with 1.3 mM citrate by AMP and $P_i$ alone and in combination. Experimental conditions same as Figure 3 (pH 7.0, 28°C, 1 mM ATP, 0.5 mM F6P.)

was greater than that of the adult; the shape of both curves was hyperbolic (Fig. 9). As the pH increased to 8.7 (data not shown), the activity of both the fetal and adult enzyme preparations decreased steadily. Saturating conditions were established as 5 mM for $Mg^{++}$ and 10 mM for $K^+$ with 3 mM ADP, and 0.5 mM phosphoenolpyruvate at pH 7.0. At levels as high as 4 mM, ATP did not change the activity of fetal or adult rhesus cardiac pyruvate kinase. Similar results have been reported by others for adult rat skeletal muscle (15, 19). This lack of ATP inhibition is probably due to the relatively high $Mg^{++}$ concentration--10 mM in these experiments; Wood (49) has reported that when the $Mg^{++}$ concentration is high enough so that all the magnesium complex species are saturated, ATP inhibition disappears. The addition of 0.1 mM fructose diphosphate to our assay medium did not change the kinetics of either fetal or adult pyruvate kinase (Table 4). However, when the protein added to the assay system was below the amount necessary for linearity, i.e., less than 8 μg $ml^{-1}$, the reaction velocity with 4 mM phosphoenolpyruvate for the 100-day fetal pyruvate kinase was less

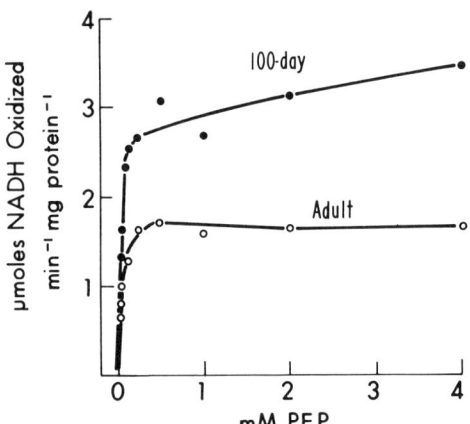

Fig. 9. Velocity curve for PK in extracts of cardiac muscle from 100-day fetal and adult rhesus monkeys. Two or more experiments were done on 2 different hearts in each series. The incubation medium was 50 mM HEPES pH 7.0 at 28°C, 10 mM $MgCl_2$, 74 mM KCl, 0.15 mM NADH, variable amounts of PEP, 3 mM ADP (saturating level) an excess of lactate dehydrogenase, and 5 mg% dialyzed BSA.

Table 4

Kinetic Data on Cardiac Pyruvate Kinase

|  | 100-Day Fetus | Adult |
|---|---|---|
| Control |  |  |
| Km µM ADP | 370 | 410 |
| Km µM PEP | 27 | 34 |
| Vmax[a] | 3.08 | 1.47 |
| 0.1 mM FDP in medium |  |  |
| Km µM PEP | 52 | 36 |
| Vmax[a] | 2.93 | 1.57 |

Cardiac extracts from a 100-day fetal and adult rhesus monkey. Experimental conditions same as Figure 9.

[a] Micromoles NADH oxidized $min^{-1}$ mg $protein^{-1}$.

than 30% of the maximum reported in Figure 9, and the addition of 0.1 mM fructose diphosphate returned the activity to the maximum control level. The apparent Km values for phosphoenolpyruvate reported by others for fetal and adult heart muscle of different species, 30-44 µM, are similar to our values (9, 16). Faulkner (9) also reported that 0.1 mM fructose diphosphate had no effect on fetal or adult muscle pyruvate kinase kinetics. Although there was no difference in our apparent Km values for fetal and adult rhesus cardiac muscle pyruvate kinase, the Vmax was higher in the fetal series. Fetal heart muscle may have a specific activator(s) for pyruvate kinase, and adult hearts may have a specific inhibitor(s). To test this possibility, we combined samples of homogenate from fetal and adult cardiac muscle in a ratio of 1.5 to 1. The activity in this mixture was proportional to the activity in the original extracts (Table 5).

Table 5

Pyruvate Kinase Activity in Mixtures of Supernatant from Homogenates of 100-Day Fetal and Adult Cardiac Muscle of the Rhesus Monkey

|  | NADH Oxidized min$^{-1}$ mg protein$^{-1}$ (μmoles) | | | |
| --- | --- | --- | --- | --- |
|  | Observed | | Calculated | |
| Control | | +0.1 mM FDP | | +0.1 mM FDP |
| Fetal | 2.62 | 2.78 | | |
| Adult | 1.52 | 1.45 | | |
| Combined | | | | |
| Fetal:adult ratio, 1.5:1 | 2.30 | 2.30 | 2.18 | 2.24 |

Experimental conditions same as Figure 9, level of PEP 2.0 mM. The PK activity was determined by the decrease in the extinction coefficient of NADH (340 nm). Protein determined by the method of Lowry et al. (26).

## DISCUSSION

Previous work on the metabolism of fetal tissues has dealt mainly with preferential substrate utilization and/ or the identification of specific enzymes and pathways. Our main objective in these experiments was to study metabolic control mechanisms in fetal cardiac muscle; that is, to determine how early in gestation the enzymes catalyzing the nonequilibrium reactions described for adult tissues, hexokinase, phosphofructokinase, and pyruvate kinase, catalyze nonequilibrium reactions in developing muscle. We found that the mass-action ratios for the glycolytic pathway in fetal and adult rhesus hearts are the same and, as has been reported for adult muscle from other species (32), hexokinase, phosphofructokinase, and pyruvate kinase are rate-limiting. Apparently these fetal nonequilibrium reactions can be identified as early as midterm, i.e., as early as we are currently able to obtain muscle. Reverse glycolysis is probably not a significant metabolic pathway in these cardiac extracts. Although some fructose diphosphatase is found in white skeletal muscle, the levels are too low to be detected in red skeletal muscle, cardiac muscle, or smooth muscle (32).

Having identified the nonequilibrium reactions in the glycolytic pathway, we began a study of these reactions in fetal cardiac muscle for possible regulatory properties (40). Phosphofructokinase is a key enzyme in the regulation of glycolysis in many adult tissues, e.g., cardiac and skeletal muscle, brain, liver, erythrocytes, and sperm. In the regulation of its activity by interaction with various compounds, phosphofructokinase is a complex enzyme. In addition to its multimolecular forms it is influenced by a multiplicity of ligands, and its kinetics are markedly influenced by pH (28, 32). Although at least four isozymes for phosphofructokinase have been identified, muscle seems to contain only one (21, 22, 46). There is no evidence in our experiments for the presence of more than one isozyme for phosphofructokinase. However, Delain et al. (8) have reported that the phosphofructokinase of the myoblast *in vitro* differs from that of the myotube. In terms of protein, overall phosphofructokinase activity increased as the fetal heart matured; in terms of percent of control activity, the sensitivity to inhibitory molecules, i.e., citrate and ATP, decreased. However, in general both fetal (midterm) and adult cardiac phosphofructokinase appear to react similarly to pH changes and effector molecules. The sensitivity of glycolytic enzymes to effector molecules is especially important since the values for maximum enzyme capacity seldom mirror their estimated activity *in vivo*. Frequently the enzyme capabilities of a given cell are used only minimally and their activity undoubtedly varies with the intracellular milieu; in erythrocytes phosphofructokinase appears to operate at about 0.5% of capacity (38).

ATP appears to be the most critical of the ligands modifying phosphofructokinase, for several reasons. ATP is present in most tissues at relatively high concentrations, and other modifiers such as inorganic phosphate appear to activate phosphofructokinase mainly by relieving ATP inhibition (28, 32). Mansour (28, 29) has reported that increasing the level of fructose-6-phosphate decreases the sensitivity to ATP of semipurified cardiac phosphofructokinase from adult guinea pig heart. When we doubled the concentration of fructose-6-phosphate in the incubation medium, the concentration of ATP necessary to inhibit the 100-day phosphofructokinase also increased; 3 mM ATP almost totally inhibited phosphofructokinase activity the amount of fructose-6-phosphate phosphorylated was tose-6-phosphate, 3 mM ATP inhibited the enzyme less than 10%.

In fetal and adult rhesus cardiac muscle, the intracellular concentration of fructose-6-phosphate was estimated to be about 0.04 to 0.09 mM and the level of ATP was greater than 3 mM, assuming uniform distribution in intracellular water. Thus at this low fructose-6-phosphate level, the ATP concentration in rhesus cardiac muscle, like that in rat heart (39) is high enough to markedly inhibit glycolysis unless there is considerable deinhibition.

We have already demonstrated that the concentration of cyclic AMP is high in fetal cardiac muscle (5). Since the level of ATP and fructose-6-phosphate at all ages ensures inhibition of phosphofructokinase, the lower phosphofructokinase activity in fetal compared to adult heart may be due to different amounts of excitatory ligands which vary with age. Our data suggest, however, that the lower phosphofructokinase activity is not due to a decrease in sensitivity of the enzyme to excitatory effector molecules. At midterm, phosphofructokinase in fetal skeletal muscle is also a nonequilibrium enzyme. However, at maximum activity the amount of fructose-6-phosphate phosphorylated was 2-3 times greater in fetal than in adult muscle, and the data suggest that this difference was not due to the presence or absence of effector molecules (2).

Citrate has long been considered a potent inhibitor of phosphofructokinase in muscle from adult animals, and inorganic phosphate was found capable of relieving this inhibition (34). This observation has been verified in many tissues (28, 36). However, to our knowledge, no one has studied the effect of citrate on fetal heart metabolism. In our study, physiological levels of citrate, 0.2 mM (11, 12) inhibited phosphofructokinase activity by heart extracts from 100-day fetal rhesus monkeys. As in adult tissues, cyclic AMP and inorganic phosphate relieved this inhibition. In view of the glucose-fatty acid cycle hypothesis originally proposed by Randle et al. (36, 37), the inhibition of glycolysis by citrate is important. According to Randle's hypothesis, the increased availability of of fatty acids and ketone bodies for oxidation in diabetes causes an increase in citric acid cycle activity, which in turn increases the citrate level in the muscle and decreases glucose uptake and utilization. Mansour and Setlow (30) comment that citrate is a ligand that appears to play a central role in the regulation of energy metabolism.

The mass-action ratio for pyruvate kinase in fetal and adult cardiac muscle indicates a reaction far removed from equilibrium. There have been far fewer studies on this reaction than on phosphofructokinase, although pyruvate kinase shares some of the regulatory properties of phosphofructokinase. The maximum activity of pyruvate kinase is so much higher in our fetal and adult cardiac muscle than that of phosphofructokinase that its activity would have to be markedly decreased to affect glycolysis (32). However, creatine phosphate, ATP, and phenylalanine inhibit pyruvate kinase, and anoxia doubles the activity in terms of flux (7, 19, 45, 48). Further work is needed to determine whether some of the regulatory effects on muscle glycolysis *in vivo* are exerted at the level of pyruvate kinase. However, pyruvate kinase cannot regulate glucose uptake; its inhibition would only cause the accumulation of glycolytic intermediates between fructose diphosphate and phosphoenolpyruvate.

The Km phosphoenolpyruvate values for pyruvate kinase reported by others (9, 15) for rat and guinea pig hearts (25 to 50 µM) were similar to the Km values for rhesus cardiac muscle in Table 4. However, the fetal Km values probably represent a composite of two or more pyruvate kinase isozymes and/or hybrids. There appear to be at least 3 basic pyruvate kinases which, upon separation, display distinctive kinetic properties (15, 16). According to Imamura and Tanaka (17) adult rat heart contains both the M1 and M2 forms of pyruvate kinase; in adult bovine cardiac muscle the M1 tetramer appears to be the predominant isozyme, with some KM hybrids (43). In his study of the electrophoretic forms of pyruvate kinase from rat tissues Susor (44) reported developmental transitions in cardiac muscle from PK4 to 3 with the appearance of electrophoretically intermediate forms--possibly hybrids--in fetal skeletal muscle. Osterman et al. (33) also reported a shift of PK4 to PK3 in both cardiac and skeletal muscle during development, with PK3 as the predominant form in adult rat cardiac muscle. Pyruvate kinase from 35- to 55-day fetal guinea pig heart was kinetically similar to adult PK3 after DEAE cellulose chromatography, although bands of activity were seen in the region of PK4 after electrophoresis (9). Electrofocusing of adult rat heart muscle extracts yielded a sharp peak at pH 7 and a smaller shoulder at 7.2 (16). PK4 or the K form is the major isozyme in the fetal hearts of rats, guinea pigs, and human beings and is a major isozyme in many adult tissues, PK3 or the M1 form is predominant in adult hearts (10, 15, 44). The human fetal isozyme pattern for cardiac muscle is comparable to that seen in rodents, which suggests a change from a predominance of PK4 to PK3 during development in both species (10). The presence of PK4 or one or more intermediate forms in our fetal heart extracts is the most probable explanation of the higher activity of fetal compared to adult pyruvate kinase.

## ACKNOWLEDGMENTS

Publication No. 852 of the Oregon Regional Primate Research Center, supported in part, by Grant RR-00163 from the National Institutes of Health. This investigation was supported by Public Health Service Research Grants HD-06069 and HD-06425 from the National Institutes of Child Health and Human Development, by General Research Support Grant RR-05694 from the National Institutes of Health, and by the Muscular Dystrophy Associations of America, Inc.

## REFERENCES

1. Atkinson, D. E. 1971. Adenine nucleotides as stochiometric coupling agents and as regulatory modifiers. In *Metabolic Pathways*, vol. 5, pp. 1-21, ed., H. J. Vogel. New York: Academic Press.

2. Beatty, C. H.; Young, M.; and Bocek, R. M. 1976. Control of glycolysis in skeletal muscle from fetal rhesus monkeys. *Ped. Res.* 10:149-153.

3. Beatty, C. H.; Bocek, R. M.; and Young, M. K. 1975. Glycolytic control mechanisms in myometrium from pregnant rhesus monkeys. *Biol. Reprod.* 12:408-414.

4. Bloxham, D. P. and Lardy, H. A. 1973. Phosphofructokinase. In *The enzymes*, vol. 8, pp. 239-278, ed., P. D. Boyer. New York: Academic Press.

5. Bocek, R. M.; Young, M.; and Beatty, C. H. 1976. Cyclic AMP in developing muscle of the rhesus monkey: effect of prostaglandin $E_2$. *Biol. Neonate* 28:92-105.

6. Bucher, T., and Pfleiderer, G. 1955. Pyruvate kinase from muscle. In *Methods in Enzymology* I, pp. 435-441. eds., S. P. Colowick and N. O. Kaplan. New York: Academic Press.

7. Carminatti, H.; Jimenez de Asua, L.; Leiderman, B.; and Rozengurt, E. 1971. Allosteric properties of skeletal muscle pyruvate kinase. *J. Biol. Chem.* 246: 7284-7288.

8. Delain, D.; Meienhofer, M. C.; Proux, D.; and Shapira, F. 1973. Studies on myogenesis *in vitro*: changes of creatine kinase, phosphorylase and phosphofructokinase isozymes. *Differentiation* 1:349-353.

9. Faulkner, A. and Jones, C. T. 1975. Pyruvate kinase isoenzymes in tissues of the developing guinea pig. *Arch. Biochem. Biophys.* 170:228-241.

10. ----. 1975. Pyruvate kinase isoenzymes in tissues of the human fetus. *FEBS Letters* 53:167-169.

11. Garland, P. B., and Randle, P. J. 1964. Regulation of glucose uptake by muscle. *Biochem. J.* 93:678-687.

12. Garland, P. B.; Randle, P. J.; and Newsholme, E. A. 1963. Citrate as an intermediary in the inhibition of phosphofructokinase in rat heart muscle by fatty acids, ketone bodies, pyruvate, diabetes, and starvation. *Nature* 200:169-170.

13. Hoskins, D. D., and Stephens, D. T. 1969. Regulatory properties of primate sperm phosphofructokinase. *Biochim. Biophys. Acta* 191:292-302.

14. Hoskins, D. D.; Stephens, D. T.; and Casillas, E. R. 1971. Enzymic control of fructolysis in the primate sperm. *Biochim. Biophys. Acta* 237:227-238.

15. Ibsen, K. H. and Trippet, P. 1973. A comparison of kinetic parameters obtained with three major noninterconvertible isozymes of rat pyruvate kinase. *Arch. Biochem. Biophys.* 156:730-744.

16. Ibsen, K. H.; Trippet, P.; and Basabe, J. 1975. Properties of rat pyruvate kinase isozymes. In *Isozymes* I, pp. 543-559, ed., C. L. Markert. New York: Academic Press.

17. Imamura, K. and Tanaka, T. 1972. Multimolecular forms of pyruvate kinase from rat and other mammalian tissues. *J. Biochem.* 71:1043-1051.

18. Johnson, R. A.; Hardman, J. G.; Broadus, A. E.; and Sutherland, E. W. 1970. Analysis of adenosine 3', 5'-monophosphate with luciferase luminescence. *Anal. Biochem.* 35:91-97.

19. Kemp, R. G. 1973. Inhibition of muscle pyruvate kinase by creatine phosphate. *J. Biol. Chem.* 248: 3963-3967.

20. Kukovetz, W. R.; Pöch, G.; and Wurm, A. 1975. Quantitative relations between cyclic AMP and contraction as affected by stimulators of adenylate cyclase and inhibitors of phosphodiesterase. In *Advances in Cyclic Nucleotide Research* 5, pp. 395-414, eds., G. I. Drummond, P. Greengard, and G. A. Robison. New York: Raven Press.

21. Kurata, N.; Matsushima, T.; and Sugimura, T. 1972. Multiple forms of phosphofructokinase in rat tissue and rat tumors. *Biochem. Biophys. Res. Commun.* 48: 473-479.

22. Layzer, R. B., and Conway, M. M. 1970. Multiple isoenzymes of human phosphofructokinase. *Biochem. Biophys. Res. Commun.* 40:1259-1265.

23. Lowry, O. H., and Lopez, J. A. 1946. The determination of inorganic phosphate in the presence of labile phosphate esters. *J. Biol. Chem.* 162:421-428.

24. Lowry, O. H., and Passonneau, J. V. 1966. Kinetic evidence for multiple binding sites on phosphofructokinase. *J. Biol. Chem.* 241:2268-2279.

25. ----. 1972. *A flexible system of enzymatic analysis.* New York: Academic Press.

26. Lowry, O. H.; Rosebrough, N. J.; Farr, A. L.; and Randall, R. J. 1951. Protein measurement with the folin phenol reagent. *J. Biol. Chem.* 193:265-275.

27. Maitra, P. K., and Estabrook, R. W. 1964. A fluoromatic method for the enzymic determination of glycolytic intermediates. *Anal. Biochem.* 7:472-484.

28. Mansour, T. E. 1972. Phosphofructokinase. In *Current topics in cellular regulation*, vol. 5, pp. 1-46, eds. B. L. Horecker and E. R. Stadtman. New York: Academic Press.

29. ————. 1963. Studies on heart phosphofructokinase: purification, inhibition, and activation. *J. Biol. Chem.* 238:2285-2292.

30. Mansour, T. E., and Setlow, B. 1972. Molecular properties of phosphofructokinase and its mechanisms of regulation. In *Biochemical regulatory mechanisms in eukaryotic cells*, pp. 58-84, eds., E. Kun and S. Grisolia. New York: Wiley Interscience.

31. Nakatsu, K., and Mansour, T. E. 1974. Reversible changes in the stability of phosphofructokinase in rat diaphragm. *Biochem. Biophys. Res. Commun.* 60:1331-1337.

32. Newsholme, E. A., and Start, C. 1973. Regulation of carbohydrate metabolism in muscle. In *Regulation in metabolism*, pp. 88-137. New York: John Wiley.

33. Osterman, J.; Fritz, P. J.; and Wuntch, T. 1973. Pyruvate kinase isozymes from rat tissues. *J. Biol. Chem.* 248:1011-1018.

34. Passonneau, J. V., and Lowry, O. H. 1963. P-fructokinase and the control of the citric acid cycle. *Biochem. Biophys. Res. Commun.* 13:372-379.

35. Pogson, C. I., and Randle, P. J. 1966. The control of rat-heart phosphofructokinase by citrate and other regulators. *Biochem. J.* 100:683-693.

36. Randle, P. J.; Denton, R. M.; and England, P. J. 1968. Citrate as a metabolic regulator in muscle and adipose tissue. In *Metabolic Roles of Citrate*, pp. 87-103, ed., T. W. Goodwin, New York: Academic Press.

37. Randle, P. J.; Garland, P. B.; Hales, C. N.; and Newsholme, E. A. 1963. The glucose fatty-acid cycle. *Lancet* i:785-789.

38. Rapoport, S. 1968. Regulation of glycolysis in mammalian erythrocytes. In *Essays in biochemistry*, vol. 4, pp. 69-102, eds., P. N. Campbell and G. D. Greville. New York: Academic Press.

39. Regan, D. M.; Young, D. A. B.; Davis, W. W.; Jack, J., Jr.; and Park, C. R. 1964. Adjustment of glycolysis to energy utilization in the perfused rat heart. *J. Biol. Chem.* 239:381-384.

40. Rolleston, F. S. 1972. A theoretical background for the use of measured concentrations of intermediates in the study of the control of intermediary metabolism. In *Current Topics in Cellular Regulation*, vol. 5, pp. 47-75, eds., B. L. Horecker and E. R. Stadtman. New York: Academic Press.

41. Staal, G. E. J.; Koster, J. F.; Banziger, C. J. M.; and van Milligen-Boersma, L. 1972. Human erythrocyte phosphofructokinase: its purification and some properties. *Biochim. Biophys. Acta* 276:113-123.

42. Stanley, P. E., and Williams, S. G. 1969. Use of the liquid scintillation spectrometer for determining adenosine triphosphate by the luciferase enzyme. *Anal. Biochem.* 29:381-392.

43. Strandholm, J. J.; Dyson, R. D.; and Cardenas, J. M. 1976. Bovine pyruvate kinase isozymes and hybrid isozymes. *Arch. Biochem. Biophys.* 173:125-131.

44. Susor, W. A. 1970. Multiple forms of pyruvate kinase during fetal development of the rat. *Fed. Proc.* 29:729.

45. Tanaka, T.; Harano, Y.; Sue, F.; and Morimura, H. 1967. Crystallization, characterization and metabolic regulation of two types of pyruvic kinase isolated from rat tissues. *J. Biochem.* 62:75-91.

46. Tanaka, T.; An, T.; and Sakaue, Y. 1971. Studies on multimolecular forms of phosphofructokinase in rat tissues. *J. Biochem.* 69:609-612.

47. Van Wagenen, G., and Catchpole, H. R. 1956. Physical growth of the rhesus monkey (*Macaca mulatta*). *Amer. J. Phys. Anthrop.* 14:245-273.

48. Williamson, J. R. 1966. Glycolytic control mechanism. II. Kinetics of intermediate changes during aerobic-anoxic transition in perfused rat heart. *J. Biol. Chem.* 241:5026-5036.

49. Wood, T. 1968. The inhibition of pyruvate kinase by ATP. *Biochem. Biophys. Res. Commun.* 31:779-785.

NOTE ON ABBREVIATIONS

The abbreviations used are: Cyclic AMP, adenosine 3': 5'-monophosphate, cAMP; isozymes of pyruvate kinase, PK1 (Type B, Type L); PK2, (Type D, Type R); PK3 (Type A, Type M1); PK4 (Type C, Type M2, Type K).

# Cardiac Output and Its Control

# 11

## Venous Return and Control of Fetal Cardiac Output

Raymond D. Gilbert
Division of Perinatal Biology
Loma Linda University
School of Medicine
Loma Linda, California

ABSTRACT

The factors that control venous return of blood to the heart were investigated utilizing a heart bypass preparation in exteriorized fetal lambs. Maximum venous return, the venous return curve, mean systemic pressure, resistance to venous return and systemic vascular compliance were measured in the control state and after: (1) increases in blood volume, (2) total spinal anesthesia, and (3) epinephrine infusion subsequent to spinal anesthesia. In the control state, the fetus maintains a high venous return (284 ± 11 ml/min/kg), predominantly through a high mean systemic pressure (15.5 ± 0.4 mm Hg), although resistance to venous return (0.069 ± 0.004 PRU·Kg) was somewhat lower than adult levels. Systemic vascular compliance (1.72 ± 0.12 ml/mm Hg) was low and probably helped support the high mean systemic pressure. Increases in blood volume caused mean systemic pressure to increase and as a result maximal venous return increased. Resistance to venous return was unchanged with the increase in volume. Total spinal anesthesia resulted in a fall in mean systemic pressure although it still remained high (10.9 ± 0.4 mm Hg) relative to adults. The lowering of mean systemic pressure coupled with an unchanged resistance to venous return caused maximal venous return to fall. The fall in mean systemic pressure could be accounted for by an increase in systemic vascular compliance. Epinephrine infusion sufficient to return arterial blood pressure to prespinal block levels returned all parameters to their control levels.

## INTRODUCTION

The fetal circulation, a high-flow, low-resistance system, functions mainly to deliver nutrients, water, and oxygen to the fetal tissues and remove wastes and carbon dioxide. Maintaining a high cardiac output is important not only for the delivery and removal of these substances but also for the support of a high fetal oxygen consumption, which may be 1.5 to 2 times that of an adult per body weight. Yet the regulation of fetal cardiac output remains poorly understood.

The circulatory system contains two major components, the heart and the peripheral vasculature, each playing a role in determining cardiac output. Both the output function of the heart and the venous return function of the peripheral vasculature must be known to understand control of cardiac output. In 1914, Patterson and Starling (10) showed that the heart increases its output with increases in right atrial pressure. This relationship, the cardiac function curve, may be shifted by changes in heart rate (14), or myocardial damage (17). Regulation of cardiac output partially depends on these changing states of cardiac function.

The same increase in right atrial pressure that causes the heart to increase its output impedes the flow of blood back to the heart from the peripheral vasculature. If right atrial pressure is raised high enough, venous return will even cease (Fig. 1, Point C). In adult dogs, 7 mm Hg constitutes the pressure needed to stop venous return (6). In the fetus, this pressure has not been determined. The relationship between venous return and right

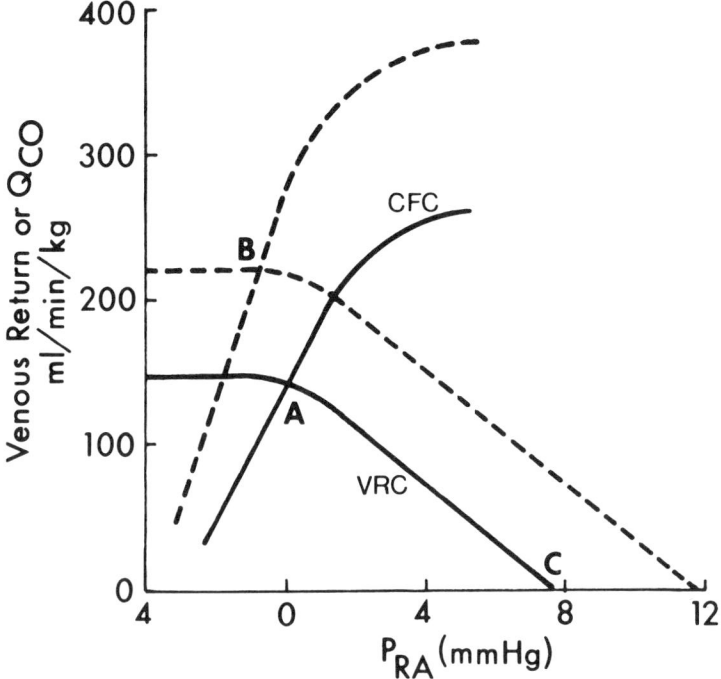

Fig. 1. Theoretical venous return curve (VRC) and cardiac function curve (CFC) for normal adults (solid lines). Dashed lines indicate the shift in both curves, which occurs with epinephrine infusion.

atrial pressure is termed the venous return function curve. Also partially responsible for determining cardiac output, the venous return function curve may be changed by various factors, including changes in sympathetic tone or circulating catecholamines.

The right atrial pressure which causes venous return to stop, corresponding to Point C on Figure 1, is called mean systemic pressure. Although it can be obtained from the venous return curve, mean systemic pressure is usually measured after stopping the circulation, allowing aortic and venous pressures to equilibrate. Being defined as the static pressure in the circulatory system, mean systemic pressure is influenced by the compliance of blood vessels and the amount of blood filling them. Mean systemic pressure constitutes the upstream driving pressure for venous return to the heart and normally venous return can be expressed as the difference between mean systemic and right atrial pressures divided by resistance to venous return.

Figure 1 shows how the two function curves interact to control cardiac output. Point A represents the resultant cardiac output and right atrial pressure in an animal with normal cardiac and venous return function curves (solid lines). Sympathetic stimulation shifts the cardiac and venous return function curves (dotted lines) resulting in a new cardiac output and right atrial pressure (Point B). If sympathetic stimulation had simply shifted the cardiac function curve, right atrial pressure would have decreased but cardiac output would have remained relatively unchanged. A shift in the venous return curve alone would have increased cardiac output but at the expense of a higher right atrial pressure. The shift of both curves allows cardiac output to increase while right atrial pressure remains normal.

Therefore, both the heart and peripheral vasculature determine cardiac output. Virtually no information is available pertaining to the role of venous return in regulating fetal cardiac output. Therefore, these studies characterize the factors that affect venous return to the heart, which in turn affect cardiac output in the fetus. The variables measured to characterize the fetal circulation in this manner were the venous return curve, mean systemic pressure, resistance to venous return and total vascular compliance.

## METHODS

The desired measurements required the use of a heart-lung bypass preparation. For that reason, all experiments were performed on exteriorized, fetal sheep. A total of 26 fetal sheep between 135 to 145 days gestation were used. Each fetus was delivered and placed on a warmed table next to the mother and wrapped in plastic to prevent heat loss. Body temperature was maintained at $38^{\circ}C$. The umbilical cord was left untouched and wrapped in cotton soaked with warm saline.

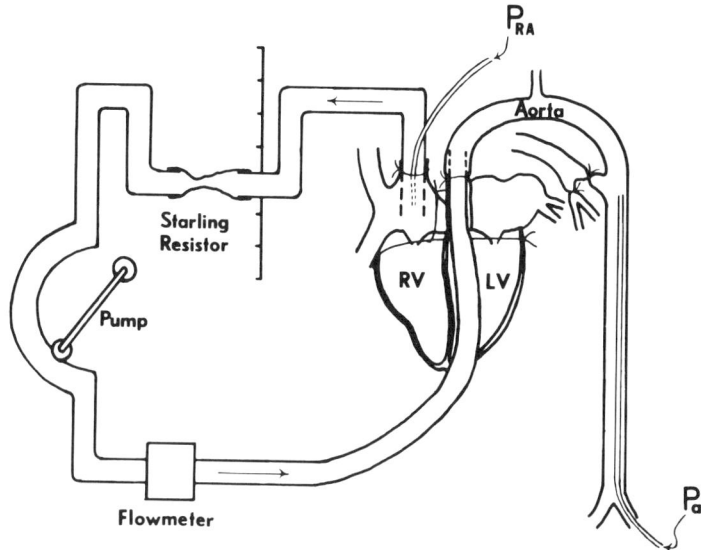

Fig. 2. Schematic of the heart bypass circuit used in the fetal animals ($P_a$ = aortic pressure, $P_{Ra}$ = right atrial pressure, RV = right ventricle, LV = left ventricle).

Figure 2 illustrates the heart-lung bypass preparation. A cannula placed in the right atrium collected all of the blood returning from the inferior and superior vena cava. Blood flowed through a Starling resistor (Penrose tubing) to an occlusive roller pump, which pumped the blood through a flow probe back into the ascending aorta. When the speed of the pump was sufficiently high to collapse the Starling resistor, right atrial pressure (PRA) could be controlled by adjusting the hydrostatic level of the Starling resistor relative to the right atrium. Under these conditions, the amount of blood flowing through the bypass circuit was dependent only on the amount of blood returning from the peripheral vasculature and was independent of the speed of the pump. The heart and lungs were isolated by ties placed around the ascending aorta, the pulmonary artery and ductus arteriosus, the atrio-ventricular junction of the heart, and around the right atrium to close the foramen ovale. Cannulae placed in the right atrium and femoral artery provided sites for pressure measurements and blood sampling. Since the heart and lungs were eliminated by the bypass circuit, all measurements reflect characteristics of the systemic vasculature alone. Thus, the bypass circuit permitted the measurement of total venous return at various right atrial pressures as well as the measurement of mean systemic pressure and total vascular compliance.

Venous return curves were obtained by slowly raising right atrial pressure from some low value, usually -10 mm Hg, until venous return ceased. This procedure took approximately 35 seconds. Venous return curves were plotted as flow ($\dot{Q}_{VR}$), measured by the flow probe in the bypass circuit, versus right atrial pressure. In six fetuses, venous return curves were also obtained using a different method for comparison. This method, reported by Guyton (6), involved rapidly elevating right atrial pressure to a predetermined level and holding it there until flow and pressures had equilibrated. When this procedure was repeated for a sufficient number of different right atrial pressures, the venous return could again be plotted as flow versus right atrial pressure. Using this method, 3-6 minutes were needed to plot the venous return curve.

Resistance to venous return was calculated from the slopes of the venous return curves. The downsloping portion of each curve (right atrial pressure $\geq$ 0) was approximated with a straight line by least-squares linear-regression analysis. The inverse of the slope of that line provided the value for resistance to venous return.

Mean systemic pressure was measured simply by stopping the pump and simultaneously clamping the tubing leading from the right atrium to exclude the Starling resistor. Having a large compliance, the Starling resistor would have introduced an error in the measurement of the fetus' mean systemic pressure. While the flow was stopped, aortic pressure ($P_a$) and right atrial pressure equilibrated (Fig. 3). The equilibrium value was taken as mean systemic pressure. The two pressures equilibrated withing 3-4 seconds and remained equal and constant until flow was restored. Flow was usually stopped for approximately 10 seconds. All pressures were referenced to the atrio-ventricular valve in the fetal right atrium.

Total vascular compliance was measured statically while the pump was stopped for the mean systemic pressure measurement. After aortic and right atrial pressures had equilibrated, a known volume of blood (usually 40 ml) was rapidly infused and the change in right atrial pressure recorded. Compliance was calculated as the change in volume divided by the change in right atrial pressure. A second method of obtaining compliance, reported by Shoukas and Sagawa (16), was also used in 10 fetuses. This method measures compliance while flow is still occurring and thus is a dynamic measure. The speed of the pump in the bypass circuit was reduced so that the Starling resistor was not collapsed. Under these conditions, venous return was constant. A known volume of blood was then infused and the change in right atrial pressure recorded. Compliance was again calculated as the change in volume divided by the change in the right atrial pressure.

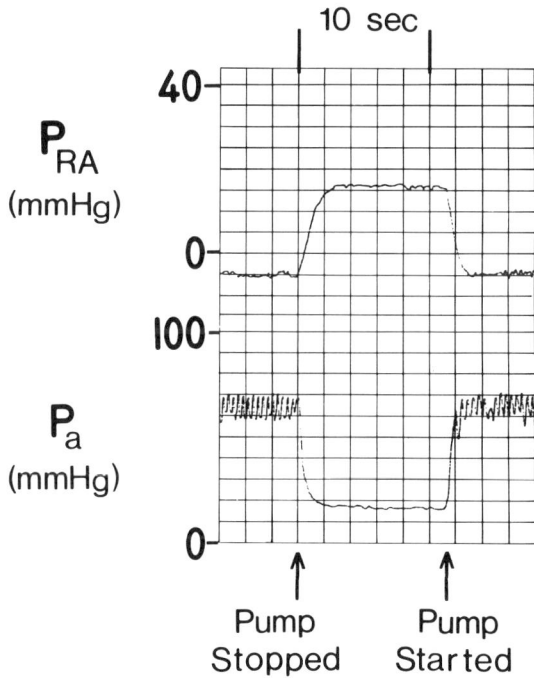

Fig. 3. Actual record of $P_{ms}$ measurement by simultaneously stopping the pump and clamping the outflow tuping of the bypass circuit ($P_{Ra}$ = right atrial pressure, $P_a$ = aortic pressure).

Once control measurements of each parameter were made, several procedures were performed to test their effect on venous return and factors controlling it. In each case, both the control and test procedure were performed in the same animal so that results could be analyzed using paired analysis.

In six fetuses, the effects of increasing fetal blood volume on venous return were investigated. After control measurements were made of the venous return curve, mean systemic pressure and resistance to venous return, 50 ml maternal blood was injected into the fetus and measurements made again within one minute after the injection to minimize loss of fluid out of the vascular compartment. Estimating total fetal blood volume as 134.7 ml/kg (1) and assuming the heart and lungs to contain no more than 15% of the fetal blood volume, the addition of 50 ml of blood produced an average 11.1% increase in fetal systemic blood volume.

In eight fetuses, sympathetic tone to the systemic vasculature was eliminated by total spinal anesthesia. Spinal anesthesia was accomplished by infiltrating the

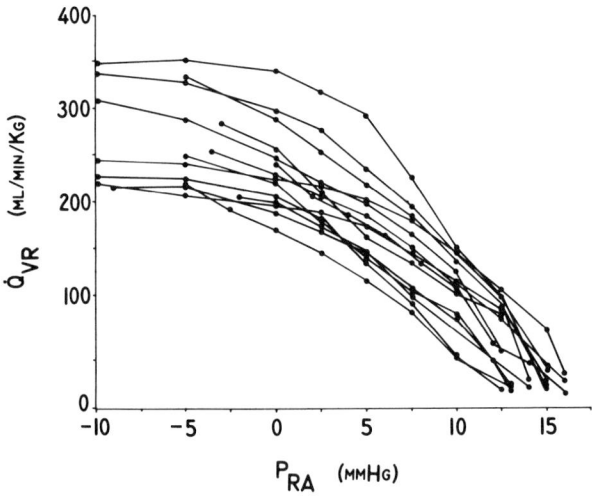

Fig. 4. Normal venous return curves of fourteen fetuses ($P_{RA}$ = right atrial pressure, $\dot{Q}_{VR}$ = venous return normalized to fetal body weight).

spinal column with 20 mg pontocaine in 7 ml isotonic saline. Measurements of venous return, mean systemic pressure, resistance to venous return, and compliance were taken before and after the spinal block. Subsequent to those measurements, epinephrine was infused at a rate sufficient to return arterial blood pressure to pre-spinal block levels and the same measurements again repeated. The rates of epinephrine infusion varied between 0.005 and 0.015 mg/kg/min.

## RESULTS

Control venous return curves for fourteen fetuses are shown in Figure 4. Values for venous return have all been normalized to fetal body weight. Figure 5 compares the venous return curves obtained by two different methods. For this comparison, the downsloping portion (right atrial pressure $\geq$ 0 mm Hg) of each venous return curve was fitted with a straight line and then the mean regression line obtained for each group. Statistically, there was no sig-

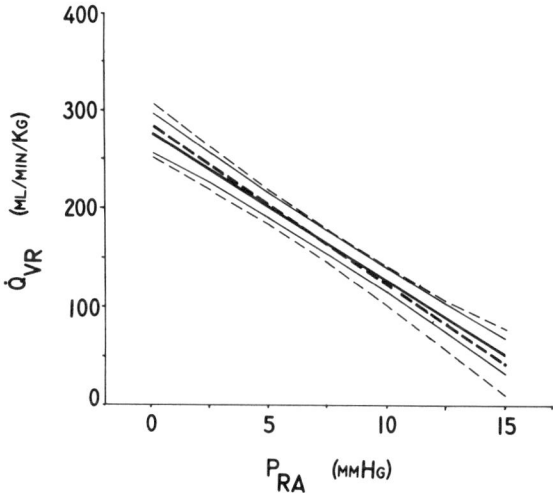

Fig. 5. Comparison of venous return curves obtained by slowly raising $P_{RA}$ (——) and by raising $P_{RA}$ to various static levels (----). Shown are the mean regression lines and 95% confidence limits for the downsloping portion of six curves in each group.

Table 1

Normal Values for Determinants of Venous Return in the Fetus and Adult

| | Fetus | Adult Value | Adult Ref. |
|---|---|---|---|
| Venous return ($\dot{Q}_{vr}$) (ml/min/kg) | 284 ± 11* | 115–123 | (2, 9)[a] |
| Mean systemic pressure ($P_{ms}$) (mm Hg) | 15.5 ± 0.4 | 6.9 | (5, 6) |
| Resistance to venous return ($R_{vr}$) (PRU·kg) | 0.069 ± 0.004 | 0.105 | (7) |
| Total vascular compliance (C) (ml/mm Hg/kg) | 1.72 ± 0.12 | 2.5–3.0 | (3, 16) |

[a] Reference for adult values.

\* Standard error; n = 26 for all fetal values

nificant difference between either the slopes or intercepts of the two groups. This figure demonstrates that venous return curves generated by slowly raising right atrial pressure are the same as those generated by raising right atrial pressure to several static levels and allowing time for equilibration.

The fetuses in this study exhibited a high maximum venous return after they were placed on the bypass circuit, compared to the adult (Table 1). Mean systemic pressure, measured by temporarily stopping flow and allowing pressures throughout the vasculature to equilibrate, was also high (Table 1). Thus, these fetuses had a high upstream pressure driving blood back towards the heart. Blood gas and pH values measured during control conditions were: $PO_2 = 21.9 \pm 0.8$ mm Hg, $PCO_2 = 44.1 \pm 1.5$ mm Hg, and pH = $7.30 \pm 0.01$.

To calculate resistance to venous return, the downsloping portion of each venous return curve (right atrial pressure $\geq 0$) was fitted with a straight line. Correlation coefficients from least squares linear regression were 0.97 or above for all of the venous return curves. The inverse of the slope of each line provided the value for resistance to venous return, which for all animals averaged $0.069 \pm 0.004$ PRU·Kg.

Control total vascular compliance, measured in 16 of the fetuses by the static method, was $1.72 \pm 0.12$ ml/mm Hg/kg. In 11 fetuses, total vascular compliance was measured dynamically while venous return was constant and averaged $1.71 \pm 0.07$ ml/mm Hg/kg, this value not being different from that obtained by the static method.

Changing fetal blood volume resulted in changes in the parameters controlling venous return. With an 11% increase in blood volume, maximum venous return increased by 48 ml/min/kg and mean systemic pressure increased by 4.2 mm Hg. As shown in Figure 6, the venous return curve was shifted up and to the right with the increased volume; however, the slope was unchanged indicating resistance to venous return did not change.

The effects of total spinal anesthesia on the factors affecting venous return are shown in Figures 7-9 for eight animals. Mean arterial blood pressure ($P_a$) and venous return decreased significantly (Figure 7) during the block;

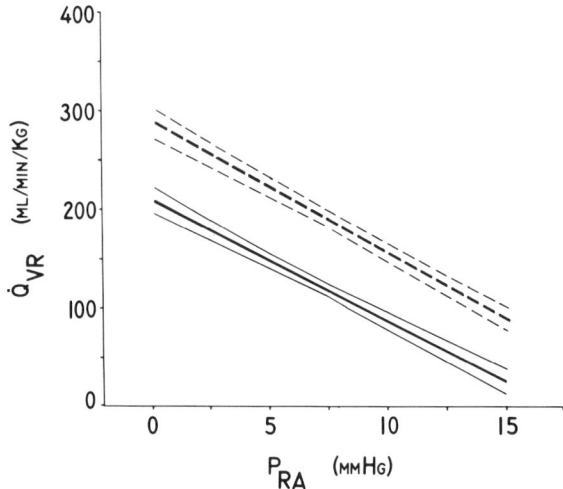

Fig. 6. Comparison of venous return curves from six fetuses before (———) and after (----) an increase in blood volume. Shown are the mean regression lines and 95% confidence limits.

however, an infusion of epinephrine, at a rate sufficient to return mean arterial blood pressure to control levels, returned venous return to control levels. Mean systemic pressure (Fig. 8) also decreased significantly during the spinal anesthesia although resistance to venous return did not change. Mean systemic pressure returned to control levels with epinephrine; however, a sufficient number of determinations of resistance to venous return were not made during epinephrine infusion for statistical comparison. Total vascular compliance (Fig. 9) increased significantly with spinal anesthesia and decreased with epinephrine infusion.

## DISCUSSION

The venous return curves obtained from these fetuses (Fig. 4) appear similar to those from adult animals (6). Below a right atrial pressure of about 0 mm Hg, venous return is relatively independent of right atrial pressure. Above 0 mm Hg, venous return decreases as right atrial pressure increases, a relationship that can be approximated with a straight line (Fig. 5). Figure 5 also indicates there was no difference in the venous return curves obtained either by raising right atrial pressure slowly or by the method reported by Guyton (6). The fact that the downsloping part of the venous return curve could be approximated well with a straight line and the comparison to Guyton's methods, devised to minimize the effects of reflexes on the measurement, both indicate that reflex changes in the cardiovascular system were probably not occurring during the measurement of the venous return curves.

Table 1 indicates that these fetuses had a high venous return (284 ± 11 ml/min/kg). This value is 2 to 3 times the normal cardiac output for an adult sheep although it is lower than values reported for cardiac output in chronic, awake lamb fetuses (2, 4, 8, 9, 12, 13). Table 1 also gives some insight into how the fetus is able to maintain a high venous return. Mean systemic pressure, which is the upstream driving pressure for venous return, is approximately twice the value measured for adult animals (5). Resistance to venous return is less than that report-

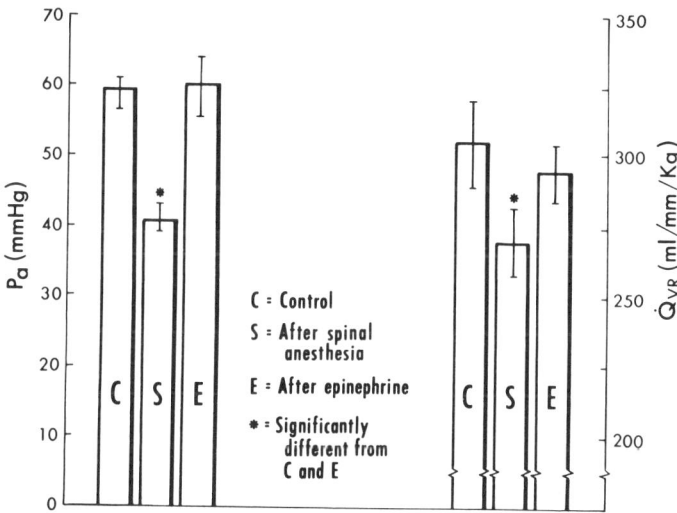

Fig. 7. Effects of spinal anesthesia and epinephrine infusion on aortic pressure ($P_a$) and venous return ($\dot{Q}_{VR}$) in fetuses.

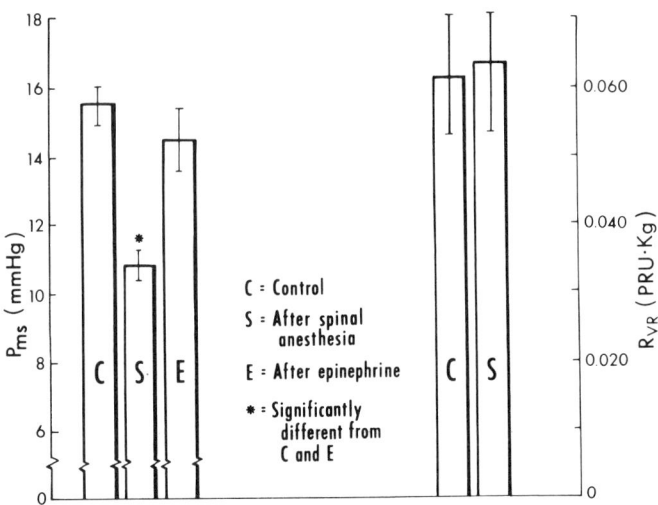

Fig. 8. Effects of spinal anesthesia and epinephrine infusion on mean systemic pressure ($P_{ms}$) and resistance to venous return ($R_{VR}$) in eight fetuses.

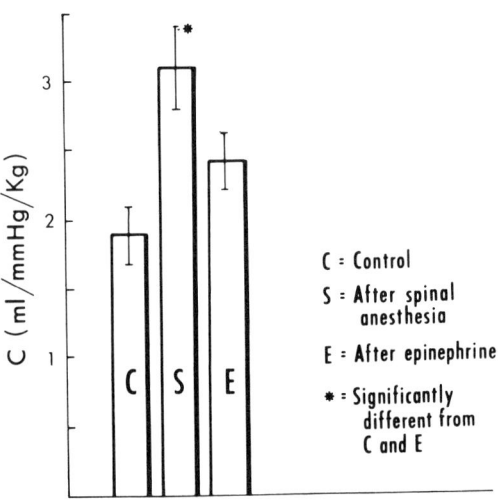

Fig. 9. Effects of spinal anesthesia and epinephrine infusion on vascular compliance ($C$).

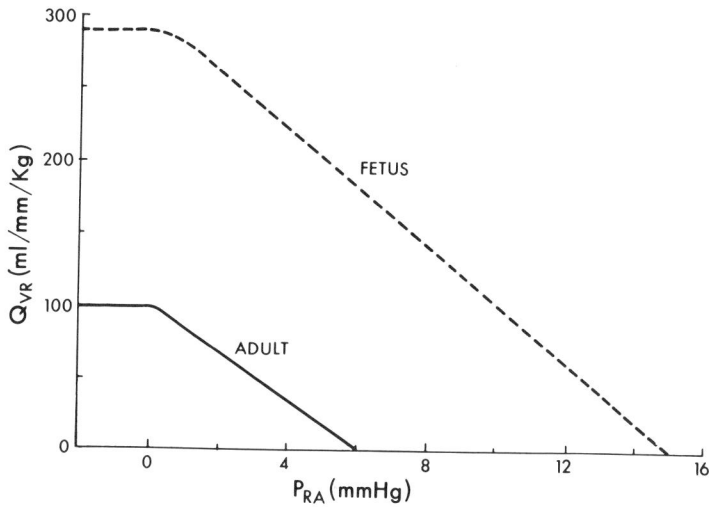

Fig. 10. Idealized venous return curves for the fetus and adult.

ed for the adult (7). The combination of a higher upstream driving pressure and lower resistance to venous return should result in a higher venous return.

This may be seen more clearly in Figure 10 which depicts idealized venous return curves of the fetus and adult. Mean systemic pressure is delineated by the intersection of each of the venous return curves with the abscissa. Mean systemic pressure in the fetus is higher and therefore its venous return curve is shifted to the right. Also, the slope of the fetal venous return curve is steeper than that of the adult, indicating resistance to venous return is less. The fact that the fetus has a higher mean systemic pressure coupled with lower resistance to venous return allows the fetal vasculature to return more blood to the heart and thus allows the fetus to have a higher cardiac output at a normal right atrial pressure of 0 mm Hg.

Venous return could be increased or decreased in these fetal animals by either increasing blood volume or eliminating sympathetic tone to the vasculature. Increasing blood volume (Fig. 6) increased maximum venous return and also increased mean systemic pressure. Resistance to venous return did not change. If the fetal heart were able to handle this increased load, then cardiac output should increase as a result of the increase in maximum venous return. Total spinal anesthesia caused maximum venous return to fall (Fig. 7). The mechanism for the decrease in venous return was a fall in mean systemic pressure with no change in resistance to venous return (Fig. 8). The fall in maximum venous return with spinal anesthesia would probably result in a fall in cardiac output.

In both of these instances, the change in venous return was controlled by a change in mean systemic pressure, while resistance to venous return remained constant. Since mean systemic pressure is the static equilibrium pressure of the systemic vasculature, it is directly proportional to distending blood volume and indirectly proportional to vascular compliance ($P_{ms} = V_D/C$). Thus, an increase in blood volume would increase mean systemic pressure and cause venous return to rise; a reduction in sympathetic tone would cause total vascular compliance to increase, resulting in a fall in mean systemic pressure and a fall in venous return.

The control systemic vascular compliance of 1.72 ml/mm Hg/kg found in these fetuses compares to values of 2.5 to 3.0 ml/mm Hg/kg reported for the adult (3, 16). This low compliance probably contributes to the high control fetal mean systemic pressure. However, the spinal anesthesia experiments indicate indirectly that the fetus may also have a higher distending blood volume. With total spinal anesthesia, mean systemic pressure fell (Fig. 8), but only to about 11 mm Hg, still rather high compared to the adult. Compliance, on the other hand, rose with spinal anesthesia (Fig. 9) to a level equal to the adult. This would indicate that not only is there tone in the fetal vessels that is maintaining a high mean systemic pressure, but that there is a relatively high distending blood volume in the fetal vasculature which, even though compliance may be the same as the adults' or even just a little greater, the mean systemic pressure can be maintained high because the vessels are filled to a greater degree than they are in the adult.

In conclusion, these studies indicate the fetus has a high upstream driving pressure forcing blood back to the heart. This high driving pressure or mean systemic pressure coupled with a relatively low resistance to venous return causes there to be a high venous return and as a result, the fetus is able to sustain a high cardiac output. The high mean systemic pressure driving blood back to the heart probably results from both a low vascular compliance and a high distending blood volume. In both experimental perturbations presented here, venous return is regulated by changes in mean systemic pressure while resistance to venous return remains unchanged.

## REFERENCES

1. Creasy, R. K.; Drost, M.; Green, M. V.; and Morris, J. A. 1970. Determination of fetal, placental and neonatal blood volumes in the sheep. *Circ. Res.* 27: 487-494.

2. Cross, K. W.; Dawes, G. S.; and Mott, J. C. 1959. Anoxia, oxygen consumption and cardiac output in newborn lambs and adult sheep. *J. Physiol.* 146:316-343.

3. Drees, J. A., and Rothe, C. F. 1974. Reflex venoconstriction and capacity vessel pressure-volume relationships in the dog. *Circ. Res.* 29:360-373.

4. Faber, J. J., and Green, T. J. 1972. Foetal placental blood flow in the lamb. *J. Physiol.* 223:375-393.

5. Guyton, A. C.; Polizo, D.; and Armstrong, G. 1954. Mean circulatory filling pressure measured immediately after cessation of heart pumping. *Am. J. Physiol.* 179:261-267.

6. Guyton, A. C.; Lindsey, A. W.; Abernathy, J. B.; and Richardson, T. 1957. Venous return at various right atrial pressures and the normal venous return curve. *Am. J. Physiol.* 189:609-615.

7. Guyton, A. C.; Lindsey, A. W.; Abernathy, B.; and Langston, J. B. 1958a. Mechanism of the increased venous return and cardiac output caused by epinephrine. *Am. J. Physiol.* 192:126-130.

8. Heymann, M. A.; Creasy, R. K.; and Rudolph, A. M. 1973. Quantitation of blood flow patterns in the foetal lamb in utero. In *Foetal and neonatal physiology*, pp. 129-135. Cambridge: Cambridge Univ. Press.

9. Metcalfe, J. and Parer, J. T. 1966. Cardiovascular changes during pregnancy in ewes. *Am. J. Physiol.* 210:821-825.

10. Patterson, S. W. and Starling, E. H. 1914. On the mechanical factors which determine the output of the ventricles. *J. Physiol.* 48:357-379.

11. Peterson, L. H. 1950. Some characteristics of certain reflexes which modify the circulation in man. *Circulation* 2:351-362.

12. Rudolph, A. M., and Heymann, M. 1967. The circulation of the fetus in utero. Methods for studying distribution of blood flow, cardiac output and organ blood flow. *Circ. Res.* 21:163-184.

13. ----. 1970. Circulatory changes during growth in the fetal lamb. *Circ. Res.* 26:289-299.

14. Rushmer, R. F. and Smith, D. A., Jr. 1959. Cardiac control. *Physiol. Rev.* 39:41-68.

15. Sarnoff, S. J. 1955. Myocardiac contractility as described by ventricular function curves. *Physiol. Rev.* 35:107-122.

16. Shoukas, A. A. and Sagawa, K. 1971. Total systemic vascular compliance measured as incremental volume-pressure ratio. *Circ. Res.* 28:277-289.

17. Stone, H. L.; Bishop, V. S.; and Guyton, A. C. 1963. Cardiac function after embolization of coronaries with microspheres. *Am. J. Physiol.* 204:16-20.

# 12

## Water Transfer Across the Placenta: Hydrostatic and Osmotic Forces and the Control of Fetal Cardiac Output

Gordon G. Power, Phillip J. Roos, and
Lawrence D. Longo

Division of Perinatal Biology
Departments of Physiology and Gynecology
Loma Linda University
School of Medicine
Loma Linda, California

The near-term fetus grows rapidly and gains an appreciable quantity of water. During the course of human gestation, about 3.5 liters of water accumulate in the fetus, the accompanying amniotic fluid, and placental tissue. Water accounts for approximately 80% of the intrauterine contents by weight. Thus, not only must a source of water be available to the fetus, but some mechanism must control the rate of water gain so that it matches changing fetal needs during gestation.

Only two sources of water are available to a growing fetus: the water in maternal blood supplying the uterus, and the fetus' own water of metabolism. While both sources are of interest, metabolic water is quantitatively less important. Barcroft (3), for example, calculated that the fetal sheep at 115 days' gestation produces about 78 g of new tissue per day. Because three-fourths of the new tissue is water, the fetus requires about 60 ml of water daily. During the same period, the fetus consumes about 15 l of oxygen. If this $O_2$ is used mainly to metabolize glucose, about 12 g of water are produced by catabolism. This equals about 20% of the total fetal requirement. Therefore, a large part of the daily fetal water requirement must be met by transfer from the mother.

As far as is currently known, water movement across the placental membranes is a passive process which follows the same hydrostatic and osmotic forces that characterize water transfer throughout the body. The combined effect of these forces is summarized in Starling's law of capillary water exchange. Figure 1 depicts this in a form adapted to placental water exchange. This straightforward relation predicts the rate of water transfer as the product of a placental filtration coefficient and the net hydrostatic and osmotic pressure differences between maternal and fetal blood. Both in vivo (3, 15, 27) and in vitro (1, 7) studies with isolated placental membranes indicate that placental water transfer follows these simple principles. In other words, no evidence has been uncovered for

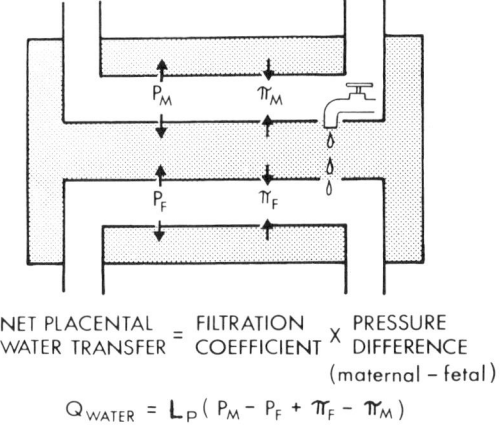

NET PLACENTAL = FILTRATION × PRESSURE
WATER TRANSFER   COEFFICIENT   DIFFERENCE
(maternal – fetal)

$$Q_{WATER} = L_p ( P_M - P_F + \pi_F - \pi_M )$$

Fig. 1. Some combinations of hydrostatic ($P_M$ and $P_F$) and osmotic ($\pi_M$ and $\pi_F$) pressure gradients are thought to explain fetal water gain throughout gestation. Starling's law relates the rate of water transfer ($Q_{water}$) to the product of the placental filtration coefficient ($L_p$) and the overall pressure difference.

*Placental Water Transfer* 319

an energy-dependent active transfer of water. In view of these findings, it seems necessary to conclude that hydrostatic or osmotic forces, singularly or in combination, must cause maternal-to-fetal water movement as the fetus grows. Short term exceptions may occur, but over the long term, the net forces must favor fetal water acquisition.

If this line of thought is accepted, however, a dilemma arises because investigators have been unable to measure hydrostatic and osmotic pressure differences that favor fetal accumulation of water. Hydrostatic forces, for instance, have been measured by several workers and reviewed by Seeds (31), who showed that while results differ among investigators and for different species, they generally indicate slightly higher hydrostatic pressures in the fetal placental exchange vessels than in maternal exchange vessels. According to the work of Reynolds and co-workers in primates (28) and more recently Moll and Künzel in guinea pigs (25), maternal intervillous lake pressure, on the order of 10 mm Hg, is less than that of fetal vessels. In sheep, hydrostatic pressures in maternal and fetal vessels are more nearly comparable, but even in that species, most estimates suggest that hydrostatic forces may cause the fetus to slowly lose water to the mother; that is to say, water moves in the opposite direction to that required for growth.

Osmotic forces could quantitatively be of far greater significance and easily counterbalance hydrostatic forces. This is true because one milliosmole is equivalent to 19.6 mm Hg at 37°C. However, several groups report osmotic pressures of maternal and fetal blood plasmas virtually identical, provided precautions are taken to avoid hypoxia (known to raise osmotic pressure (4), and to compare samples of equal $CO_2$ contents. Meschia, Battaglia, and Barron (24), for example, have shown in sheep and goats that the osmotic pressure as measured by freezing-point depression (i.e., total solute concentration) of fetal plasma is equal to or less than that of maternal plasma. Meschia (21) also has measured a higher colloid osmotic pressure in maternal sheep and goat plasma throughout the last half of pregnancy.

In humans, the colloidal osmotic pressures of fetal plasma is equal to or slightly lower than that of maternal plasma (5, 11, 20). Hinkley and Blechner (14) calculated the mean difference of pressure favoring maternal plasma to equal 51 mm of $H_2O$. On the other hand, Scoggin et al.

(29) found human fetal blood to be in osmotic equilibrium across the placental barrier with maternal blood in the placental sinus. We also have compared maternal and fetal plasmas drawn from mothers and their infants during cesarean sections. Using a specially constructed differential osmometer and cellulose acetate membranes impermeable even to small ions, we found maternal total osmotic pressure very nearly equal to fetal total pressure (the difference of 0.14 ± 0.02 SEM mOsm, equivalent to 2.6 ± 0.5 mm Hg, favoring the mother). These results are based on an average of 3 determinations in each of 8 patients. Carbon dioxide levels were identical in maternal and fetal samples for these measurements.

There is, then, no apparent difference in total solute or colloid concentration across the placenta to explain the intrauterine accumulation of water during pregnancy. The mechanism of fetal water accumulation without measurable hydrostatic or osmotic gradients has remained an enigma.

In an effort to resolve this problem, a number of investigators are searching systematically for a substance that might create and maintain small osmotic gradients. This substance would presumably generate gradients too small to detect with present methods, but ones which, over the course of 24 hours, could account for the 30 to 50 ml of water the term fetus accumulates daily. Faber and co-workers have proposed that amino acids and glucose could draw water osmotically into the fetus. Other workers (Effros, personal communication) propose that potassium or other solutes might be actively accumulated intracellularly as the fetus grows. Then as the number of cells increased during growth, water would be drawn passively into the cells and, in turn, from the fetal circulation and thence from the mother.

Another hypothesis was advanced by Longo and Power (18), implicating $CO_2$ and bicarbonate in the exchange vessels of the placenta as the substances creating osmotic gradients for water movement. Because the fetus continuously forms $CO_2$ and since this ultimately is eliminated through the maternal lungs, it follows that mean fetal levels must exceed mean maternal levels in the placental exchange vessels. Carbon dioxide is carried in the blood largely as bicarbonate ion. Both bicarbonate ion and probably $CO_2$ are also osmotically active across placental membranes (see below). Fetal placental mean osmotic pressure must therefore exceed maternal placental mean osmotic pressure, although presumably by the end of the capillary, exchange may have progressed to the point of virtual equilibrium. Thus, although an investigator would measure the same osmotic pressures in venous blood samples, the osmotic pressure need not necessarily have been equal along the entire capillary length.

The essence of our theory would require the following chain of events (Fig. 2): As the fetus grows, additional $CO_2$ is produced which is transported in fetal blood largely as bicarbonate ion. Along the length of the placental exchange vessels, bicarbonate ion exerts an osmotic force that gradually diminishes as it converts to $CO_2$, in which form it crosses the placental membranes and enters the maternal circulation. Because mean fetal levels must exceed maternal levels for excretion to take place, the overall $CO_2$ osmotic force would tend to move water into the fetus. Fetal water space and blood volume would expand

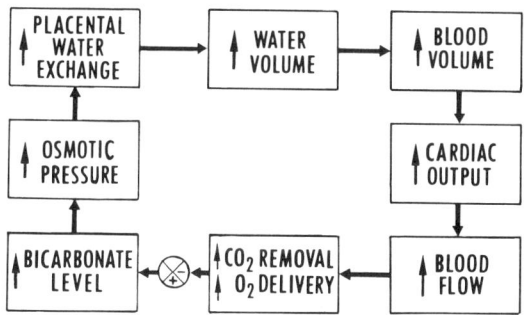

Fig. 2. $CO_2$ hypothesis to explain water gain during fetal growth. The following sequence of events is proposed: ↑fetal $O_2$ consumption, ↑$CO_2$ production and ↑venous $CO_2$ and bicarbonate content; ↑fetal total osmotic pressure; ↑ rate of maternal to fetal water movement across the placenta; ↑fetal water and blood volume, ↑mean systemic pressure; ↑cardiac output; ↑tissue blood flow; ↓veno-arterial $CO_2$ content difference; and hence restored venous $CO_2$ content. This hypothesis would link metabolic demands and tissue blood flow on a physical basis.

and fetal cardiac output would increase (Starling's law of the heart). As a result, tissue blood flow would increase and the arteriovenous difference for $CO_2$ would diminish. This would tend to restore the $CO_2$ level towards normal and complete a negative feedback loop whereby fetal $CO_2$ production indirectly would maintain fetal cardiac output in line with fetal needs.

We designed the present studies to investigate several aspects of placental water exchange and the $CO_2$ hypothesis. First, they were designed to measure water movement in response to hydrostatic forces. Second, they were designed to measure water flux in response to osmotic forces caused by adding or removing carbon dioxide to fetal blood perfusing the placenta. Although water flux across the placenta in response to osmotic changes has been well studied in humans (5), other primates (6, 8), rabbits (10), rats (2, 32), and other species, the present studies were designed to quantitate the volume of water moving per unit $CO_2$ difference between the maternal and fetal placental exchange vessels. We used an isolated perfused cotyledon in order to control and measure the osmotic and hydrostatic pressures and the $CO_2$ levels as closely as possible in both arteries and veins. This permitted a reasonable estimate to be made of capillary values. (Elsewhere in this volume studies from our group, report fluid shifts among various body compartments and the changes in fetal cardiac output caused when water is added to the fetal circulation.) Third, these studies were designed to establish the relation of $CO_2$ levels to osmotic pressure in maternal and fetal blood. Finally, we compared the importance of $CO_2$, glucose, amino acids, and other factors that move water across the placenta, using a mathematical model of placental water exchange.

## EXPERIMENTAL PROCEDURES

Pregnant sheep in the last month of gestation were given spinal anesthesia (tetracaine, 4 mg) and barbiturate sedation (pentobarbital, 5-10 mg/kg, I.V., repeated as required). Maternal femoral arterial and venous catheters were inserted in order to monitor hydrostatic pressures and to collect blood samples. The abdomen and uterus were opened through a midline incision and the uterine wall opposite the incision was partially invaginated so as to expose one of the attached cotyledons. We catheterized (PE

90) inflowing and outflowing fetal vessels to the cotyledon and tied off collateral vessels to surrounding membranes. Then we perfused the cotyledon with whole, heparinized filtered maternal blood at a rate of 1.6 ml/min from a syringe pump. Inflow and outflow pressures were monitored (Statham transducers P23Db) and recorded (Brush polygraph). All pressures were referenced to the height of the perfused cotyledon and no corrections were made for amniotic fluid pressure since the uterus was not tightly sealed. The uterine wall and abdominal incision were loosely approximated, leaving the isolated cotyledon and fetus *in situ*.

The outflow drainage was passed through a fraction collector. Timed 1-min collections of outflowing blood were made into tared test tubes. At the conclusion of the experiment, blood was collected directly from the pump (without passing through the cotyledon) as a calibration procedure. The weight difference between the experimental and the calibration collections was assumed to be the weight of water added to or removed from the blood as it passed through the cotyledon.

Fetal inflow pressure could be varied by changing the rate of perfusion. The height of the outflow drainage line could be changed so as to vary outflow pressure. All pressures were corrected for the resistance of the tubing leading to and from the placenta.

After fetal blood samples were weighed, those samples collected during the last 3 min of an experimental interval were pooled. Fetal and maternal samples were analyzed for total and colloid osmotic pressure using a freezing-point depression apparatus (Advanced), in triplicate, and a membrane osmometer with cutoff at about 10,000 m.w., respectively.

## RESULTS

### Effect of Varying Fetal Arterial and Venous Pressures on Net Water Transfer

To study transplacental water movement in response to hydrostatic forces, we perfused the umbilical artery of a sheep cotyledon at constant flow rate with maternal blood and varied inflow pressure from 30 to 150 mm Hg and outflow pressure from 0 to 40 mm Hg. Since the same maternal blood flowed through both sides of the placenta, we could avoid water movements caused primarily by osmotic forces. We collected and weighed outflowing blood samples and estimated water movement by weight changes during passage through the placenta.

Figure 3 illustrates the entire procedure for one experiment. Initially, there was a control period with inflow and outflow pressures near their normal values, during which time only a little water was transferred. Then we raised umbilical venous (outflow) pressure to 45 mm Hg. After a 5-min delay for approach to a new steady state, blood collections were repeated. Water transfer rate was calculated to have increased to 0.4 ml/min from fetus to mother. This returned essentially to zero during the next

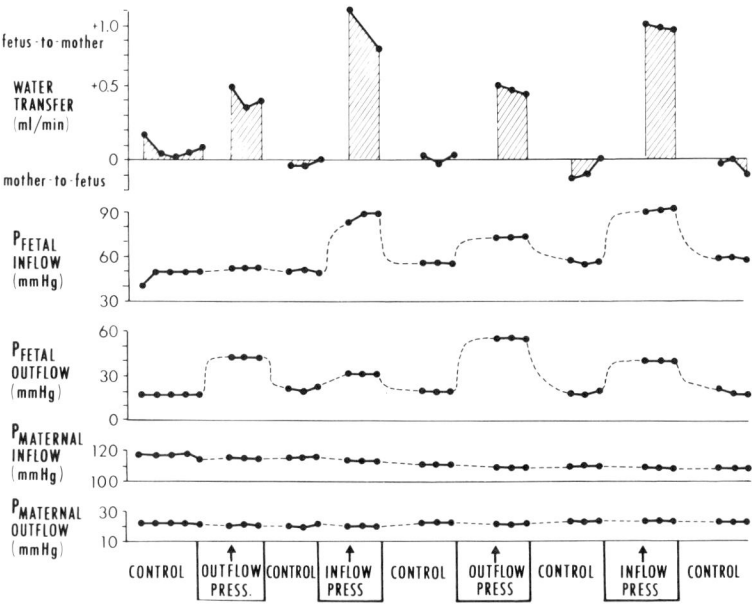

Fig. 3. Experimental procedure to study transplacental water flux across a single isolated cotyledon of a sheep placenta perfused *in situ*. Water moved briskly from fetus to mother when either fetal inflow or outflow pressures were elevated and returned to zero when pressures were lowered to control range.

control period. Next, fetal placental inflow pressure was raised to about 90 mm Hg (venous pressure being noted to rise concomitantly to about 30 mm Hg). Water moved briskly from fetus to mother at about 0.9 ml/min. This procedure was followed by another control period, and then the entire cycle was repeated, pressure perturbations always being bracketed with control periods.

To summarize the results and to compare different preparations, we expressed water movement per unit pressure difference between maternal and fetal capillaries and per mass of placental tissue. We estimated maternal and fetal pressures in the exchange areas on the basis of measured arterial and venous pressures, using the procedure described by Pappenheimer and Soto-Rivera (26). We assumed arterial-to-venous resistance ratios of 2 for maternal vessels and 1 for fetal vessels (ratios selected by computer fitting of the present data to minimize the sum of squares of deviations from Starling's law of water exchange). That is, we assumed that mean capillary pressure lay two-thirds of the way between inflow and outflow pressures on the maternal side and midway on the fetal side, and then plotted water transfer versus maternal-fetal capillary pressure difference. Figure 4 shows 64 average results based on 354 collections in 4 sheep. We estimated a filtration coefficient from the slope of the line of best fit for each cotyledon. We found the placental filtration coefficient ($L_p$) averaged 0.059 (± 0.015 SEM) ml/min/100 grams placental tissue/mm Hg maternal-to-fetal capillary pressure difference for the individual cotyledons, an estimate considered better than that from the pooled data. Problems in its interpretations will be discussed below.

The value of 0.059 for the placental filtration coefficient may be compared with filtration coefficients measured for different organs in various species in Table 1, modified from Guyton, Taylor, and Granger (12), where many original references are conveniently listed. The intact cotyledon is relatively freely permeable to water, a finding that accords with many studies of placental membranes. For example, Seeds (30) reports that the osmotic permeability for term human chorion laeve is over 100-fold greater than tracer-water diffusion permeabilities when studied in the same *in vitro* chamber. This non-diffusional flow is typically characterized as bulk streaming through membrane pores.

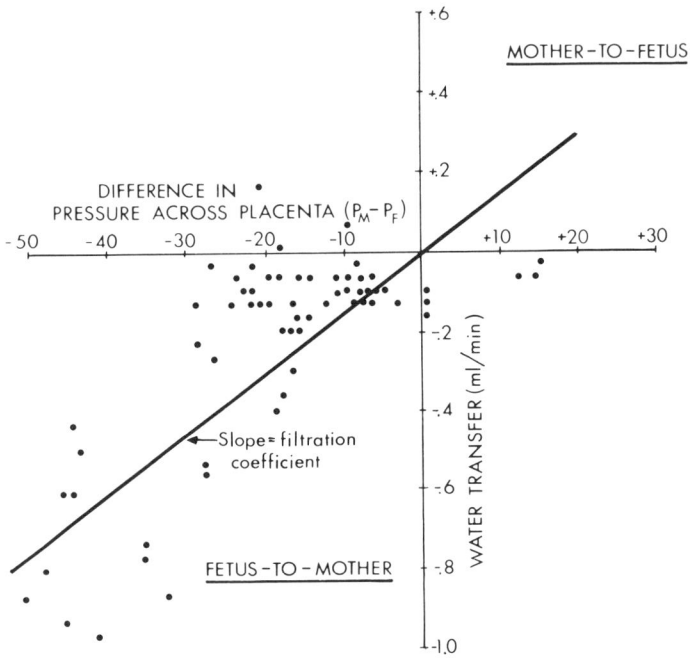

Fig. 4. Summary results shown as 64 average water flux determinations, based on 354 collections in 4 cotyledons. The slope is an index of the placental filtration coefficient.

Several problems arise in the measurement of the filtration coefficient and presentation of a single value for overall function: (1) A tendency was noted for calculated filtration coefficients to decrease when venous pressure (and to a lesser extent arterial pressure) was elevated. This has been demonstrated in other tissues. In isolated perfused intestine, for example, the filtration coefficient decreases five-fold as venous pressure is elevated from 5 to 30 mm Hg (see Ref. 12 for further references). (2) The sheep placenta is actually a combination of several membranes in series. The filtration coefficient measured in a multimembrane system may not represent a single membrane function but, instead, a complex function of the different membranes, as discussed by Curran and

Table 1

Coefficients for Various Capillary Beds[a]

| Tissue | Filtration Coefficient (ml/min x 100 g x mm Hg) |
|---|---|
| Muscle | |
| Cat | 0.0105 |
| Dog | 0.0104 |
| Rat | 0.0330 |
| Forearm of man | 0.0057 |
| Lung | |
| Dog | 0.070 |
|  | 0.260 |
| Sheep | 0.030 |
| Intestine | |
| Cat | 0.08-0.44[b] |
|  | 0.09-0.40[b] |
| Dog | 0.08-0.44 |
| Placenta[c] | |
| Sheep | 0.059 |

[a] Modified from Guyton, Taylor, and Granger (12), whose work lists individual references.

[b] Variation with venous pressure.

[c] Present study.

MacIntosh (9). (3) Very likely, a larger value for the filtration coefficient would have been calculated if water fluxes could have been measured before any diluting or concentrating effects of water movements tended to retard the further movement of water. This "osmotic buffering" no doubt occurred in the present experiments and complicates an interpretation of steady-rate results of water flux. Osmotic buffering is thought to cause underestimates of the filtration coefficient in other vascular beds of up to 20%. (4) The measured filtration coefficient probably increases several-fold throughout gestation (13) and present results take no account of this change. (5) Finally, the measured filtration coefficient is a composite result for an entire cotyledon. Because variations in maternal-to-fetal blood flow ratios probably markedly affect water transfer and these ratios may differ widely in different locations within a single cotyledon (Power and Nelson, unpublished), the observed value may not accurately reflect membrane function in any given location (Wilbur, Power, and Longo, submitted for publication).

On several occasions, we performed control experiments to test for possible deterioration in the cotyledon's function with time. Twelve to 18 collections were carried out at one-minute intervals and water flux calculated for each interval. The filtration coefficient for the cotyledon remained constant or diminished no more than 10 to 20%, indicating that the permeability of the placental membranes was changing only slightly with time. Furthermore, since in some instances the total amount of water exchange approached the total mass of the cotyledon, it was apparent that the water was exchanged between maternal and fetal capillaries and not merely between fetal capillaries and the surrounding placental tissues.

These limitations present a serious barrier in attempts to understand basic membrane function. Nonetheless, the measured filtration coefficient is useful because it indicates the changes that would result in fetal water volume as a consequence of changing vascular hydrostatic pressures. Said another way, it indicates how much water the placenta would actually transfer per min, given the vascular pressures, initial osmotic equilibrium, and other conditions during these experiments. It is an operational index of placental permeability and capillary surface area.

In summary, these initial experiments confirm the work of many investigators. They suggest the placenta is freely permeable to water, possibly because of bulk streaming through membrane pores. A minimal estimate for the filtration coefficient is 0.059 ml/min/100 g placental tissue/mm Hg capillary gradient. Regardless of the mechanism of water movement, however, and irrespective of the precise numerical value of the filtration coefficient, these results have not indicated why the fetus acquires water from the mother.

## $CO_2$ and Osmotic Pressure

Up to this point, we have considered the role of hydrostatic forces in transplacental water movement. Before we turn our attention to osmotic forces, it is first useful to examine the osmotic activity of $CO_2$ and bicarbonate ion *in vitro*. Inasmuch as osmotic pressure is a colligative property of a solution, it would be expected to depend on the number of particles present rather than on their type. Thus, osmotic pressure would be predicted to relate to the sum of bicarbonate ions and physically dissolved $CO_2$ molecules. Variations in $CO_2$ in the form of carbamino compounds would not alter osmotic pressure appreciably because the total number of particles in solution remains unchanged when $CO_2$ combines directly with hemoglobin. Nor are changes in unbuffered hydrogen ions and undissociated carbonic acid large enough to be significant.

*In vitro*, there is little question that $CO_2$ exerts an osmotic force just as do other solutes. Margaria (19) first noted that blood osmotic pressure was proportional to its $CO_2$ content. Meschia and Barron (22, 23) demonstrated that plasma osmotic pressure increased 0.9 mOsm/kg water per mM $CO_2$/kg·water added to whole blood; and that during the course of transit through pulmonary vessels with loss of $CO_2$ and uptake of $O_2$, plasma osmotic pressure decreased 1.2 mOsm/kg water. We also determined the relation of osmotic pressure to $CO_2$ for whole, heparinized human maternal and fetal blood equilibrated with various gas mixtures. In blood from near-term human newborns and their mothers, osmotic pressure, $\pi$(in mOsm), in both maternal and fetal blood, was found to be

$$\pi = 270.5 \ (\pm 0.97 \text{ SEM}) + 1.23 \ (\pm 0.11) \ ([HCO_3^-] + [CO_2]) \quad (1)$$

where $[HCO_3^-]$ and $[CO_2]$ are the millimolar concentrations of bicarbonate ion and dissolved $CO_2$, respectively.

Nor is there much question that bicarbonate exerts an osmotic force across the membranes of a functioning sheep placenta (17). Its reflection coefficient, $\sigma$, is relatively high and according to our estimate is about 0.54 for human placental membranes. But some would question the assertion that dissolved molecular $CO_2$ can exert an osmotic force *in vivo*. This is predicted, nonetheless, on the basis of a new set of equations recently derived by Wilbur, Power, and Longo (submitted for publication) for water fluxes across membranes. These equations are analogous to those derived by Kedem and Katchalsky (16) for total volume flux. They relate transmembrane water movement to a modified reflection coefficient, $\sigma'$, which differs from $\sigma$ by a term that is essentially the ratio of the rates at which solute and solvent penetrate the membrane. $\sigma'$ for $CO_2$, in fact, is close to 1.0, and the gas is predicted to exert practically its full potential osmotic force. Overall, this study suggests that significant transplacental water flow may be caused by lipid soluble, freely penetrating molecules, such as $CO_2$, and that there is theoretical justification to expect water movement in response to $CO_2$ differences between maternal and fetal blood.

### Effect of Varying Fetal $CO_2$, Bicarbonate, and Osmotic Pressure on Net Water Transfer

In these studies we varied osmotic pressure of the perfusing blood by changing $CO_2$ levels and measured water exchange across a single cotyledon of the sheep placenta. We administered spinal anesthesia to near-term pregnant ewes (4 mg tetracaine hydrochloride) and performed a tracheostomy under local anesthesia. The ewe was allowed to breathe spontaneously a gas mixture containing sufficient oxygen to maintain arterial oxygen tension of 100 to 150 torr. Next, we withdrew 200 ml of maternal blood and used it to perfuse the umbilical circulation after equilibrating it was a gas mixture containing either a high (18% $CO_2$, 4.5% $O_2$, balance $N_2$) or a negligible (4.5% $O_2$, balance $N_2$) carbon dioxide content.

We prepared a placental cotyledon for perfusion as described above. Following a 3-min period for the preparation to stabilize, timed collections of 10-min duration were begun. We perfused the cotyledon alternately with blood of high and blood of low $CO_2$ content. Upon completion of a 10-min perfusion, outflow volume was measured and samples were drawn from the maternal carotid artery and from the inflow and outflow vessels of the cotyledon. We determined $P_{O_2}$, $P_{CO_2}$, and pH on each sample with microelectrodes (Radiometer, Model BMS3), and calculated $HCO_3^-$ with the Henderson-Hasselbalch equation. The osmotic pressure of each sample was determined in duplicate by freezing-point depression (Precision Systems Osmette osmometer, Model 5002).

We performed a total of 31 paired perfusions (both high and low $P_{CO_2}$) in 9 experimental animals. High $P_{CO_2}$ in the inflowing blood, associated with increased [$HCO_3$] and osmotic pressure, resulted in an increased volume of blood from the outflow catheter. This increase in volume indicated a transplacental flux of water from the maternal-to-fetal circulation. Figure 5 shows the levels of $P_{CO_2}$

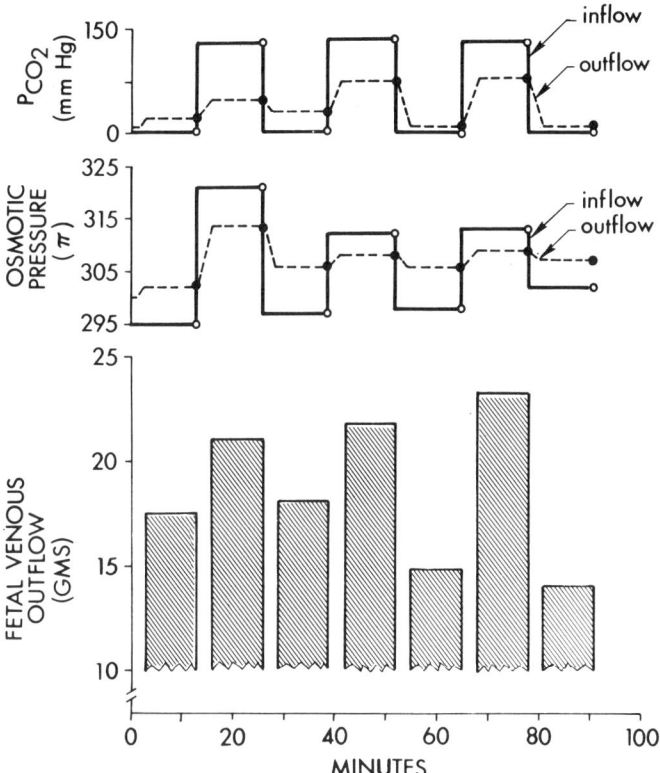

Fig. 5. Placental water transfer during alternating perfusions with blood of high and low $CO_2$ content in a typical experiment. $P_{CO_2}$ and osmotic pressure are shown for the inflowing blood (solid line) and outflowing blood (interrupted line) during successive periods of perfusion. Open and closed circles indicate points at which blood samples were taken. Since inflow rate was constant, variations in venous outflow reflect water added to or removed from blood while passing through the placenta.

and osmotic pressure in inflowing blood (solid lines) and outflowing blood (interrupted lines) during successive periods of perfusion in one experiment. Alternating changes in outflow volume are indicated. Similar changes were noted in the other 8 animals studied.

Figure 6 summarizes the results for all animals. Average water transfer rate increased 1.8 ml/min/100 g placental tissue in the maternal-to-fetal direction when fetal osmotic pressure was raised from 304 to 325 milliosmoles. Water transfer rate, accordingly, averaged 0.086 ml/min/

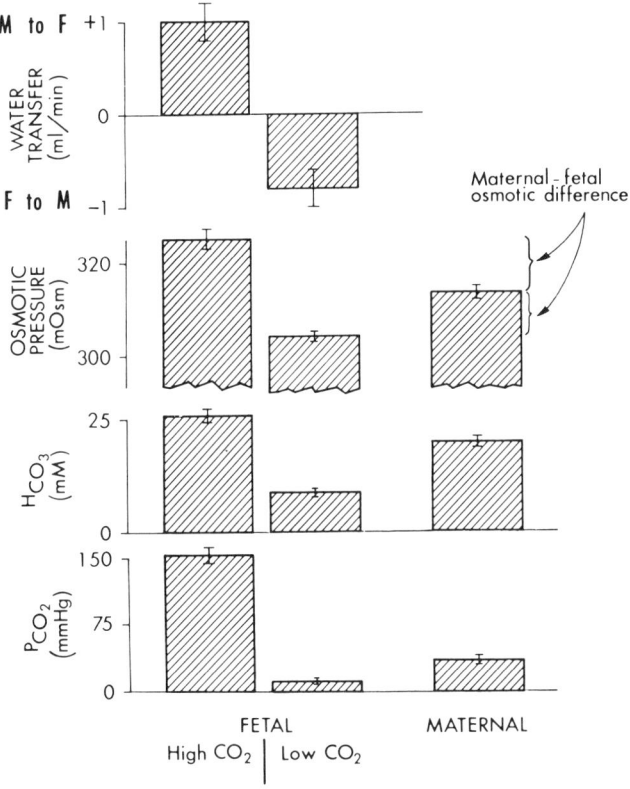

Fig. 6. Summary of placental water transfer during perfusion with blood of high and low $CO_2$ content. Results are averages of 31 comparisons in 9 sheep. Water transfer is given as ml/min/100 g placental tissue.

100 g placental tissue/milliosmole difference between maternal and fetal arterial blood. This is equivalent to 0.0044/ml/min/100 g tissue/mm Hg pressure difference. In terms of $CO_2$, water transfer averaged 0.013 ml/min/100 g/torr $P_{CO_2}$ difference.

These values relate water flux to initial arterial differences. They indicate, for example, that a fetal-to-maternal arterial difference of only 0.08 milliosmole (undetectable with present methods) would over the course of a day provide the 50 or so milliliters of water the near-term lamb needs for growth. Said another way, a sustained rise in $CO_2$ of only 0.53 torr would supply fetal water requirements.

These results indicate that water moves freely across placental membranes in response to $CO_2$-induced osmotic gradients, but several qualifying statements are appropriate. First, water transfer has been related to initial arterial differences rather than to mean capillary differences, although the latter would be preferred in assessing membrane function. Although the mean capillary gradient cannot be measured experimentally, it is no doubt far smaller than the arterial gradient and depends on how rapidly water exchanges during the capillary transit. If water flux were related to the capillary gradient, placental permeability would appear far larger (we estimate from mathematical modeling at least an order of magnitude difference). The placental membranes would also appear more permeable if water flux could be measured before any diluting or concentrating effects take place which in turn would tend to retard further movement.

And finally, the greater part of the $CO_2$-induced water flux may have resulted from its hydrostatic effect rather than from its osmotic effect. When we perfused the cotyledon with blood with high $CO_2$ content rather than low $CO_2$ content, the perfusion pressure fell to 90 (±4) mm Hg from 143 (±10) mm Hg. We estimated that this change lowered fetal mean capillary pressure about 26 mm Hg, assuming an arterial/venous resistance ratio of 1. Using the earlier reported filtration coefficient of 0.059 ml/min/100 g/mm Hg pressure gradient, we further estimated this decrement to have caused 85% of the observed water movement. Thus, $CO_2$-induced hydrostatic effects supplement the $CO_2$-induced osmotic effects and can be more important quantitatively, at least when studied over the wide range of $P_{CO_2}$ examined in these experiments. That the hydrostatic effects of $CO_2$ can so greatly augment its osmotic effects is important to an understanding of total placental function. However, the physiologic consequences for water balance and the fetal cardiovascular system are similar regardless of whether $CO_2$ moves water by hydrostatic or by osmotic effects. The fact is that it can do so in relatively large quantities. The result substantiates one of the critical links in the $CO_2$ hypothesis presented earlier.

### Relative Importance of Various Factors for Placental Water Transfer

In an effort to better understand the forces affecting human placental water exchange, Wilbur, Power, and Longo (submitted for publication) recently developed a mathematical model of the process. It consists of a system of differential equations describing the flows of water, glucose, bicarbonate ion, amino acids, $CO_2$, $O_2$, passive cations ($Na^+$, $K^+$, etc.), and chloride ion transplacentally and across the erythrocytes in maternal and fetal placental exchange vessels. The equations are based largely on the flux equations of irreversible thermodynamics, although Goldman's hypothesis and equation are employed to treat the ionic currents across the erythrocyte membrane. Figure 7 shows the predicted influence of different factors on placental water exchange at different locations along the length of the exchange vessels. Bicarbonate ion and dissolved carbon dioxide were indicated as likely major forces acting early during a capillary transit to produce a large flow of water toward the fetus. Near the end of the capillary an almost equal amount of water would return to the mother, due predominantly to the effect of glucose. Amino acids would tend to move water towards the fetus throughout the length of the exchange vessels. The mean concentration gradients, weighted by their reflection coefficients, for the individual factors that affect transplacental water exchange are indicated on the right hand side of Figure 7. Overall, the importance of these various factors would be determined by the percent of their contribution to water flux. Under the steady-state conditions assumed for the model, glucose would contribute about 38% of the overall effective osmotic force; bicarbonate ion and dissolved carbon dioxide, about 31%; and amino acids, 21%. Passively diffusing ions, including sodium, potassium, and chloride and additional plasma solutes taken

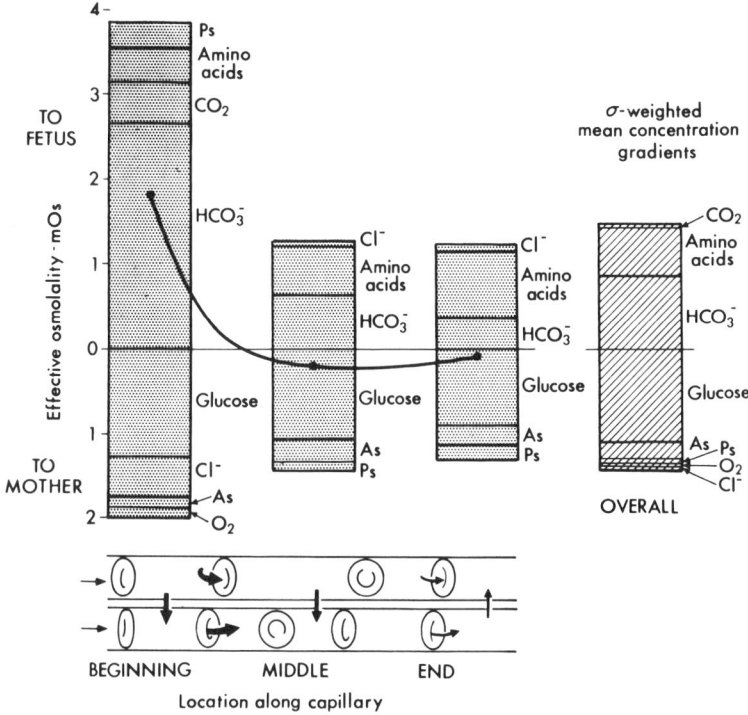

Fig. 7. Transplacental osmotic forces at the beginning, middle, and end of human exchange vessels predicted by the mathematical model. The gradients of the different components are shown. The gradient created by plasma proteins is of the order of 0.03 mOsm toward the mother and is too small to include in the diagram. Gradients due to $CO_2$ and $O_2$ rapidly disappear and are negligible by the middle of the capillary. Changes observed in the gradients of the passively diffusing cations ($P_s$) and additional plasma solutes ($A_s$) are due almost exclusively to water fluxes across the red cells shown at the bottom of the figure. The bar at the right depicts the contribution of the various factors to the overall osmotic force during a single capillary transit, although this may vary under different circumstances, e.g., blood flow rates, and so forth.

together, would contribute about 10% of the effective osmotic force. The importance of each factor would vary, of course, depending upon the blood flow rate and the time available for the many different processes to proceed toward equilibrium. In the conditions assumed in the model, the effect of glucose moving water towards the mother would be just slightly more than offset by the osmotic force of bicarbonate and amino acids moving water towards the fetus, the net result being sufficient to explain the small daily water gain by the fetus.

The model also made possible predictions for changes in water flux induced by small deviations from accepted normal values, such as might occur if the fetus were to grow. Using the model, we varied each factor independently to assess its effects alone, without compensatory changes occurring in the rest of the system. Water transfer was found to be most sensitive to changes in sodium, potassium, and chloride ions, followed by bicarbonate ion and $CO_2$, plasma solutes (including lactate, glucose, and amino acids) and hydrostatic pressure. Membrane characteristics, including the permeabilities and reflection coefficients of glucose, bicarbonate ion, and $CO_2$ and amino acids, pointed generally toward a lesser effect.

The water filtration coefficient, $L_p$, was of special interest. The rate of net water transfer would rise slowly from zero as the filtration coefficient increased, approaching normal transfer at about 20% of the standard $L_p$ value, and reach a maximum at about 90% of the standard value. With further rises in $L_p$, net water transfer increasingly would favor the mother, an effect based on the assumption that if $L_p$ were small, water transport would not be sufficient to significantly affect osmotic gradients, and water transport would then depend on the average gradient along the capillary, thus favoring the fetus. In the capillary, water movement would be roughly proportional to $L_p$, thereby accounting for the slow, almost linear rise in the early part of the curve. On the other hand, for larger values of $L_p$, net water transport would depend principally on the end-capillary distribution of osmotically active particles, a process favoring the mother.

These results indicate only the sensitivity of water exchange to a given factor; they do not indicate the range over which that factor might vary physiologically or the changes it might induce elsewhere in the system to augment

## Placental Water Transfer

or suppress water transfer. It is the product of sensitivity and percentage variation that would establish the importance of a given factor for placental water movement. All this leads to an important deduction that can be made from the modeling studies. Since in normal circumstances, glucose, $CO_2$, and amino acids have comparable effects in moving water and since placental water transfer is about equally sensitive to changes in these three constituents, one might at first thought conclude that increasing fetal utilization of glucose (or amino acids) would offset the carbon dioxide produced. Such is clearly not the case, however, since for every mole of glucose the fetus completely metabolizes, six moles of bicarbonate and $CO_2$ will be formed. Thus, metabolism effectively raises total osmotic pressure, and the net effect is a movement of water towards the fetus. Recalling the colligative character of osmotic force and the fact that the reflection coefficients of $CO_2$ and bicarbonate are relatively large and comparable to metabolic substrates, we believe this 6 to 1 ratio requires that $CO_2$ be several-fold more important than other constituents in causing osmotic changes during growth.

If $CO_2$ does indeed have a pivotal role in water balance, then there are also important implications in the regulation of the fetal cardiovascular system. As the fetus grew and produced additional quantities of $CO_2$, a chain of events would follow that would tend to restore $CO_2$ levels to baseline. Additional $CO_2$ production would lead to a rise in venous $CO_2$ levels, $HCO_3^-$ concentration, and fetal osmotic pressure. These changes, in turn, would lead to an increased rate of maternal-to-fetal water movement across the placenta, followed in order by an increase in fetal blood volume, fetal mean systemic pressure, and fetal cardiac output. This increased flow of blood to the tissues would lower veno-arterial $CO_2$ content differences and hence restore venous $CO_2$ content. Coincidentally, this increased flow of blood would also provide additional oxygen, glucose, amino acids, and other nutrients to fetal tissues. This would set the stage for the next increment of growth. Thus, $CO_2$ would link fetal cardiac output to growth.

This chain of events would not require any new physiologic mechanisms. It is based on a number of established mechanisms, including water exchange under osmotic and hydrostatic pressure gradients, the sensitive dependence of cardiac output on circulating blood volume, and the relation of osmotic pressure to $CO_2$ levels in the blood.

The role of $CO_2$ provides a chemomechanical explanation of the increase in fetal blood flow that answers the metabolic demands of growing tissue. It can do this early in gestation before cardiovascular reflex controls are active.

Finally, the dilemma concerning fetal water gain without demonstrable osmotic or hydrostatic differences can be resolved. The seeming paradox grew up because some early investigators compared maternal and fetal osmotic pressures using the freezing-point methods, in which small plasma samples were exposed momentarily to air.

To test the extent to which $CO_2$ diminishes during this exposure and thereby leads to falsely low osmotic pressure readings, the following experiment was done. Plasma was obtained by centrifugation from freshly drawn, heparinized human blood and divided into two parts, one of which was equilibrated with 2% $CO_2$ and the other with 8% $CO_2$, balance air. $P_{CO_2}$ and pH were measured and bicarbonate ion calculated by the Henderson-Hasselbalch relation for each sample. The results are shown in Table 2 as initial readings. Total solute osmotic pressure was then measured by freezing-point depression (Advanced osmometer, 0.3 ml samples) with no special efforts made to prevent $CO_2$ from being lost, such as might normally occur. After each freezing-point determination, the plasma sample was immediately reclaimed from the apparatus without further exposure to air, and $P_{CO_2}$, pH, and bicarbonate ion were measured again. Table 2 also gives the post-freezing-point values, showing that $CO_2$ and bicarbonate of both samples fell, as might be expected and that, more importantly, the fall was greater in the sample with the initially higher $CO_2$ level. The $CO_2$ difference of 32.3 mm Hg initially diminished to 5.2 mm Hg between post-freezing point samples. The bicarbonate difference decreased from 5.3 to 2.3 mM. The post-freezing-point values probably truly reflect the values prevailing when $\pi$ is measured. Thus, an investigator must exercise caution not to compare $\pi$ in maternal (low $CO_2$) and fetal (high $CO_2$) samples in which $CO_2$ have fallen to low and similar levels.

As a further test, a plasma sample equilibrated with 8% $CO_2$ was analyzed repeatedly for osmotic pressure. Readings, in order, were: 301, 300, 297, 295, 294, and 291. The sample was reclaimed and final readings of $P_{CO_2}$ = 4.8 mm Hg, pH = 8.04, and $HCO_3^-$ = 12.5 mM obtained. Clearly the sample had progressively lost $CO_2$. In this instance an

Table 2

Changes in Measured Total Solute Osmotic Pressure ($\pi$)
Caused by $CO_2$ Loss During Freezing-Point Determinations

| Reading | Low $CO_2$ Blood | | | | High $CO_2$ Blood | | | |
|---|---|---|---|---|---|---|---|---|
| | $P_{CO_2}$ (torr) | pH | $HCO_3^-$ (mM/L) | $\pi$ (mOsm) | $P_{CO_2}$ (torr) | pH | $HCO_3^-$ (mM/L) | $\pi$ (mOsm) |
| Initial | 22.5 | 7.56 | 19.6 | 297±1* | 54.8 | 7.28 | 24.9 | 302±1[a] |
| Post-freezing point | 17.0 | 7.69 | 19.7 | | 22.2 | 7.62 | 22.0 | 301 |
| | | | | | | | | 300 |
| | | | | | | | | 297 |
| | | | | | | | | 295 |
| | | | | | | | | 294 |
| Post-repeated analyses | | | | | 4.8 | 8.04 | 12.5 | 291 |

[a] Mean and SEM, n = 5.

investigator might have concluded that $\pi$ in fetal (high CO2) plasma was greater than, equal to, or even less than maternal plasma, depending upon the extent of $CO_2$ loss from the sample. When plasma was merely laid aside undisturbed in a beaker, $P_{CO_2}$ and $\pi$ also fell, although the drop was not as rapid.

We think the resolution of the apparent paradox of fetal water accumulation without demonstrable osmotic gradients is as follows: (1) nearly identical, total (crystalloid), osmotic pressures are usually measured in maternal and fetal plasmas because $CO_2$ levels are low and similar in the freezing-point apparatus (and the blood samples are obtained usually from the post-exchange vessels, where the $CO_2$ values are almost the same); (2) in the body's placental exchange vessels, however, fetal $CO_2$ is higher than maternal; (3) osmotic pressure is directly proportional to $CO_2$ content; and therefore, (4) fetal osmotic pressure exceeds maternal pressure in the body. This difference would favor water movement into the fetal circulation. With the data in this report, one may calculate that a $P_{CO_2}$ difference of less than 0.6 mm Hg would provide the water a term human fetus needs for growth.

## NOTE

Very recently, Conrad and Faber (*Am. J. Physiol.* 233: H474-H487, 1977) presented a detailed analysis of placental water exchange in the sheep. Using measured levels of osmotic pressures in maternal and fetal plasma (Armentrout, et al. *Am. J. Physiol.* 233:H466-H474, 1977), they concluded that mean maternal capillary pressure would be required to exceed mean fetal pressure by 20 to 60 mm Hg to explain fetal water acquisition. To the contrary, we have concluded from the data described here and predictions from a mathematical model that this would not be required and that $CO_2$-bicarbonate differences would be sufficient explanation. To resolve this controversy definitively, direct measurements of capillary hydrostatic pressures are needed.

## REFERENCES

1. Abramovich, D. R.; Page, K. R.; and Jandial, L. 1976. Bulk flows through human fetal membranes. *Gynec. Invest.* 7:157-164.

2. Adolph, E. F., and Hoy, P. A. 1963. Regulation of electrolyte composition of fetal rat plasma. *Am. J. Physiol.* 204:392-411.

3. Barcroft, Sir J. 1947. *Researches on prenatal life.* Springfield, Ill.: Thomas.

4. Battaglia, F. C.; Meschia, G.; Hellegers, A.; and Barron, D. H. 1958. The effects of acute hypoxia on the osmotic pressure of the plasma. *Exptl. Physiol.* 43:197-208.

5. Battaglia, F.; Prystowsky, H.; Smisson, C.; Hellegers, A.; and Bruns, P. 1960. Fetal blood studies. XIII. The effect of the administration of fluids intravenously to mothers upon the concentrations of water and electrolytes in plasma of human fetuses. *Pediat.* 25:2-10.

6. Behrman, R. E.; Seeds, A. E.; Battaglia, F. C.; Hellegers, A.; and Bruns, P. D. 1964. The normal changes in mass and water content in fetal rhesus monkey and placenta throughout gestation. *J. Pediat.* 65:38-44.

7. Boyd, E. M. 1936. Lipid composition of blood in newborn infants. *Am. J. Dis. Child.* 52:1319-1324.

8. Bruns, P. D.; Hellegers, A. E.; Seeds, A. E.; Behrman, R. E.; and Battaglia, F. C. 1964. Effects of osmotic gradients across the primate placenta upon fetal and placental water contents. *Pediat.* 34:407-411.

9. Curran, P. F. and MacIntosh, J. R. 1962. A model system for biological water transport. *Nature* 193:347-348.

10. Dancis, J.; Worth, M.; and Schneidau, P. B. 1957. Effect of electrolyte disturbances in the pregnant rabbit on the fetus. *Am. J. Physiol.* 188:535-537.

11. Delivoria-Papadopoulos, M.; Battaglia, F. C.; and Meschia, G. 1969. A comparison of fetal versus maternal plasma colloidal osmotic pressure in man. *Proc. Soc. Exptl. Biol. Med.* 131:84-87.

12. Guyton, A. C.; Taylor, A. E.; and Granger, H. J. 1975. *Circulatory physiology II: dynamics and control of the body fluids*, p. 106. Philadelphia, Pa.: Saunders.

13. Hellman, L. M.; Flexner, L. B.; Wilde, W. S.; Vosburgh, G. J.; and Proctor, N. K. 1948. The permeability of the human placenta to water and the supply of water to the human fetus as determined with deuterium oxide. *Amer. J. Obstet. Gynec.* 56:861-868.

14. Hinkley, C. M., and Blechner, J. N. 1969. Colloidal osmotic pressures of human and maternal and fetal blood plasma. *Am. J. Obstet. Gynec.* 103:71-72.

15. Hutchinson, D. L.; Hunter, C. B.; Nelsen, E. D.; and Plentl, A. A. 1955. The exchange of water and electrolytes in the mechanism of amniotic fluid formation and the relationship to hydramnios. *Gynec. Obst.* 100:391-396.

16. Kedem, O. and Katchalsky, A. 1958. Thermodynamic analysis of the permeability of biological membranes to non-electrolytes. *Biochim. Biophys. Acta* 27:229-246.

17. Longo, L. D.; Delivoria-Papadopoulos, M.; and Forster, R. E., II. 1974. Placental $CO_2$ transfer after fetal carbonic anhydrase inhibition. *Am. J. Physiol.* 226:703-710.

18. Longo, L. D. and Power, G. G. 1973. Long-term regulation of fetal cardiac output. *Gynec. Invest.* 4:277-287.

19. Margaria, R. 1931. On the state of $CO_2$ in blood and haemoglobin solutions, with an appendix on some osmotic properties of glycine in solution. *J. Physiol.* 73:311-330. London.

20. McCarthy, E. F. 1946. The osmotic pressure of human foetal and maternal sera. *J. Physiol.* 104:443-448. London.

21. Meschia, G. 1955. Colloidal osmotic pressures of fetal and maternal plasmas of sheep and goates. *Am. J. Physiol.* 181:1-8.

22. Meschia, F., and Barron, D. H. 1956. Freezing point depression of arterial and venous plasma *in vivo*. *Yale J. Biol. Med.* 29:54-59.

23. ----. 1956. The effect of $CO_2$ and $O_2$ content of the blood on the freezing point of the plasma. *Quart. J. Exp. Physiol.* 41:180-194.

24. Meschia, G.; Battaglia, F. C.; and Barron, D. H. 1957. A comparison of the freezing points of fetal and maternal plasmas of sheep and goats. *Quart. J. Exptl. Physiol.* 42:163-170.

25. Moll, W., and Künzel, W. 1973. The blood pressure in arteries entering the placentae of guinea pigs, rats, rabbits, and sheep. *Pflugers Arch.* 338:125-131.

26. Pappenheimer, J. R., and Soto-Rivera, A. 1948. Effective osmotic pressure of the plasma proteins and other quantities associated with the capillary circulation in the hindlimbs of cats and dogs. *Am. J. Physiol.* 152:471-491.

27. Paul, W. M.; Enns, T.; Reynolds, S.; and Chinnard, F. P. 1956. Sites of water exchange between the maternal system and the amniotic fluid of rabbits. *J. Clin. Invest.* 35:634-640.

28. Reynolds, S. R. M.; Freese, U. E.; Bieniarz, J.; Caldeyro-Barcia, R.; Mendez-Bauer, C.; and Escarcena, L. 1968. Multiple simultaneous intervillous space pressure recorded in several regions of the hemochorial placenta in relation to functional anatomy of the fetal cotyledon. *Am. J. Obstet. Gynec.* 102:1128-1134.

29. Scoggin, W. A.; Harbert, G. M.; Anslow, W. P., Jr.; Van't Riet, B.; and McGaughey, H. S. 1964. Fetomaternal exchange of water at term. *Amer. J. Obstet. Gynec.* 90:7-16.

30. Seeds, A. E. 1970. Osmosis across term human placental membranes. *Am. J. Physiol.* 219:551-554.

31. ----. 1965. Water metabolism of the fetus. *Am. J. Obstet. Gynec.* 92:727-745.

32. Seller, M. J. 1963. The effect of maternal overhydration on the rat foetus. *Quart. J. Exptl. Physiol.* 48:292-297.

33. Vosburg, G. J.; Flexner, L. B.; Cowie, D.; Hellmen, L.; Proctor, N.; and Wilde, W. 1948. The rate of renewal in woman of the water and sodium of the amniotic fluid as determined by tracer techniques. *Am. J. Obstet. Gynec.* 56:1156-1159.

# 13

# The Interrelations of Blood and Extracellular Fluid Volumes and Cardiac Output in the Newborn Lamb

Lawrence D. Longo, William W. Allen, Jerome W.H. Niswonger, Kirk D. Pagel, Howard C. Wieland, and Raymond D. Gilbert

Division of Perinatal Biology
Departments of Physiology and Obstetrics and Gynecology
Loma Linda University
School of Medicine
Loma Linda, California

INTRODUCTION

In the fetus and newborn infant the relation of cardiac output to blood volume and, in turn, the relation of of extracellular fluid volume to blood volume are of interest both physiologically from a consideration of the basic mechanisms of the control of cardiac output, and clinically with regard to the effects of blood or fluid administration on cardiac function.

The physiologic control of cardiac output in the fetus and newborn infant is mediated by several short-term regulators, including venous return, baroreceptors, chemoreceptors, humoral agents, and perhaps myogenic influences (18, 21, 22). The long-term control of this overall process and the interrelation of these various factors are understood only poorly. As noted in another chapter in this volume (32), the interrelations of cardiac output, blood volume, and extracellular fluid volume probably are important factors in this process.

Clinically, cardiac output in the newborn infant and young child is an important determinant for optimum delivery of oxygen and other nutrients to the brain, heart, and other vital structures. Because cardiac output can be altered by infusions of blood (or its constituents) and intravenous fluids, some understanding of how such infusions influence the whole-blood volume and the volumes of erythrocytes, plasma, and extracellular fluid would be of value. Few studies have explored these interrelations.

In the present study we investigated in young lambs the relation of cardiac output to circulating blood volume and extracellular fluid volume, varying these volumes by infusing or removing whole blood or by infusing Tyrode's solution. Initially, our studies attempted to explore these interrelations in the fetus. Unfortunately, the rapid placental loss of fetal radioactive volume markers (particularly $^{82}$Br for extracellular space, and to a lesser extent $^{125}$I-radioiodinated serum albumin for plasma volume) precluded such determinations. We therefore pursued the studies in young lambs in an attempt not only to determine these interrelations per se, but to use this data to extrapolate back to the fetal state.

## METHODS

The principle of the method was to measure blood and extracellular fluid volume using radionuclides, and cardiac output using indicator-dilution techniques. Lambs from 1 to 3 weeks of age were kept off food or water for 2 hr, after which we administrered barbiturate sedation (10 mg/kg pentobarbital) and thereafter gave small doses of barbiturate as needed. Supplemental oxygen was administered through a face mask to maintain arterial oxygen tension at about 100 torr. We placed small tygon catheters (1.8-mm o.d.) in one carotid artery and jugular vein and connected them to an extracorporeal shunt. We also placed catheters in a femoral artery and vein and connected them to syringes for injection of dye and withdrawal of blood for cardiac-output measurements. A catheter was also placed in the other jugular vein and its tip advanced to the superior vena cava in order to monitor central venous pressure.

Through a midline abdominal incision, we ligated the renal artery and vein of both kidneys, performing a functional nephrectomy to eliminate urine formation. In addition, we ligated the major vessels supplying the spleen in order to preclude variable sequestering of erythrocytes (25). We placed the lamb on a heating pad to maintain its body temperature at about 39°C and continuously monitored its core temperature (Tele-thermometer, Yellow Springs Instrument Co., Yellow Springs, Ohio).

### Preparation of Radioactive-Labeled Compounds

We used $^{51}$chromium-labeled erythrocytes to determine red cell volume according to previously described methods (2, 10, 20). Eight to twelve ml of blood from an adult sheep were labeled with 50 microcuries (µCi) of $^{51}$Cr. We washed the cells two or three times in an effort to remove the free label. The $^{51}$Cr-labeled erythrocytes were drawn into a syringe which was weighed both before and after their injection into the experimental animal. We prepared standards from the same blood. We used $^{125}$I-radioiodinated serum albumin (RISA) (Albumotope, E. R. Squibb and Sons, New York) to determine the plasma volume (13, 25, 26, 42). Five µCi of $^{125}$I RISA were injected into the animal for this measurement. We used $^{82}$bromine ($^{82}$Br) ions to measure extracellular fluid volume (ECFV) (5, 31, 39, 40), injecting into the circulation 50 µCi of sodium bromide with 15 mg of unlabeled sodium bromide as carrier. Because $^{82}$Br has a half-life of only 35.9 hr, we corrected for its decay during the experiment. (Incidentally, we originally attempted to measure extracellular fluid volume, using $^{99}$technetium-labeled inulin, but abandoned this approach, as $^{99}$Tc failed to remain attached to the inulin molecule.) The above three radionuclides collectively were diluted in 10 ml of blood and administered to the lamb after withdrawal of a similar volume of blood. The isotopes were present in sufficient concentration to give counts 10 or more times background when distributed in the lamb's circulation.

### Measurement of Blood and Fluid Volumes

We continuously monitored the radioactivity in the lamb's blood, using an extracorporeal shunt (26) containing a plastic coil (3) fixed in a deep scintillation well

(Baird Atomic, Model 810c), connected to an analyzer and scaler with digital readout (Integrated Medical Spectrometer, Model INS 15). Corrections for overlap of radioactivity were made easily, as the three radionuclides have widely differing energy peaks. We calibrated the scaler to achieve maximum count rate for $^{125}I$ at 28 keV, $^{51}Cr$ at 280 keV, and $^{82}Br$ at 777 keV. With this calibration, no $^{125}I$ or $^{51}Cr$ counts appeared in the $^{82}Br$ window, and no $^{125}I$ counts appeared in the $^{51}Cr$ window. But 2.4 (±0.1)% of the $^{51}Cr$ and 3.9 (±0.4)% of the $^{82}Br$ counts appeared in the $^{125}I$ window; and 18.3 (±2.1)% of the $^{82}Br$ counts appeared in the $^{51}Cr$ window. The injection syringes were weighed before and after injection of each isotope to determine the amount of radioactivity injected. Amounts used for standards were also weighed on the analytic balance. Blood was propelled through the shunt at a rate of 10 ml/min by means of a peristaltic pump (Masterflex Model No. 7014, Cole-Parmer the shunt was about 14 ml with about 10 ml of blood contained in the coil itself. We attempted to minimize blood coagulation by heparinizing the animal (150 units/kg) initially, and later as needed at an approximate rate of 35 units/(kg x hr), and siliconizing the tubing prior to use.

## Cardiac-Output Determination

We determined cardiac output by the indicator-dilution technique, using indocyanine green dye and a densitometer (Model D401, Waters Instruments, Inc., Rochester, Minn.), injecting 0.5 ml of indocyanine green (5 mg/ml) followed by 3 ml of saline; the time interval required for injection was less than 0.5 sec. Ten to fifteen ml of blood were withdrawn over a period of 15 to 20 sec for dye-concentration measurement and replaced following each determination. Determinations were repeated 3 times at 3-min intervals at each level of volume expansion (see below). On an average, 24 cardiac-output determinations were made in each of the 21 lambs.

We calculated the cardiac output and vascular volume, using standard techniques (44) with an on-line minicomputer (Univac 60).

## Blood Pressure and Blood Gas Measurements

We continuously monitored the lamb's arterial and central venous blood pressures with a pressure transducer

(Model P23Db, Statham Laboratories, Inc., Hato Rey, Puerto Rico) and continuously monitored the heart rate. At 30- to 45-min intervals blood samples were withdrawn anaerobically into siliconized glass syringes, the dead space of which was filled with heparin. We determined blood $PO_2$, $PCO_2$, and pH values, using microelectrodes (Radiometer Model BMS3, The London Co., Westlake, Ohio) and plasma osmotic pressure using a freezing point osmometer (Digimatic, Model 30, Advanced Instruments, Inc., Needham Heights, Mass.). Plasma albumin, globulin, and total protein were determined by electrophoresis and the Biuret reaction, and blood hemotocrit was determined by a microhematrocrit method (Clay-Adams, Model MB micro-hematocrit).

## Experimental Procedures

A typical experiment consisted of a 45- to 60-min control period following the introduction of the $^{51}$Cr-labeled erythrocytes, $^{125}$I-labeled albumin, and $^{82}$Br. When the radioactivity counts stabilized, we recorded the counts in each channel 10 to 15 times, sampled the blood, for qualitating each radioactive nuclide, and measured cardiac output 3 times. In an effort to minimize clotting in the extracorporeal circuit, every 30 min we stopped the pump and flushed the entire system with heparinized Tyrode's solution (25).

In 5 lambs we infused isotonic Tyrode's solution, previously warmed to 39°C, into the circulation at a rate of 6 to 10 ml/min. Typically, 10 to 20 ml/kg body weight was infused at a given time. We continuously monitored radioactivity in the plastic coil for 30 to 45 min until stable readings were recorded, at which time we withdrew blood samples and repeated the cardiac output determinations. Then we repeated the infusion of Tyrode's solution and waited another 30 to 45 min to achieve a quasi steady rate. In all, 7 to 10 such infusions were repeated in each animal studied.

In another 11 lambs we infused 10 to 15 ml/kg warmed maternal whole blood under conditions similar to those described above. From most lambs we also withdrew 30 to 45 ml/kg of blood before commencing the blood infusion in an effort to span both sides of the normal blood volume range.

At the end of a given study, we reweighed the lamb and measured the volume and level of radioactivity in several body fluids, including the urine (usually there was none), and gastric, bowel, and tracheal secretions.

In order to determine the steady-state rate of isotope disappearance from the circulation (see below) in the absence of changes in blood or fluid volumes, we performed a functional nephrectomy and splenectomy in 6 lambs, as in the other animals. Following the injection of radioactive isotopes, we withdrew 5-ml blood samples every 30 min for 4 to 5 hr, and measured radioactivity of $^{125}I$, $^{82}Br$, and $^{51}Cr$, but otherwise did not add or remove fluid or blood.

## Calculations

We determined the blood volume, plasma volume, and extracellular volume using standard techniques (25, 26).

## RESULTS

### Control Values

The baseline values of the infant lamb variables were as follows: weight = 9.1 (±0.7 SEM) kg; heart rate = 159 (±6) $min^{-1}$; mean arterial pressure = 76.6 (±4.1) mm Hg; central venous pressure = 2.9 (±0.3) mm Hg; $PO_2$ = 87.5 (±15.5) torr; $PCO_2$ = 39.3 (±1.4) torr; pH = 7.386 (±0.007) units; cardiac output = 106.1 (±10.2) ml/(min x kg); erythrocyte volume = 27.4 (±1.3) ml/kg; plasma volume = 54.0 (±2.2) ml/kg; whole blood volume = 81.5 (±3.0) ml/kg; extracellular fluid volume = 337.1 (±9.5) ml/kg; whole-body hematocrit = 33.3 (±1.0) ml/100 ml; large-vessel hematocrit = 41.8 (±0.7) ml/100 ml; F cell (whole-body hematocrit/large-vessel hematocrit) = 0.8 (±0.02); blood osmotic pressure = 308.9 (±2.3) mOsm; serum albumin = 2.35 (±0.11) g/100 ml; serum globulin = 3.08 (±0.23) g/100 ml; albumin/globulin = 0.83 (±0.08).

### Control Equilibration Times of Radiosotopes

We measured the time course of changes in mean values of the radionuclides in 6 lambs to determine the control rate of isotope disappearance from the circulation.

The disappearance rates were determined by linear extrapolation after the first 1 hr and for up to 4.5 hr following isotope injection. The disappearance half-times were: $^{51}$Cr (RBCs) = 26.5 (±2.0 SEM) hr; $^{125}$I (plasma) = 13.2 (±0.9) hr; $^{82}$Br (ECFV) = 111.6 (±9.3) hr. The average rate of loss of each radioactive nuclide per hr (λ) was: $^{51}$Cr = 2.6 (±0.2)%; $^{125}$I RISA = 5.3 (±0.3)%; $^{82}$Br = 0.6 (±0.1)%. During the first 10 to 60 min of a study the disappearance half-time for $^{125}$I-labeled RISA was 3.8 (±0.1) hr, while the disappearance rate was 20.3% per hr.

In two experiments (otherwise not included) we measured the isotope disappearance times in fetal lambs in which the placental circulation was intact. In these cases the disappearance half-time of $^{51}$Cr was essentially the same as in the newborn lambs, while the half-times for $^{125}$I and $^{82}$Br were 6.5 and 7.5 hr, respectively.

## Infusion of Tyrode's Solution

We infused Tyrode's solution repeatedly into the lamb circulation in amounts of 100 to 250 ml (10 to 70 ml/kg, mean = 19.7 [±2.4 SEM] ml/kg lamb weight). We repeated the infusions an average of 6.5 (±0.7) times. Figure 1 shows the change in volume of extracellular fluid, whole blood, plasma, and erythrocytes, and cardiac output following these infusions in a single experiment. Typically, the radioactive counts of $^{51}$Cr (RBCs), $^{125}$I (plasma), and $^{82}$Br (ECFV) became relatively constant 30 to 40 min following the infusion period. The half-time for these changes in plasma volume and ECFV averaged about 6 min. Central venous pressure (CVP) usually, but not always, rose during and immediately following the infusion period, particularly after the lamb had previously received 50 to 60 ml/kg Tyrode's solution. Following the transient increase, the central venous pressure returned to pre-infusion level, but it increased slowly as the total amount of solution infused increased. The least-squares linear-regression equation for this relation is CVP (mm Hg) = 4 (±0.6 SEM) + 0.1 (ml/kg Tyrode's solution). Lamb cardiac output initially increased following the blood volume increase, but after several repeated infusions (when the volume of Tyrode's solution infused exceeded 40 to 50 ml/kg), cardiac output decreased dramatically and the animal deteriorated. Unfortunately, we had no way of decreasing extra-cellular fluid volume independently of blood volume; therefore, we could not explore these interrelations below normal values.

Figure 2 shows the interrelations of the volumes of erythrocytes, plasma, whole blood, and extracellular fluid as a function of the amount of Tyrode's solution added. As expected, total erythrocyte volume did not change sig-

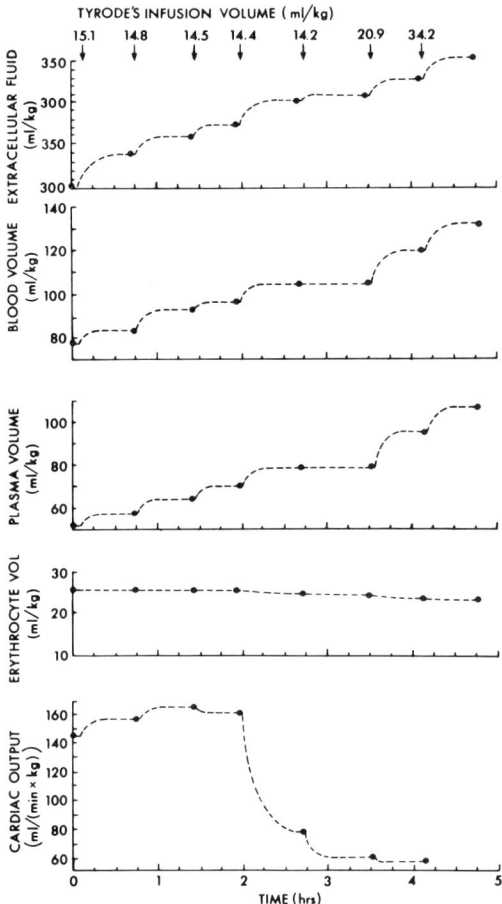

Fig. 1. The volume of extracellular fluid, whole blood, plasma, and erythrocytes, and cardiac output following repeated infusions of Tyrode's solution in a lamb. (See text for details.)

nificantly following infusion of Tyrode's solution. Plasma volume increased 0.29 ml/kg per ml/kg Tyrode's solution infused ($\underline{Y}$ = 51.8 + 0.3$\underline{X}$; $\underline{R}$ = 0.67); total blood volume also increased 0.29 ml/kg per ml/kg Tyrode's infused ($\underline{Y}$ = 77.0 + 0.3$\underline{X}$; R = 0.62), while extracellular fluid volume increased 0.91 ml/kg per ml/kg Tyrode's infused ($\underline{Y}$ = 317.8 + 0.9$\underline{X}$; $\underline{R}$ = 0.80). The $p$ value of each of these relations was < 0.005.

Figure 3 shows the relation of erythrocyte, plasma, and whole-blood volumes to extracellular fluid volumes following infusion of Tyrode's solution. The regression equations for the various volumes as a function of extracellular fluid volume are: for plasma, $\underline{Y}$ = 23.8 + 0.2$\underline{X}$ ($\underline{R}$ = 0.65); for whole blood, $\underline{Y}$ = -5.41 + 0.3$\underline{X}$ ($\underline{R}$ = 0.65). There

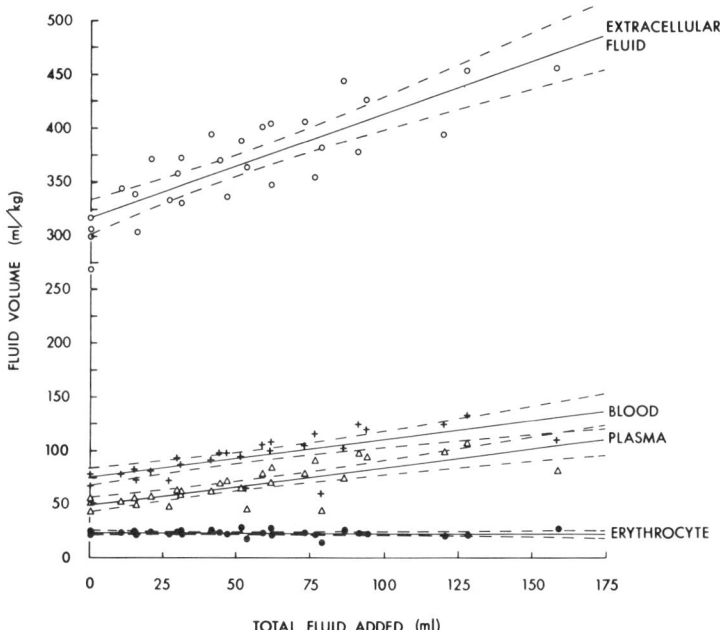

Fig. 2. The interrelations of volumes of erythrocytes (•), plasma (Δ), whole blood (+), and extracellular fluid (o) as a function of the volume of Tyrode's fluid infused. The solid lines indicate the fitted linear regressions, and the broken lines indicate the 95% confidence limits of the mean.

was no relation to erythrocyte volume. Although both whole body and peripheral hematocrits decreased as a function of the amount of Tyrode's solution infused, the F cell ratio remained essentially unchanged.

Lamb cardiac output remained relatively constant or increased slightly until the volume of Tyrode's infused exceeded 40 to 50 ml/kg, at which point cardiac output decreased substantially. Unfortunately, we had no relatively simple way of decreasing extracellular fluid volume independently of blood volume, and could not, therefore, explore these interrelations below normal values.

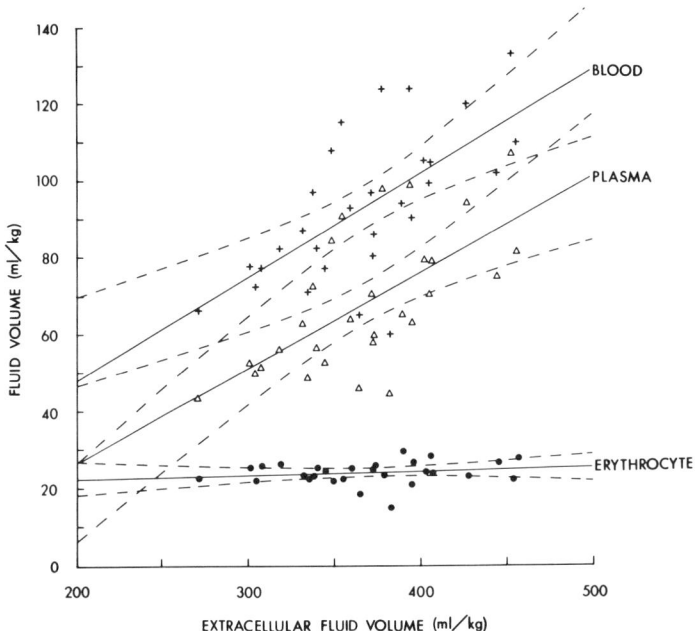

Fig. 3. The interrelations of erythrocyte, plasma, and whole-blood volumes to extracellular fluid volume with the addition of Tyrode's solution. The symbols are the same as in Figure 2. The solid lines indicate the linear regressions and the broken lines indicate the 95% confidence limits of the mean.

## Infusion of Whole Blood

We both withdrew whole blood from the lamb's circulation and infused maternal blood following the readministration of the previously withdrawn lamb blood. The volume of blood infused at one time ranged from 3.6 to 16.2 (mean = 6.7 (±0.7 SEM) ml/kg. We infused blood an average of 4.7 (±0.4) times. Figure 4 shows the changes in volume of erythrocytes, plasma, whole blood, and extracellular fluid as a function of the amount of whole blood added to or taken

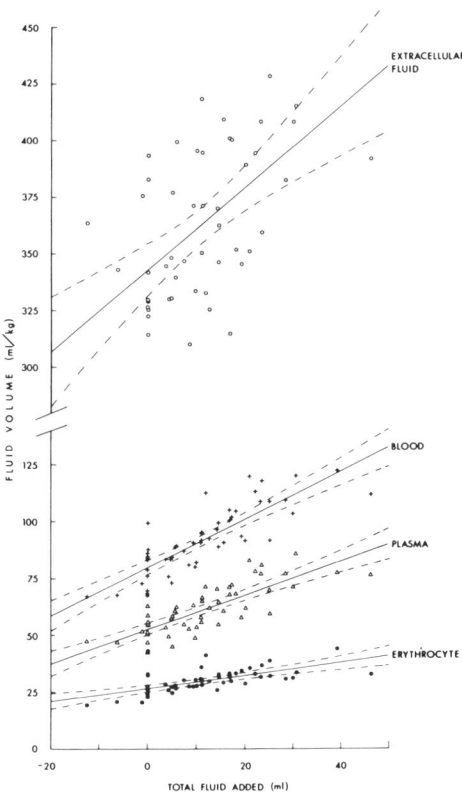

Fig. 4. The interrelations of volumes of extracellular fluid, whole blood, plasma, and erythrocytes to the volume of whole blood infused. The symbols are the same as in Figure 2. The linear regressions and 95% confidence limits of the mean are shown.

from the newborn circulation. The appropriate regression equations are: for erythrocyte volume, $\underline{Y} = 26.9 + 0.3\underline{X}$ ($\underline{R} = 0.63$); for plasma volume, $\underline{Y} = 52.7 + 0.7\underline{X}$ ($\underline{R} = 0.80$); for whole-blood volume, $\underline{Y} = 79.9 + 1.1\underline{X}$ ($\underline{R} = 0.83$); for extracellular fluid volume, $\underline{Y} = 343.2 + 1.8\underline{X}$ ($\underline{R} = 0.58$).

Figure 5 shows the relation of total blood volume to the extracellular fluid volume following the withdrawal or infusion of whole blood. On the basis of a linear relation, the regression equation is $\underline{Y} = 13.5 + 0.2\underline{X}$ ($\underline{R} = 0.53$). A power function probably fits the data as well (see "Discussion"). Unfortunately, when we attempted to study these relations much beyond the physiologic range, the animal preparation deteriorated.

Figure 6 demonstrates newborn lamb cardiac output (ml/min x kg) as a function of blood volume (ml/kg) when

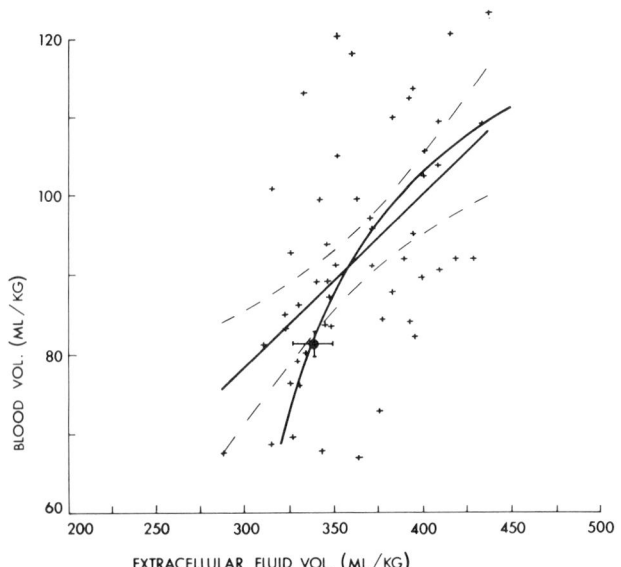

Fig. 5. The relation of blood volume to extracellular fluid volume. The linear regressions and 95% confidence limits of the mean are shown. The closed circle indicates the normal value (±SEM) prior to blood infusion or withdrawal. A hyperbolic function fits the data as well as the linear function.

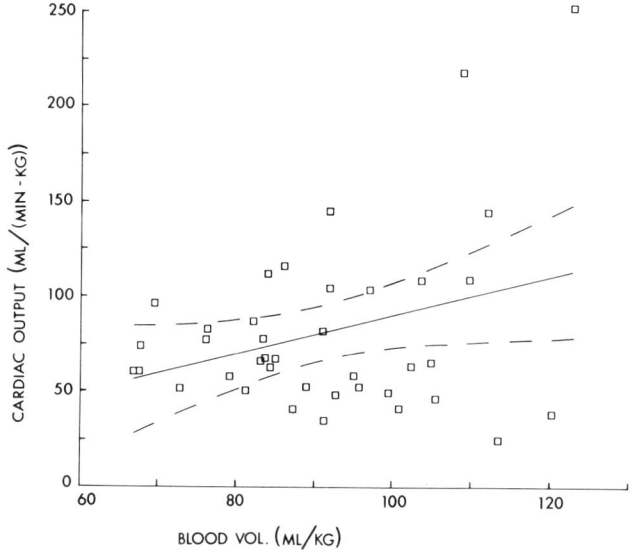

Fig. 6. Cardiac output in young lambs as a function of blood volume. The linear regression is $\underline{Y} = 11.5 \pm 1.0\underline{X}$. The 95% confidence limits of the mean are shown.

blood volume was altered with whole blood expansion. The regression equation for this relation is $\underline{Y} = 11.5 + 1.0\underline{X}$ ($\underline{R} = 0.32$). The relation of cardiac output to total whole blood added is $\underline{Y} = 57.5 + 4.5\underline{X}$ ($\underline{R} = 0.52$).

## DISCUSSION

### Validity of the Methods

The methods for calculation of blood volumes, extracellular fluid volume, and cardiac output using the various radionuclides should meet certain criteria as nearly as possible (31): the labeled compounds should be nontoxic; they should be distributed uniformly within a given compartment; they should not be metabolized; they should not be excreted; and their measurement should be relatively straightforward.

## Red Blood Cell Volume

The accuracy of erythrocyte volume measurements by the $^{51}$Cr method depends on several factors (1, 2), including the reproducibility of successive blood radioactivity determinations and the ability of erythrocytes to remain intact and uniformly mixed. Following a period of mixing in the circulatory system, the calculated red cell volume remained relatively constant with a disappearance half-time of about 27 hr. The functional splenectomy precluded intermittent or variable sequestering of erythrocytes in that organ (25), an effect which has been shown to affect calculated erythrocyte volume (24). The lack of change in the ratio of the whole body hematocrit (determined from the erythrocyte and plasma radioactivity) to the central venous hematocrit (F cell ratio) (34, 35) during the course of the experiment suggested little sequestering of erythrocytes in slowly equilibrating vascular pools. Periodically, we checked blood samples for hemolysis and found that despite the use of the peristaltic pump in the extracorporeal shunt, no hemolysis was detectable.

## Plasma Volume Measurements

The accurate determination of plasma volume by use of RISA requires that the albumin not be lost from the vascular system. As noted by several workers (1, 13, 19, 42), albumin slowly leaks from the vascular bed into tissue spaces at rates varying from 10 to 30% per hr. We corrected all plasma volume measurements for both the relatively rapid initial loss of RISA from the circulation, and the slower rate of loss after the first hour.

## Whole-Blood Measurements

Because total blood volume is calculated as the sum of the red cell and plasma volumes, errors in these determinations obviously will affect the calculated blood volume.

## Extracellular Fluid Volume

Two main groups of substances have been used for the determination of extracellular fluid volume, but none are entirely satisfactory. The first group includes substances such as bromide and thiocyanate ions, having a distribution volume similar to that of chloride ion, i.e., the so-called chloride space. The volume measured by these molecules is probably greater than the true extracellular space, as they are readily secreted into the fluid of the gastrointestinal tract (31, 39, 40). These ions also enter cells to a certain extent, erythrocytes being particularly permeable (27, 31, 39, 40). The second group of substances have a volume distribution similar to that of inulin. These include sucrose, dulcitol, mannitol, thiosulfate, and sulfate (6, 37). The volume of distribution of these molecules is considerably less than the chloride space and probably less than the extracellular fluid volume, because they do not penetrate connective tissue readily. Furthermore, because of their slow rates of diffusion, these substances require prolonged periods for equilibration with extracellular fluids. The true extracellular fluid volume thus probably lies between the "chloride" and "inulin" space measurements. Any determination of extracellular fluid volume may be criticized, therefore, as not representing the true extracellular fluid volume (5, 39).

As noted above, we chose $^{82}Br$ for the determination of extracellular fluid volume as the best compromise. Because of its relatively short half-life, we corrected for radioactive decay during the course of each experiment. We also corrected for the entry of bromide ion into the erythrocytes (27, 31, 39, 40). Red cell bromine has been shown to amount to from 2% (31) to 5% (40) of the total amount of nuclide present. In young lambs this amounted to 3.5%. Because of these losses of $^{82}Br$, we sampled fluids from several compartments at the end of each study and corrected for the loss of isotopes.

The prolonged time to equilibrate in body tissues is an additional problem with bromide ion. Although complete equilibration probably requires 24 hr (40), equilibration is 90 to 95% complete by 3 hr (31). In the lamb, after the initial period of 2 to 3 hr, re-equilibration following the injections of Tyrode's solution or whole blood was 90% complete within 40 min, as determined with continuous radioactive counting in the external shunt.

As noted above, we initially attempted these studies in fetal lambs *in utero*. However, the rapid passage of $^{82}Br$ and, to a lesser extent, the slower transfer of $^{125}I$

across the placenta and into the amniotic fluid made any meaningful interpretation of the data difficult, if not impossible. Therefore we abandoned that approach for the one reported here.

### Cardiac-Output Determination

The problems associated with the determination of cardiac output by using indicator-dilution methods have been discussed by several authors (38, 44). With adequate mixing of dye in the blood and the absence of shunts, the technique is applicable in young animals (4), and may be used even in the presence of bidirectional shunts in newborns (9, 15, 33). The reproducibility of the dye curves in the present study compared favorably to that in older subjects by densitometry (38). The variation between successive measurements equaled about 17% in our study; thus the value of cardiac output obtained represents only an approximation of the true value. In six instances we observed significant left-to-right shunts and discarded the data in those cases.

### Comparative Aspects of Blood and Extracellular Fluid Volumes

Table 1 lists some comparative values for extracellular fluid volume, blood volume, and cardiac output in lambs and human infants.

### Physiologic and Clinical Significance

The relation of blood volume to extracellular fluid volume, and in turn of cardiac output to blood volume, is of interest for several reasons. From a physiologic standpoint, these functional relations are important for an understanding of the control of cardiac output. Guyton and his colleagues (21, 22) have stressed the importance of blood volume and extracellular fluid volume in affecting venous return and in determining the long-term control of cardiac output and blood pressure in adults. These relations are probably also important in the long-term regulation of cardiac output and blood pressure in newborn infants and the fetus *in utero* (28). Figure 5 illustrates

Table 1

Blood and Extracellular Fluid Volumes

| Species | Age | Erythrocyte Volume (ml/kg) | Plasma Volume (ml/kg) | Whole-Blood Volume (ml/kg) | Extracellular Fluid Volume (ml/kg) | Intracellular Fluid Volume (ml/kg) | Total Body Water (ml/kg) | Ref. |
|---|---|---|---|---|---|---|---|---|
| Sheep | Newborn | 31.6 (±2.3)[a] | 72.4 (±2.6) | 104 (±4.2) | | | | 10 |
| | Young lambs | 27.4 (±1.3) | 54.0 (±2.2) | 81.5 (±3.0) | 337.1 (±9.5) | | | Present study |
| | 20- to 60-days old | | | | | | 623 (±10.3) | 23 |
| Human | Newborn | | 45.1 (±1.3) | | 430 | 310 | 725 | 41 |
| | | 32.1 to 50.6[b] | 45.9 to 48.0[b] | 78.0 to 98.6[b] | 360 (±7.0) | 356 (±10.1) | 715 (±10.3) | 30 |
| | | | | | | | | 43 |
| | | | 43.1 (±1.5) | 98.3 (±1.9) | 358 (±6.5) | | | 7 |
| | | | | | | | | 14 |
| | | | | | 445 | 339 | 784 | 16 |
| | | | | | 387 (± 9.8) | 398 (±12.4) | 774 (±8.9) | 8 |

a SEM

b Value in newborns in which clamping of the umbilical cord was either early or delayed after delivery.

the relation of blood volume to extracellular fluid volume. The mean values (±SEM) of these parameters under control conditions, prior to the infusion of blood, are shown by the closed circle with bars. As extracellular fluid volume increases, the blood volume increases about one-tenth to one-half as much, depending on the extent to which the vascular system is filled. Both a linear and hyperbolic curve are fitted to the data (Fig. 5). As noted with the curvilinear plot, beyond a certain point essentially all of the additional fluid that enters the extracellular fluid compartment fills the interstitial spaces, and essentially none remains in the blood, a result predicted by Guyton (21).

An increase in blood volume would predict an increase in mean systemic pressure, and this in turn would increase cardiac output in the newborn or fetus, as Gilbert (18) discusses elsewhere in this volume. As observed in Figure 1, cardiac output increased somewhat when blood volume was increased a small amount. However, further increases in blood volume resulted in a deterioration of the preparation with a dramatic fall in cardiac output. Figure 6 shows cardiac output as a function of blood volume for all the data from these experiments. When the data for each experiment is plotted separately this trend is even more dramatic.

From a clinical standpoint, the relation of cardiac output to blood and extracellular fluid volumes is important in infants with disorders of hydration. The present studies may be useful in understanding the pathophysiology of disorders such as overhydration with edema and/or congestive heart failure, polycythemia, anemia, and related disorders. The therapeutic goals in caring for these patients perhaps may be more readily achieved through awareness of basic physiological mechanisms and interrelations of these factors.

In conclusion, newborn (and probably fetal) blood and plasma volumes and extracellular fluid volume are related closely to the volume of Tyrode's solution or whole blood infused. The present report depicts some of these interrelations in the newborn lamb. These relations are important to an understanding of the long-term control of cardiac output, as this is a function of mean systemic pressure, which in turn is a function of blood volume, itself a function of extracellular fluid volume. Extracellular fluid volume in turn is a complicated function of water intake and output. These interrelations are probably of profound importance in the long-term control of cardiac output and the delivery of oxygen and other nutrients in the newborn and fetus. It remains to be determined to what extent changes in blood volume affect the mean systemic pressure, and to what extent the distribution of blood flow to the various organs (or placenta in the case of the fetus) is altered with changes in cardiac output.

## ACKNOWLEDGMENT

U. S. Public Health Service Grant HD 03807 supported this study.

## REFERENCES

1. Albert, S. N.; Gavel, Y.; Turmel, Y.; and Albert, C. A. 1965. Pitfalls in blood volume measurement. *Anesth. Analg.* 44:805-814.

2. Albert, S. N.; Shibuya, J.; Economopoulos, B.; Radice, A.; Cuevo, N.; Varrone, E. V.; and Albert, C. A. 1968. Simultaneous measurement of erythrocyte, plasma and extracellular fluid volumes with radioactive tracers. *Anesthesiology* 29:908-916.

3. Albert, S. N.; Spencer, W. A.; Finkelstein, M.; Shibuya, J.; Alpert, S.; and Coakley, C. S. 1956. A plastic coil simplifying radioactive liquid phase counting. *J. Lab. Clin. Med.* 48:471-475.

4. Arcilla, R. A.; Oh, W.; Wallgren, G.; Hanson, J.S.; Gessner, I. H. and Lind, J. 1967. Quantitative studies of the human neonatal circulation. II. Hemodynamic findings in early and late clamping of the umbilical cord. *Acta Paediat. Scand.* 179:23-42.

5. Berson, S. A. and Yalow, R. S. 1955. Critique of extracellular space measurements with small ions: $Na^{24}$ and $Br^{82}$ spaces. *Science* 121:34-36.

6. Calcagno, P. L.; Husson, G. S.; and Rubin, M. I. 1951. Measurement of "extracellular fluid space" in infants by equilibration technique using inulin and

sodium ferrocyanide. *Proc. Soc. Exper. Biol. Med.* 77:309-311.

7. Cassady, G. 1966. Plasma volume studies in low birth weight infants. *Pediatrics* 38:1020-1027.

8. Clapp, W. M.; Butterfield, L. J.; and O'Brien, D. 1962. Body water compartments in the premature infant, with special reference to the effects of the respiratory distress syndrome and of maternal diabetes and toxemia. *Pediatrics* 29:883-889.

9. Cotton, R. B.; Lindstrom, D. P.; Kanarek, K. S.; Sundell, H. and Stahlman, M. T. 1977. Quantitation of cardiac output components in the presence of bidirectional shunts. *J. Appl. Physiol.: Respirat. Environ. Exercise Physiol.* 43:352-356.

10. Creasy, R. K.; Drost, M.; Green, M. V.; and Morris, J. A. 1970. Determination of fetal, placental and neonatal blood volumes in sheep. *Circ. Res.* 27:487-494.

11. Cross, K. W.; Dawes, G. S.; and Mott, J. C. 1959. Anoxia, oxygen consumption and cardiac output in newborn lambs and adult sheep. *J. Physiol.* 146:316-343. London.

12. Dyrbye, M. O. and Kragelund, E. 1970. Simultaneous determination of the apparent $^3$HOH, $^{82}$Br, $^{125}$I human albumin and $^{51}$Cr red cell volumes in human subjects. *Scand. J. Clin. Lab. Invest.* 26:61-66.

13. Fine, J., and Seligman, A. M. 1943. Traumatic shock. IV. A study of the problem of the "lost plasma" in hemorrhagic shock by the use of radioactive plasma protein. *J. Clin. Invest.* 22:285-303.

14. Fink, C. W. and Cheek, D. B. 1960. The corrected bromide space (extracellular volume) in the newborn. *Pediatrics* 26:397-401.

15. Friedman, P. J. and Downing, S. E. 1968. Estimation of systemic blood flow by the indicator-dilution technique in the presence of central shunts. *Am. J. Cardiol.* 22:672-677.

16. Friis-Hansen, B. 1961. Body water compartments in children: changes during growth and related changes in body composition. *Pediatrics* 28:169-181.

17. Friis-Hansen, B. J.; Holiday, M.; Stapleton, T.; and Wallace, W. M. 1951. Total body water in children. *Pediatrics* 7:321-327.

18. Gilbert, R. D. 1978. Venous return and control of fetal cardiac output. In *Fetal and newborn cardiovascular physiology*, eds., L. D. Longo and D. D. Reneau. New York: Garland Publishing.

19. Gitlin, D. 1957. Distribution dynamics of circulating and extravascular $I^{131}$ plasma proteins. *Ann. New York Acad. Sci.* 70:122-136.

20. Gray, S. J. and Sterling, K. 1950. The tagging of red cells and plasma proteins with radioactive chromium. *J. Clin. Invest.* 29:1604-1613.

21. Guyton, A. C. and Coleman, T. G. 1967. Long-term regulation of the circulation: interrelationships with body fluid volumes. In *Physical basis of circulatory transport: regulation and exchange*, eds., E. B. Reeve and A. C. Guyton. Philadelphia: Saunders.

22. Guyton, A. C.; Coleman, T. G.; and Granger, H. J. 1972. Circulation: overall regulation. *Ann. Review Physiol.* 34:13-46.

23. Hansard, S. L. and Lyke, W. A. 1956. Measurement of total body water in sheep using $I^{131}$ labeled 4-iodoantipyrine. *Proc. Soc. Exper. Biol.* 93:263-266.

24. Hodgetts, V. E. 1961. The dynamic red cell storage function of the spleen in sheep. III. Relationship to determination of blood volume, total red cell volume, and plasma volume. *Austr. J. Exp. Biol.* 39:187-196.

25. Leonard, J. I. 1971. Dynamics of plasma-interstitial fluid distribution following intravenous infusions in nephrectomized dogs. Ph.D. dissertation, Univ. of Michigan, Ann Arbor.

26. Leonard, J. I. and Abbrecht, P. H. 1974. A method for continuously monitoring blood volume. *J. Appl. Physiol.* 36:506-508.

27. Leth, A., and Binder, C. 1970. The distribution volume of $^{82}Br^-$ as a measurement of the extracellular fluid volume in normal persons. *Scand. J. Clin. Lab. Invest.* 25:291-297.

28. Longo, L. D. and Power, G. G. 1973. Long-term regulation of fetal cardiac output. A hypothesis on the role of carbon dioxide. *Gynec. Invest.* 4:277-287.

29. McCance, R. A. and Widdowson, E. M. 1951. A method of breaking down the body weights of living persons into terms of extracellular fluid, cell mass and fat, and some applications of it to physiology and medicine. *Proc. Royal Soc., Series B* 138:115-130.

30. MacLaurin, J. C. 1966. Changes in body water distribution during the first two weeks of life. *Arch. Dis. Child.* 41:286-291.

31. Nicholson, J. P. and Zilva, J. F. 1960. Estimation of extracellular fluid volume using radiobromine. *Clin. Sci.* 19:391-398.

32. Power, G. G.; Roos, P.; and Longo, L. D. 1978. Water transfer across the placenta: hydrostatic and osmotic forces and the control of fetal cardiac output. In *Fetal and newborn cardiovascular physiology*, eds., L. D. Longo, and D. D. Reneau. New York: Garland Publishing.

33. Prec, K. J. and Cassels, D. E. 1955. Dye dilution curves and cardiac output in newborn infants. *Circulation* 11:789-798.

34. Reeve, E. B.; Gregersen, M. I.; Allen, T. H.; and Sear, H. 1953. Distribution of cells and plasma in the normal and splenectomized dog and its influence on blood volume estimates with $P^{32}$ and T-1824. *Am. J. Physiol.* 175:195-203.

35. Reeve, E. B.; Gregersen, M. I.; Allen, T. H.; Sear, H.; and Walcott, W. W. 1953. Effects of alteration in blood volume and venous hematocrit in splenecto-

mized dogs on estimates of total blood volume with $P^{32}$ and T-1824. *Am. J. Physiol.* 175:204-210.

36. ----. 1953. Validity of cell and blood volume measurements in the bled splenectomized dog. *Am. J. Physiol.* 175:211-217.

37. Rosenberg, L. E.; Downing, S. J.; and Segal, S. 1962. Extracellular space estimation in rat kidney slices using $C^{14}$ saccharides and phlorizin. *Am. J. Physiol.* 202:800-804.

38. Sleeper, J. C.; Thompson, H. K., Jr.; McIntosh, H. D.; and Elston, R. C. 1962. Reproducibility of results obtained with indicator-dilution technique for estimating cardiac output in man. *Circ. Res.* 11:712-720.

39. Spears, C. P.; Hyatt, K. H.; Vogel, J. M.; and Langfitt, S. B. 1974. Unified method for serial study of body fluid compartments. *Aerospace Med.* 45:274-278.

40. Staffurth, J. S. and Birchall, I. 1960. The measurement of the extracellular fluid volume with radioactive bromine. *Clin. Sci.* 19:45-53.

41. Stearns, G. 1939. The mineral metabolism of normal infants. *Physiol. Rev.* 19:415-438.

42. Storaasli, J. P.; Krieger, H.; Friedell, H. L.; and Holden, W. D. 1950. The use of radioactive iodinated plasma protein in the study of blood volume. *Surg. Gynec. Obstet.* 91:458-464.

43. Usher, R.; Shephard, M.; and Lind, J. 1963. Blood volume of the newborn infant and placental transfusion. *Acta Paediatr.* 52:497-512.

44. Zierler, K. L. 1962. Circulation times and the theory of indicator-dilution methods for determining blood flow and volume. In *Handbook of physiology, Sec. 2, Circulation*, vol. 1, pp. 585-615. Washington, D.C.: Amer. Physiological Soc.

# 14

## Myocardial Determinants of Fetal Cardiac Output

### Stanley E. Kirkpatrick and William F. Friedman
Division of Pediatric Cardiology
Department of Pediatrics
University of California, San Diego
La Jolla, California

Studies from our laboratory concerning the intrinsic mechanical properties of the developing heart have shown that cardiac muscle isolated from fetal lambs close to term is significantly less capable of generating isometric active tension per unit of cross-sectional area when compared to the adult (5, 14). Significant reductions were also noted when fetal muscle was compared to adult in both the extent and velocity of shortening at any load (5). A systematic evaluation of the fetal heart with respect to active (17) and passive stiffness (20), energy production (26), the rate of turnover of energy-yielding processes (5), and of the utilization of energy by the contractile apparatus (5, 10) did not yield an explanation for the age-related differences in the heart's mechanical properties.

Both scanning and transmission electron microsopic studies of the ultrastructural development of cardiac myocytes in fetal and postnatal sheep provided an explanation for the depression of forced generation and the ability to shorten in fetal cardiac muscle (5, 23). It was quite clear that myofibrillar development was quite sparse in fetal life, when compared to the adult, so that less contractile tissue existed per unit volume in fetal myocardium. Thus, although sarcomere length-tension relations appeared identical in the fetus and adult, the fetal heart contained fewer sarcomeres/unit volume.

In further studies from our laboratory, dynamic changes were observed in the anatomic, biochemical, and physiological disposition of cardiac catecholamines in the perinatal period(5, 6, 8, 9). Fetal lambs and quite young neonates were supersensitive to norepinephrine, the sympathetic neurotransmitter. Histochemical studies demonstrated that these hearts were partially innervated with a good deal of the visualized norepinephrine residing in preterminal nerve trunks, rather than in terminal nerve endings. Beta receptor sensitivity was similar in fetal and adult myocardium, indicating that catecholamine receptor sites were functional before the complete development of an extrinsic nerve supply. At a comparable stage of development, the adrenal glands, unlike the heart, were found to contain abundant catecholamine stores. Puppy studies indicated that cardiocirculatory control in the young depended upon extracardiac adrenergic support to a remarkably greater degree than in the adult (9). It was of interest that in both the late fetal and early neonatal stages of development when a reduction existed in sympathetic nerves and in cardiac norepinephrine stores and in the intraneuronal enzymes concerned with norepinephrine biosynthesis and degradation, there were no age-related differences in parasympathetic innervation or in responsiveness of cardiac tissue to acetylcholine, the cholinergic neurotransmitter (5, 6). Thus, one could speculate that whenever an imbalance exists between the cardiac output and the perfusion requirements of peripheral tissues in the fetal or neonatal organism, there is a reduced capability from a mechanical point of view for a cardiac response as vigorous as in the adult, and that in the perinatal period the interaction between a supersensitive myocardium and the adrenal release of catecholamines plays a more critical compensatory role in maintaining ventricular contractility than in the adult.

In an effort to analyze more clearly the determinants of fetal cardiac output, we have turned most recently to utilization of a chronically instrumented fetal lamb preparation that allows the serial evaluation of fetal cardiac function (7, 12, 13). The *in situ* study of the latter was considered of major importance in view of the evidence presented above that the intrinsic physiological properties, as well as the ultrastructure of the heart, may vary considerably with age. In addition, it should be recognized that the assessment of ventricular performance in the fetus *in situ* offers important insights to understanding the differences in the responses of the fetal, neonatal, and adult circulations to disease states and to alterations in the physical and chemical environment.

Accordingly, the remainder of this report describes the results of chronically instrumented fetal lamb experiments in which three factors determining fetal cardiac output have been examined: (1) heart rate and poststimulation potentiation, (2) pacemaker location and its effect on redistribution of blood flow across the foramen ovale, and (3) the effectiveness of the Frank-Starling relationship in maintaining a stable fetal cardiac output.

The general experimental situation is shown diagrammatically in Figure 1. Data are obtained from continuously monitoring internal left ventricular dimensions and pressures in the intact, undisturbed fetal lamb *in utero*. Time-dated, pregnant ewes were operated upon at known fetal gestational ages of 103-115 days (total term, 150 days). Under spinal anesthesia, hysterotomy was performed and the instrumentation accomplished. Systemic arterial pressure was recorded with a catheter inserted in the fetal carotid artery. A balloon catheter with lumen was placed in the superior vena cava to measure central venous pressure and also to produce acute alterations in loading conditions by inflation of the balloon. In addition, after a left thoracotomy, bipolar pacing wires were sutured to the left atrial appendage for cardiac pacing or for recording the electrocardiogram. In selected experiments, pacing wires were sutured to both right and left atrial appendages. A left atrial catheter was inserted both for recording pressure and for volume infusion to allow the study of pressure-volume relations. In order to obtain left ventricular cardiac output, indicator dilution curves were obtained with injection of dye into the left atrium and sampling from the carotid artery. Since injection of dye was performed in left atrium with sampling from carotid artery, it was important that indicator did not cross from left to right across the foramen ovale. This potential problem was as-

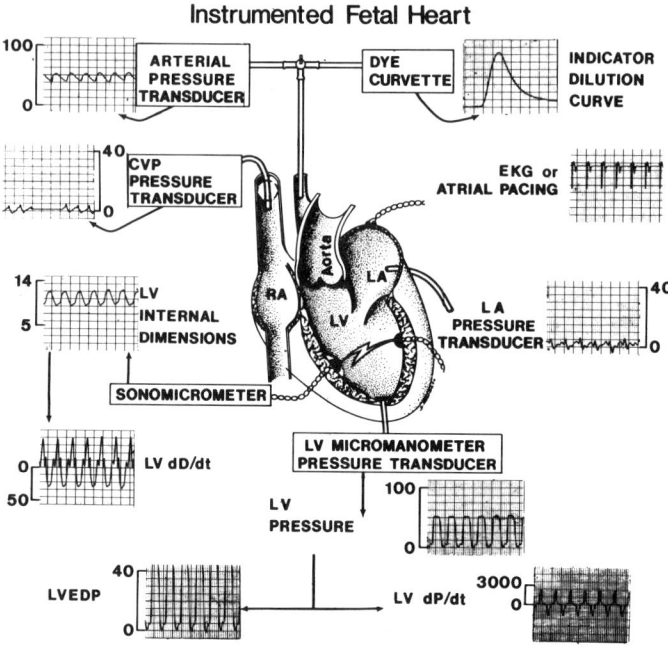

Fig. 1. The instrumentation of the fetal lamb heart is illustrated. Arterial pressure is obtained via a catheter in the carotid artery. This catheter is also used to withdraw blood for indicator dilution curves. A catheter in the left atrium (LA) records pressure and serves as the injection site for indicator dilution curves. Central venous pressure (CVP) is recorded from a superior vena caval (SVC) balloon catheter with lumen positioned above the SVC-right atrial (RA) junction. Bipolar pacing wires are sutured to the left atrial appendage to obtain the fetal electrocardiogram or to achieve a fixed heart rate. Sonomicrometer piezoelectric crystals on the endocardial surface of the left ventricle (LV) continuously record internal dimensions. This signal is differentiated to obtain the velocity of dimension change (LV dD/dt). A solid-state micromanometer pressure transducer obtains high fidelity LV pressure which is amplified to give LV end diastolic pressure (LVEDP) and differentiated to derive LV dP/dt. Source: Am. J. Obstet. Gynec.

sessed in selected experiments by having a catheter in the main pulmonary artery with injection of indicator in left atrium. Early appearance of indicator in the main pulmonary artery was not observed. Moreover, in selected experiments, dye outputs were compared after injection into left atrium and into left ventricle. Since no significant differences existed in these values there was no suggestion that indicator was lost into the right heart after left-sided injection.

A solid-state, micromanometer pressure transducer was inserted into the cavity of the left ventricle. This transducer was calibrated using the systemic arterial and left atrial pressures. The zero pressure reference was provided by a catheter sutured to the fetal precordium or by a catheter placed into the fetal trachea. The same zero reference catheter also allowed the measurement of intrauterine pressure which was subtracted from all fetal pressure data. Lastly, two 3.5-mm piezoelectric crystals were placed opposite one another on the endocardial surface of the left ventricle, perpendicular to the longitudinal axis of the left ventricle and across its greatest internal diameter. The sonomicrometer used to measure left ventricular internal diameter was a modification of the one described previously by Stegall, et al. (24). In order to measure distance linearly to 3 mm with a 1 percent error, one endocardial ultrasonic transducer was shock-excited at a high repetition rate (5,000/sec) and the time required for each ultrasonic burst to pass from one transducer to the other was converted into a voltage suitable for recording. Since sound velocity in blood is $1.5 \times 10^3$ M/sec, the readings may be converted into distance. Differentiating circuits were employed to derive the first derivatives of both left ventricular pressure and left ventricular internal dimensions.

After operation the uterus was replaced and all leads and catheters exteriorized to the flank of the ewe. Fetal hemodynamics stabilize 1-2 weeks postoperatively and physiological studies were performed up to 3 times weekly thereafter. During the study of fetal hemodynamics, the ewe stood quietly watching or eating in a mobile cage.

## POSTSTIMULATION POTENTIATION AND HEART RATE

It has been suggested that the potentiating effect of extra systoles on myocardial contractility may be poorly developed in the young mammalian heart (1). Moreover, questions have arisen concerning the influence on cardiac performance in the conscious animal of the relation between contraction frequency and contractility (18, 25).

Utilizing our chronic fetal lamb preparation, recordings during the basal state and at paced heart rates per minute of 180, 210, 240, 270, and 300 were obtained during quiescent periods in the fetus when there were no respiratory movements (14). Pacing was maintained for 30 sec at each level in all animals. Steady-state data were chosen for analysis at the end of each of the paced heart rates, and sufficient recovery time was allowed after each pacing period for heart rate and blood pressure to return to control levels. In order to evaluate changes in inotropic state induced by changes in contraction frequency, beats with similar R-R intervals were analyzed immediately prior to and after pacing that were matched with regard to systemic arterial pressure, left ventricular end-diastolic pressure, and left ventricular end-diastolic diameter (Fig. 2). The first or second beats after pacing were analyzed prior to any expected decline in a frequency-induced enhancement of contractility. Paired t tests were used for statistical comparisons.

Thirteen studies, beginning 14 days postoperatively, were performed in seven chronically instrumented fetal lambs (Fig. 3). Basal heart rate averaged 150/min (range 135-165). As the heart rate was increased with left atrial pacing, there was a progressive decline in the internal left ventricular dimensions with a significantly greater reduction in end-diastolic diameter than in end-systolic diameter. At heart rates of 240 and behond there is a slight, but significant, fall in peak systemic pressure. During the increase in heart rate induced by pacing, there were no significant changes observed either in the directly measured velocity of shortening, dD/dt, or in the calculated rate of mean circumferential fiber shortening, VCF. In contrast, dP/dt increased significantly and progressively up to a heart rate of 270 min and returned to near basal values at 300 beats/min.

In comparing control beats and the first or second beats immediately after pacing, there is a striking and stepwise increase in both dD/dt and mean VCF (Fig. 4). Thus, at a heart rate of 300/min, dD/dt increased 47%, and mean VCF increased 39%.

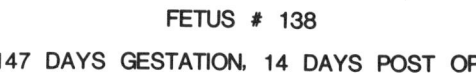

## FETUS # 138
### 147 DAYS GESTATION, 14 DAYS POST OP

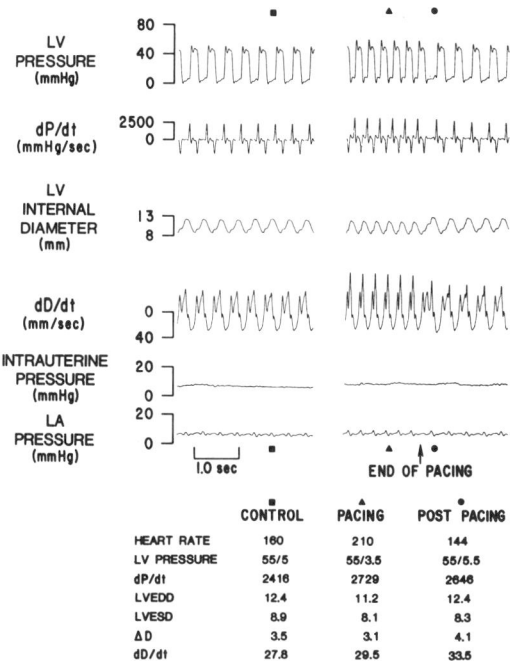

|  | CONTROL | PACING | POST PACING |
|---|---|---|---|
| HEART RATE | 160 | 210 | 144 |
| LV PRESSURE | 55/5 | 55/3.5 | 55/5.5 |
| dP/dt | 2416 | 2729 | 2646 |
| LVEDD | 12.4 | 11.2 | 12.4 |
| LVESD | 8.9 | 8.1 | 8.3 |
| ΔD | 3.5 | 3.1 | 4.1 |
| dD/dt | 27.8 | 29.5 | 33.5 |

Fig. 2. Fetus 138 at 147 days gestation, 14 days postoperative. A representative pacing study at 210 beats/min. Square, triangle, and circle, mark beats used to obtain control, pacing, and postpacing data, respectively. Note similarly in left ventricular (LV) pressure and left ventricular end-diastolic diameter (LVEDD) pre- and post-pacing. Heart rate is slightly slower following pacing. On first beat after pacing with LV pressure and LVEDD the same, there is a significant increase in rate of LV pressure rise (dP/dt). Postpacing beat was comparable to control with respect to LV systolic pressure and end-diastolic diameter but showed a marked augmentation in rate of rise of LV pressure (dP/dt) and in velocity (dD/dt) and extent (ΔD) of LV shortening. LA = left atrium. Source: *American Journal of Physiology*.

The results of these studies demonstrate that increased frequency of contraction exerts a significant, positive inotropic effect upon the fetal heart. The augmentation was observed best after fiber length increased immediately after pacing, but was masked during pacing by the

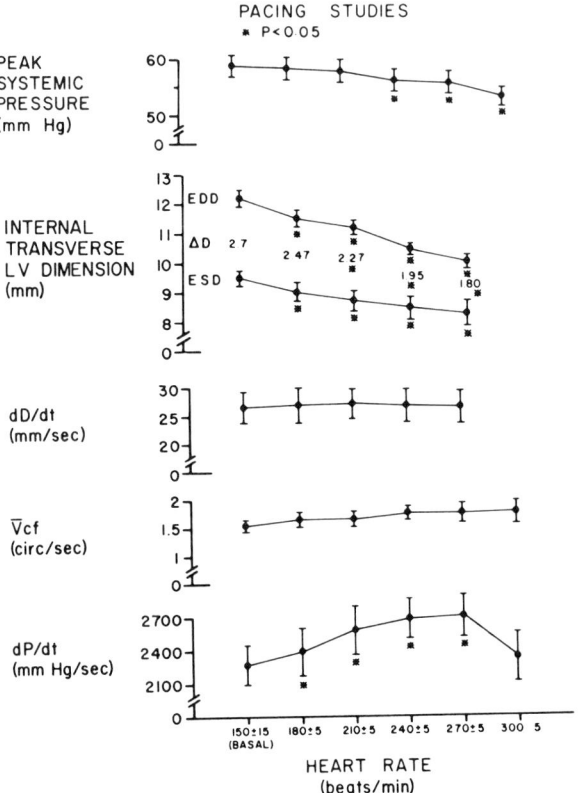

Fig. 3. Measurements of peak systemic pressure, internal transverse left ventricular dimension (EDD, end-diastolic diameter; EDS, end-systolic diameter; ΔD, EDD - ESD), maximum rate of left ventricular diameter change (dD/dt), mean circumferential fiber shortening (Vcf), maximum rate of rise of left ventricular pressure (dP/dt) at basal heart rate (150 ± 15 beats/min) and at each of the indicated paced heart rates. Asterisk denotes statistically significant differences from basal values ($P < 0.05$). Source: Am. J. of Physiol.

rate-induced reduction in fiber length. These results are quite comparable to findings in the chronically instrumented adult dog (16). As in the latter study, it did not appear in the present investigation that the increase in contractility was related to beta-adrenergic stimulation, since no differences in the responses to pacing were observed after pretreatment with the beta-adrenergic blocking agent, propranolol.

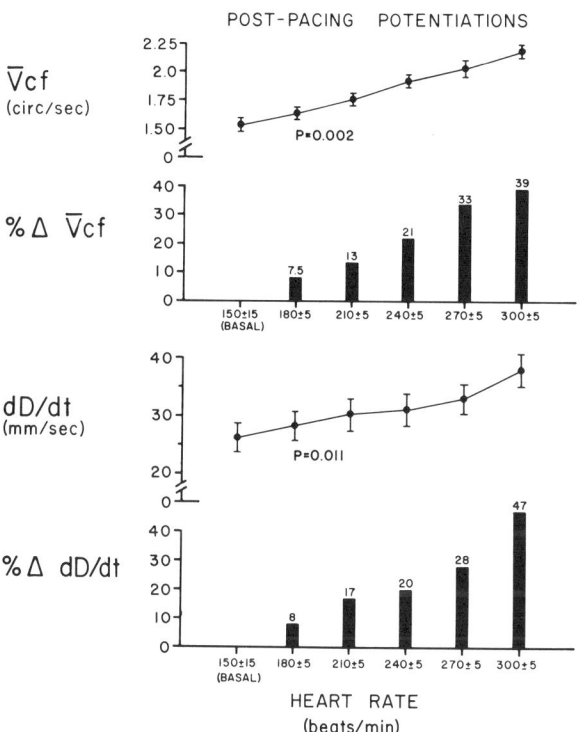

Fig. 4. Absolute values of mean $\overline{V}cf$ and LV dD/dt are shown in first and third panels at control heart rate and at indicated paced heart rates. In second and fourth panels the percentage increase from control is shown. At and beyond a paced heart rate of 180 ± 5 beats/min, velocity parameters were significantly increased over basal values. Source: *Am. J. of Physiol.*

## PACEMAKER LOCATIONS

It has been suggested that heart rate is the primary regulator of fetal cardiac output (21). Rudolph and Heymann and their associates have shown that *right* ventricular output increases substantially as increases in heart rate were produced by left atrial pacing (22). Studies from our laboratory have shown a reduction in *left* ventricular output associated with *left* atrial pacing (14) over a comparable range of heart rates. Since the above analyses of individual ventricular outputs show opposite directional changes with alterations in heart rate, simultaneous measurements of right and left ventricular output and of total cardiac output with selective pacing of each atrium was considered of importance in order to examine the latter interactions (19).

Five fetuses were studied acutely; the four remaining fetuses were returned to the uterus after their catheter and wires were exteriorized. The full two-week recovery period was allowed before any studies were performed in the latter group, and a total of eight chronic studies were performed.

In accord with our previous results (12), fetal arterial pH, $pO_2$, $pCO_2$, and hematocrits were monitored during all studies, and averaged $7.35 \pm .01$, $24.8 \pm 1.5$, $50.6 \pm 1.6$, and $37.2 \pm 1.6$, respectively. During acute studies, intravascular pressures were obtained using Statham P-23 dB strain gauge transducers and recorded on a Clevite-Brush multichannel oscillograph and on magnetic tape. In all studies, the indocyanine green dye technique was employed to measure cardiac outputs (19). Cardiogreen dye was injected (0.5 cc in a concentration of 1.25 mg/cc) into the right atrium while blood for sampling was withdrawn simultaneously from the fetal ascending aorta and main pulmonary artery. Catheter volumes were 1.1 cc, and sampling rate was 8 cc/min from each great vessel. The total sampling period averaged 40 sec, and therefore approximately 12 cc were removed for each study. Sampled blood was reinfused and indicator was then injected into the left atrium with simultaneous sampling from the great arteries. These paired studies were done in the basal state, again when the right atrium was paced at 270 stimuli/min, and a third time when the left atrium was paced at the latter rate.

Ventricular outputs were calculated from the areas under the appearance curves on a standard program employing a PACE-EAI analog digital computer.

The present study identified the necessity of considering the location of the atrial pacemaker site before drawing conclusions concerning the role of heart rate per se in regulating fetal cardiac output. We found that total cardiac output was unchanged statistically from control values in both acute and chronic experiments with both left and right atrial pacing. Despite the fact that total cardiac output did not change significantly with selective atrial pacing, important changes occurred in the contribution of each individual ventricle to total cardiac output. Further, selective atrial pacing resulted in important differences in atrial pressure relations that were related

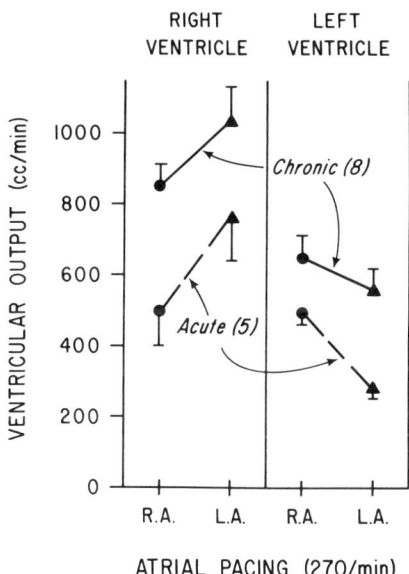

Fig. 5. Right and left ventricular outputs are shown under conditions of right atrial (RA) and left atrial (LA) pacing at a heart rate of 270 beats/min in both acute and chronic experiments. Each point and the vertical bars represent the mean ±SEM. The numbers in parentheses reflect the number of experiments performed. Source: Am. J. of Physiol.

not to heart rate, but rather to the location of the pacemaker site since output of either ventricle was reduced significantly when the ipsilateral atrium was paced, when compared to either ventricle's output with contralateral atrial pacing (Fig. 5). This study suggests that the higher initial left atrial pressure during left atrial pacing diminishes the magnitude of normal right to left shunting across the foramen ovale, resulting in augmentation of right ventricular filling and a relative increase in right ventricular output. Conversely, right atrial pacing maintains the normal fetal right atrial pressure dominance, promotes right to left flow across the foramen ovale, and increases left ventricular filling and output. Although changing the articifial pacemaker site alters individual ventricular outputs, no important changes occur in total cardiac output.

## THE FRANK-STARLING RELATION

In previously reported studies (21) it was stated that the Frank-Starling mechanism is of little importance in the fetal sheep, and that heart rate is the primary regulator of cardiac output. Evidence for this latter view emanated from chronically instrumented fetal lamb experiments in which intravascular volume infusions did not increase right ventricular output substantially (11) and studies in which left atrial pacing significantly increased right ventricular output (22). In the latter experiments right ventricular output was shown to increase progressively from low heart rates produced by vagal stimulation to higher rates produced by left atrial pacing until a decline occurred beyond heart rates of 300 beats/min (15).

Thus, although past studies utilizing isolated cardiac muscle (5), chick embryo (4), acute intact fetal lamb (2), or neonatal lamb (3) models have shown that an intrinsic relationship exists between resting fiber length and cardiac performance, questions have arisen concerning the presence and importance of the Frank-Starling mechanism in the intact fetal heart. The purpose of the studies described below was to examine the effects of natural and experimentally induced changes in left ventricular end diastolic diameter on the extent of left ventricular shortening in the intact, undisturbed lamb fetus *in utero*. In addition, the relationship was examined between left ventricular end diastolic pressure and the extent of myocardial fiber shortening.

Smaller left ventricular end diastolic diameters and lesser end diastolic pressures were obtained by inflation of a balloon in the superior vena cava for no longer than 30 sec. Greater end diastolic pressures and larger end diastolic diameters were obtained by infusion of heparinized fetal blood (15-20 cc over 15-20 sec) directly into the left atrium. The extent of shortening of the left ventricular internal diameter was measured as the distance from beginning to end of ejection on the diameter trace. During balloon inflation or blood infusions, data were accumulated during quiescent periods in the fetus when there were no spontaneous respiratory movements. However, changes in left ventricular pressures and dimensions were also recorded during the course of frequently occurring periods of respiratory activity, as well as during spontaneous fluctuations in heart rate. Data were obtained from a total of 21 experiments on 7 fetal lambs.

The relationship between fetal left ventricular stroke volume and the change in left ventricular internal diameter

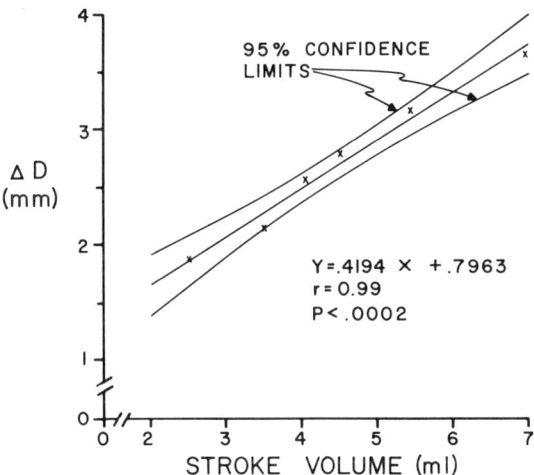

Fig. 6. The relationship between the extent of LV shortening ($\Delta D$) measured directly from the internal dimension signals and LV stroke volume obtained by dividing heart rate into left ventricular output determined by the indocyanine green dye technique. Each data point is the average value obtained from four animals.
Source: *Am. J. of Physiol.*

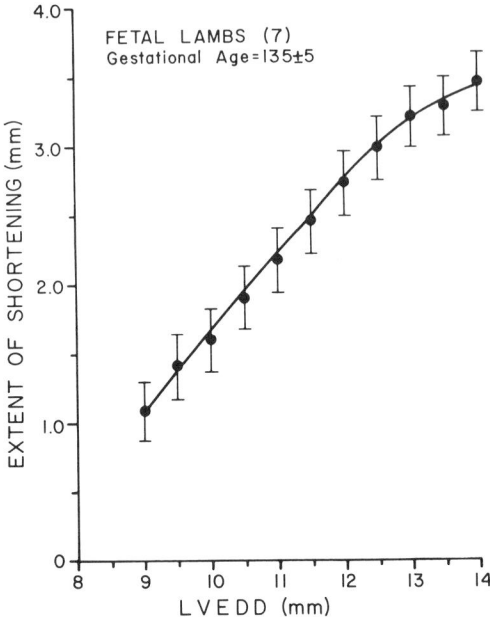

Fig. 7. Relationship between left ventricular end diastolic diameter (LVEDD) and extent of LV shortening in 7 experiments from 4 fetal lambs. Each point and the vertical bars represent the mean ±S.E. Source: *Am. J. of Physiol.*

is depicted in Figure 6. A highly significant relationship was observed with a correlation coefficient of + 0.99. Similarly, a direct correlation existed between the extent of shortening and the end diastolic diameter (Fig. 7). Thus, a 250% increase in left ventricular shortening was observed in association with a 5-mm increase in end diastolic diameter.

The relationship between end diastolic pressure and the extent of left ventricular shortening from four fetal lambs is shown in Figure 8. Over the physiological range of end diastolic dimensions (10.5-13 mm) and pressures (2.5-8 mm Hg) (16), the change in extent of left ventricular shortening averaged 68%. Beyond a left ventricular end diastolic pressure of 10 mm Hg there was little further increase in the extent of left ventricular shortening.

Intermittently, in the course of observing fetuses in the basal state, spontaneous respiratory activity occurred (Fig. 9). Thus, it was possible to observe beat-to-beat fluctuations in end diastolic pressure and dimensions while systemic pressure and heart rate remained essentially unchanged. During an inspiratory effort, intratracheal and left atrial pressure fell sharply while end diastolic diameter increased variably with the augmentation in ventricular filling (arrow, Fig. 9). The latter increase in preload resulted immediately in substantial enhancement of the extent of shortening and, hence, left ventricular stroke volume. Left ventricular end diastolic pressure appeared lower on oscillographic tracings recorded during an inspiratory effort but it must be pointed out that the appropriate subtraction of intratracheal pressure from left ventricular pressure would have resulted in an

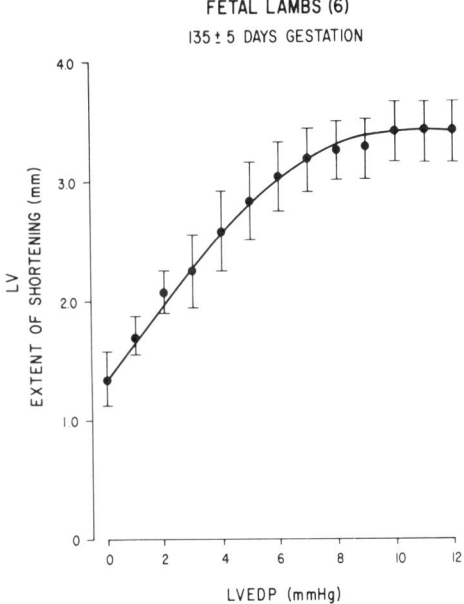

Fig. 8. Relationship between left ventricular end diastolic pressure (LVEDP) and extent of LV shortening in six studies on four fetal lambs. Each point and the vertical bars represent the mean ±S.E. Source: Am. J. of Physiol.

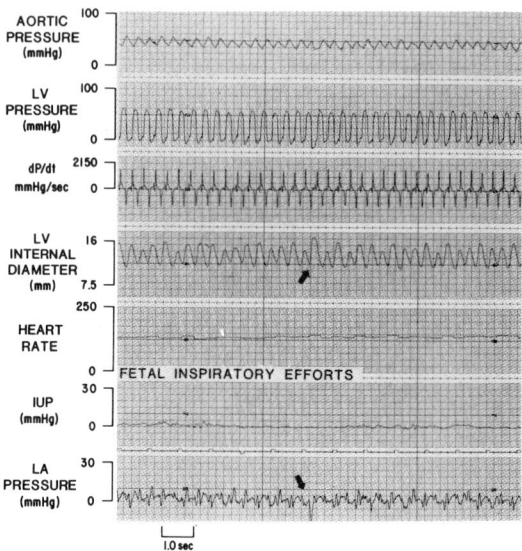

Fig. 9. Representative tracings from a fetal lamb (128 days gestation, 15 days postoperative) showing spontaneous fetal respiratory movements and their major beat-to-beat effects upon left ventricular internal dimensions with only minimal changes in aortic pressure and heart rate. The arrow marks one of the larger negative deflections caused by an inspiratory effort of the fetus. Note that no changes occur in intrauterine pressure (IUP). Source: *Am. J. of Physiol.*

increase in end diastolic pressure. Thus, during respiratory efforts beat-to-beat changes in ventricular filling result in alterations of stroke volume ($\Delta D$) which maintain cardiac output stable while heart rate remains unchanged.

In order to evaluate the dependence of the left ventricle on the Frank-Starling mechanism to maintain its output constant during respiratory-induced variations in end-diastolic diameter the average extent of shortening per unit time was calculated in four basal studies of three fetal lambs. In any single fetus, while heart rate varied no more than ± 2 beats/min, the average extent of shortening/sec was not significantly different during varied rates of respiratory activity (Fig. 10). Thus, at a constant heart rate, while fetal respiratory rate varied from 50 to 100/min there is no significant alteration in left ventricular output/time by virtue of beat-to-beat stroke volume changes appropriate to changes in diastolic fiber length.

An additional approach to this same aspect of cardiovascular homeostasis was provided in four studies of two fetal lambs (gestational age 131-141 days) in which heart rate varied spontaneously from 114 to 180 beats/min (for prolonged periods) in the absence of respiratory activity. In each fetus left ventricular outputs measured over this wide range of spontaneous frequencies of contraction remained constant (Fig. 11) because of adjustments in the extent of shortening as a function of left ventricular end diastolic diameter.

The results of the present investigation demonstrate clearly that the Frank-Starling relationship is both operative and effective in the fetal lamb heart. Measurement of left ventricular end diastolic diameter appeared to be

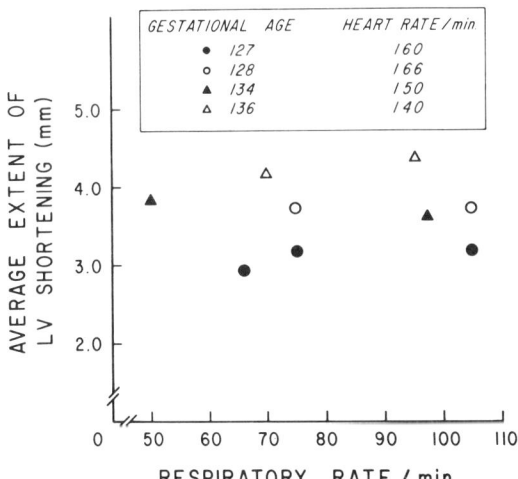

Fig. 10. In the absence of a change in heart rate the constancy of LV cardiac output in any individual fetus is illustrated over a wide range of rates of respiration (four studies of three fetal lambs). Each like symbol represents a value from the same animal. Source: *Am. J. of Physiol.*

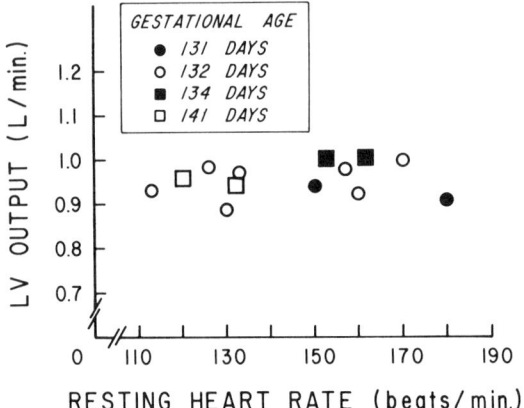

Fig. 11. The constancy of left ventricular output in two fetuses studied at various gestational ages is illustrated over a wide range of naturally occurring heart rates. Each like symbol represents a value from the same animal. Source: Am. J. of Physiol.

a substantially more reliable indicator of resting fiber length than measurements of end diastolic pressure. In this regard, our results indicate that the extent of left ventricular shortening changes little beyond end diastolic pressures of 10 mm Hg and support the notion that the fetal heart may be overstretched beyond that level. Over the narrow, but physiological, range of end diastolic pressures from 2.5 to 8 mm Hg there was a profound augmentation in the extent of left ventricular shortening. Furthermore, in the absence of experimental interventions designed specifically to reduce or enhance left ventricular end diastolic volume, significant beat-to-beat changes occurred in end diastolic diameter and in the extent of left ventricular shortening during spontaneous respiratory activity in the fetus. During fetal respirations, in the absence of significant changes in the heart rate or systemic pressure, appropriate changes in the extent of left ventricular shortening up to and often exceeding 100% maintained ventricular output constant. Moreover, prolonged changes in spontaneous heart rates (114 to 180 beats/min) were accompanied by changed in left ventricular output. Thus, in contrast to the vagal stimulation studies of Rudolph and Heymann (21), the fetus with spontaneously occurring slow heart rates is capable of maintaining cardiac output. Thus, it would appear that while changes in heart rate can influence the inotropic state of the myocardium (14) there is no convincing evidence that such contraction frequency changes are the only determinant of ventricular output in the fetus. Rather, the observations generated by our chronic studies provide strong support for the concept that changing myocardial fiber length is a fundamental determinant of fetal cardiac output along with the heart rate and inotropic state.

## ACKNOWLEDGMENTS

Supported by U.S. Public Health Services Grants HL 12373 and HL 05846.

Permission has been granted by the *American Journal of Obstetrics and Gynecology* for the use of Figure 1 and the *American Journal of Physiology* for the use of Figures 2-11.

## REFERENCES

1. Arcilla, R. A.; Lind, J.; Zetterquist, P.; and Oh, W. 1966. Hemodynamic features of extrasystoles in newborn and older infants. *Am. J. Cardiol.* 18:191.

2. Brinkman, C. R., II; Johnson, G. H.; and Assali, N. S. 1972. Hemodynamic effects of bradycardia in the fetal lamb. *Am. J. Physiol.* 223:1465-1469.

3. Downing, S. E.; Talner, N. S.; and Gardner, T. H. 1965. Ventricular function in the newborn lamb. *Am. J. Physiol.* 208:931-937.

4. Faber, J. J.; Green, T. J.; and Thornburg, K. L. 1974. Embryonic stroke volume and cardiac output in the chick. *Dev. Biol.* 41:14-21.

5. Friedman, W. F. 1972. The intrinsic physiologic properties of the developing heart. *Prog. Cardiovasc. Dis.* 15:87.

6. ----. 1972. Neuropharmacologic studies of perinatal myocardium. *Cardiovasc. Clinics* vol. 4, no. 3:44-57.

7. Friedman, W. F. and Kirkpatrick, S. E. 1975. *In situ* physiological study of the developing heart. In *Recent advances in studies on cardiac structure and metabolism*, vol. 5, p. 497, eds., A. Fleckenstein and N. Dhalla. Maryland: Univ. Park Press.

8. Friedman, W. F.; Pool, P. E.; Jacobowitz, D.; Seagren, S.; and Braunwald, E. 1968. Sympathetic innervation of the developing rabbit heart: biochemical and histochemical comparisons of fetal, neonatal, and adult myocardium. *Circ. Res.* 23:25-32.

9. Geis, W. P.; Tatooles, C. J.; Prioloa, D. V.; and Friedman, W. F. 1975. Factors influencing neurohumoral control of the heart in the newborn. *Am. J. Physiol.* 228:1685.

10. Henry, P.; Ahumada, G.; Friedman, W. F.; and Sobel, B. 1972. Simultaneous isometric tension and ATP hydrolysis in glycerinated fibers. *Circ. Res.* 31:740.

11. Heymann, M. A. and Rudolph, A. M. 1973. Effects of increasing preload on right ventricular output in fetal lambs *in utero*. *Circulation* 48:IV-37.

12. Kirkpatrick, S. E.; Covell, J. W.; and Friedman, W. F. 1973. A new technique for the continuous assessment of fetal and neonatal cardiac performance. *Am. J. Obstet. Gynec.* 116:963.

13. Kirkpatrick, S. E., and Friedman, W. F. 1974. Advances in the study of the fetal heart. *Contemporary Ob. Gyn.* 3:21.

14. Kirkpatrick, S. E.; Naliboff, J.; Pitlick, P. T.; and Friedman, W. F. 1975. The influence of post stimulation potentiation and heart rate on the fetal lamb heart. *Am. J. Physiol.* 229:318-323.

15. Kirkpatrick, S. E.; Pitlick, P. T.; Naliboff, J.; and Friedman, W. F. 1976. The importance of the Frank-Starling relationship as a determinant of fetal cardiac output. *Am. J. Physiol.* (In press.)

16. Mahler, F.; Yoran, C.; and Ross, J., Jr. 1974. Inotropic effect of tachycardia and post-stimulation potentiation in the conscious dog. *Am. J. Physiol.* 227:569-575.

17. McPherson, R. A.; Kramer, M. F.; Covell, J. W.; and Friedman, W. F. 1976. A comparison of the active stiffness of fetal and adult cardiac muscle. *Ped. Res.* 10: 660-664.

18. Noble, M. I. M.; Wyler, J.; Milne, E. N. C.; Trenchard, D.; and Guz, A. 1969. Effect of changes in heart rate on left ventricular performance in conscious dogs. *Circ. Res.* 24:285-295.

19. Pitlick, P. T.; Kirkpatrick, S. E.; and Friedman, W. F. 1976. Distribution of fetal cardiac output-importance of pacemaker location. *Am. J. Physiol.* 231:204-208.

20. Romero, T.; Covell, J.; and Friedman, W. F. 1972. A comparison of the pressure-volume relations of the fetal, newborn, and adult heart. *Am. J. Physiol.* 222:1285.

21. Rudolph, A. M., and Heymann, M. A. 1973. Control of the foetal circulation. In *Foetal and neonatal physiology, proceedings of the Sir Joseph Barcroft centenary symposium*, pp. 89-111. Cambridge: Cambridge Univ. Press.

22. ----. 1974. Fetal and neonatal circulation and respiration. *Ann. Rev. Physiol.* 36:187-207.

23. Sheldon, C. A.; Friedman, W. F.; and Sybers, H. D. 1976. Scanning electron microscopy of fetal and neonatal lamb cardiac cells. *J. Molec. Cell. Cardiol.* (In press.)

24. Stegall, H. F.; Kardon, M. B.; Stone, H. L.; and Bishop, V. S. 1967. A portable, simple sonomicrometer. *J. Appl. Physiol.* 23:289.

25. Vatner, S. R., and Braunwald, E. 1973. A comparison of the inotropic effects of ouibain and tachycardia in the normal and failing right ventricles of the conscious dog. *Clin. Res.* 21:456.

26. Wells, R. J.; Friedman, W. F.; and Sobel, B. E. 1972. Increased oxidative metabolism in the fetal and newborn lamb heart. *Am. J. Physiol.* 222:1488-1493.

# 15

## Force-Frequency Relations of the Neonatal Cat Heart

J.G. Maylie, K.L. Thornburg, and J.J. Faber

Department of Physiology
University of Oregon
Health Sciences Center
Portland, Oregon

The role of the sarcoplasmic reticulum in the generation of contractile force in the mammalian heart has been studied extensively. The sarcoplasmic reticulum is divided into two segments: the longitudinal sarcoplasmic reticulum and the sarcoplasmic reticulum cisternae. The longitudinal sarcoplasmic reticulum consists of a fine network of tubules that enmesh the myofibrils. The sarcoplasmic reticulum cisternae are specialized portions of the sarcoplasmic reticulum which make contact with the cell surface and the T-tubular membranes. Excitation-contraction coupling is now believed to be due to intracellular release of calcium for the cisternae during the initial phase of the action potential, and a transsarcolemmal influx of calcium during the plateau phase of the action potential (2, 4). The amount of calcium derived from these two sources determines the intracellular calcium concentration during systole, and hence the force of contraction.

At the same time, two mechanisms compete for the removal of calcium. The longitudinal sarcoplasmic reticulum actively sequesters calcium and a sarcolemmal sodium/calcium exchange carrier extrudes calcium into the extracellular space. Relaxation ensues when the calcium concentration begins to decline. The calcium sequestered into the longitudinal sarcoplasmic reticulum is transported back into the cisternae, where it becomes available for release in subsequent beats. This partial recycling of stored calcium explains such phenomena as post-extrasystolic potentiation, and the subsequent return to a steady state tension only after 5-8 beats.

An extrasystole in the adult heart potentiates the following beat by 60 to 120% above the control tension (Fig. 1). The hypothesis that post-extrasystolic potentiation is a function of the sarcoplasmic reticulum is also supported by studies on the frog heart which ultrastruc-

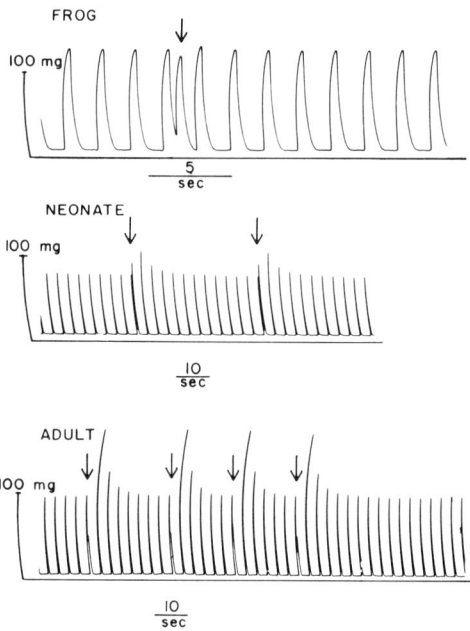

Fig. 1. Post-extrasystolic potentiation in electrically driven frog, neonatal cat and adult cat hearts. Arrows mark the extrasystoles.

turally lacks the well organized sarcoplasmic reticulum
and T-tubular system of the mammalian heart (6). Post-
extrasystolic potentiation in the frog heart is only zero to
five percent of the control tension and the return to steady
state occurs within one beat (Fig. 1).

Orkand (5) demonstrated that 1- to 2-day-old neonatal cats lack T-tubules and a well developed network of
sarcoplasmic reticulum. We used a modified Karnovsky meth-

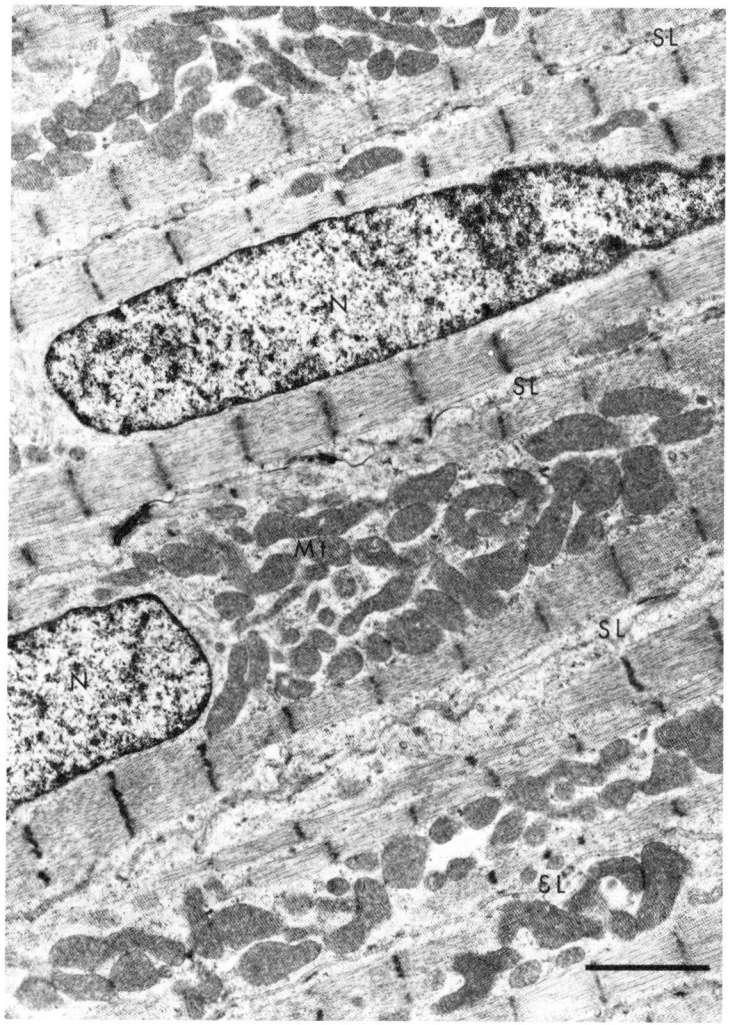

Fig. 2. Electron micrograph of papillary
muscle for a 2-day-old kitten: mitochondria
(Mt); sarcolemma (SL); nucleus (N). Horizontal bar represents 2 microns.

od to perfuse kitten hearts with 2.5% gluteraldehyde buffered in 0.1 M phosphate buffer and post-fixed in 2% osmium tetroxide. Thin plastic sections were photographed in a Joel 100S electron microscope. We confirmed that the neonatal kitten heart lacked T-tubules (Fig. 2). Neonatal heart cells averaged 3-5 microns in diameter with a single bundle of sarcomeres along the edge of the cells and the nucleus and mitochondria in the center of the cell. There was an occasional appearance of saccules in contact with the sarcolemma, suggestive of developing sarcoplasmic reticulum cisternae (Fig. 3). We saw membrane bound structures between sarcomeres, suggestive of developing longitudinal sarcoplasmic reticulum, but the amount present was very sparse in comparison to that found in the adult cat heart (3). The neonatal kitten heart is very similar to the frog heart with respect to the organization of sarcomeres and the amount of sarcoplasmic reticulum present. Thus a study of the neonatal cat heart was performed to see if it behaved as one would expect, considering its sparse sarcoplasmic reticulum system.

Right ventricular papillary muscles were excised from 1 to 3 day old neonatal and adult cat hearts under oxygen-

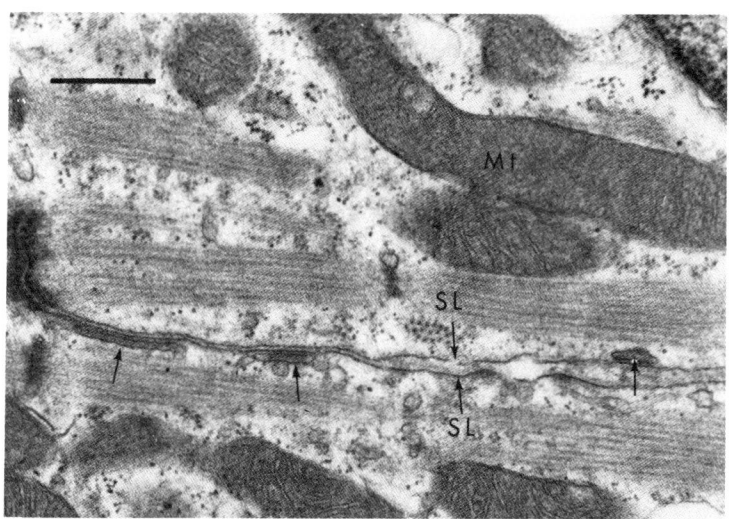

Fig. 3. Electron micrograph of papillary muscle from a 2-day-old kitten. Arrow marks saccules believed to be developing sarcoplasmic reticulum cisternae adjacent to the sarcolemna (SL); mitochondria (Mt). Horizontal bar represents 0.5 micron.

ated Tyrode solution. The muscles were transferred to a chamber perfused with oxygenated Tyrode at 37°C. Tension was recorded at a basal stimulation rate of 20 beats/minute. An extra stimulus was applied at a set interval following a regular beat. The interval was chosen so as to produce maximum potentiation of the following beat. In neonates this potentiation ranged from 15 to 40% as shown in Fig. 1. The potentiation in the adult heart is 3 to 5 times greater. The potentiated state decayed back to steady state control tension in 5 to 8 beats in both neonatal and adult hearts. Functionally, this potentiation is an example of the change in contractility associated with changes in the rate of contraction ("rate inotropism").

It has been reported (1) that the change in peak tension associated with a change in frequency of stimulation is composed of two time components as shown in Figure 4. For a change in stimulation rate from 20 to 60 beats/minute peak tension rapidly increased for 5 to 8 beats followed by a slow rise in peak tension until it eventually reached a new steady state. The increase in potentiation associated with the fast component was much greater in the

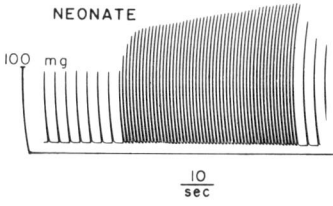

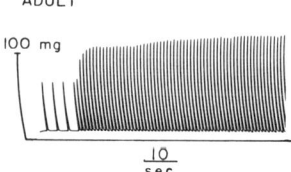

Fig. 4. Ascending frequency staircases for neonatal and adult cat hearts. Rate increase is from 20 to 60 beats/minute.

adult than the neonate. The fast component in the neonate produced only 40% of the total potentiation and 75% in the adult.

The fact that the rate inotropism consists of a fast and a slow component is also demonstrated in Figure 5. An extrasystole was applied during the slow component of increasing peak tension. The following beat was potentiated but the potentiation decayed back to the level of tension that would have existed had no extrasystole been applied. Thus the slow component of tension increase was not affected by the extrasystole suggesting independence between the fast and the slow components. This result was observed for both neonates and adults (Fig. 5). As can be seen, the degree of potentiation from the extrasystole applied during the frequency staircase was much greater in the adult than in the neonate. In fact, if an extrasystole was applied during the early phase of the slow component the peak potentiation in the neonate was less than the new steady-state tension at the end of the staircase. In the adult peak potentiation from an early extrasystole was much greater than the new steady-state level of tension (Fig. 5).

These findings support the hypothesis that there are two independent compartments from which calcium is released upon excitation. Furthermore, the fast component of the

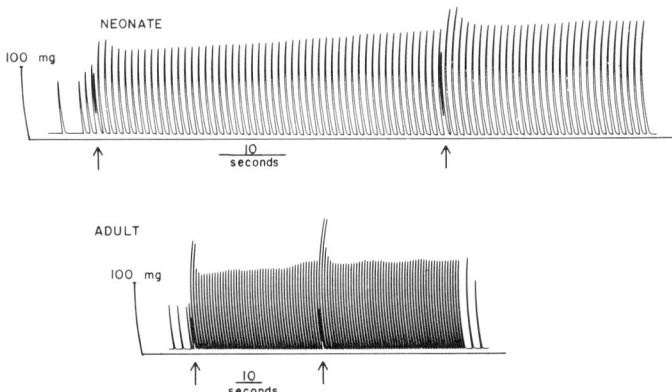

Fig. 5. Ascending frequency staircases for neonatal and adult cat hearts. Rate increase is from 20 to 60 beats/minute. Arrows mark application of extrasystolic stimuli.

frequency staircase is very likely to be the same component as the one associated with post-extrasystolic potentiation. Evidence for this is shown in Figure 6. Plotted is the decay of potentiation above control tension versus the beat number following the potentiating stimulus. Both post-extrasystolic potentiation and the fast component of a descending[1] frequency staircase for the adult and the neonate fall virtually on the same decay curve with an exponential beat constant of about 1.2.

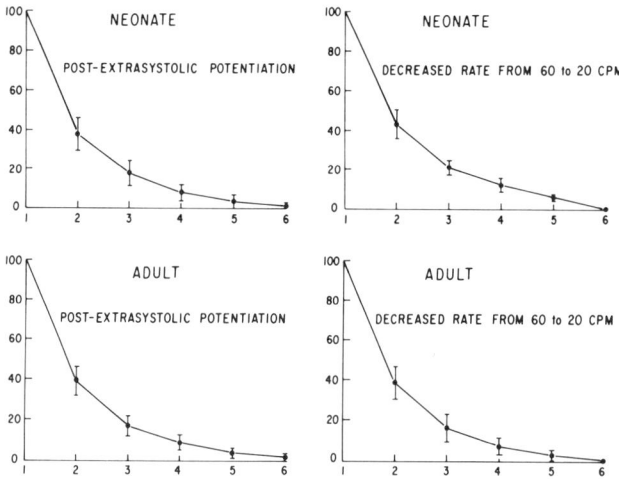

Fig. 6. Beat-dependent decay of potentiation in right ventricular papillary muscle from adult and neonatal cats. Potentiation as a percent of maximum increase above control tension is plotted on the ordinate versus beat number in the decay train on the abscissa. The descending staircase for a decreased rate from 60 to 20 beats/minute is obtained for the fast component. The decay of potentiation is essentially the same in all four experiments. The vertical bars at each point represent one standard deviation. Average number of determinants for each point ranged from 6 to 13.

---

[1] The descending frequency staircase is the decay of tension following a change from a high to a low frequency of stimulation.

This study demonstrates that the ability of the fast component of rate inotropisms to add to the tension generated in systole is much less in the neonate than in the adult. We feel that the fast component is associated with calcium release and re-accumulation in the sarcoplasmic reticulum, which is not as well developed in the neonate as in the adult.

## ACKNOWLEDGMENTS

We would like to thank Carla Baltrusch Dow for her expert technical assistance.

Supported by NIH Training Grant 5T01 GM00538 and grants from the Oregon Heart Association.

## REFERENCES

1. Anderson, P. A. W.; Manring, A.; Sommer, J. R.; and Johnson, E. A. 1976. Cardiac muscle: an attempt to relate structure to function. *J. Mol. Cell. Cardiol.* 8:123-143.

2. Bassingthwaighte, J. B., and Reuter, H. 1972. Calcium movements and excitation contraction coupling in cardiac cells. In *Electrical phenomena in the heart*, ed., W. DeMello. New York: Academic Press.

3. Fawcett, D. W. and McNutt, S. N. 1969. The ultrastructure of the cat myocardium. *J. Cell. Biol.* 42:1-45.

4. Morad, M. and Goldman, Y. 1973. Excitation-contraction coupling in heart muscle: membrane control of development of tension. In *Prog. Biophys. Mol. Biol.* vol. 27, eds., A. J. V. Butler, and D. Noble. Oxford and New York: Pergamon Press.

5. Orkand, P. M. 1964. Light and electron microscopical studies of skeletal and cardiac muscle in the normal state and in drug induced myopathies in the cat. Ph.D. dissertation. Univ. of London.

6. Page, S. G. and Niedergerke, R. 1972. Structures of physiological interest in the frog heart ventricle. *J. Cell. Sci.* II:179-203.

# 16

## Functional Interaction of Both Ventricles at Birth and the Changes During the Neonatal Period in Relation to the Changes of Geometry

**Adrian Versprille and Jos R.C. Jansen**
Clinical Physiological Laboratory/Department of Pulmonary Diseases
University Hospital/Erasmus University/Rotterdam/The Netherlands

**Eric Harinck**
Department of Pediatric Cardiology/Wilhelmina Children's Hospital
University of Utrecht/Utrecht, The Netherlands

**Cornelis J. van Nie**
Department of Anatomy/Free University/Amsterdam/The Netherlands

**Karel J. de Neef**
CNS — Pharmacology
R&D Laboratories
Organon International/OSS/The Netherlands

Long-term studies have indicated that left ventricular hypertrophy eventually occurs after the main pulmonary artery in dogs has been banded (13, 14). Left ventricular hypertrophy has been described also in patients with congenital stenosis of the pulmonary valve (3, 7, 8). Herbert and Yellin (11) observed an increased left ventricular end diastolic pressure ($P_{lved}$) in patients with this disease. In acute experiments, elevation of the left ventricular end diastolic pressure was observed along with increased end diastolic pressure in the right ventricle ($P_{rved}$) when the pulmonary artery pressure was raised (1,

2, 5, 6, 7, 15, 16, 17, 18). Thus, changes in right ventricle hemodynamics can affect that of the left ventricle.

Whether the left ventricle changes influence the function of the right ventricle in a similar manner is still a matter of dispute. According to Henderson and Prince (10), stroke volume of the cat right ventricle is decreased more by an increasing filling pressure of the left ventricle than the stroke volume of the left ventricle is decreased by increased right ventricular filling pressure. Although Elzinga et al. (6) also found a left-to-right (l-r) effect in cats, Von Capeller (4) and Bucher and Von Capeller (2) observed little, if any, reaction in the right ventricle when the left ventricle was loaded.

A key to the causal mechanism of the right-to-left ventricular effect is implied in the observations of Urschell et al. (17) and Bemis et al. (1). These authors described a flattening of the left ventricle produced by an increase in diastolic filling pressure of the right ventricle. This flattening consists of an increase of the anterior-to-posterior distance and a decrease of the septum-to-free wall distance. We believe, therefore, that the functional interaction results from a flattening of the contra-lateral ventricle. In a balloon, flattening will increase the radius (r) for a large part of the wall, which will change the pressure (P) to tension (T) relation according to Laplace's law, $T = Pr/2d$, in which d represents the wall thickness. A flattened ventricular shape, thus, is less able to build up pressure than is a sphere.

It seems reasonable to expect that the anatomic and functional relations must be linked. Hence, changes of the geometry may be expected to interfere with function. As geometric changes of the ventricles occur in the early postnatal period (9, 12), we studied the functional interaction and geometry of both ventricles in hearts of neonatal pigs from birth to 37 days of age.

Because an increased load on the right ventricle of young piglets increased shunt flow through the foramen ovale, we also measured the dimensions of the foramen ovale and its changes after birth, as well as the influence of a changing right-to-left shunt on the right-to-left effect.

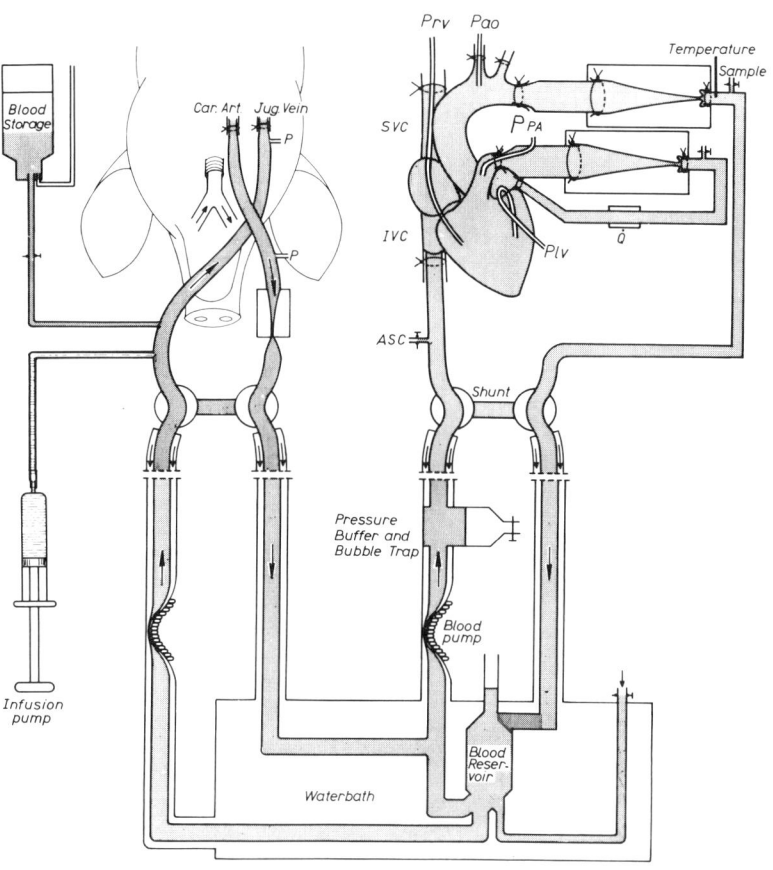

Fig. 1. Schematic representation of the isolated perfused, supported, pig heart. Abbreviations: $P_{rv}$, $P_{ao}$, $P_{pa}$, and $P_{lv}$ are blood pressures of the right ventricle, aorta, pulmonary artery, and left ventricle respectively; SVC and IVC are superior and inferior vena cava, ASC is injection of ascorbinate for right-to-left shunt detection with a platinum electrode at the top of the aortic pressure catheter; $\dot{Q}$ is the volume flow through the pulmonary artery.

## METHODS

### The Isolated, Perfused Heart Preparation

The experiments were performed with an isolated, perfused heart preparation of newborn pigs (Yorkshire) up to the age of 37 days (Fig. 1). The isolated hearts were supported by donor pigs of the same race, 8 to 10 weeks old, with a weight of 15-20 kg. A Sigma finger pump transported the blood from a reservoir through a pressure buffer/bubble trap into the inferior vena cava of the newborn piglet. The right and left ventricles pumped blood against arterial pressures, which were controlled by Starling resistors connected respectively to the main truncus of the pulmonary artery and the aortic arch.

Blood was pumped from the reservoir into the right external jugular vein of the "donor" pig by a Sigma finger pump. The blood was driven by the arterial pressure of this pig through a Starling resistor into the "venous line" of the isolated heart. Starling resistor, bubble trap, and reservoir consisted of perspex, the tubes of Tygon R3603, and the Starling resistor was a thin latex tube 50-70 mm long with an internal diameter of 13 mm.

### The Perfusion Blood

Blood with 2 ml heparin (Organon Oss, 5000 IU ml$^{-1}$) per 500 ml was obtained from the slaughter house and stored one day at 4°C. The compatability of the blood was tested by injection of 10 ml into the "donor" pig. The blood was rejected if a transient decrease of the arterial blood pressure was observed. The blood was kept at 39.0 ± 0.1°C by a warm water bath surrounding the artificial system. The arterial $PCO_2$ was maintained at 40 to 45 mm Hg and pH at 7.35 to 7.45 by respiring the donor pig with oxygen (a tidal volume of 250 ml, a frequency dependent of $PaCO_2$ and pH, and an end expiratory pressure of +5 cm $H_2O$). The oxyhemoglobin saturation was above 95% in all experiments.

### Measurements of Hemodynamic Variables

The right ventricular output ($\dot{Q}_{rv}$) was measured electromagnetically with a cannulating probe (Skalar, Delft,

Type SP8) distal to the Starling resistor. After each
experiment the finger pump of the isolated heart and the
flow meter were calibrated. The accuracy is given by the
mean values of all individual standard deviations (SD)
from our 1973 experiments: pump $\overline{SD}$ = 0.41% (±0.35%, n =
31); flow-meter $\overline{SD}$ = 0.45% (±0.30%, n = 25). The heart
rate was kept constant by a pacemaker on the right auricle
at a rate of 3.5 or 4 per sec. The ECG was measured between two electrodes, one near the apex of the heart (midsternal) and one on the right foreleg. Pressures (P) were
measured with Statham pressure transducers (Type P23De)
through polyethylene catheters (internal diameter = 1 mm;
length = 10 to 12 cm). These catheters were introduced
into the right ventricle through the right external jugular vein; into the aorta through the right common carotid
artery or the mouthpiece of the Starling resistor; into
the left ventricle through the left auricle, and into the
pulmonary artery through the mouthpiece of the Starling
resistor. Right-to-left shunts through the foramen ovale
of the young piglets ($\dot{Q}_{fo}$) were detected by the ascorbinate dilution method.

## Experimental Procedures

We studied the effect of left ventricle on right ventricle by stepwise changes (5 to 10 mm Hg) with aortic
pressure as an independent variable and with left ventricular end diastolic pressure and right ventricular end
diastolic pressure as dependent variables. The left-to-right effect was quantified by the slope of the regression
line through the plot of right ventricular end diastolic
pressure versus left ventricular end diastolic pressure.
Pump flow ($Q_s$) and pulmonary artery pressure ($P_{pa}$) were
parameters. The right ventricle to left ventricle effect
was studied by stepwise changes of pulmonary artery pressure with aortic pressure and pump flow as parameters.
Measurements were done about 3 minutes after the stepwise
change of afterload, when the heart function was in steady-state condition.

## Coronary Circulation

When the right-to-left shunts are zero, output of
the left ventricle and output of the right ventricle equal
the sum of the pump flow and coronary flow ($Q_{coron}$) (i.e.,

coronary flow, if the flow through the Thebesian veins is neglected). Under these conditions output of the right ventricle and thus output of the left ventricle showed a slight decrease during the procedure of stepwise increase of pulmonary artery pressure. The stepwise increase of aortic pressure increased the output of the right ventricle only slightly. We disregarded the possible influences of these small changes of right ventricular output and left ventricular output on the end diastolic pressures during the observations of the left-to-right and the right-to-left effects respectively.

RESULTS

Postnatal Changes of the Ventricle

Figure 2 shows the postnatal changes of the pig heart in transversal sections through the midpoint between apex and base. The right and left ventricle appear to be almost symmetrical at birth; the septum is not curved. By the third day of life the left ventricle shows the circular geometry. This shape is maintained at older ages

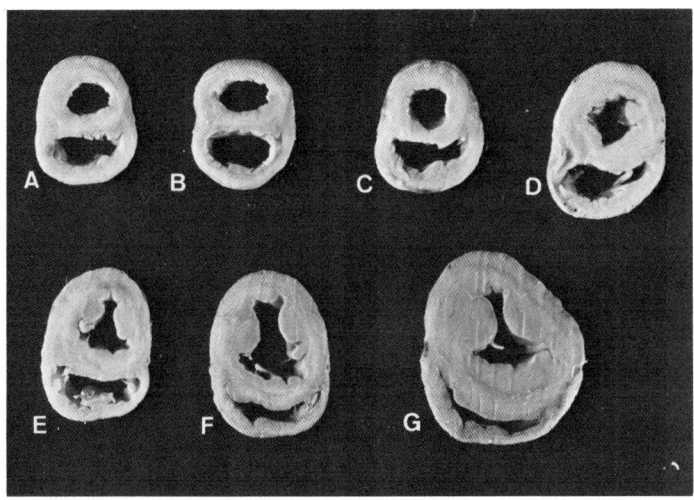

Fig. 2. Horizontal sections across the ventricles of pig hearts from 1 day to 9 weeks old. $A$=1 day; $B$=2 days; $C$=3 days; $D$=9 days; $E$=18 days; $F$=37 days; and $G$=9 weeks.

while the wall thickness increases. From birth the shape of the right ventricle changes from the more balloon-like geometry into the shell-like structure; wall thickness remains more or less consistent.

## The Foramen Ovale

In piglets the foramen ovale is a funnel-like canal through the septum with the widest portion on the right atrial side. A membrane attached at both sides to the septum at the left atrial side arises from about half the muscular edge of this opening. Thus a narrowing sheath is formed opening into the left atrium. The free rim of this slit is slightly thickened.

In 40 piglets this slit was measured as if it had a round shape, and its size was expressed as the diameter of this circle. Then the opening was sounded carefully with a calibrated conical probe. Figure 3 plots these diameters against age. During the first few days of life this diameter is about 5 mm. Then the free rim starts to grow as well as the muscular septum where it is attached. Within about two weeks after birth, the slit is closed completely in most cases. In these experiments four animals older than 15 days had a patent foramen ovale of about 2 mm. In one animal 36 days old the patent foramen ovale was as large as at birth; however, in this case the hemodynamics were aberrant and the data were not used in this study.

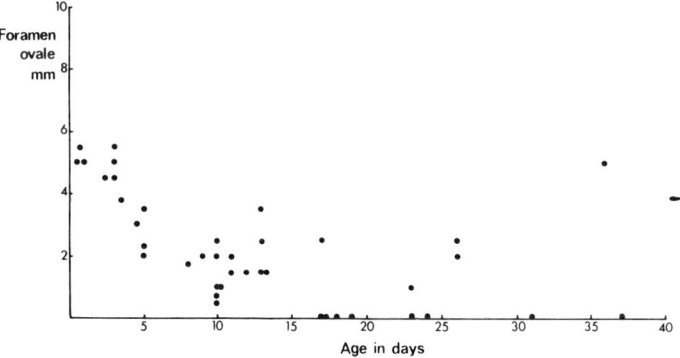

Fig. 3. Foramen ovale diameter as a function of newborn age.

## Direct Interaction of Both Ventricles

Stepwise increase of the pulmonary artery pressure from 15 to about 50 mm Hg caused a progressive increase of both right ventricular end diastolic pressure and left ventricular end diastolic pressure (Fig. 4A). A plot of left ventricular end diastolic pressure versus right ventricular end diastolic pressure yields a straight line with a slope s (= $\Delta P_{lved}/\Delta P_{rved}$) which quantifies the influence of the right on the left ventricle (Fig. 4B). The influence of the left on the right ventricle was studied by the stepwise increase of aortic pressure with 5-10 mm Hg within the range from 50 to 130 mm Hg. At birth the left-to-right ventricular effects equaled those of the right-to-left effects (Figs. 5A and B). However, at older ages, the left-to-right effect was essentially nil. In these experiments the right ventricular end diastolic pressure changed only slightly if at all, while the left ven-

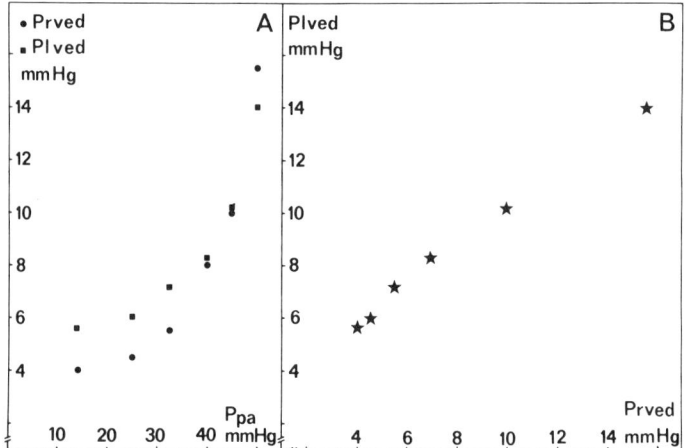

Fig. 4. A: The effect of elevation of pulmonary artery blood pressure on right ventricular end diastolic pressure (●) and left ventricular end diastolic pressure (■). B: The relation between left ventricular end diastolic pressure versus right ventricular end diastolic pressure, derived from 4A. The age of the pig was 12 days; aortic blood pressure = 70 mm Hg; heart rate 3.5 $s^{-1}$; $\dot{Q}_{rv}$ = 4.8 to 4.9 ml $s^{-1}$.

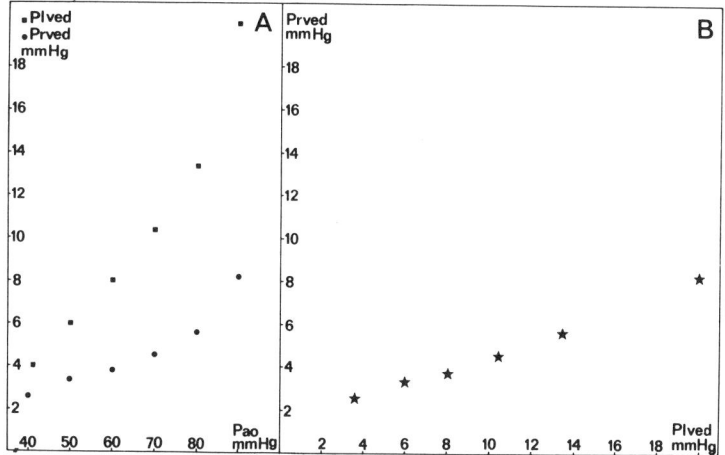

Fig. 5. A: The effect of elevation of aortic blood pressure on left ventricular end diastolic pressure (■) and right ventricular end diastolic pressure (●). B: The relation right ventricular end diastolic pressure versus left ventricular end diastolic pressure, as derived from 4B. The age of the pig was 1 day; pulmonary artery blood pressure = 20 mm Hg; heart rate 3.5 s$^{-1}$; $\dot{Q}_{rv}$ = 3.4 ml s$^{-1}$.

tricular end diastolic pressure increased progressively. The age dependency of the left-to-right effect is presented in Fig. 6, where the slope of the linear relation is plotted against age. During the first month of life the left-to-right effect decreased gradually. The right-to-left effect appeared to be independent of age (Fig. 7). The mean value of the right-to-left effect was 0.60 (SD, ±0.19) for 36 animals.

Hemodynamic Influence of the Foramen Ovale

The stepwise increase of pulmonary artery pressure during the studies of the right-to-left effect induced a progressive right-to-left shunt through the foramen ovale of the piglets less than one week old. This shunt changed the progressive increase of the right and left end diastolic pressures into an S-shaped relation with a plateau at the higher levels of pulmonary artery pressure when foramen ovale output was maximal (Fig. 8A). Notwithstanding this different pattern of end diastolic pressure rise, the interaction yielded a straight line (Fig. 8B), with a slope which appeared to be independent of the flow through the heart.

## DISCUSSION

Our results on the right-to-left effect agree with other results reported from the literature. A stepwise

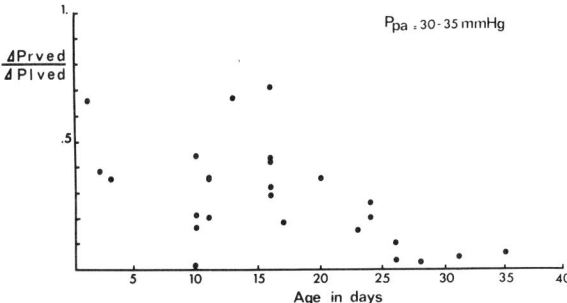

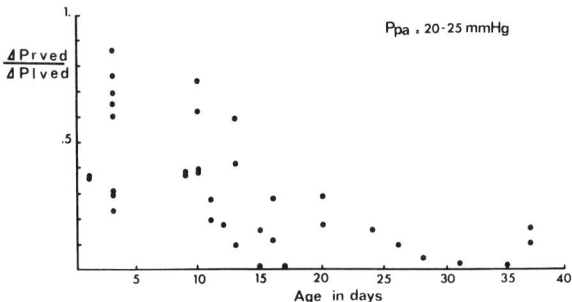

Fig. 6. The left-to-right effect, expressed as the regression line of the relation between right ventricular end diastolic pressure versus left ventricular end diastolic pressure, is plotted against age. The effect was studied at two levels of pulmonary artery blood pressure, 30 to 35 mm Hg (A) and 20 to 25 mm Hg (B) respectively.

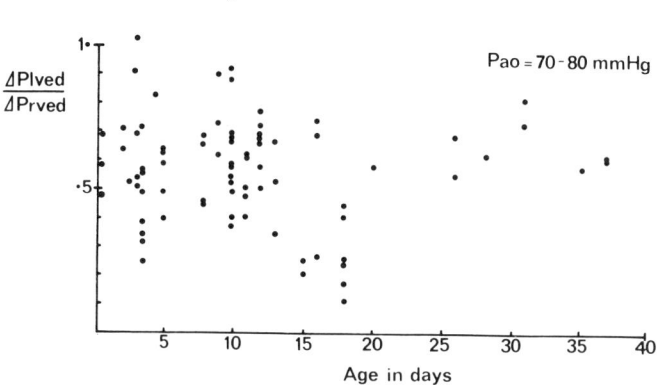

Fig. 7. The right-to-left effect is plotted against age. The level of aortic blood pressure was 70 to 80 mm Hg.

increase of pulmonary artery pressure causes a proportional increase of left ventricular end diastolic pressure along with right ventricular end diastolic pressure. Our mean value of 0.60 (SD, 0.19) for the slope of the straight line agrees closely with the value of 0.56 (SD, 0.19) presented in a preliminary paper by Urschell et al. (17), who worked with 6 dogs. In their publication the interaction was described as $P_{lved} = 0.45\ P_{rved} + 0.68$ (1). They did not explain this difference.

As shown in Figure 8, the slope of the left ventricular end diastolic pressure versus the right ventricular end diastolic pressure appears independent of the flow through the heart. Only the regression line through the points is displaced vertically. Therefore, we believe that the value 0.68 in the equation of Bemis et al. (1) will not have any significance for the interaction itself. Hence it seems reasonable to express the amount of right-to-left interaction by the slope of the relation, not only because of its linearity, but also because of the independency of this linearity on hemodynamic changes at the right side of the heart; that is to say, at the side where the changes of end diastolic pressure was established. A decrease of flow through the right ventricle caused by an increase of foramen ovale output in the piglets produces a progressive but proportional change of

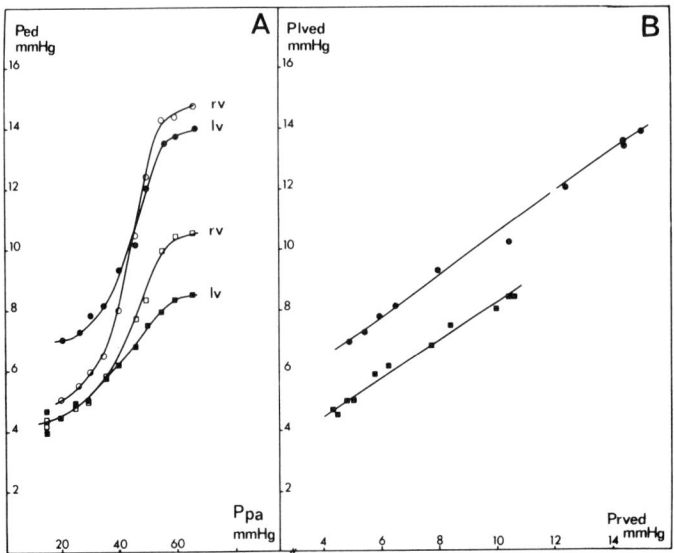

Fig. 8. A: The effect of pulmonary artery blood pressure on right ventricular end diastolic pressure and left ventricular end diastolic pressure with patent foramen ovale at two levels of left ventricular output, (●) 4.0 and (■) 2.9 ml s$^{-1}$, respectively. In the upper curves, right ventricular output changed from 4.0 to 0.8 ml s$^{-1}$, in the lower curves, from 2.9 to 0.7 ml s$^{-1}$. B: The plot of left ventricular end diastolic pressure versus right ventricular end diastolic pressure as derived from the points in Fig. 8A. The age of the pig was 4 hours; aortic blood pressure = 70 ml s$^{-1}$; heart rate 4 s$^{-1}$.

$\Delta P_{lved}$ with $\Delta P_{rved}$ without changing the linearity (Fig.8). This fortifies our opinion that $\Delta P_{lved}$ only depends on $\Delta P_{rved}$ and that both changes need not be seen as parallel mechanisms based on the same cause as are, for instance, changes in coronary flow. Therefore, changes in coronary flow should not be considered as a common underlying mechanism because of the opposite changes in the left-to-right effect as compared to the right-to-left effect. Moreover, it would be difficult to explain the age-dependency of the left-to-right effect.

Another consideration has been the development of a progressive subvalvular stenosis in the heterolateral ventricle. However, a systolic pressure gradient over the similunar valve did not develop with increasing end diastolic pressure. Again, it would be difficult to explain the age-dependency of the left-to-right effect. The age-dependency of the left-to-right effect (reported in a preliminary paper (18)) includes approximately the same period after birth as the right ventricular geometric changes from the balloon-like shape into a definite shell-like structure. This parallel phenomenon strengthens our opinion that the interaction of both ventricular functions is based on functional geometric changes induced by loading one side of the heart. Anatomically, the left ventricle retains its balloon-like shape; therefore the right-to-left effect probably does not change, which means that the effect of flattening is consistent. The right ventricle. however, becomes a "shell" around the "balloon-like" left ventricle. Increasing the left ventricle load increases the diameter of this balloon, presumably uniformly; thus the round shape may be expected to be unchanged. In that case, the edges of the shell will be displaced in the outward direction from the center of the left ventricle, but without changing the radius of the right ventricular free wall noticeably. This probably explains why the left-to-right effect is smaller at older ages.

The observed left ventricular hypertrophy in cases of increased outflow resistance at the right side of the heart (3, 7, 13, 14) might result from increased tension in the wall of the left ventricle according to the mechanism mentioned earlier. Thus, clinicians should not only check the hemodynamics of the right ventricle, but also those of the left ventricle in patients with pulmonary valvular stenosis and intact ventricular septum.

### ACKNOWLEDGMENT

This work was supported by grants from the Netherlands Organization for the Advancement of Pure Research (ZWO) received through the Foundation for Basic Medical Research (FUNGO).

REFERENCES

1. Bemis, C. E.; Serur, J. R.; Borkenhage, D.; Sonnenblick, E. H.; and Urschel, C. W. 1974. Influence of right ventricular filling pressure on left ventricular pressure and dimension. *Circ. Res.* 34:498.

2. Bucher, K., and Von Capeller, D. 1954. Mechanismen der gegenseitigen Anpassung von Lungen und Körperkreislauf. *Helv. Physiol. Acta* 12:253.

3. Becu, L.; Somerville, J.; and Gallo, A. 1976. "Isolated" pulmonary valve stenosis as part of more widespread cardiovascular disease. *Brit. Heart J.* 38:472.

4. Von Capeller, D. 1954. Mechanismen der gegenseitigen Anpassung von Lungen-und Körperkreislauf. *Helv. Physiol. Acta* 12:23.

5. Elzinga, G. 1972. Cross talk between right and left heart. M.D. Thesis, Amsterdam, Free University.

6. Elzinga, G.; Van Grondelle, R.; Westerhof, N.; and Van den Bos, G. C. 1974. Ventricular interference. *J. Physiol.* 226:941.

7. Harinck, E. 1974. Wederzijdse beinvloeding van de beide hartshelften en veranderingen hiervan na de geboorte (Interaction of the right and left side of the heart and postnatal changes thereof). M.D. Thesis, Leiden.

8. Harinck, E.; Becker, A. E.; Gittenberger-de Groot, A. C.; Oppenheimer-Dekker, A.; and Versprille, A. 1976. The left ventricle in congenital valvular pulmonary stenosis. *Eur. J. Cardiol.* 4:267.

9. Van Harreveld and Russell, F. E. 1956. Postnatal development of a left-right atrial pressure gradient. *Am. J. Physiol.* 186:521.

10. Henderson, Y. and Prince, A. L. 1914. The relative systolic discharges of the right and left ventricles and their bearing on pulmonary congestion and depletion. *Heart* 5:217.

11. Herbert, W. H. and Yellin, E. P. D. 1969. Left ventricular diastolic pressure elevation consequent to pulmonary stenosis. *Circulation* 40:887.

12. Keen, J. A. 1942. Note on the closure of the foramen ovale and the postnatal changes of the ventricle in the human heart. *J. Anatomy* 77:104. London.

13. Laks, M. M.; Morady, F.; and Swan, H. J. C. 1969. Canine right and left ventricular cell and sarcomere lengths after banding the pulmonary artery. *Circ. Res.* 24:705.

14. Laks, M. M.; Morady, F.; Garner, D.; and Swan, H. J. C. 1972. Relation of ventricular volume, compliance and mass in the normal and pulmonary arterial banded canine heart. *Cardiovasc. Res.* 6:187.

15. Moulopoulos, D.; Sarcas, A.; Stamatelopoulos, S.; and Arealis, E. 1965. Left ventricular performance during bypass or distension of the right ventricle. *Circ. Res.* 17:484.

16. Ullrich, K. J.; Riecker, G.; and Kramer, K. 1954. Das Druckvolumdiagramm des Warmblüterherzens. Isometrische Gleichgewichtskurven. *Pflüger Arch.* 259:481.

17. Urschel, C. W.; Bemis, C. E.; Serur, J. R.; Borkenhagen, D.; and Sonnenblick, E. H. 1971. The influence of right ventricular filling pressure on left ventricular pressure and dimension. *Circulation* 43-44, suppl. 2, abs. 125.

18. Versprille, A.; Harinck, E.; van Nie, C. J.; Jansen, J. R. C.; and de Neef, K. J. 1974. Cross-talk between right and left ventricle during the neonatal period. *Arch. Intern. Physiol. Bioch.* 82:329.

# 17

## The Fetal Renin-Angiotensin System

Joan C. Mott
University of Oxford
Nuffield Institute for Medical Research
Headley Way, Headington
Oxford, U.K.

INTRODUCTION

Until the last decade or so the renin-angiotensin system has been largely neglected by mainstream circulatory physiologists inclined to regard it as a pathological preserve unfit for students of the normal.

Why should the fetus be expected to have a renin-angiotensin system and if it has, why should it be of any particular interest even if its adult counterpart has now attained physiological respectability? First of all, evidence of a renin-angiotensin system is not confined to mammals but has been reported in bony fish, amphibia, reptiles and birds (40). Second, as long ago as 1942 renin activity was found in the fetal hog kidney and a pressor response obtained on injection of renin and angiotensin into fetal rats (28, 16). Later incidental observations in fetal lambs by various workers confirmed that angiotensin has a pressor action (2) and that angiotensin II is a more potent pressor agent than either noradrenaline or adrenaline (29). Thus the enzyme is there and fetal tissues are sensitive to one of the most potent peptides resulting from its action. Thirdly, a systematic investigation of the consequences of bleeding newborn, infant and adult rabbits showed that, in terms of equi-proportionate reduction of blood volume, immature rabbits maintained arterial pressure significantly better than adults (31) and that their capability in this respect, unlike the adult, depended on the presence of the kidneys and was not impaired by various maneuvers modifying autonomic nervous activity. Simultaneously it became evident that in very young rabbits especially anemic ones, even maintenance of resting arterial pressure depended on the presence of the kidneys (32).

Measurement of angiotensin II - like activity in circulating blood of rabbits of different ages added weight to the inference that high activity of the renin-angiotensin system is responsible for the superior ability of infant rabbits to stand bleeding (12). The question then naturally arose as to the age dependency and other characteristics of the renin-angiotensin system in fetal life.

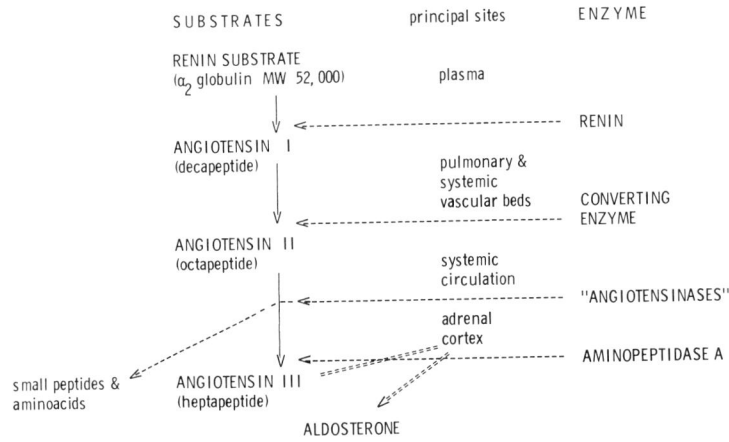

Fig. 1. Outline of the renin-angiotensin system.

## The Fetal Renin-Angiotensin System

Most of the information about the fetal renin-angiotensin system (other than that obtained from human umbilical cord blood) pertains to the sheep. In the absence of contrary evidence there is a tacit assumption that there are at least no qualitative differences between ewe and fetus in the sequence of enzymatic actions outlined in Figure 1.

It is clear that, taken literally, the phrase "activity of the renin-angiotensin system" is ambiguous or imprecise. In different contexts interest may center on tissue concentrations of angiotensins I, II or III all of which depend on specific enzymatic activity for formation and destruction. The most widely employed measurement is that of plasma renin activity (PRA or endogenous velocity) determined as the initial rate of production of angiotensin-I by endogenous renin from endogenous substrate during incubation of plasma *in vitro* at body temperature and pH in the presence of inhibitors of converting enzyme and of "angiotensinases." It is assumed that this provides a reasonable assessment of the ability of plasma renin *in vivo* to detach the decapeptide angiotensin I (AI) from renin substrate (RS). Kinetic investigation (38) shows that at the substrate concentrations found *in vivo*, the reaction is first order (i.e., substrate dependent). Thus plasma renin activity does not represent enzyme concentration though relative activities determined at a fixed substrate concentration will reflect relative enzyme concentrations. Plasma concentration of renin substrate in the sheep is much less than that at the $K_m$ (2μM or 2μg/ml); the extent to which changes of renin substrate contribute to changes of plasma renin activity in experimental or developmental situations requires further investigation. Some investigators measure plasma renin activity in the presence of added substrate (e.g. Table 1a). Unless renin destruction by the liver and other tissues can be assumed constant, renin secretion can only be accurately assessed with knowledge of renal plasma flow and a-v renin difference. These considerations make it clear that caution may be necessary when comparing measurements of plasma renin activity from different sources.

Measurement of angiotensins in fresh plasma naturally depends on an appropriate extraction procedure. Angiotensins are now usually measured by radioimmunoassay. Previously angiotensin I produced by incubation of plasma was assayed in terms of its pressor action (34). Changes of angiotensin II concentration can be estimated from its ability to cause contraction of the isolated ascending colon of a rat, superfused with blood in an extracorporeal circuit (45).

## Renal Renin

A long-lasting pressor response was observed in nephrectomized fetal lambs on injection of a saline extract of fetal kidney in contrast to the relatively short-lived responses to angiotensin II (Fig. 2). The youngest lamb from which an active kidney extract was prepared was only 75/147 term. Granules identified as renin by a positive reaction to Bowie's stain have been described in the juxtaglomerular apparatus from lambs of less than 100 days gestation (39).

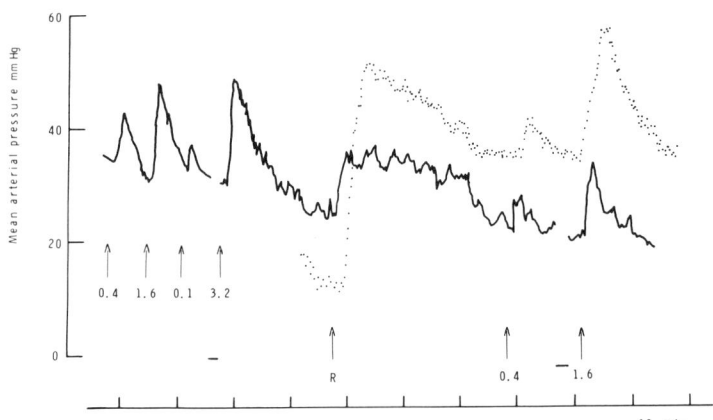

Fig. 2. *Solid line:* arterial pressure of acutely nephrectomized fetal lamb of ewe (133/147 term) under general anesthesia. *Dotted line:* length of rat colon (contraction upwards) superfused with fetal blood in extracorporeal circuit. *Arrows:* intravenous injection of angiotensin II amide (Hypertensin Ciba, dose in µg) or 1-ml saline extract of kidney (R) prepared from another lamb of 123 days gestation.

### Fetal Bilateral Nephrectomy

Plasma renin activity and plasma angiotensin II (MW 1045) fell to very low levels in two bilaterally nephrectomized fetal lambs; both of these had twins with plasma renin activity in the normal range (11). In six bilaterally nephrectomized fetal lambs the half-life calculated for the initial decrements of plasma renin activity was 24 to 48 minutes and fetal plasma renin activity was unaffected by maternal nephrectomy (33a). In another investigation the half-life of fetal plasma renin activity after bilateral nephrectomy was 28 min (G. C. Liggins, personal communication). In accordance with expectation there is no evidence that maternal renin (M.W. $\sim$ 40,000) or renin substrate (M.W. 52,000) can cross the placenta.

### Plasma Renin Activity in Unanesthetized Fetal Lambs

#### 111-144 Days Gestation

Table 1 summarizes data obtained not less than 4 days post-catheterization and published almost simultaneously from three laboratories (11, 39, 22). It shows that fetal plasma renin activity (ng/ml·hr$^{-1}$) on average substantially exceeds maternal activity. Large day to day variations of unidentified origin occur and fetal renin activity is rarely less than maternal; Col. a (Table 1) conceals a rise of fetal plasma renin activity from 9.3 ± 2.0 S.E.M. at 105-119 days to 12.6 ± 2.6 S.E.M. after 130 days gestation (17). No circadian rhythm in phase with other fetal circadian rhythms has been detected (11).

### Immature Lambs

#### 97-103 Days Gestation

In 6 lambs in which arterial $P_{O_2}$ was 19-27 mm Hg plasma renin activity did not exceed 2.1. In 3 lambs $P_{O_2}$ was 12-20 mm Hg and plasma renin activity was 5.7-71. The negative correlation of paired measurements of plasma renin activity and $P_{O_2}$ for the whole group (n = 7 for the more highly oxygenated lambs and n = 5 for the remainder) was significant (r = -0.735, P < 0.01). One lamb of 98 days gestation with an arterial $P_{O_2}$ of 24 mm Hg had a plasma renin activity of 0.9 while that of its twin with a $P_{O_2}$ of 12 mm Hg was 71.

Table 1
Renin Activity in Plasma from Unanesthetized Fetal Lambs and Their Mothers Measured as Angiotensin I ng/ml·hr$^{-1}$ (± S.E.M.)

| Group | a | b | c | d |
|---|---|---|---|---|
| Reference | 11 | 11 | 39 | 22 |
| Gestation age (days) | 111 – 144 | 114 – 144 | 114 – 122 | 128 – 140 |
| Days postcatheterization | ≥ 5 | ≥ 5 | ≥ 6 | ≥ 4 |
| Added substrate | + | 0 | 0 | 0 |
| Plasma renin activity | | | | |
| Ewe | 1.5 ± 0.2 | 0.9 ± 0.3 | 1.0 ± 0.4 | 2.5 ± 0.1 |
|  | (40) | (7) | (3) | (13) |
| Fetus | 10.7 ± 1.1 | 4.7 ± 1.7 | 2.9 ± 0.1 | 7.8 ± 0.6 |
|  | (55) | (7) | (3) | (13) |
| Fetal Mean | 7.1 | 5.2 | 2.9 | 3.3 |
| Maternal Mean | | | | |

Figures in parentheses are numbers of samples, except in last column, where they are numbers of daily means obtained from 15 pregnancies.

### 104-110 Days Gestation

The lowest plasma renin activity encountered in this group was 9.3, in a lamb with an arterial $P_{O_2}$ of 22. It appears that at or shortly before 104/147 (0.71) term plasma renin activity increases abruptly independently of any hypoxemic stimulus (Fig. 3). The average $P_{O_2}$ for the lambs of Table 1a was 22.6 ± 0.5 S.E.M. mm Hg.

## ANGIOTENSIN II

### Converting Enzyme Activity

The C terminal His-Leu moiety of angiotensin I (AI) generated by renin from renin substrate is detached by converting enzyme with formation of the octapeptide angiotensin II (AII). Ng and Vane (33) showed that conversion of angiotensin I to angiotensin II was not, as previously had been held, confined to plasma but in adult animals takes place predominantly in the pulmonary vascular bed

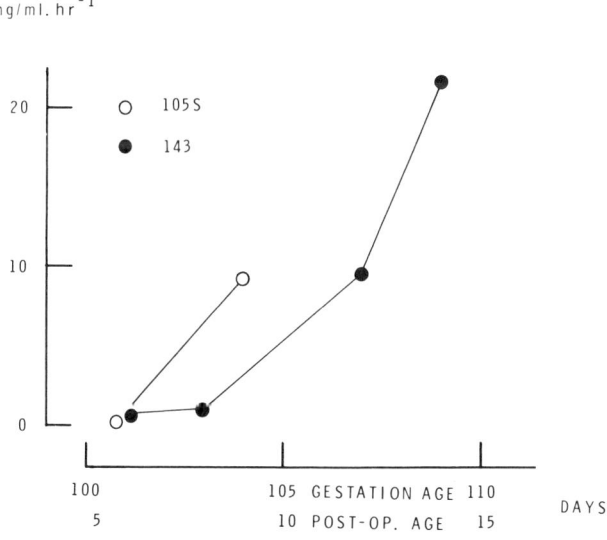

Fig. 3. Plasma renin activity in two immature fetal lambs.

and also in various systemic vascular beds (1). So far as the fetus is concerned the questions that arise are first, is angiotensin II formed in the absence of lung ventilation? And second, is converting enzyme activity quantitatively similar to that in the adult?

The ability of angiotensin II to contract smooth muscle is about 10-fold that of angiotensin I; angiotensin II passes through the pulmonary circulation unchanged. Pulmonary conversion of angiotensin I to angiotensin II has the consequence that less angiotensin I is required to evoke a given systemic pressor response when injected intravenously than when given in the ascending aorta. In ewes the intravenous dose required averaged 56% of the arterial dose; the corresponding figures in fetal and newborn lambs (with the ductus arteriosus ligated) were 82% and 78% respectively. Thus there is no indication that lung ventilation is responsible for any change in converting enzyme activity at birth (26).

The activity of pulmonary converting enzyme is greater in infant than in adult rabbits (9). It is therefore a little surprising that the contrary appears to be the case in the sheep. However, this investigation (26) was based on the estimation of the dose required in lambs and ewes to produce a rise of arterial pressure of 20 mm Hg at all ages, regardless of resting pressure. It would be desirable to repeat these experiments using the superfused rat colon preparation or another method of estimating angiotensin II independent of the animal under investigation.

## Angiotensin II in Fetal Plasma

Plasma concentrations of angiotensin II in 5 mature fetal lambs 122-139 days gestation (57 ± 12 S.E.M. pg/ml) were similar to those in 4 ewes (47.3 ± 6.6 S.E.M. pg/ml) (11), despite much greater fetal plasma renin activity. Concentrations of angiotensin up to 10-fold of these values have been found both in some younger fetal lambs and their mothers (17); the correlation between maternal and fetal values was 0.761 (n = 16, $P < 0.001$, Fig. 4). Instances have occurred of high plasma concentrations of angiotensin II coexisting with low plasma renin activity in several lambs of less than 111 days gestation, while in older lambs plasma renin activity is normally high and angiotensin II concentration lower. It was of interest to enquire if the high plasma angiotensin II found in some immature lambs was related to high substrate concentrations.

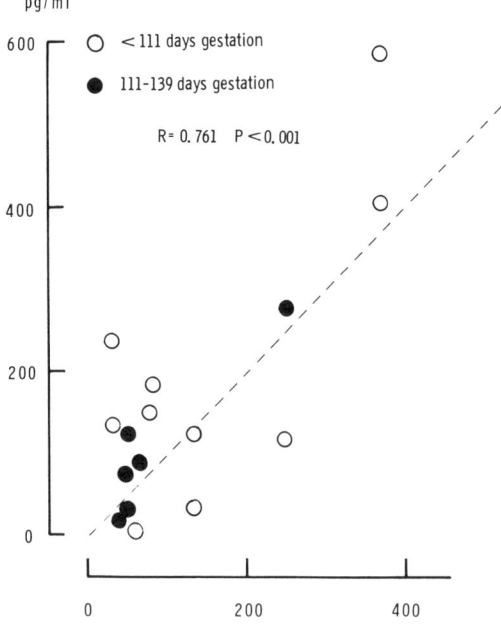

Fig. 4. Correlation of fetal and maternal angiotensin II. The line of equality is shown dotted and is not significantly different from the linear regression.

## Renin Substrate

The plasma concentration of renin substrate (which is measured as the total amount of angiotensin I liberated from plasma by excess renin) averaged 366 ± 21 S.E.M. ng/ml in 21 samples from 13 pregnant ewes and 211 ± 26 S.E.M. ng/ml for 22 samples from 14 fetal lambs at 97-142 days gestation. The corresponding figure for other normal sheep is 410 ± 30 S.E.M. ng/ml (38). In only one of 17 maternal-fetal pairs was fetal substrate concentration greater than maternal (paired t = 5.67, $P < 0.001$) (16a, 17a). Thus neither the greater plasma renin activity of the mature fetal lamb compared with the ewe, nor the higher concentrations of angiotensin II found in many younger lambs depend on high plasma concentration of renin substrate. Indeed 9 fetal and 9 maternal pairs of angiotensin II and renin substrate measurements showed an inverse correlation ($r = -0.492$, $P < 0.05$).

## Metabolism of Angiotensin II

In adult animals, about two thirds of the angiotensin II present in arterial blood is inactivated on each passage through the systemic circulation (27). In infant rabbits, such inactivation is significantly less in muscle, kidney and liver compared with the adult (12, 8, 9). A potentially important though presumably quantitatively inconsiderable site of degradation is the adrenal cortex where formation of angiotensin III is believed to play a part in aldosterone secretion (25).

No information is available about inactivation of angiotensin II in the fetus, though clearly it is possible that the high concentrations of angiotensin II found in some immamature fetal lambs may be the consequence of slower destruction. A recent report (14) that umbilical venous angiotensin II exceeds umbilical arterial angiotensin II in human infants at delivery suggests that in this species any destruction of angiotensin II in the fetal placental circuit is outweighed by other factors.

The apparent paradox of relatively high angiotensin II coexisting in immature fetal lambs with low plasma renin activity may be a consequence of the suppression of renin secretion by angiotensin II (37, 43) but this possibility requires experimental verification.

## FACTORS WHICH CHANGE FETAL PLASMA RENIN ACTIVITY

### Drugs

### Anesthetics

Trimper and Lumbers (42) were the first to measure plasma renin activity in fetal lambs of ewes anesthetized with thiopentone and chloralose. The mean renin activity in control plasma samples from the fetal lambs exceeded that of the ewe in only 3 of 13 pregnancies. This was due however to high maternal ($13.4 \pm 2.0$ S.E.M. $ng/ml/hr^{-1}$) rather than low fetal activity ($7.3 \pm 2.0$ S.E.M. $ng/ml \cdot$

$hr^{-1}$, of Table 1a). Non-pregnant ewes did not show this high plasma renin activity although a sham abdominal operation was performed. Three other mature fetal lambs from ewes anesthetized with sodium pentobarbitone had renin activities in the same range (7.2 ± 4.2) (10). The mean fetal/maternal ratio of plasma renin activity for the established preparations of Table 1a was greater than that (8.3/2.9) found during the first four post-operative days (11). It therefore seems that fetal plasma renin activity is comparatively little affected by maternal anesthesia despite the increase (often large) in maternal plasma renin activity during and after anesthesia and operation. Measurements of renin activity in such anesthetized fetal lambs are perhaps less vulnerable to criticism than might have been supposed.

Measurements of plasma renin activity in fetal lambs under maternal spinal anesthesia from the same source as the chronic preparations of Table 1c were also not systematically different. However, the rapid increase in plasma renin activity at about 104 days of age described earlier is less conspicuous in this small series under spinal anesthesia; in the absence of simultaneous blood gas measurements the cause of the difference must remain uncertain.

### Furosemide

This natriuretic is well known to stimulate renin release in adults and Trimper and Lumbers (42) showed that as early as 110 days gestation age it also increased plasma renin activity in the fetal lambs of ewes anesthetized with thiopentone and α-chloralose. In unanesthetized fetal lambs (Table 1d) 6 of 9 fetal lambs (128-136 days gestation age) responded to furosemide with substantial (3.2 - 17.8 fold) increases of plasma renin activity. Three lambs in which resting plasma renin activity exceeded 8 $ng/ml \cdot hr^{-1}$ were unresponsive (22).

### Hemorrhage

#### Unanesthetized Normoxic Lambs

Removal of 4 ml of blood at zero time and 30 min later caused plasma renin activity to rise (11) in 4 of 5 lambs of 126-144 days gestation age ($P < 0.01$ for mean % change from control). Hematrocrit (PCV) also fell and calculation of the apparent change of blood volume (assuming an initial fetoplacental volume of 116 ml/kg) at 30 and 60 min suggested that plasma renin activity was doubled when blood volume decreased by 6.4 ml/kg (5.5%). The lamb in which plasma renin activity had slightly decreased was calculated to have slightly increased its blood volume (Fig. 5a). Fetal arterial pressure was unchanged in these experiments.

When 4 ml blood was withdrawn 2 days postoperatively in unanesthetized lambs of 93-98 days gestation age plasma renin activity 15 min later was increased in 6 of 8 even though the initial activity in all but two of these lambs relatively soon after catheterization was higher than that

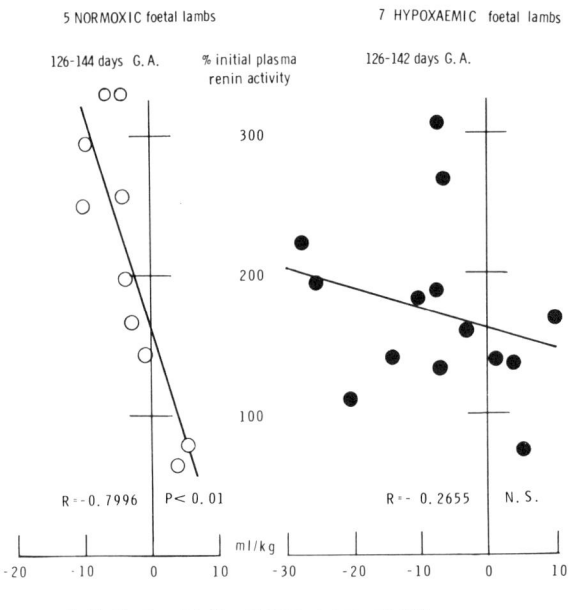

Fig. 5. Percentage change in plasma renin activity 30 and 60 min after control measurement plotted against change of blood volume (ml/kg) calculated from hematocrit (assuming initial blood volume of 116 ml/kg bodyweight, no undetected loss of red cells or change in $F_{cells}$ in 5 normoxic and 7 hypoxemic fetal lambs.

of established preparations illustrated in Fig. 3. After bleeding the heart rate increased from 200 ± 3 to 215 ± 5 beats/min on average but the fall of arterial pressure was not significant (16a).

The great sensitivity of the renin-angiotensin system to blood loss renders control experiments imperative in any investigation involving removal of even small volumes of blood.

## Hypoxemia

Fetal lambs of ewes caused to breathe a gas mixture containing 9% $O_2$ and 2.8% $CO_2$ in nitrogen become hypoxemic without accompanying hypocapnia or change of pH. In seven such experiments, the change of plasma renin activity during blood sampling was unrelated to the calculated change of blood volume (Fig. 5b). It therefore appears that hypoxemia in some way obscured the relation of the increase of plasma renin activity to the apparent reduction of blood volume. This may have been due to the accompanying rise of arterial pressure or possibly to liberation of antidiuretic hormone (36) which is known to depress renin release (37). Plasma angiotensin II was measured in 4 of these experiments and increased in parallel with plasma renin activity (11).

## Anesthetized Lambs

Removal of 25% of blood volume for 30 min in fetal lambs of ewes anesthetized with pentobarbitone increased the angiotensin II concentration in their circulating blood (rat colon bioassay) by 480 ± 210 pg/ml (10). This degree of hemorrhage increased plasma renin activity in 3 lambs from 7.2 ± 4.2 to 15.6 ± 3.1 S.E.M. $ng/ml \cdot hr^{-1}$ (10). Following removal of 8-10% of blood volume Smith et al. (39) recorded a 7-10 fold increase in plasma renin activity with the ewe under spinal anesthesia. They also observed an increase in plasma renin activity following aortic constriction sufficient to reduce abdominal/arterial and renal pressure to 10-12 mm Hg below carotid pressure. Thus, mechanisms augmenting fetal plasma renin activity do not seem to be impaired by maternal anesthesia.

## Volume Addition to the Circulation

### Single Large Injections of Maternal Blood

Infusion of 30-75 ml/kg bodyweight of freshly drawn maternal blood into fetal lambs 103-118 days gestation age reduced plasma renin activity for 1-2 hours (Table 2a). Blood sampling alone (109-122 days gestation) produced a small increase or no change of plasma renin activity over the same period. These infusions were made in a few minutes and arterial pressure rose so the reduction of renin activity cannot be solely attributed to increase of blood volume. Blood pressure was already returning towards the initial level by 60 min but the hematocrit was raised.

### Prolonged Infusion of Sodium Chloride Solution

Infusion of osmotic diuretics is known to reduce experimentally raised renin secretion (44). Infusion of isotonic or hypertonic sodium chloride 29.2-65.4 ml/kg for 60 min on five occasions in 4 fetal lambs (120-125 days gestation age) had no effect on arterial or blood gas pressures, caused only a small fall of hematocrit, but heart rate rose slightly. Although plasma renin activity sometimes fell below control activity during and as long as 24 hr after the infusion, the time course of the responses varied (Table 2b).

## PLASMA SODIUM AND POTASSIUM

It is well know that in adult humans, low plasma $[Na^+]$ is associated with high plasma renin activity. In fetal lambs plasma $[Na^+]$ falls very little with gestation age (97-144 days) though the fall in fetal plasma $[Na^+]/[K^+]$ is statistically significant (n = 73, P < 0.02, 31 lambs). Of more interest is the independence of fetal and maternal plasma $[Na^+]$ from 120 days gestation, whereas in younger lambs plasma $[Na^+]$ is correlated with the maternal level (Table 3); plasma $[Na^+]$ in mature fetal lambs tends to be lower than that of the ewe (17, 30). The fetal adrenal produces aldosterone (47) and it is possible that the angiotensin II-angiotensin III pathway (25) is implicated in this. Some observations suggest that to maintain growth the fetal lamb may need to conserve sodium ion as term approaches (4, 15, 30). As indicated earlier, wholly con-

Table 2

Plasma Renin Activity (±SEM) Following Volume Addition to the Circulation Expressed as a Percentage of Control

| Group | Number of Experiments / Number of Lambs | Time (min) 60 | 150 |
|---|---|---|---|
| (a) Rapid infusion of freshly drawn maternal blood (30-75 ml/kg) | | | |
| Infusion | 4/3 | $33 \pm 9^*$ | $58 \pm 27$ |
| No infusion | 3/2 | $128 \pm 35$ | $102 \pm 1$ |
| (b) 60 min infusion of iso- or hypertonic NaCl solution (29.2-65.4 ml/kg) | 5/4 | $45 \pm 16^*$ | $99 \pm 19$ |

$^*$ $P < 0.05$ (paired t-test for difference from control)

vincing experimental suppression of fetal plasma renin by administration of sodium has not been achieved but it is possible that lambs nearer term would present conditions more conducive to the success of this maneuver.

## PARTURITION

Newborn lambs delivered vaginally under farm conditions and less than 10 hr old had substantial concentrations of angiotensin II in circulating blood whereas in similar lambs delivered by cesarean section under maternal epidural anesthesia angiotensin II was undetectable (10). Comparable differences have been found in newborn infants (Table 4).

Which feature(s) of parturition are responsible for the difference is not known, though hypoxemia per se may be excluded with some confidence (11). Any mechanism tending to maintain intermittently obstructed placental blood flow is likely to have immediate survival value and may influence placental transfusion at birth.

Table 3

Correlation of Maternal and Fetal Plasma Sodium and Potassium

| Days of Gestation | No. of pairs of samples | Correlation Coefficients $[Na^+]$ | $[K^+]$ |
|---|---|---|---|
| < 120 | 26 | 0.541* | 0.283 |
| > 119 | 44 | 0.154 | 0.246 |

*$P < 0.01$

## DISCUSSION

### The Renin-Angiotensin System

In the fetal lamb, renal renin is detectable at least from 75/147 (0.51) term, and plasma renin activity and angiotensin II from before 100/147 (0.68) term. There is no evidence that maternal renin, renin substrate, or angiotensin can cross the placenta. Plasma renin activity is normally low until about 104/147 (0.71 term) (Fig. 3) though plasma angiotensin II is often high at this age (Fig. 4). Plasma renin activity can be increased by bleeding from 93/147 (0.63) term. In mature fetal lambs, plasma renin activity has consistently been reported to be several-fold maternal (Table 1) but plasma angiotensin II concentration is lower than at earlier ages. Fetal renin substrate concentration averages only 58% maternal. It is more likely that the high concentration of angiotensin II found in some immature fetal lambs reflect limited ability of immature tissues to destroy angiotensin II than increased production. Angiotensin II can inhibit renin release (37, 43) and the low plasma renin activity/high angiotensin II situation present in some immature fetal lambs may be due to this phenomenon. In older lambs increased metabolism of angiotensin II would then allow increased renin secretion and higher plasma renin activity. The inverse relationship between plasma angiotensin II and renin substrate invites more attention to substrate concentration under varying circumstances.

Table 4

The Renin-Angiotensin System and Mode of Delivery

|  | Lamb (< 10 hr old) | Human cord blood | | |
|---|---|---|---|---|
|  | AII | AII | AII | PRA |
| Reference | 10 | 14 | 35 | 35 |
|  | rat colon | radioimmunoassay | | |
| Vaginal | 839 | 185 | 40 | 4.3 |
| Vaginal (epidural anesthesia) |  |  | 67 | 6.4 |
| Caesarean | < 123 | 73 | 7.5 | 2.4 |

Angiotensin II (AII) in pg/ml and plasma renin activity (PRA) as AI ng/ml·hr$^{-1}$.

Attempts to manipulate fetal arterial pressure or blood volume have produced changes of plasma renin activity or angiotensin II consistent with one of the accepted mechanisms of the renal baroreceptor hypothesis of renin release (18). It may be that the response of the fetal renin-angiotensin system to blood loss is particularly important before the autonomic nervous system is fully developed. When the newborn rabbit is bled, massive angiotensin II release quickly follows a small release of pressor amines, but in adult rabbits much larger concentrations of pressor amines are produced and smaller angiotensin II release is delayed until the concentration of pressor amines has fallen (12). Adrenaline causes renin release in the isolated kidney (18) but this mechanism was not apparent in the fetus (11). The possibility that limitation of sodium supply promotes renin release through stimulation of the macula densa is an attractive one in relation to fetal growth. Direct evidence for this mechanism is lacking even in adults. Recent work on superfused glomeruli *in vitro* shows that renin retention within the cell is an active process and suggests the possibility that a common pathway based on active volume regulation of renin-containing cells is implicated in renin release whether the stimulus is initiated through renal baroreceptors or by the ionic and osmotic environment of the macula densa (3, 24).

The coincidence of falling sodium ion concentration in amniotic and allantoic fluids and in urine (30), the decreasing permeability of the skin to active influx of sodium (23) and the increasing absolute growth rate in the last third of gestation all suggest that progressive adjustments to meet fetal sodium requirements may induce the high plasma renin activity characteristic of the mature fetal lamb.

Consequences of Bilateral Nephrectomy in Fetal Lambs

The well-known growth retardation associated with congenital absence of kidneys (Potter's syndrome) has been mimicked by removal of the fetal kidneys at 0.57 term in sheep (41). This procedure was without significant effect on plasma electrolytes. Body weight measured at 0.92 term implied a growth rate half that of normal siblings. It has been suggested that this growth retardation may be due to elimination of renal degradation of growth hormone to a somatomedin or other active substance (41, 46).

The longer-term cardiovascular consequences of fetal nephrectomy have not been systematically examined. However, our observations and those of Faber et al. (21) indicate that arterial pressure was low in a few nephrectomized fetal lambs. Moreover arterial $O_2$ saturation was consistently slightly lower in nephrectomized lambs than in their intact sibs (paired t-test n = 9, P < 0.02, 3 lambs) though $P_{CO_2}$ and pH were not significantly different. The constrictor action of angiotensin II on umbilical vessels is weak, at least *in vitro* (20). The tonic action of angiotensin II on fetal systemic vessels may therefore make a contribution to the maintenance of umbilical blood flow which over a long period helps to ensure adequate $O_2$ uptake by the fetus for metabolism and growth.

It is also significant that Berton (5) found that bilateral nephrectomy of fetal rabbits at 25-29 days gestation (0.8-0.93 term) was invariably fatal. If arterial pressure at that age (19) falls as much after bilateral nephrectomy as in the newborn rabbit (7), it is likely that arterial pressure would then be less than 10 mm Hg and incompatible with survival to term (31 days). The newborn rabbit is far more dependent on the renin-angiotensin system than on the autonomic nervous system for the maintenance of arterial pressure and resistance to bleeding than the adult (32). The much greater maturity of the lamb than the rabbit at birth and the fact that bilateral nephrectomy at 84 days gestation is not fatal implies that in this species any crucial contribution of renal renin to fetal survival may occur at an even younger age. However, the occurrence of Potter's syndrome also implies that fetal development cannot be uniquely dependent on fetal renal renin in all species.

## CONCLUSION

The fetal renin-angiotensin system is clearly autonomous though there is no evidence that it is qualitatively different from the adult. Quantitatively, differences of plasma renin activity and renin substrate are substantial, but the correlation of maternal and fetal plasma angiotensin II concentrations (Fig. 4) may be indicative of a placental role in the metabolism of angiotensins. The responses of the fetal renin-angiotensin system to stimuli resemble those found in adults. In a few instances (e.g. the failure of furosemide to increase an already high fetal plasma renin activity and the association of high plasma renin activity and low $O_2$ in immature lambs) unusual combinations of circumstances are likely and their detailed explanation will no doubt prove intriguing.

The evidence suggests that activity of the fetal renin angiotensin system makes a significant or, in the rabbit, perhaps crucial contribution to the control of the fetal cardiovascular system. More documentation of its interaction with other neural and hormonal systems is needed. In particular reproductive hormones (18) can influence substrate concentration. It is unlikely that this ancient mechanism is merely an atavism; in its absence both growth and distribution of the circulation may be impaired, with serious consequences for development in general.

## ACKNOWLEDGMENT

This work was supported by the Medical Research Council and the Spastics Society. Gifts of renin from Dr. E. Haas and of antisera to angiotensins I and II from Dr. E. R. Lumbers are gratefully acknowledged.

REFERENCES

1. Aiken, J. W. and Vane, J. R. 1972. Inhibition of converting enzyme of the renin angiotensin system in kidneys and hindlegs of dogs. *Circ. Res.* 30:263-273.

2. Assali, N. S.; Holm, L. W.; and Seghal, N. 1962. Regional blood flow and vascular resistance of the foetus *in utero*. Action of vasoactive drugs. *Am. J. Obstet. Gynecol.* 83:809-819.

3. Baumbach, L.; Leyssac, P. P.; and Skinner, S. L. 1976. Studies on renin release from isolated superfused glomeruli: effects of temperature, urea, ouabain and ethacrynic acid. *J. Physiol.* 258:243-256. London.

4. Bernstine, R. L. 1970. A chronic renal model for the fetus. *Lab. Anim. Sci.* 20:949-956.

5. Berton, J-P. 1970. Effects de la néphrectomie bilatérale chez foetus de lapin (survie et metabolisme hydrique). *C.R. Acad. Sci.* (Paris), 271:219-222.

6. Blair-West, J. R.; Coghlan, J. P.; Denton, D. A.; Funder, J. W.; Scoggins, B. A.; and Wright, R. D. 1971. Inhibition of renin secretion by systemic and intrarenal angiotensin infusion. *Am. J. Physiol.* 220:1309-1315.

7. Broughton Pipkin, F. 1971. Cardiovascular responses in rabbits of different ages to hypertensin and adrenaline. *Quart. J. Exp. Physiol.* 56:210-220.

8. ----. 1972. Hepatic inactivation of val$^5$ angiotensin II amide (Hypertensin), val$^5$ angiotensin II free acid and adrenaline in immature and adult rabbits. *J. Physiol.* 225:35-36. London.

9. ----. 1973. Circulating vasoactive hormones in immature animals. Ph.D. dissertation, Oxford University.

10. Broughton Pipkin, F.; Kirkpatrick, S. M. L.; Lumbers, E. R.; and Mott, J. C. 1974. Renin and angiotensin-like levels in foetal, newborn and adult sheep. *J. Physiol.* 241:575-588.

11. Broughton Pipkin, F.; Lumbers, E. R.; and Mott, J. C. 1974. Factors influencing plasma renin and angiotensin II in the conscious pregnant ewe and its foetus. *J. Physiol.* 243:619-636. London.

12. Broughton Pipkin, F.; Mott, Joan C.; and Roberton, N. R. C. 1971. Angiotensin II-like activity in circulating arterial blood in immature and adult rabbits. *J. Physiol.* 218:385-403. London.

13. Broughton Pipkin, F. and Smales, O. R. C. 1975. Blood pressure and angiotensin II in the newborn. *Arch. Dis. Child.* 50:330.

14. Broughton Pipkin, F. and Symonds, E. M. 1975. Angiotensin II and the placenta in normal pregnancy and in pregnancy complicated by hypertension. *J. Physiol.* 256:121-122P. London.

15. Buddingh, F.; Parker, H. R.; Ishizaki, G.; and Tyler, W. S. 1971. Long term studies of the functional development of the fetal kidney in sheep. *Am. J. Vet. Res.* 32:1993-1998.

16. Burlinghame, P.; Long, J. A.; and Ogden, E. 1942. The blood pressure of the foetal rat and its response to renin and angiotonin. *Am. J. Physiol.* 137:473-484.

16a. Carver, J. G. 1977. The renin-angiotensin system of the foetus and newborn. Master's thesis, Oxford University.

17. Carver, J. G., and Mott, J. C. 1974. Plasma renin, [$Na^+$] and [$K^+$] in immature foetal lambs with indwelling catheters. *J. Physiol.* 245:73-75. London.

17a. ----. 1978. Renin substrate in plasma of unanaesthetized pregnant ewes and their foetal lambs. *J. Physiol.* 276:395-402. London.

18. Davis, J. O., and Freeman, R. H. 1976. Mechanisms regulating renin release. *Physiol. Rev.* 56:1-56.

19. Dawes, G. S.; Handler, J. J.; and Mott, J. C. 1957. Some cardiovascular responses in foetal, newborn and rabbits. *J. Physiol.* 139:123-136. London.

20. Dyer, D. C. 1971. Effect of metabolic inhibitors on responses to serotonin and angiotensin on isolated sheep umbilical arteries. *Eur. J. Pharmacol.* 16:357-360.

21. Faber, J. J.; Green, T. J.; and Thornburg, K. L. 1974. Arterial blood pressure in the unanaesthetized fetal lamb after changes in fetal blood volume and haematocrit. *Quart. J. Exp. Physiol.* 59:241-256.

22. Fleischman, A. R.; Oakes, G. K.; Epstein, M. F.; Catt, K. J.; and Chez, R. A. 1975. Plasma renin activity during ovine pregnancy. *Am. J. Physiol.* 228:901-904.

23. France, V. M. 1976. Active sodium uptake by the skin of the foetal sheep and pigs. *J. Physiol.* 258:377-392. London.

24. Frederiksen, O.; Leyssac, P. P.; and Skinner, S. L. 1975. Sensitive osmometer function of juxtaglomerular cells *in vitro*. *J. Physiol.* 252:669-680. London.

25. Goodfriend, T. L., and Peach, M. J. 1975. Angiotensin III: (Des-aspartic acid[1]) angiotensin II. Evidence and speculation for its role as an important agonist in the renin-angiotensin system. *Circ. Res.* (suppl.) 1:38-48.

26. Hébert, F.; Fouron, J. C.; Boileau, J. C.; and Biron, P. 1972. Pulmonary fate of vasoactive peptides in fetal, newborn and adult sheep. *Am. J. Physiol.* 232:20-23.

27. Hodge, R. L.; Ng, K. K. F.; and Vane, J. R. 1967. Disappearance of angiotensin from the circulation of the dog. *Nature* 215:138-141.

28. Kaplan, A., and Friedman, M. 1942. Studies concerning the site of renin formation in the kidney III. The apparent site of renin formation in the tubules of the mesonephros and metanephros of the hog fetus. *J. Exp. Med.* 76:307-316.

29. Lumbers, E. R.; Reid, G. C.; and Lee-Lewes, J. 1976. A comparison of the effects of noradrenaline, angiotensin II, adrenaline and isoprenaline on the fetal cardiovascular system. *Proc. Soc. Aust. Physiol. Pharmacol.* 7:38.

30. Mellor, D. J., and Slater, J. S. 1971. Daily changes in amniotic and allantoic fluid during the last three months of pregnancy in conscious unstressed ewes, with catheters in their foetal fluid sacs. *J. Physiol.* 217:573-604. London.

31. Mott, J. C. 1965. Haemorrhage as a test of the function of the cardiovascular system in rabbits of different ages. *J. Physiol.* 181:728-752. London.

32. ----. 1969. The kidneys and arterial pressure in immature and adult rabbits. *J. Physiol.* 202:25-44. London.

33. Ng, K. K. F., and Vane, J. R. 1967. The conversion of angiotensin I to angiotensin II. *Nature* 216:762-766.

33a. Oakes, G. K.; Fleischman, A. R.; Catt, K. J.; and Chez, R. A. 1977. Plasma renin activity in sheep pregnancy after fetal or maternal nephrectomy. *Biol. Neonate.* 31:208-212.

34. Peart, W. S. 1955. A new method of large scale preparation of hypertensin with a note on its assay. *Biochem. J.* 59:300-302.

35. Reid, G. C., and Lumbers, E. R. 1975. The effect of parturition on the renin-angiotensin system of the human fetus. *Proc. Soc. Aust. Physiol. Pharmacol. Soc.* 6:102-103.

36. Rurak, D. W. 1975. Plasma vasopressin in foetal lambs. *J. Physiol.* 256:36-37P. London.

37. Shade, R. E.; Davis, J. O.; Johnson, J. A.; Gotshall, R. W.; and Spielman, W. S. 1973. Mechanism of action of angiotensin II and antidiuretic hormone on renin secretion. *Am. J. Physiol.* 224:926-929.

38. Skinner, S. L.; Dunn, J. R.; Mazzetti, J.; Campbell, D. J.; and Fidge, N. H. 1975. Purification, properties and kinetics of sheep and human renin substrates. *Aust. J. Exp. Biol. Med. Sci.* 53:77-88.

39. Smith, F. G.; Lupu, A. N.; Barajas, L.; Bauer, R.; and Bashore, R. A. 1974. The renin-angiotensin system in the fetal lamb. *Pediatr. Res.* 8:611-620.

40. Sokabe, H.; Ogawa, M.; Oguri, M.; and Nishimura, H. 1969. Evolution of the juxtaglomerular apparatus in vertebrate kidneys. *Tex. Rep. Biol. Med.* 27:868-885.

41. Thorburn, G. D. 1974. The role of the thyroid gland and kidneys in fetal growth. In *Size at birth,* Ciba Foundation Symposium, London, 27.

42. Trimper, C., and Lumbers, E. R. 1972. The renin-angiotensin system in foetal lambs. *Pfluegers. Arch.* 336:1-10.

43. Vander, A. J., and Geelhoed, G. W. 1965. Inhibition of renin secretion by angiotensin II. *Proc. Soc. Exp. Biol. Med.* 120:399-403.

44. Vander, A. J., and Miller, R. 1964. Control of renin secretion in the dog. *Am. J. Physiol.* 207:537-545.

45. Vane, J. R. 1969. The release and fate of vasoactive hormones in the circulation. *Br. J. Pharmacol.* 35:209-242.

46. Wallace, A. L. C.; Stacy, B. B.; and Thorburn, G. D. 1970. The effect of kidney ligation on the release of plasma free fatty acids following the injection of growth hormone. *J. Endocrinol.* 48:297-298.

47. Wintour, E. M.; Brown, E. H.; Denton, D. A.; Hardy, K. J.; McDougall, J. G.; Oddie, C. J.; and Whipp, G. T. 1975. The ontogeny and regulation of corticosteroid secretion by the ovine foetal adrenal. *Acta Endocrinol. (Kbh)* 79:301-316.

# Prostaglandins and the Developing Cardiovascular System

# 18

## Role of Prostaglandins in Control of Fetal and Neonatal Pulmonary Circulation

Sidney Cassin, Thom Tyler, Charles Leffler, and Richard Wallis

Department of Physiology
College of Medicine
University of Florida
Gainesville, Florida

PERINATAL PULMONARY CIRCULATION

Regulation of blood flow through the lungs plays a key role in control of the fetal circulation. If pulmonary vascular resistance were low, a large arteriovenous shunt between the two sides of the heart would exist. A large proportion of the right ventricular output would enter the left heart. Such a diversion of the cardiac output could only be detrimental to the fetus by increasing the work of the left heart (12). Normally, however, blood flow through fetal lungs is quite low. Rudolph (33) has suggested that pulmonary blood flow in mature fetal lambs (130-140 days gestation) is 35-45 ml/min/kg body weight.

Since postnatal survival of the fetus, after elimination of the placental circulation, depends on establishment of appropriate alveolar ventilation and pulmonary blood flow for adequate gaseous exchange, a remarkable transformation must occur in the pulmonary circulation at birth.

With ventilation of the lungs, pulmonary blood flow increases by a factor of 5 or 10 in the lamb. The mechanisms responsible for the dramatic decrease in pulmonary vascular resistance have been studied extensively (6, 13, 17). This decrease in pulmonary vascular resistance, seems at least in part, due to to local vasodilator effects of increased $PO_2$ and decreased $PCO_2$. Observations by Rudolph and Yuan (34) in newborn calves suggest in addition that changes in pulmonary vascular resistance with hypoxia are dependent on pH. Some decrease in pulmonary vascular resistance is probably also related to the development of a gas-liquid interface on the alveolar wall. Surface forces (15) would tend to collapse the alveoli with a resulting subatmospheric pressure which would be exerted on alveolar septa and small vessels coursing between the alveoli. Thus, alveolar vessels would be distended. The basic mechanism by which oxygen, carbon dioxide, and hydrogen ion exert their effects on pulmonary arterioles is not known. It is conceivable that an increase in $PO_2$ releases or activates some substance in the lung which exerts a dilator influence on the vascular supply of that organ. The elevation in arterial $PO_2$ with ventilation is by far the most important means of dilating the pulmonary bed of the fetus. In all experiments reported, improved oxygenation, whether by arterial blood, gaseous expansion (6), or liquid (oxygenated) expansion (27) caused pulmonary vasodilation.

## RESPONSES OF PERINATAL PULMONARY CIRCULATION TO DRUGS

The lungs of fetal lambs and goats are not only sensitive to changes in blood gas tensions and hydrogen ion, but also to a variety of drugs. Small doses of histamine, acetycholine, isoproterenol (7), and bradykinin (3) cause marked vasodilatation of the pulmonary circulation. After birth, dilator responses to drugs are reduced, probably because relative to unventilated fetal lungs the vasculature of air-ventilated lungs is very dilated. However, acetycholine and bradykinin will continue to produce vasodilatation in lambs 2-4 months after birth (11) if they exhibit a moderate pulmonary vasoconstrictor tone. Although there are many other vasoactive materials which may alter fetal pulmonary vascular resistance, it is not our intention to review them.

There are few drugs that currently occupy more widespread interest in biological circles than do the prostaglandins (14, 18). Until recently, however, little if anything had been published on the effects of prostaglandins in the fetal and neonatal pulmonary circulations. It is the intent of this chapter to review the work done in our laboratory on the effects of the prostaglandins, as well as inhibitors of their synthesis, on the perinatal pulmonary circulation.

Prostaglandin $E_1$ ($PGE_1$) has been shown to decrease pulmonary and systemic vascular resistance of several mammalian species (21, 22, 23) including man (1, 4). However, in contrast, Kadowitz, Joiner, and Hyman have reported vasoconstrictor properties for prostaglandin $E_2$ ($PGE_2$) on lobar arteries and veins of dog, swine, and sheep (24). Prostaglandins of the F series (PGF) are generally accepted as vasoconstrictors of the pulmonary vasculature (20). The actions of prostaglandins are not blocked by ganglionic blocking agents, botulinus toxins, adrenergic inhibitors, cholenergic inhibitors, or antihistamines (45).

In adult mammals, 90% or more of prostaglandins of the E series and 35-70% of the F series are inactivated in a single pass through the pulmonary circulation (2, 16, 19). Thus, the effects of exogenous PGE and PGF are greatly influenced by the route of administration and metabolism by the lung (35, 43). Smith (35) has suggested that the E and F series of prostaglandins which are known to be produced and metabolized by lungs may be of importance in the regional control of balance between ventilation and perfusion so that arterial blood gas content is maintained constant.

## EFFECTS OF $PGE_1$ AND $PGE_2$ ON
## FETAL PULMONARY VASCULAR RESISTANCE

Pregnant goats close to term were anesthetized with chloralose and their fetuses were delivered by cesarean section with umbilical circulation intact, as described in detail elsewhere (8). Briefly, the left pulmonary artery was perfused, at constant flow with blood from fetal inferior vena cava (level of liver). Pulmonary arterial perfusion pressure and flow as well as pulmonary venous pressure were monitored. Pulmonary venous blood was collected in a reservoir and returned to the fetus via the femoral

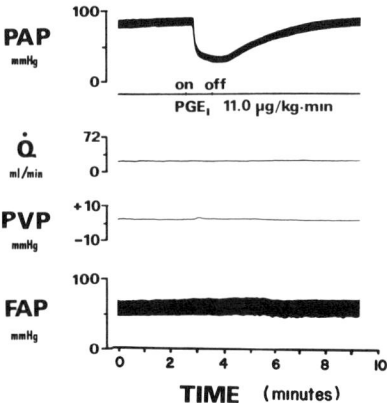

Fig. 1. Infusion of $PGE_1$ directly into left pulmonary artery on pulmonary artery perfusion (PAP), lower left lobe pulmonary venous pressure (PVP) and femoral arterial pressure (FAP) of slightly hypertensive fetal unventilated goat. Pulmonary inflow ($\dot{Q}$) was maintained constant. (From Cassin, Tyler, Wallis *Proc. Soc. Exper. Biol. Med.*, 1975.)

artery. With pulmonary arterial inflow held constant, infusion of $PGE_1$ (11.0 µg/kg·min) into the left pulmonary artery of fetal goats (Fig. 1) decreased perfusion pressure and pulmonary vascular resistance in an isolated unventilated lobe. The decrease in vascular resistance reflects a reduction in tonic smooth muscle contraction and resultant vasodilatation. These results are consistent with reports of active vasodilatation and decreased pulmonary vascular resistance with $PGE_1$ in other species (18, 21, 22). In this preparation, infusions for 1 minute in doses as high as 25 µg/kg·min did not alter systemic pressure. These data suggested to us that $PGE_1$ was inactivated in the fetal lung. The same dose of $PGE_1$ injected into the left ventricle in a 2 day old goat produced a systemic hypotension. Olley, Coceani, and Kent (31) reported an average inactivation of $PGE_1$ by fetal lamb lung of 72%, although up to 89% inactivation occurred with doses of $PGE_1$ ranging from 0.38 to 9.50 µg/kg·min. In man (1, 4) infusions of $PGE_1$ greater than 0.18 µg/kg·min produce reductions not only in pulmonary vascular resistance, but also in systemic arterial pressure.

We suggested (8) that systemic hypotension consequent to intravenous infusion of $PGE_1$ may be due to saturation of the pulmonary inactivating mechanism with subsequent delivery of noninactivated $PGE_1$ to the left heart. However, in our preparation pulmonary venous blood from the left lung was not returned directly to the heart, but was collected in a reservoir and passed thru a large extracorporeal circuit before returning to the descending aorta. Clearly, the lack of systemic effect also may have been due to inactivation in the extracorporeal circuit. In order to evaluate this possibility we modified our circuitry so that pulmonary venous blood returned normally to the left atrium without intervening extracorporeal circuitry. Under these conditions, continuous infusions of $PGE_1$ (3.47

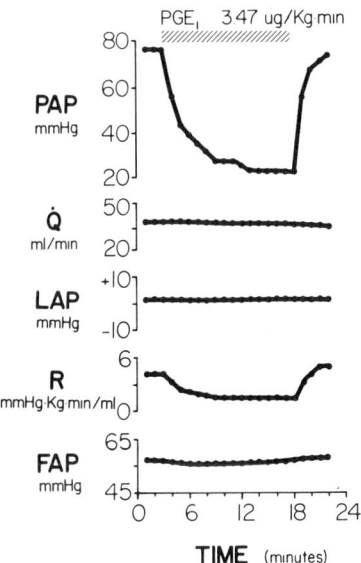

Fig. 2. Infusion for 15 minutes of $PGE_1$ directly into left pulmonary artery of unventilated fetal goat with intact umbilical circulation. Measurements were made every minute of pulmonary perfusion pressure (PAP), pulmonary flow (Q), left atrial pressure (LAP), and systemic mean pressure (FAP). Resistance (R) was calculated as $\frac{PAP - LAP}{\dot{Q}/kg\ body\ wt}$ (From Tyler, Wallis, Cassin IRCS 2:1662, 1974.)

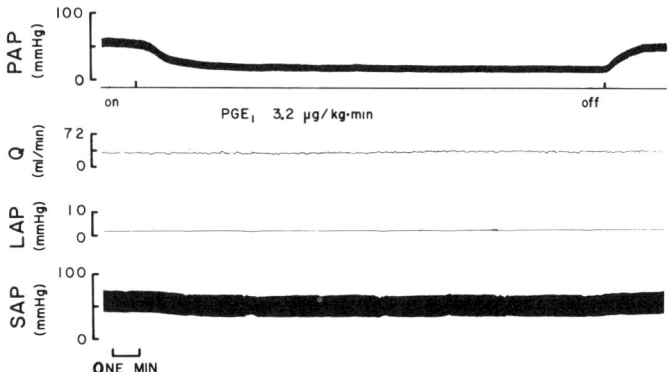

Fig. 3. Photograph of actual record of 17 minute infusion of $PGE_1$ into left pulmonary artery of unventilated fetal goat. Records were made of pulmonary arterial pressure (PAP), pulmonary inflow (Q), left atrial pressure (LAP), and systemic arterial pressure (SAP). (From Tyler, Leffler, Wallis, and Cassin, submitted to Amer. J. Physiol., 1976.)

µg/kg·min) for 15 minutes (Fig. 2) decreased pulmonary vascular resistance without a concomitant systemic effect. An actual trace of another experiment is presented in Figure 3. In this experiment, $PGE_1$ was infused into the pulmonary artery for about 17 minutes at the rate of 3.2 µg/Kg·min. There is a remarkable decrease in pulmonary arterial perfusion pressure at constant inflow while systemic pressure is not depressed significantly.

Experiments using similar animal preparations were also carried out with $PGE_2$ which in most species and in most vascular beds is thought to be a potent vasodilator (14). Data described in Figure 4 clearly demonstrate that $PGE_2$ injected directly into the pulmonary artery of fetal goats causes a pulmonary vasodilatation. $PGE_2$ is not as effective a dilator as $PGE_1$; approximately 10 times as much $PGE_2$ as $PGE_1$ is needed to produce the same response.

Dose response (39) curves for effects of $PGE_1$ and $PGE_2$ on pulmonary vascular resistance as well as heart rate and systemic blood pressure are presented in Figure 5. Half of the maximal decrease in pulmonary vascular resistance

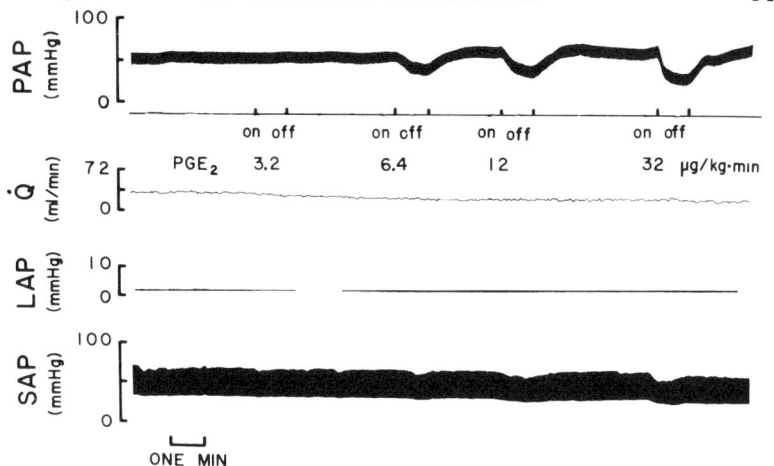

Fig. 4. Dose response curves for 1 minute infusion of 3.2, 6.4, 12, and 32 µg/kg·min $PGE_2$ directly into left pulmonary artery of unventilated fetal goat. Measurements made of left pulmonary artery perfusion pressure (PAP), pulmonary arterial inflow (Q), left atrial pressure (LAP), and systemic arterial pressure (SAP). (From Tyler, Leffler, Wallis, and Cassin, submitted to *Amer. J. Physiol.*, 1976.)

was achieved with a dose of 1.9 µg/kg (infused for 1 minute). Tyler, et al. (39) have described the effects of $PGE_1$ and $PGE_2$ on the fetal cardiovascular system as a function not only of dose but rate of delivery. Figure 6, taken from their work, is based on 187 determinations on 29 fetal goats. Clearly, as the duration of infusion increases, the dose necessary to achieve a 50% decrease in pulmonary vascular resistance, at the same rate of infusion, is diminished. These data are extremely important and should be useful to anyone wishing to use $PGE_1$ and $PGE_2$ clinically. Whereas injection of a single bolus of $PGE_1$ or rapid infusion of an effective dose might saturate the receptor system and produce systemic effects, a smaller dose at the same rate of infusion for a longer duration would produce the desired effect without saturating the receptors. Since exogenous $PGE_1$ infused directly into the pulmonary artery of slightly hypertensive fetal goats dilated the pulmonary vasculature without systemic effects, this drug may have therapeutic value in perfusion-ventilation imbalances with elevated pulmonary vascular resistance, particularly in instances where other measures have failed to improve blood gases.

A specific role for endogenous prostaglandins in the regulation of pulmonary blood flow is not yet clear. However, the data presented above clearly indicate that exogenous $PGE_1$ and $PGE_2$ are dilators of the fetal pulmonary circulation.

The prostaglandins are among the most widely distributed hormone-like substances known to man and have been detacted in almost every tissue and body fluid. Their production seems to increase in response to a multitude of stimuli including distension and changes in circulating hormones. Since they produce a broad spectrum of effects in remarkably minute amounts, it is conceivable that endogenous prostaglandins might be involved in the regulation of fetal and neonatal pulmonary blood flow as well as the changes in pulmonary vascular resistance with the onset of air breathing.

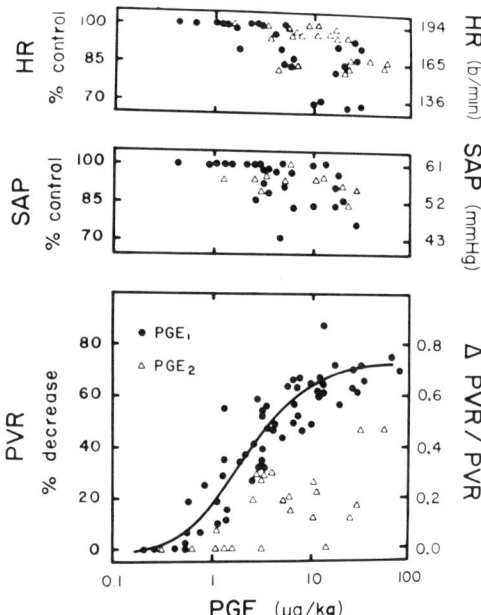

Fig. 5. Dose-response curves for $PGE_1$ and $PGE_2$ in fetal (25) goats on pulmonary vascular resistance (PVR), systemic arterial pressure (SAP), and heart rate (H.R.) (From Tyler, Leffler, Wallis, and Cassin, submitted to Amer. J. Physiol., 1976.)

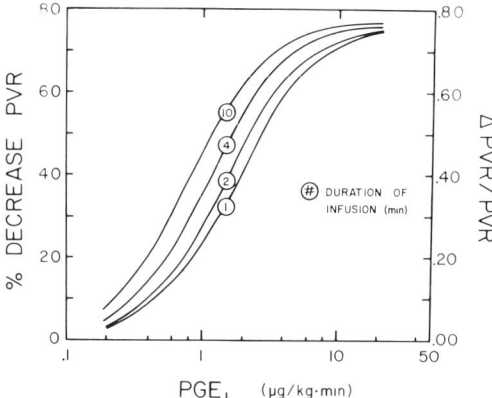

Fig. 6. Computer generated regression curves for decrease in pulmonary vascular resistance (PVR) in fetal goats given $PGE_1$ directly into pulmonary artery. Numbers in circles are durations of infusions. (From Tyler, Leffler, Wallis, and Cassin, submitted to Amer. J. Physiol., 1976.)

## EFFECTS OF AEROSOLIZED $PGE_1$

Although prostaglandins of the E series are effective pulmonary dilators if injected into the pulmonary artery, administration by aerosolization would be a much more convenient method of achieving reductions in pulmonary vascular resistance. To this end, Leffler, Tyler, and Cassin (28) examined the effects of aerosolized $PGE_1$ on the perinatal pulmonary vasculature. Figure 7 summarizes the results of experiments on newborn hypoxic (to produce pressor response) goats given $PGE_1$ in various doses to lower pulmonary vascular resistance. Although infusions of $PGE_1$ caused a reduction in pulmonary vascular resistance to approximately 60% of control, inhalation of an aerosol reduced the pulmonary vascular resistance minimally and appeared to be dose independent up to 50 μg/kg·min. Additionally, there was a marked systemic hypotensive effect (Fig. 8). In fact, aerosolization resulted in a response which was not statistically significantly different from infusion of the same dose of $PGE_1$ into the left ventricle. These experiments demonstrated that administration of $PGE_1$ by aerosol (1) produced small reductions in pulmonary vascular resistance, and (2) had pronounced effects on the systemic circulation. We concluded that a systemic effect occurred with $PGE_1$ given by aerosol because major sites of activity and catabolism of $PGE_1$ in the neonatal goat, are upstream from where aerosolized $PGE_1$ enters the bloodstream. If sites of activity and catabolism of $PGE_1$ in human and goat lung are the same, aerosol administration of $PGE_1$ for treatment of pulmonary hypertension would not produce the therapeutic effect of decreasing pulmonary vascular resistance without reducing systemic arterial pressure.

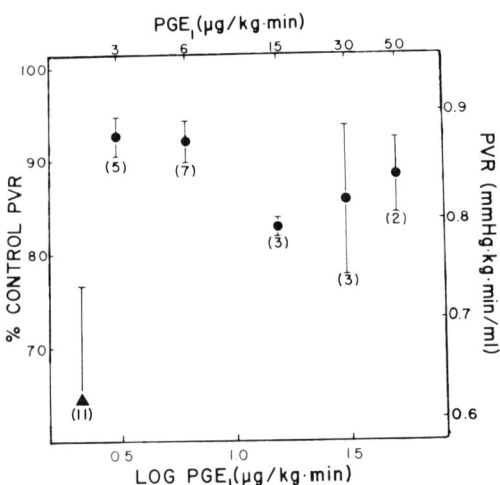

Fig. 7. Effect of $PGE_1$ administered as an aerosol on pulmonary vascular resistance (PVR) of newborn goats. All animals received indomethacin (3 mg/kg) 60 minutes prior to beginning experiments. Solid circles are means ± SE of means for reduction in PVR with aerosol. Solid triangle is response to infusion of 2 µg/kg·min into left pulmonary artery. (From Leffler, Tyler, and Cassin *Proc. Soc. Exper. Biol. Med.* 155:19-22, 1977.)

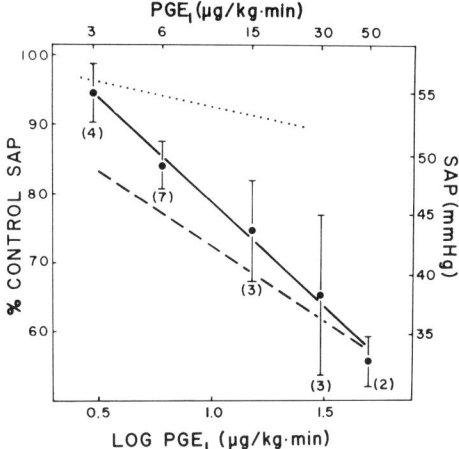

Fig. 8. Effect of PGE₁ administered as an aerosol (·-mean ± SE), infused into left pulmonary artery (·····), or infused into left ventricle (----) on mean systemic blood pressure of newborn goats. Numbers in parenthesis are sample sizes for aerosol PGE₁ treatment. Eleven goats were used for left pulmonary artery infusion with 22 points between 3.2 and 25 µg/kg·min. Four goats were used for left ventricular infusion. (16 points 1.03 - 70 µg/kg·min. (From Leffler, Tyler, and Cassin *Proc. Soc. Exper. Biol. Med.* 155:19-22, 1977.)

## SEGMENTATION OF CHANGES IN PULMONARY VASCULAR RESISTANCE DUE TO PGE₁

Over the years various models have been utilized to describe flow through adult pulmonary vasculature. Several years ago we described a model for the fetal and neonatal pulmonary circulation (17) which suggested that it behaved as if it were controlled by vessels acting as Starling resistors. A Starling resistor is an easily collapsible tube through which flow is related to the pressure drop between inflow and surrounding pressure rather than the pressure drop between inflow and outflow, as long as the surrounding pressure is greater than the outflow pressure. The present studies use a modification of the MacDonald, Butler (30) model and our previous model of Starling resistors, which permits us to calculate resistances in two segments of the pulmonary vasculature. Resistances were calculated in a segment proximal to the presumed Starling resistors as well as in the segment distal to them. Details of the methodology for these calculations and a complete discussion of Starling resistors have been presented elsewhere (5, 17).

Data on blood gases from animals subjected to infusion of $PGE_1$ (10 µg/kg·min) are presented in Table 1. Average values for resistances proximal (Rp) and resistances distal (Rd) to the presumed Starling resistors as well as pressure surrounding the Starling resistors (Ps) are shown in Figure 9. Infusions of $PGE_1$ decreased proximal resistance, distal resistance and surrounding pressure significantly ($P < .01$). Proximal resistance decreased from a value of 5.99 to 1.53 mm Hg·Kg·min/ml. Distal resistance decreased from 0.73 to 0.23 mm Hg·kg·min/ml. Surrounding pressure dropped from 15.02 to 9.19 mm Hg. After infusion was stopped, proximal resistance, distal resistance, and surrounding pressure returned to approximately preinfusion levels. Although we do not know the exact location of the segments of the Starling resistor in the pulmonary vasculature, indirect evidence (17) suggests that they include vessels both pre and post capillaries. Experiments presented previously in this chapter on aerosolization suggest that the major sites of action and inactivation of prostaglandins are precapillary in ventilated animals. If this

Table 1

Arterial Blood

| Condition | $P_{O_2}$ mmHg | $P_{CO_2}$ mmHg | pH |
|---|---|---|---|
| Before Drug | 25.1 ± 1.1 | 48.6 ± 2.7 | 7.28 ± .03 |
| During Drug | 23.0 ± 1.1 | 49.0 ± 3.6 | 7.22 ± .03 |
| After Drug | 23.5 ± 0.7 | 50.0 ± 2.6 | 7.23 ± .02 |

Values are means ± SE.

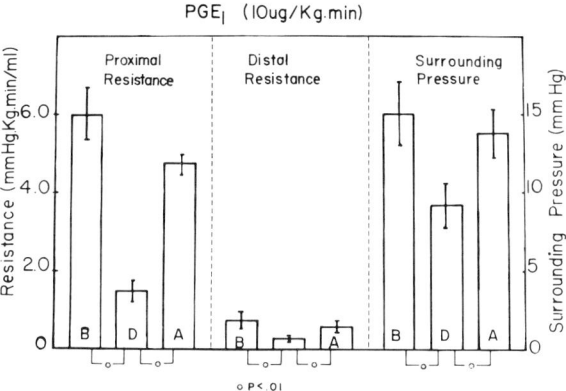

Fig. 9. Calculated changes in proximal and distal resistance as well as surrounding pressure before, during and after PGE$_1$ infusion into left pulmonary artery (N = 16).

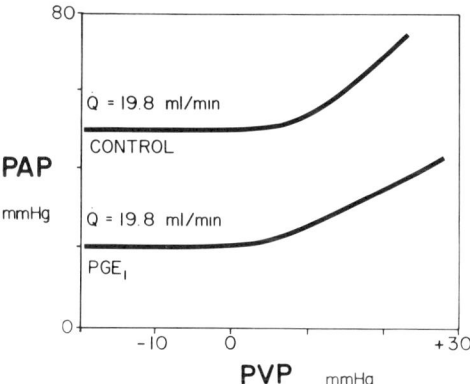

Fig. 10. Record of effect of elevation of pulmonary venous pressure (PVP) on pulmonary arterial pressure (PAP) at constant pulmonary arterial inflow (19.8 ml/min) before and after infusion of PGE$_1$ (10 µg/kg·min) in a fetal goat of 135 days gestation.

also applies to the unventilated fetus, it is conceivable that proximal resistance, which normally constitutes about 87% of the total resistance across the lungs, is situated on the arterial side of the capillaries. In contrast, distal resistance, which contributes about 11% of total resistance, may be located in the venules. This is substantiated by the fact that $PGE_1$ presented by aerosol does in fact produce some decrease in pulmonary vascular resistance, though not much of it must enter the vasculature upstream of the capillaries. Finally, the change in proximal resistance which occurred was not due to a passive dilatation. Figure 10 shows that for these experiments there was a remarkable decrease in pulmonary arterial pressure (Ppa) at the same flow. Also these data indicate that pulmonary arterial pressure decreased substantially more than can be accounted for by a decrease in surrounding pressure. Thus, it is clear that although a decrease in surrounding pressure may have contributed to the reduction in pulmonary arterial pressure at constant flow, there was a decrease in resistance to flow as well.

## HYPOXIA AND INDOMETHACIN

As indicated previously the perinatal pulmonary circulation is responsive not only to hypoxia but prostaglandins. Inhibition of synthesis and release of prostaglandins by indomethacin leads to an elevation of pulmonary vascular resistance which appears to be potentiated by hypoxia. The response (42) to indomethacin appears to be more pronounced in premature (ventilated fetus 133-135 days gestation) than in mature newborn (less than 2 weeks of age) goats. In contrast, the response to hypoxia (5% oxygen) is less in the immature than in the mature perinatal circulation. These results are consistent with the hypothesis that there is a greater net dilator influence in the premature than in the mature neonatal lung. This is further substantiated by the fact that the augmentation in pressor response to hypoxia following indomethacin in the premature animals is considerably greater than in the mature goats (Figs. 11, 12).

Elevation of pulmonary vascular resistance as a result of hypoxia (6) at parturition may lead to a ventilation perfusion imbalance by promoting right to left shunting through the ductus arteriosus and/or foramen ovale (3). Thus, the possibility that exogenous $PGE_1$ and $PGE_2$ might

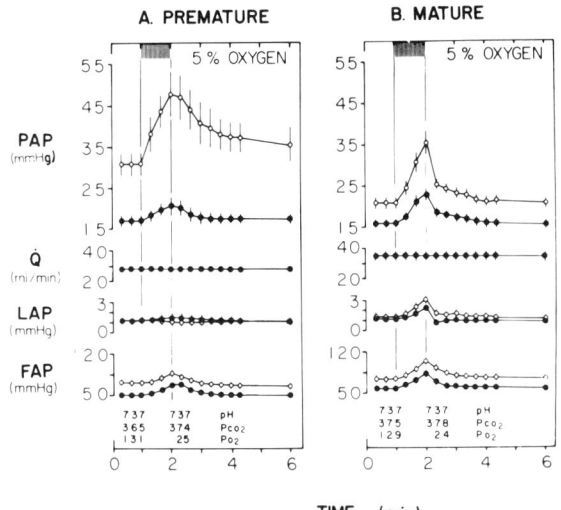

Fig. 11. Effect of ventilation of premature (133-135 days gestation) and mature (less than 2 weeks) goats with 5% oxygen in nitrogen on pulmonary artery pressure (PAP), left atrial pressure (LAP), and femoral arterial pressure (FAP) at constant pulmonary arterial inflow before (solid circles) and after (open circles) indomethacin (2 mg/kg). Results are means ± SE. Duration of hypoxia indicated by solid bar (1 minute). (From Tyler, Wallis, Leffler, and Cassin *Proc. Soc. Exper. Biol. Med.*, 1975.)

be useful in abating the pulmonary pressor response to hypoxia was studied in our laboratory (40) in newborn goats. Figure 13 shows the response of the pulmonary and systemic blood pressure in newborn goats made hypoxic (5% oxygen in nitrogen). Infusion of $PGE_1$ (2 µg/kg·min) during the hypoxic response completely eliminated the pulmonary pressor response. However, probably due to the fact that the $PGE_1$ was inactivated in the lung, the systemic pressor response was not altered. In contrast to $PGE_1$, the same dose of $PGE_2$ diminished the pulmonary pressor response minimally and has no effect systemically. $PGE_2$ is by far a less potent pulmonary vasodilator than is $PGE_1$.

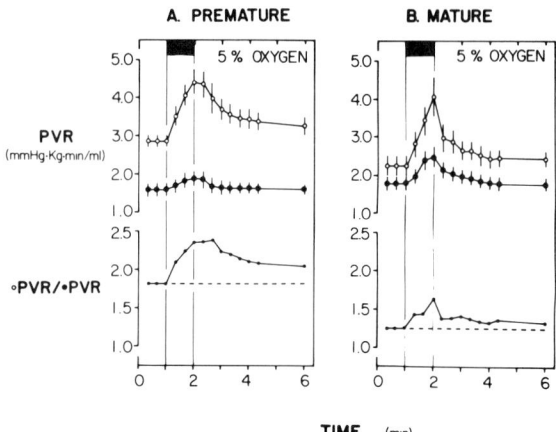

Fig. 12. Effect of ventilation with 5% oxygen in nitrogen on pulmonary vascular resistance of premature and mature newborn goats before (solid circle) and after (open circle) indomethacin. Duration of hypoxia indicated by solid bar (1 minute). The fractional increase in resistance after indomethacin is indicated in the lower portion of the figure. Results are means ± SE. (From Tyler, Wallis, Leffler, and Cassin Proc. Soc. Exper. Biol. Med., 1975.)

## VASOCONSTRICTOR PROSTAGLANDINS

Prostaglandins of the F series have been shown to increase pulmonary vascular resistance in adult mammals (20). Recently, $PGF_{2\alpha}$ has been shown (9, 10, 25) to increase in umbilical cord blood and amniotic fluid during labor. Since $PGF_{2\alpha}$ is suspected of being involved in uterine contractions, exogenous $PGF_{2\alpha}$ is being utilized clinically to induce labor (29, 36). Thus, we (38) investigated the effects of $PGF_{2\alpha}$ as well as $PGF_{1\alpha}$ on the perinatal circulation. Although infusions of either $PGF_{1\alpha}$ or $PGF_{2\alpha}$ produced pronounced pulmonary pressor responses in premature and mature newborn goats, most of our effort was directed at evaluation of $PGF_{2\alpha}$ because of its clinical trial. Figure 14 shows the pressor response to 25 µg/kg·min $PGF_{2\alpha}$ in mature and premature newborn goats before and after infusion of indomethacin. Although there is clearly a potentiation of

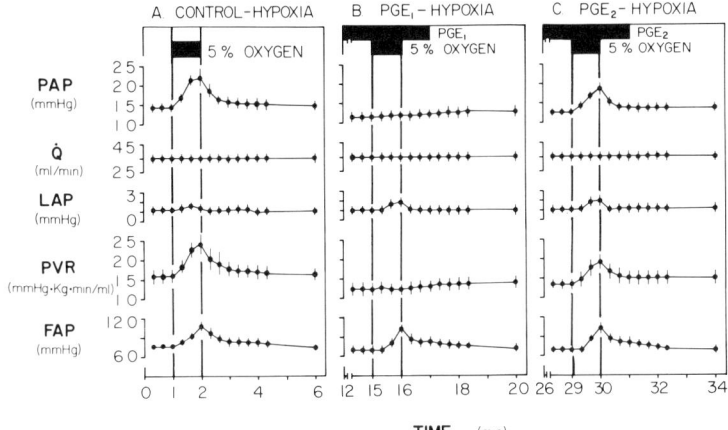

Fig. 13. Changes in pulmonary arterial pressure (PAP), left atrial pressure (LAP), pulmonary vascular resistance (PVR) and femoral arterial pressure (FAP) as a result of breathing 5% oxygen for 1 minute. Panels are for before (A) and during intrapulmonary infusion of $PGE_1$ (B) or $PGE_2$ (C) at 2 µg/kg·min. Results are means ± SE. (From Tyler, Leffler, Wallis, and Cassin *Prostaglandin*, 1975.)

effect in both groups after treatment with indomethacin, there appears to be little difference in responsiveness of the premature and mature neonatal goats to $PGF_{2\alpha}$ following indomethacin. Since there seems to be a greater dilator activity in the premature fetal animal, removal of this influence with indomethacin might be expected to produce a greater pressor response.

## PROSTAGLANDIN PRECURSORS

As indicated previously, many tissues of the body (26) are capable of synthesizing, releasing and metabolizing prostaglandins. A particularly large source of production is found in the lungs (32). Synthesis of prostaglandins is carried out (32) from certain polyunsaturated fatty acids by the formation of a five-member ring and incorporation of 3 oxygen atoms at certain positions in the 20-carbon carboxylic acid in the presence of prostaglandin synthetase. Dihomo-γ-linolenic acid and arachidonic acid are precursors of $E_1$-$F_1$ and $E_2$-$F_2$ series of prostaglandins, respectively. Thus, we decided to evaluate these precursors (41) and analogs of an endoperoxide intermediate ($PGH_2$) on the control of perinatal circulation. Infusions of arachidonic acid directly into the pulmonary circulation (<100 µg/kg·min) of fetal and (<20 µg/kg·min) newborn goats (Fig. 15) resulted in dose related increases in pulmonary arterial pressure at constant inflow. Doses of arachidonic acid in excess of 100 or 20 µg/kg·min respectively in fetal and neonatal animals resulted in systemic hypotension. When drug infusion was stopped, both pulmonary and systemic effects were eliminated. Dihomo-γ-linolenic acid was less effective than arachidonic acid as a pulmonary vasopressor agent in both fetal and neonatal goats. It was of interest to us that left ventricular infusions of arachidonic acid (>300 µg/kg·min) resulted in immediate systemic hypotension. Infusions directly in the fetal and neonatal pulmonary

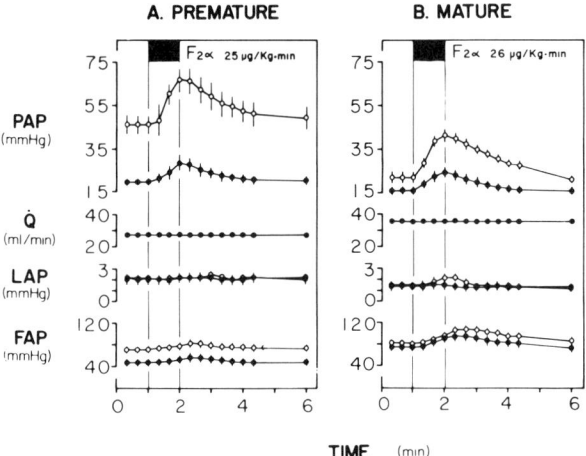

Fig. 14. Effects of a one minute infusion of $PGF_2\alpha$ directly into left pulmonary artery of 6 premature and 6 mature newborn goats on pulmonary artery pressure (PAP), left atrial pressure (LAP), and femoral artery pressure (FAP) before (solid circle) and after (open circle) indomethacin (2 mg/kg). Results are means ± SE. (From Tyler, Leffler, and Cassin, submitted to Amer. J. Physiol., 1976.)

circulation of endoperoxide I or II (in contrast to the effects of arachidonic acid and dihomo-γ-linolenic acid) resulted in an increase in pulmonary arterial pressure and systemic arterial pressure (Fig. 16). Pulmonary pressor responses to arachidonic acid, dihomo-γ-linolenic acid and endoperoxides were not abolished by alpha adrenergic blockade. In agreement with data recently presented by Wicks et al. (44) on vascular responses of perfused canine lungs to precursors indomethacin abolished the circulatory responses to arachidonic acid and dihomo-γ-linolenic acid without altering the responses to the endoperoxide analogs of $PGH_2$.

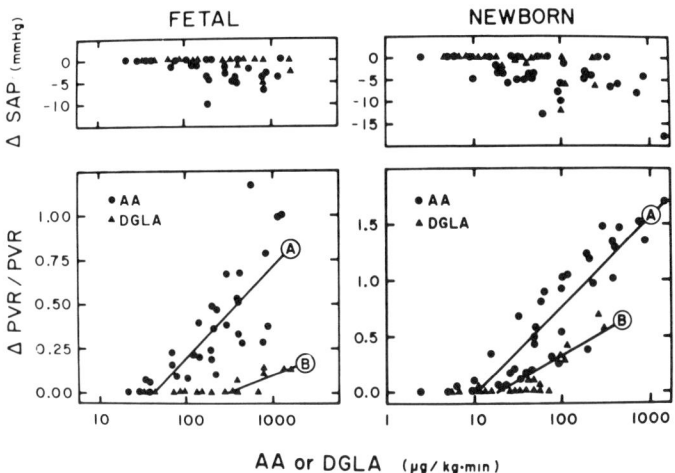

Fig. 15. Effects of one minute infusions of arachidonic acid (AA) or dihomo-γ-linolenic acid (DGLA) on pulmonary vascular resistance (PVR) and systemic arterial pressure (SAP) of fetal and neonatal goats. *Left panel*: Least-square linear regressions are indicated for fetal AA data (Line A, R=0.78, y intercept=-0.86, m=0.53, n=8) and DGLA data (Line B, R-0179, y intercept =-0.50, m=0.20, n=4). Changes in PVR and SAP are referenced to mean PVR (3.54 ± 0.06) or SAP (68 ± 2.0) at zero change. (From Tyler, Leffler, and Cassin *Chest* 715:2715-2735, 1977.)

## SUMMARY AND CONCLUSIONS

Prostaglandins $E_1$ and $E_2$ are both dilators of the perinatal pulmonary circulation. In addition, both of these compounds produce a diminution of the pulmonary pressor response to hypoxia. However, $PGE_1$ is by far a much more potent dilator of the perinatal pulmonary circulation during normoxia as well as during hypoxia. Studies to segment and localize resistance changes due to PGE's suggest that they act primarily upstream of the capillaries although there is some postcapillary activity. Although we

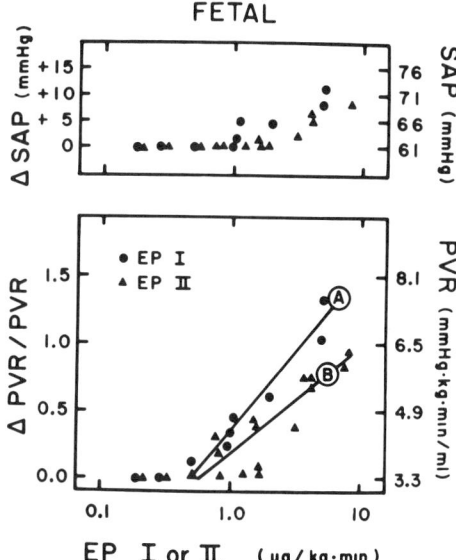

Fig. 16. Dose related effects of one minute infusions of analogs of endoperoxide $PGH_2$ (EP I or EP II; see text) on pulmonary vascular resistance (PVR) and systemic arterial pressure (SAP) of fetal goats. Least-square linear regressions are indicated for EP 1 date (Line A, R = 0.96, y intercept = -3.12, m = 1.16, n = 3) and EP II data (Line B, R = 0.87, y intercept = -2.15, m = 0.77, n = 4). Changes in PVR and SAP are referenced on the right ordinate to mean PVR (3.25 ± 0.10) or SAP (61 ± 2.0) at zero change. (From Tyler, Leffler, and Cassin *Chest* 715:2715-2735, 1977.)

have been able to produce systemic hypotension with intrapulmonary infusion of large doses of $PGE_1$, judicious choice of a proper dose and rate of infusion permits lengthy administration without consequent systemic hypotension. Decreased sensitivity to $PGE_1$ with prolonged infusion did not occur. Thus, we suggest that prostaglandin $E_1$ should be considered for possible use as a therapeutic agent in newborn humans with persistent pulmonary hypertension, and ventilation perfusion imbalances--especially in cases where ventilatory assistance and other maneuvers fail to improve blood gases.

Prostaglandins of the F series are powerful vasoconstrictors of the pulmonary circulation. The F prostaglandins, just as the E prostaglandins, are produced and metabolized by the lungs. The natural occurrence and release of these compounds along with their marked activity suggests to us that they may be important in the local control of the pulmonary vasculature in health and disease.

Increase in pulmonary vascular resistance following administration of indomethacin suggests to us that there is a net dilator influence present in perinatal lungs. The hypoxic pulmonary pressor response in perinates is augmentted with indomethacin. In contrast, $PGE_1$ is most effective in diminishing the pressor response to hypoxia. These results indicate that use of prostaglandin synthetase inhibitors (aspirin, meclofenamate and indomethacin) to control ductal shunting or to prolong labor may be detrimental to the fetus. Although they may be effective in their intended use, it is possible that these drugs will produce an increase in pulmonary vascular resistance as well as an increased pulmonary pressor response to hypoxia.

Arachidonic acid and dihomo-γ-linolenic acid, precursors of prostaglandins, produced marked pulmonary pressor responses and systemic hypotensive effects when infused into the pulmonary artery. The responses are dependent on prostaglandin synthetase activity, but independent of the alpha adrenergic system. Endoperoxide $H_2$ analogs produced a pulmonary pressor response, but instead of a systemic hypotension, resulted in systemic hypertension. The pulmonary and systemic circulatory system effects of the precursors used are not explainable on the basis of actions of PGE, PGF, or endoperoxide $H_2$ alone. Thus, we have suggested that in addition to newly synthesized E and F series prostaglandins intermediates in their synthesis and non-prostanoate vasoactive compounds of alternate synthetic pathways may be responsible for circulatory events described.

## ACKNOWLEDGMENT

Work presented in this chapter which was carried out in our laboratory, was supported in part by NIH-HL 10834, Florida Heart Grant No. 74-AG-2, 75-AG-231 and The Florida Lung Association.

## REFERENCES

1. Bergstrom, S.; Carlson, L. A.; Ekelund, L. G.; and Oro, L. 1965. Cardiovascular and metabolic response to infusion of prostaglandin $E_1$ and to simultaneous infusions of noradrenaline and prostaglandin $E_1$ in man. *Acta Physiol. Scand.* 64:332-339.

2. Bloor, C. M.; White, F. C.; and Sobel, B. E. 1973. Coronary and systemic thermodynamic effects of prostaglandins in unanesthetized dog. *Card. Res.* 7:156-166.

3. Campbell, A. G. M.; Dawes, G. S.; Fishmen, A. P.; Hyman, A. L.; and Perks, A. M. 1968. The release of bradykinin like pulmonary vasodilator substance in foetal and newborn lambs. *J. Physiol.* 195:83-96. London.

4. Carlson, L. A.; Ekelund, L. G.; and Oro, L. 1969. Circulatory and respiratory effects of different doses of $PGE_1$ in man. *Acta Physiol. Scand.* 75:161-169.

5. Cassin, S. 1973. The Starling resistor model in the foetal and neonatal pulmonary circulation, in foetal and neonatal physiology. In *Proceedings of the Sir Joseph Barcroft Centenary Symposium*, pp. 112-128. Cambridge: Cambridge Univ. Press.

6. Cassin, S.; Dawes, G. S.; Mott, J. I.; Ross, B. B.; and Strang, L. B. 1964. The vascular resistance of

the foetal and newly ventilated lung of the lamb. *J. Physiol.* 171:61-79. London.

7. Cassin, S.; Dawes, G. S.; and Ross, B. B. 1964. Pulmonary blood flow and vascular resistance in immature foetal lambs. *J. Physiol.* 171:80-89.

8. Cassin, S.; Tyler, T.; and Wallis, R. 1975. The effects of prostaglandin $E_1$ on fetal pulmonary vascular resistance. *Proc. Soc. Exper. Biol. and Med.* 148:584-587.

9. Challis, J. R. G.; Osathanondh, R.; Ryan, K. J.; and Tulchinsky, D. Maternal and fetal plasma prostaglandin levels at vaginal delivery and caesarean section. *Prostaglandins* 6:281-287.

10. Craft, I. L.; Scrivener, R.; and Dewhurst, C. J. 1973. Prostaglandin $F_{2\alpha}$ levels in the maternal and fetal circulation in late pregnancy. *J. Obstet. Gynec. Brit. Cmvlth.* 80:616-618.

11. Dawes, G. S. 1969. Control of the pulmonary circulation in the fetus and newborn. In *The Pulmonary circulation and interstitial space*, eds., P. P. Fishman and H. A. Hecht, pp. 293-304. Univ. of Chicago Press.

12. ----. 1968. *Foetal and neonatal physiology*. Year Book Medical Publishers.

13. Dawes, G. S.; Mott, J. C.; and Widdicombe, J. G. 1954. The foetal circulation in the lamb. *J. Physiol.* 126:563-587. London.

14. Douglas, W. W. 1975. Polypeptides--angiotensins, plasmakinins, and other vasoactive agents; prostaglandins. In *The Pharmacological basis of therapeutics*, 5th ed., eds., L. S. Goodman and A. Gilman, pp. 640-652.

15. Enhörning, G.; Adams, F. H.; and Norman, A. 1966. Effect of lung expansion on the fetal lamb circulation. *Acta Pediatrica Scand.* 55:441-451.

16. Ferreira, S. H., and Vane, J. R. 1967. Prostaglandins: their disappearance from and release into the circulation. *Nature* 216:886-873.

17. Gilbert, R. D.; Hessler, J. R.; Eitzman, D. V.; and Cassin, S. 1972. Sites of pulmonary vascular resistance in fetal goats. *J. Appl. Physiol.* 32:47-53.

18. Horton, E. W. 1969. Hypothesis on physiological roles of prostaglandins. *Physiol. Rev.* 49:122-161.

19. Horton, E. W. and Jones, R. L. 1960. Prostaglandins $A_1$, $A_2$ and 19-hydroxy $A_1$, their actions on smooth muscle and their inactivation on passage through the pulmonary and hepatic portal vasuclar beds. *Brit. J. Pharmacol.* 37:705-722.

20. Kadowitz, P. J.; Joiner, P. D.; Greenberg, S.; and Hyman, A. L. 1976. Comparison of the effects of prostaglandins A, E, F, and B on the canine pulmonary vascular bed. *Advances in prostaglandin and thromboxane research*, vol. 1, eds., B. Samuelsson and R. Paoletti, pp. 403-415. New York: Raven Press.

21. Kadowitz, P. J.; Joiner, P. D.; and Hyman, A. L. 1974. Effects of prostaglandins $E_1$ and $F_{2\alpha}$ in the swine pulmonary circulation. *Proc. Soc. Exper. Biol. Med.* 145:53-56.

22. ----. 1974. Influence of prostaglandins $E_1$, and $F_{2\alpha}$ on pulmonary vascular resistance in the sheep. *Proc. Soc. Exper. Biol. Med.* 145:1258-1261.

23. ----. 1975. Physiological and pharmacological roles of prostaglandins. *Ann. Rev. Pharmacol.* 15:255-306.

24. ----. 1975. Effect of prostaglandin $E_2$ on pulmonary vascular resistance in intact dog, swine and lamb. *European J. Pharmacol.* 31:72-80.

25. Karim, S. M. M. 1966. Identification of prostaglandins in human amniotic fluid. *J. Obstet. Gynec. Brit. Cmwlth.* 73:903-908.

26. Katz, R. L., and Katz, G. J. 1974. Prostaglandins--basic and clinical observations. *Anesthesiology* 40:471-493.

27. Lauer, R. M.; Evans, J. A.; Aoki, H.; and Kittle, C. F. 1965. Factors controlling pulmonary vascular resistance in fetal lambs. *J. Pediat.* 67:568-577.

28. Leffler, C. W.; Tyler, T.; and Cassin, S. 1977. Vascular responses of newborn goats to aerosolized prostaglandin $E_1$: pulmonary vascular sites of action and catabolism on an E-series prostaglandin. *Proc. Soc. Exper. Biol. Med.* 155:19-22.

29. Lindmark, G.; Zador, G.; and Wilsson, B. A. 1975. The induction of labor with prostaglandin $F_{2\alpha}$ by intravenous infusion. *Acta Obstet. Gynec. Scand.* 37: 17-26.

30. McDonald, F. G., and Butler, J. 1967. Distribution of vascular resistance in the isolated perfused dog lung. *J. Appl. Physiol.* 23:463-474.

31. Olley, P. M.; Coceani, F.; and Kent, G. 1974. Inactivation of prostaglandin $E_1$ by lungs of the foetal lamb. *Experentia* 30:58-59.

32. Pike, J. 1971. Prostaglandins. *Sci. Amer.* 225:84-97.

33. Rudolph, A. M. 1968. *Congenital heart disease*, p. 30. Year Book Medical Publishers.

34. Rudolph, A. M., and Yuan, S. 1966. Responses of the pulmonary vasculature to hypoxia and $H^+$ ion concentration changes. *J. Clin. Invest.* 45:399-411.

35. Smith, A. P. 1971. *Prostaglandins*, ed., P. W. Ramwell, p. 203. London, Plenum Press.

36. Spellacy, W. N.; Gall, S. A.; Shevach, A. B.; and Holsinger, K. K. 1973. The induction of labor at term. *Obstet. Gynecol.* 41:14-21.

37. Strang, L. B., and MacLeish, M. H. 1961. Ventilatory failure and right to left shunt in newborn infants with respiratory distress. *Ped.* 28:17-27.

38. Tyler, T. L.; Leffler, C. W.; and Cassin, S. 1976. Pulmonary and systemic circulatory responses to prostaglandins of the F series in fetal, premature fetal and mature neonatal goats. Submitted to *Am. J. Physiol*.

39. Tyler, T.; Leffler, C.; Wallis, R.; and Cassin, S. 1976. Circulatory responses of perinatal goats to prostalandins $E_1$ and $E_2$. Submitted to *Am. J. Physiol*.

40. ----. 1975. Effects of prostaglandins of the E-series on pulmonary and systemic circulations of newborn goats during normoxia and hypoxia. *Prostaglandins* 10:963-970.

41. Tyler, T. L.; Leffler, C. W.; and Cassin, S. Effects of prostaglandin precursors, prostaglandins, and prostaglandin metabolites in pulmonary circulation of perinatal goats. Aspen Lung Conf., June 1976, accepted *Chest*.

42. Tyler, T.; Wallis, R.; Leffler, C.; and Cassin, S. The effects of indomethacin on the pulmonary vascular response to hypoxia in the premature and mature newborn goat. *Proc. Soc. Exper. Biol. Med.* 150:695-698.

43. Vane, J. R. 1969. The release and fate of vasoactive hormones in the circulation. *Brit. J. Pharmacol.* 35: 209-242.

44. Wicks, T. C.; Rose, J. C.; Johnson, M.; Ramwell, P. W.; and Kot, P. A. 1976. Vascular responses to arachidonic acid in the perfused canine lung. *Circ. Res.* 38:167-171.

45. Weeks, J. R. 1972. Prostaglandins. *Ann. Rev. Pharmacol.* 12:317-336.

# 19

## Pharmacologic Closure of Patent Ductus Arteriosus in the Premature Infant

William F. Friedman and Stanley E. Kirkpatrick
Division of Pediatric Cardiology
Department of Pediatrics
University of California at San Diego
School of Medicine
San Diego, California

INTRODUCTION

Significant left-to-right shunting across a patent ductus arteriosus (PDA) commonly complicates the clinical course of prematurely born infants. The ductal shunt has been implicated especially in the deterioration of pulmonary function in infants with respiratory distress syndrome in whom severe congestive failure is often unresponsive to digitalis and diuretics (18, 28, 30).

Since infusion of prostaglandins $PGE_1$ and $PGE_2$ have been shown recently to dilate the already constricted ductus arteriosus (4, 8, 19, 20) the aim of the present study was to employ an inhibitor of prostaglandin synthesis to reduce or abolish this physiologically disadvantageous shunt. In our chronic fetal lamb preparation described in greater detail in Chapter 14 "Myocardial Determinants of Fetal Cardiac Output," we placed catheters in the main pulmonary artery (proximal to the ductus), ascending aorta, and descending aorta and a Doppler flow probe around the main pulmonary artery (Fig. 1). Indomethacin 2.5 mg/kg was given intravenously to the fetus and within one hour there was an elevation of all pressures with the main pulmonary artery pressure being significantly higher than ascending or descending aorta. At 90 minutes a 20 mm Hg gradient existed between peak main pulmonary artery pressure and aortic pressure (ascending and descending aortic pressures remaining equal throughout the study) with no change in mean velocity of flow. An infusion of $PGE_1$ at this point resulted in prompt reversal of the gradient

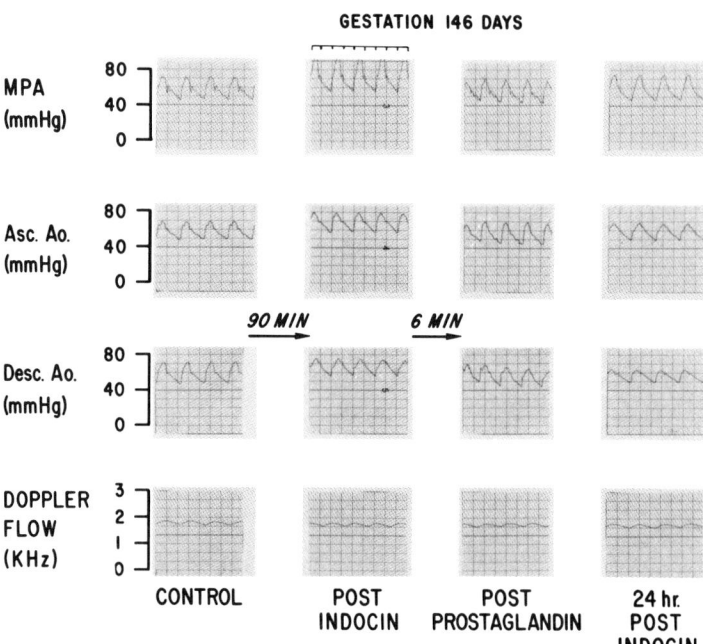

Fig. 1. Actual pressure tracings from main pulmonary artery (MPA), ascending aorta (Asc. Ao.) and descending aorta (Desc. Ao.) along with MPA Doppler flow velocity in a 146-day fetal lamb. Recordings illustrated are control values, 90 min after 2.5 mg/kg of indomethacin, 6 min after prostaglandin E, to reverse the effects of indomethacin and 24 hr later.

## Table 1
## Patient Data

| Patient | Gestational Age (weeks) | Sex | Birth Weight (g) | Indomethacin (mg/kg) |
|---------|-------------------------|-----|------------------|----------------------|
| 1 | 32 | F | 1332 | 2.5 p. o. |
| 2 | 33 | M | 2020 | 5.0 p. r. |
| 3 | 36 | M | 2320 | 5.0 p. r. |
| 4 | 32 | F | 1247 | 2.5 p. o. |
| 5 | 30 | M | 1060 | 2.5 p. o. |
| 6 | 29 | M | 1100 | 2.5 p. o. |

Source: *New England Journal of Medicine.*

with a return to control values. Within 30 minutes of stopping the prostaglandin infusion the gradient reappeared and was still present twenty-four hours later, although to a lesser degree. We did not see any significant change in main pulmonary artery mean flow velocity during the study. These findings provide evidence of ductal constriction by inhibiting synthesis of prostaglandins and ductus dilation by infusion of prostaglandins.

Accordingly, indomethacin, a potent inhibitor of prostaglandin synthetase (5) was administered to 6 consecutive premature infants who would have otherwise undergone surgical ligation of their patent ductus arteriosus. The magnitude of left-to-right shunt in each of these infants was assessed serially by the ultrasound calculation of left atrial dimension (2, 22, 25). This study was presented previously in preliminary form (6).

## MATERIALS AND METHODS

Six prematurely born infants were admitted with clinical, arterial blood gas and pH, and radiographic features typical of severe respiratory distress syndrome (12). Gestational age, sex, and weight are provided in Table 1. Each infant required assisted ventilation by respirator, as well as positive end expiratory alveolar pressure. Between 3 and 9 days of age each of these infants showed bounding peripheral pulses, a hyperactive precordium, cardiomegaly, a systolic infraclavicular and precordial murmur, progressive pulmonary vascular engorgement and edema roentgenographically, and an increase in the ultrasound-determined left atrial dimension, and in the left atrium/aortic root dimension ratio. Despite fluid restriction, digitalis, and diuretics, 4 our of 6 of these infants could not be weaned from the ventilator and required higher ventilator rates and inspiratory pressures and higher inspired oxygen tension to maintain systemic arterial oxygen tension in the range of 45-50 torr. In the two infants who no longer required assisted ventilation, spells of apnea and bradycardia and pronounced cardiomegaly, pulmonary edema, and hepatomegaly indicated that cardiac decompensation was refractory to medical therapy. Within 72 hr after the onset of decongestive therapy each infant was considered a candidate for surgical ligation of the patent ductus arteriosus. Echocardiographic studies are performed daily on infants with respiratory distress syndrome admitted to our nursery by previously described methods (7, 21) using a commercially available ultrasonoscope Picker 103, interfaced with a Honeywell 1856 fiberoptic strip-chart recorder and a 5-megahertz, quarter-inch, nonfocused transducer. Left atrial and aortic root dimensions are determined at end systole during the same or adjacent cardiac cycle. Since errors introduced by small differences in measurement of left atrium or aortic root absolute dimensions are minimized when the data is expressed as a left atrial/aortic root ratio, it is our practice to employ the ratio to serially analyze changes in left to right ductal shunting in individual infants. We, and others, have established the validity of this approach (2, 22, 25).

After obtaining consent from the parents of each infant, indomethacin was administered in a dose of 5 mg/kg per rectum in two infants and in a dose of 2.5 mg/kg by nasogastric tube to 4 infants (Table 1). Prior to drug administration laboratory studies included hemoglobin, hematocrit, white blood count, platelet count, serum bilirubin, blood urea nitrogen, serum creatinine, stool guiaic, prothrombin time, and partial thromboplastin time. These laboratory studies were repeated at 24, 48, and 72 hr after drug administration. Arterial blood gases and pH were monitored frequently. Evidence of GI bleeding, hyperbilirubinemia in excess of 10 mg %, reduced platelets, or abnormal coagulation studies was considered contraindications to indomethacin administration.

After a single dose of indomethacin each infant was evaluated by physical examination and by ultrasound analysis of cardiac and great vessel dimensions at 3, 6, 12, and 24 hr, and daily thereafter.

## RESULTS

Dramatic clinical improvement was observed in each infant within 24 hr of indomethacin administration. Each infant receiving ventilator support was extubated without difficulty, and those physical findings attributed to patency of the ductus arteriosus and cardiac decompensation disappeared completely in 5 of the 6 infants. In the sixth infant a precordial systolic murmur persisted for an additional 48 hr, although peripheral pulses and precordial activity had returned to normal and signs of heart failure had disappeared within 24 hr of indomethacin administration. In one instance there was evidence of recurrence of left-to-right ductal shunting after initial drug-induced ductal closure. At this writing, patients have been followed from 14 to 18 months since drug administration.

Figure 2 illustrates the mean (±SEM) values for the left atrial/aortic root dimension ratio observed in 110 normal premature and full-term newborns described previously by our laboratory (7, 12). The normal value of $0.77 \pm 0.1$ is significantly lower ($p < 0.001$) than the ratio ($1.09 \pm 0.5$) determined immediately prior to ligation of the ductus arteriosus in 21 premature infants evaluated prior to the present clinical investigation of indomethacin. The individual pre- and 24-hr post-indomethacin values of LA/Ao ratio in the study group shown in Figure 1 provide unequivocal evidence of drug-induced ductal closure. A similar conclusion may be reached by examining absolute and relative left atrial dimension changes in the study group, compared to normal values obtained in a previous study by others (Figure 3) (2).

Hemoglobin, hematocrit, prothrombin time and partial thromboplastin time values, and platelet counts, remained normal from the time of indomethacin administration to sev-

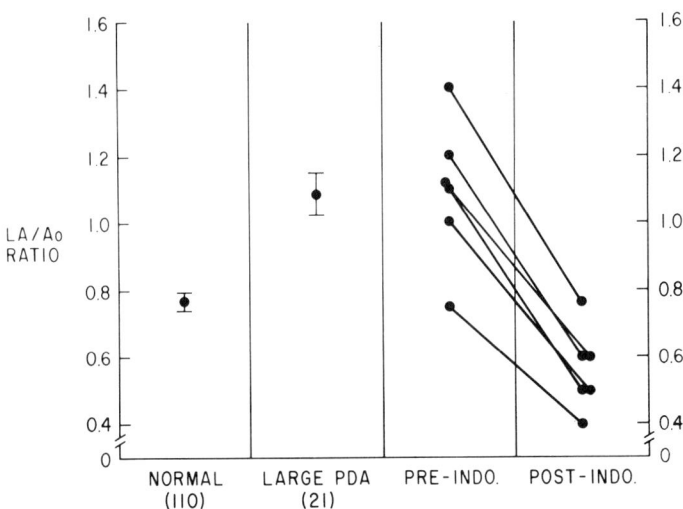

Fig. 2. The echocardiographically derived ratio between left atrium and aortic root dimension is significantly less in normal premature and full-term newborns than in those with a large patent ductus arteriosus ($p < 0.001$). The numbers in brackets refer to the number of infants from whom values were obtained. The individual echo ratios of the infants in the study group are significantly different before and twenty-four hours after indomethacin administration ($p < 0.001$). Source: New England Journal of Medicine.

eral weeks thereafter. No evidence of abnormal bleeding was observed. Total serum bilirubin ranged from 7.3 mg-% to 9.5 mg-% at the time of drug administration. An average reduction of 1.2 mg-% in total serum bilirubin occurred within 24 hr of indomethacin administration and subsequent values diminished progressively towards normal during the ensuing follow-up.

In two patients, beginning 12 hr after indomethacin administration, a significant reduction was noted in urine output. In one infant (weight = 2320 g, indomethacin dose = 5 mg/kg per rectum) oliguria persisted for 36 hr and was accompanied by a significant rise in blood urea nitrogen

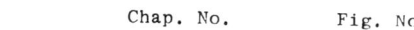

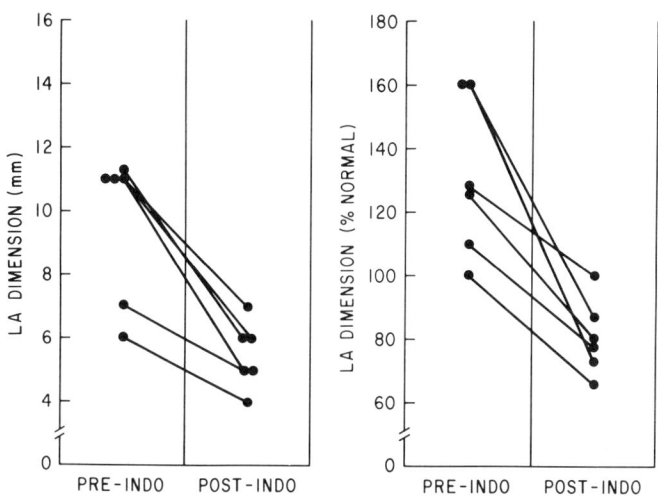

Fig. 3. The reduction in left atrial dimensions immediately prior to and twenty-four hours after indomethacin administration are expressed in absolute terms (*left panel*) and as a percentage of normal (*right panel*).
Source: New England Journal of Medicine

(50 mg-%) and serum creatinine (3.2 mg-%), and a substantial reduction in urinary sodium concentration (7-10 meg/liter). In the second infant (weight = 1130 g, indomethacin dose = 2.5 mg/kg p.o.) urine flow diminished less markedly and persisted for only 24 hr. The oliguria was associated with an elevation of blood urea nitrogen (26 mg-%) and creatinine (2.5 mg-%) and a reduced urine sodium concentration (24 meg/liter). Within 72-96 hr in each of these infants a return to normal occurred in urinary output, BUN, creatinine, and urinary sodium concentration. Through out their episodes of oliguria serum electrolytes remained normal and excessive weight gain was prevented by appropriate restriction of fluid intake.

## DISCUSSION

Prostaglandins are virtually ubiquitous in animal tissues and appear to be formed from polyunsaturated fatty acids and released in most human organs in response to many varied stimuli (29). It seems likely that these lipids act throughout the body as local mediators or modulators of physiological functions. It is now clear that prostaglandins of the A, E, and F-types are extremely potent vasoactive substances (11). Recently, strong suggestive evidence has emerged that the prostaglandins play a role in modulating the smooth muscle tone of the ductus arteriosus. Thus, inhibition of prostaglandin synthesis in a variety of animals has been shown to constrict the ductus arteriosus *in utero* (9, 23, 24, 26) and infusions of $PGE_1$ and $PGE_2$ have been shown to increase arterial blood $pO_2$ and oxygen saturation, presumably by dilating the ductus arteriosus, in infants with marked or complete right ventricular outflow tract obstruction in whom ductal shunting provides pulmonary blood flow (4, 8, 19, 20).

The results of the present study were quite dramatic. The oral or rectal administration of a single dose of indomethacin, a potent prostaglandin synthetase inhibitor, resulted, 12 to 24 hr later, in the permanent disappearance of all of those clinical symptoms and physical, radiographic, and cardiac ultrasound signs attributable to significant left to right shunting through a patent ductus arteriosus, presumably by constricting the latter vessel. The study group consisted of premature infants in whom a deterioration in cardiopulmonary function followed the classical features of severe respiratory distress syndrome. Thus, infants who would ordinarily have undergone ductal ligation were spared a thoracotomy.

A paucity of information exists concerning the actions of indomethacin or other prostaglandin synthetase inhibitors in the newborn period although indomethacin is being employed increasingly in obstetrical practice to delay parturition (27, 31). It would appear that the drug crosses the placenta readily and that its half-life in the newborn is remarkably prolonged when compared to the adult (27). It is important to recognize that premature closure of the ductus arteriosus *in utero* has been reported in one infant whose mother chronically ingested large doses of aspirin (1) and that administration of aspirin (9) and indomethacin

(32) to the pregnant sheep has been shown to constrict fetal lamb ductus arteriosus *in utero* (Fig. 1). In our judgement the latter observations raise serious questions concerning the obstetrical practice of inhibiting synthesis of prostaglandins, although, thus far, no complications have been reported in those human newborns whose mothers have received this form of therapy (27).

Since indomethacin is bound tightly to albumin (13) and since platelet function may be altered by inhibition of prostaglandin synthesis (14), infants were chosen for the current study only if their serum bilirubin levels did not exceed 10 mg-% and if no evidence existed of a bleeding tendency or disorder. We did not observe any complications in the current study that might have been attributed to a sudden increase in tissue bilirubin levels and none of the infants in the study group experienced a coagulation disorder.

A significant observation in two infants in the study group was the transient appearance of oliguria, azotemia, creatininemia, and reduced sodium concentration in the absence of systemic hypertension. Thus, it would appear that indomethacin altered renal function significantly. $PGE_2$, PFG, and $PGA_2$ have been isolated from the renal medulla and it would appear that $PGE_2$, in particular, plays a conspicuous role in defending renal function against excessive activity of the salt and water conserving system (the adrenergic-nervous-renin-angiotensin-antidiuretic hormone system) and supports renal blood flow while regulating its intrarenal distribution (16, 17). We would presume that the two infants whose renal function was altered transiently in the current study excreted indomethacin principally via the gastrointestinal tract since enterohepatic recirculation of the drug has been demonstrated in man (3, 15). Whether or not the acute renal changes that were observed are dose related, or related, in part, to an age-dependent immaturity of the kidney remains conjectural at the present time. It is likely that the doses employed in the present feasibility study were excessive since Heymann, Rudolph and Silverman have described successful pulmonary ductus arteriosus closure with markedly lesser doses of indomethacin (0.1-0.3 mg/kg) (10). In the same regard, it remains to be seen if other inhibitors of PG synthesis will preferentially influence the smooth muscle tone of the ductus arteriosus without an influence on renal function.

The observation that inhibition of the synthesis of prostaglandins causes constriction and closure of the patent ductus arteriosus in premature infants raises important possibilities for the improved treatment of the respiratory distress syndrome. Our findings support the need for a detailed analysis in the fetus and newborn of the relative organ system effects of various prostaglandin inhibitors, as well as the interaction between such compounds and those agents capable of selectively increasing renal blood and urine flow.

The extremely high morbidity and mortality in infants with the respiratory distress syndrome and patency of the ductus arteriosus provides abundant justification for pursuing non-surgical methods to alter the course of this common complication of prematurity. However, it is clear that knowledge of the potential complications of inhibition of prostaglandin synthesis, as well as those described herein, should provide a deterrent to uncontrolled clinical applications.

## NOTE

Since submission of this report we have detailed in the literature a more complete description of the *in utero* effects of indomethacin on the fetal ductus arteriosus (32). Moreover, we have expanded our clinical experience significantly (32, 33, 34). We now have constricted and closed the PDA in 31 of 35 preterm infants using 0.2 mg/kg indomethacin. Approximately 60% of the infants have required more than one dose.

## ACKNOWLEDGMENTS

The authors wish to acknowledge the advice of Drs. Stanley Mendoza and William Griswold in the conduct of this study, and especially the counsel and encouragement of Dr. Louis Gluck, Head of Perinatal Medicine UCSD, and his associate, Dr. T. Allan Merritt.

## REFERENCES

1. Arcilla, R. A.; Thalaneous, O. G.; and Ranniger, K. 1969. Congestive heart failure from suspected ductal closure in utero. *J. Pediat.* 75:74-78.

2. Baylen, B. G.; Myer, R. A.; Kaplan, S.; Ringenberg, W. E.; and Korfhagen, J. 1975. The critically ill premature infant with patent ductus arteriosus and pulmonary disease--An echocardiographic assessment. *J. Pediat.* 86:423-431.

3. Duggan, D. E.; Hogans, A. F.; Kwan, K. C.; and McMahon, F. G. 1972. The metabolism of indomethacin in man. *J. Pharmacol. Exper. Ther.* 181:563-575.

4. Elliott, R. B.; Starling, M. B.; and Neutz, J. M. 1975. Medical manipulation of the ductus arteriosus. *Lancet* 1:140-142.

5. Flower, R. J. and Vane, J. R. 1974. Inhibition of prostaglandin biosynthesis. *Biochem. Pharmacol.* 23: 1439-1450.

6. Friedman, W. F.; Hirschklau, M. J.; Printz, M. P.; Pitlick, P. T.; and Kirkpatrick, S. E. 1976. Pharmacological closure of patent ductus arteriosus in the premature infant. *Pediatr. Res.* 10:312.

7. Hagan, A. D.; Deely, W. J.; Sahn, D. J.; and Friedman, W. F. 1973. Echocardiographic criteria for normal newborn infants. *Circulation* 48:1221-1226.

8. Heymann, M. A. and Rudolph, A. M. 1976. Dilatation of the ductus arteriosus by prostaglandin $E_1$ in infants with pulmonic atresia. *Pediatric Res.* 10:313.

9. ----. 1976. Effects of acetylsalicylic acid on the ductus arteriosus and circulation in fetal lambs in utero. *Circ. Res.* 38:418.

10. Heymann, M. A.; Rudolph, A. M.; and Silverman, N. H. 1976. Closure of the ductus arteriosus in premature infants by inhibition of prostaglandin synthesis. *N. Engl. J. Med.* 295:530.

11. Higgins, C. B., and Braunwald, E. 1972. Prostaglandins: biochemical, physiological and clinical consideration. *Am. J. Med.* 53:92-112.

12. Hunt, C. E.; Matalon, S.; Wangensteen, O. D.; and Leonard, A. S. 1974. Mass spectrometer evaluation of ventilation-perfusion abnormalities in respiratory distress syndrome. *Pediatric Res.* 8:621-627.

13. Koch-Weser, J., and Sellers, E. M. 1976. Drug therapy: binding of drugs to serum albumin. *N. Eng. J. Med.* 294:311-316.

14. Kocsis, J. J.; Hernandovich, J.; Silver, M. J.; Smith, J. B.; and Ingerman, C. 1973. Duration of inhibition of platelet prostaglandin formation and aggregation by ingested aspirin or indomethacin. *Prostaglandins* 3:141-144.

15. Kwan, K. C.; Breault, G. O.; Umbenhauer, E. R.; McMahon, F. G.; and Duggan, D. E. (In press.) The kinetics of indomethacin absorption, elimination, and enterhepatic circulation in man. *J. Pharmacokinetics and Biopharmaceuticals*.

16. McGiff, J. C.; Crowshaw, K.; and Itskovitz, H. D. 1974. Prostaglandins and renal function. *Fed. Proc.* 33:39-47.

17. McGiff, J. C. and Nasjletti, A. 1976. Tynan's renal function and blood pressure regulation. *Fed. Proc.* 35:172-174.

18. Neal, W. A.; Messenger, F. B., Jr.; Hunt, C. E.; and Lucas, R. V. 1975. Patent ductus arteriosus complicating respiratory distress syndrome. *J. Pediat.* 86:127-131.

19. Olley, P. M. 1975. Non-surgical palliation of congenital heart malformations. Editorial. *N. Eng. J. Med.* 292:1292-1294.

20. Olley, P. M.; Coceani, F.; and Bodach, E. 1976. E-type prostaglandins: a new emergency therapy for certain cyanotic congenital heart malformations. *Circulation* 53:728-731.

21. Sahn, D. J.; Deely, W. J.; Hagan, A. D.; and Friedman, W. F. 1974. Echocardiographic assessment of left ventricular performance in normal newborns. *Circulation* 49:232-236.

22. Sahn, D. J.; Vaucher, Y.; Williams, D. E.; Allen, H. D.; Goldberg, S. J.; and Friedman, W. F. (In press.) Echocardiographic detection of large left to right shunts and cardiomyopathies in infants and children. *Am. J. Cardiol.*

23. Sharpe, G. L.; Thalme, B.; and Larsson, K. S. 1974. Studies on closure of the ductus arteriosus II. Ductal closure *in utero* by a prostaglandin synthetase inhibitor. *Prostaglandins* 8:363-368.

24. Sharpe, G. L.; Larsson, K. S.; and Thalme, B. 1975. Studies on closure of the ductus arteriosus XII. *In utero* effect of indomethacin and sodium salicylate in rats and rabbits. *Prostaglandins* 9:585-596.

25. Silverman, N. H.; Lewis, A. B.; Heymann, M. A.; and Rudolph, A. M. 1974. Echocardiographic assessment of ductus arteriosus shunts in premature infants. *Circulation* 50:821-825.

26. Starling, M. B. and Elliott, R. B. 1974. The effects of prostaglandins, prostaglandin inhibitors and oxygen on closure of the ductus arteriosus, pulmonary arteries and umbilical vessels *in vitro*. *Prostaglandins* 8:187-203.

27. Tareger, A.; Noschel, H.; and Zaumseil, J. 1973. Zur Pharmakokinetik von Indomethatin bei schwangeren Kreissenden un deren Neugeborenen. *Zentralblatt Gynakol.* 95:635-641.

28. Thiebeault, D. W.; Emmanoulides, G. C.; Nelson, R. J.; Lachman, R. S.; Rosengard, R. M.; and Oh, W. 1975. Patent ductus arteriosus complicating the respiratory distress syndrome in pre-term infants. *J. Pediat.* 86: 120-126.

29. Weeks, J. R. 1972. Prostaglandins. *Ann. Rev. Pharmacol.* 12:317-336.

30. Zachman, R. D.; Steinmetz, G. P.; Botham, R. J.; Graven, S. N.; and Ledbetter, M. K. 1974. Incidence and treatment of the patent ductus arteriosus in the ill premature neonate. *Am. Heart J.* 87:697-703.

31. Zuckerman, H.; Uziel, R.; Rubinstein, I. 1974. Inhibition of human premature labor by indomethacin. *Am. J. Obstet. Gynecol.* 44:787-792.

32. Friedman, W. F.; Printz, M. P.; and Kirkpatrick, S. E. 1978. Blockers of prostaglandin synthesis: A novel therapy in the management of the premature human infant with patent ductus arteriosus. In *Advances in prostaglandins and thromboxanes*, vol. 4. Eds., F. Coceani and P. M. Olley, pp. 373-381. New York: Raven Press.

33. Friedman, W. F.; Hirschklau, M. J.; Printz, M.; Pitlick, P. T.; and Kirkpatrick, S. E. 1976. Pharmacological closure of patent ductus arteriosus in the premature infant. *New England Journal of Medicine* 295 (10):526.

34. Friedman, W. F.; Heymann, M. A.; and Rudolph, A. M. 1977. Commentary: new thoughts on an old problem--patent ductus arteriosus in the premature. *J. Pediat.* 90:338-340.

# 20

## Prostaglandins in the Regulation of the Placental Blood Flows

John H.G. Rankin
Departments of Physiology and Gynecology-Obstetrics
University of Wisconsin
School of Medicine, and
Wisconsin Perinatal Center
Madison, Wisconsin

The uneven distribution of the ventilation-perfusion ratios in the lung is known to be one of the determinants which affects respiratory gas transfer, and in the mammalian placenta Faber (6) has shown that unevenly distributed perfusion-perfusion ratios can severly depress the placental transfer of respiratory gases and nutrients. The uneven ventilation-perfusion ratios in the lung are primarily a function of the effects of gravity in an air-filled organ surrounded by a fluid. This does not apply to the placenta, and we have shown that the perfusion-perfusion ratios at the macroscopic and microscopic levels are remarkably evenly distributed over the surface of the sheep placenta in the awake and unanesthetized condition (14, 19).

The ventilation-perfusion ratios in the lung are partially regulated by some form of chemical mediation at the local level. Several reports have implicated prostaglandins in this regulation (22, 27, 29). Prostaglandin $E_2$

may be involved in the maintenance of ventilation-perfusion ratios in the lung, and this substance has been shown to have a constricting action on the pulmonary vascular bed (9) and a dilating action on airway smooth muscle (13).

Our search for the mechanism whereby the perfusion-perfusion ratios in the placenta are stabilized was first directed to a description of the characteristics of a possible chemical mediator (15). The chemical mediator should act at the local level and should act in such a direction as to maintain the stability of each placental blood flow and the ratio between these blood flows. It immediately became apparent that the need to preserve the stability of the perfusion-perfusion ratios produced some difficulties. If no local mechanism existed, then external vasoactive

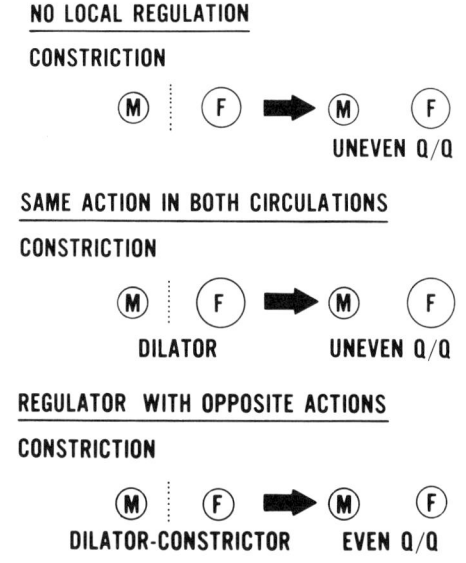

Fig. 1. The effect of an external vasoconstricting stimulus on the maternal-fetal placental perfusion-perfusion ratios where no independent control exists (*top*), where the local homeostatic agent has similar effects in both circulations (*center*), and where the local homeostatic agent has opposite effects in the two circulations (*bottom*).

agents could not affect one placental blood flow without affecting the other. This would produce a deviation from the normal perfusion-perfusion ratios. The most logical first step would be to postulate the existence of a substance which, in response to an external stimulus, would oppose that stimulus at the local level. Should such a substance be able to cross the placenta and affect the adjacent circulation then the potential for massive disruption of perfusion-perfusion ratios exists. An example of this is shown in the center section of Figure 1 in which an external constricting stimulus in the maternal circulation stimulates the production of an opposing local vasodilator. Should the vasodilator have access to the adjacent umbilical circulation, then the net result would be a vasoconstriction of the uterine circulation and a vasodilation of the umbilical circulation which would move the perfusion-perfusion ratios further from the normal value. The solution to this difficulty is found if the locally produced chemical were to have opposite actions in the two placental circulations. An example of this is shown in the lower section of Figure 1. In this example an external constricting stimulus in the uterine circulation would induce the elaboration of a locally produced vasodilator to oppose the external stimulus. Should the locally produced substance have a vasoconstricting action in the umbilical circulation then the perfusion-perfusion ratios would be maintained at a relatively normal value.

We were aware that prostaglandin $E_2$ had opposite actions on the two sides of the pulmonary exchanging surfaces and instituted a series of experiments to determine whether or not this substance had analogous actions in the placental circulation. Using the chronically catheterized near-term sheep as our model we were able to demonstrate that prostaglandin $E_2$ produced a vasoconstriction in the umbilical placental circulation (16). This finding confirmed the earlier results of Novy et al. (12).

Several investigators have postulated that prostaglandin $E_2$ should have a vasodilating action in the maternal placental circulation. All of these experiments were indirect. Terragno et al. (25) showed that the pregnant uterus synthesizes a large quantity of prostaglandin $E_2$ and that the rate of synthesis and the uterine blood flow decreased after treatment with indomethacin. Ryan et al. (21) postulated that the uterine hyperemia secondary to the administration of estrogen is mediated by the formation of prostaglandins of the E series and Venuto et al. (28), working with the pregnant rabbit, found results similar to those of Terragno's group.

We attempted to demonstrate the vasodilating properties of prostaglandin $E_2$ in the maternal placental circulation but found that when prostaglandin $E_2$ is injected into the maternal left ventricle the uterus responds with a strong vasoconstriction (18). This result, which was at variance with the indirect results in the literature, was attributed to the action of prostaglandin $E_2$ initiating a uterine contraction by acting on the myometrium. The responses of the placental vasculature were masked by the occlusion of the input and output vessels of the placenta. We therefore administered prostaglandin $E_2$ to the fetal circulation of these animals and observed significant vasodilatation in the uterine vascular bed.

The administration of prostaglandin $E_2$ to the maternal circulation induced changes in the fetal circulation which were essentially the same as those observed when prostaglandin $E_2$ was administered to the fetal circulation. The fact that prostaglandin $E_2$ in the mother can affect the fetus and when administered to the fetus can affect the maternal circulation supports the proposition that prostaglandin $E_2$ can cross the placenta. It therefore appeared that prostaglandin $E_2$ was produced by the pregnant uterus, could probably cross the placenta, and had opposite actions on both sides of the exchanging area. In addition, the fact that indomethacin causes a decrease in prostaglandin $E_2$ synthesis and a fall in the uterine blood flow is confirmatory evidence that prostaglandin $E_2$ indeed plays a role in the maintenance of the placental blood flows.

The finding that prostaglandin $E_2$ caused hypertension, bradycardia, and vasoconstriction of the fetal renal and umbilical circulations is somewhat anachronous because prostaglandin $E_2$ is known to be a potent vasodilator in the adult animal. In a recent series of experiments we have examined the fetal response to prostaglandin $E_2$ and have attempted to determine whether the substance acts directly or whether it acts on the fetal circulation by releasing other vasoactive substances (17).

In the following experiments we used chronically catheterized awake sheep 48 hr after surgery. We placed polyvinyl catheters in the maternal left ventricle and a femoral artery and in the fetal hindlimb artery and vein. Our basic procedure was to measure the distribution of the subdiaphragmatic fetal blood flows by injecting radioactive microspheres into the catheter in the fetal vein and withdrawing an integrated arterial blood sample from the catheter in the fetal artery. We then injected 15 µg/kg of prostaglandin $E_2$ into the fetal vein and again measured the distribution of the fetal blood flows after 1.5 min. In most preparations, it was possible to repeat these procedures after a 4 hr period of stabilization. All microspheres were 25 microns in diameter and labeled with one of the following isotopes: $^{46}Sc$, $^{85}Sr$, $^{141}Ce$, or $^{125}I$. At the end of the experiment the mother was killed and the fetal and placental tissues were assayed for radioactivity. Details of the surgical procedures, tissue preparations and gamma spectroscopy are provided elsewhere (16, 18). We had previously determined that systematic errors due to the geometry of the samples in the counting well, the environment of the samples, and the spillover of the isotopes into other channels affected the results by less than 5%.

Blood samples were drawn from the fetal artery in all experiments, the average pH was 7.36, and the average $PO_2$ was 17.6 mm Hg. The lowest pH that we observed in any experiment was 7.31. All fetuses appeared to be in good physiological condition as determined by blood pressure and blood gases.

## CONTROL SERIES

A control series, in which prostaglandin $E_2$ was given to the fetus, has been reported elsewhere (16) and is summarized in Table 1. It can be seen that the administration of prostaglandin $E_2$ to the fetus increased the cotyledonary resistance by a factor of 3.03, the resistance of the placental membrane by a factor of 3.03 and the resistance of the fetal kidney by a factor of 1.68. The fetal blood pressure, which was defined as fetal arterial minus fetal venous pressure, increased by a factor of 1.40.

## ALPHA RECEPTOR BLOCKADE

We tested the possibility that the prostaglandin $E_2$ was acting as a pressor agent in the fetus by releasing

Table 1

Effect of Pretreatment with Phenoxybenzamine (250 mg to the mother, 10 mg to the fetus) on the Fetal Arterial Minus Venous Blood Pressures (Blood Pressure) and Resistances of the Cotyledons, Fetal Placental Membranes, and Kidneys in the Control (C) State and 1.5 Minutes After the Administration of 50 µg $PGE_2$ to Fetal Lambs (T).

| Number | Blood Pressure (mm Hg) | | Fetal Placental Resistance [mm Hg/ (ml/min x kg fetus)] | | | | | |
|---|---|---|---|---|---|---|---|---|
| | | | Cotyledons | | Fetal Placental Membranes | | Fetal Kidneys | |
| | C | T | C | T | C | T | C | T |
| 9a | 51 | 69 | 0.290 | 0.784 | 4.64 | 13.80 | 6.38 | 11.50 |
| 9b | 43 | 46 | 0.439 | 2.421 | 6.14 | 23.00 | 6.14 | 9.20 |
| 10a | 40 | 40 | 0.255 | 2.353 | 5.71 | 13.33 | 5.71 | 40.00 |
| 10b | 30 | 40 | 0.309 | 1.212 | 2.73 | 8.00 | 3.75 | 20.00 |
| 11 | 46 | 66 | 0.301 | 0.555 | 15.33 | 33.00 | 4.60 | 11.00 |
| 15a | 38 | 44 | 0.106 | 0.312 | 1.90 | 6.29 | 4.75 | 4.89 |
| 15b | 35 | 42 | 0.141 | 0.282 | 1.94 | 3.00 | 3.50 | 3.50 |
| 24a | 42 | 50 | 0.275 | 0.256 | 5.25 | 7.14 | 7.00 | 8.33 |
| 24b | 37 | 43 | 0.222 | 0.642 | 3.08 | 7.17 | 5.29 | 10.75 |
| MEAN | 40 | 49 | 0.260 | 0.980 | 5.19 | 12.75 | 5.24 | 13.24 |
| | $P < .002$ | | $P < .003$ | | $P < .001$ | | $P < .008$ | |

endogenous catecholamines by administering 250 mg of phenoxybenzamine via the maternal left ventricular catheter and 10 mg phenoxybenzamine via the catheter in the fetal vein. After a delay of 2 hr the response of the fetal circulation to prostaglandin $E_2$ was tested by measuring the blood pressures and blood flows with radioactive microspheres before and after the injection of prostaglandin $E_2$. Nine observations were made in 5 near-term fetuses and the results obtained are shown in Table 1. It can be seen that the fetal pressor response is still apparent. The prostaglandin $E_2$ produced an increase in the resistance of the umbilical circulation as seen by the response of the cotyledons and fetal placental membranes. The fetal renal circulation also responded with vasoconstriction. These results are essentially the same as those seen without alpha receptor blockade. A summary of these results is given in Table 1. We concluded from this series that the pressor effects of prostaglandin $E_2$ in the fetal circulation are not due to the secondary release of catecholamines.

## ANGIOTENSIN RECEPTOR BLOCKADE

Several investigators (5, 26) have shown that the renin-angiotensin system is functional in the sheep fetus. As endogenous catecholamines were not the cause of the fetal vasoconstrictor responses to prostaglandin $E_2$ we decided that prostaglandin $E_2$ may have activated the renin-angiotensin system in the fetus. In order to test this a series of experiments was performed in which the fetal and maternal circulations were treated with $[Sar^1, Ile^8]$ angiotensin II, which is a potent angiotensin inhibitor (3). The substance was infused into the maternal left ventricle at a rate of 20 µg/min and the fetal venous catheter at the rate of 8 µg/min. After 15 min of the infusion the fetal responses to prostaglandin $E_2$ were again tested. Nine observations were made in 7 fetuses and the results are given in Table 2. It can be seen that the fetal pressor response is still present. Cotyledonary vascular resistance also increased as did that of the fetal placental membranes. The vascular resistance of the fetal kidneys did not change. These results are also summarized in Table 4. We conclude that in this series the fetal pressor response and vasoconstriction of the umbilical circulation were not due to the secondary release of angiotensin into the fetal circulation. An important inference to be drawn from these studies is that the vasoconstriction seen in the fetal kidneys after the injection of prostaglandin $E_2$ to the fetus appears to be due to the secondary release of endogenous angiotensin or to the potentiation of angiotensin binding in the fetal kidney. This result is interesting because the adult kidney does not respond to prostaglandin $E_2$ with vasoconstriction (10). It is difficult to postulate a mechanism whereby prostaglandin $E_2$ can be vasoconstricting in the fetal kidney and vasodilating in the adult kidney, and it is interesting to note that the fetal renal vasoconstriction seen after prostaglandin $E_2$ is probably indirect and due to angiotensin II.

## ALPHA AND ANGIOTENSIN RECEPTOR BLOCKADE

The possibility exists that in the sheep blocked with phenoxybenzamine we were observing vasoconstriction due to endogenous angiotensin and in the sheep treated with angiotensin antagonist we were observing vasoconstrictor responses due to endogenous catecholamines. In an attempt to solve this problem we performed a further series of experiments in which phenoxybenzamine blockade in the mother and fetus was induced as described above. After a 2 hour delay we then induced angiotensin receptor blockade as described above. We then observed the response of the fetal circulation to prostaglandin $E_2$. Eleven observations were made in 6 near-term sheep and the results are shown in Table 3. It can be seen that the fetal pressor response was still present. We continued to observe vasconstriction in the circulation of the fetal cotyledons and membranes. The resistance of the fetal renal circulation did not change. These results are summarized in Table 4 and are essentially the same as those seen with angiotensin receptor blockade alone.

The efficacy of the angiotensin and alpha receptor blockades was tested in all animals by observing the pressor responses to 10 μg of angiotensin or 10 μg of norepinephrine injected into the fetal venous catheters before and after administering the appropriate blocking agent. We observed a strong pressor response in all animals after these maneuvers in the control condition and we could obtain no pressor response in these animals after blockade with phenoxybenzamine or after the administration of $[Sar^1, Ile^8]$ angiotensin II. We conclude that the blockade used in these experiments was sufficient to suppress the effects of endogenous catecholamines and angiotensin II. It there-

## Table 2

Effect of Pretreatment with [Sar$^1$, Ile$^8$]Angiotensin II (20 μg/min to the Mother, 8 μg/min to the Fetus) on Fetal Arterial Minus Venous Blood Pressures (Blood Pressure) and Resistances of the Cotyledons, Fetal Placental Membranes, and Kidneys in the Control (C) State and 1.5 Minutes After the Administration of 50 μg PGE$_2$ to Fetal Lambs (T).

Fetal Placental Resistance [mm Hg/(ml/min × kg fetus)]

| Number | Blood Pressure (mm Hg) C | T | Cotyledons C | T | Fetal Placental Membranes C | T | Fetal Kidneys C | T |
|---|---|---|---|---|---|---|---|---|
| 16a | 40 | 47 | 0.412 | 0.810 | 4.44 | 11.75 | 3.64 | 4.27 |
| 16b | 30 | 38 | 0.330 | 0.826 | 3.33 | 7.60 | 2.73 | 6.33 |
| 17 | 40 | 48 | 0.263 | 1.043 | 5.00 | 24.00 | 6.67 | 5.33 |
| 18a | 36 | 37 | 0.237 | 0.435 | 4.00 | 7.40 | 7.20 | 7.40 |
| 18b | 32 | 33 | 0.264 | 0.635 | 4.57 | 5.50 | 4.57 | 4.13 |
| 27 | 41 | 48 | 0.209 | 0.565 | 4.10 | 16.00 | 13.67 | 12.00 |
| 28 | 38 | 47 | 0.197 | 1.306 | 2.38 | 7.83 | 4.22 | 5.22 |
| 29 | 53 | 60 | 0.340 | 1.500 | 2.94 | 6.67 | 13.25 | 15.00 |
| 30 | 55 | 63 | 0.313 | 2.739 | 13.75 | 15.75 | 18.33 | 15.75 |
| MEAN | 41 | 47 | 0.285 | 1.095 | 4.95 | 11.39 | 8.25 | 8.38 |
| | $P < .001$ | | $P < .001$ | | $P < .001$ | | NS | |

Table 3

Effect of Pretreatment with Phenoxybenzamine and [Sar$^1$, Ile$^8$] Angiotensin II on Fetal Arterial Minus Venous Blood Pressures (Blood Pressure) and Resistance of the Cotyledons, Fetal Placental Membranes, and Fetal Kidneys in the Control (C) State and 1.5 Minutes After the Administration of 50 μg PGE$_2$ to Fetal Lambs (T)

| Number | Blood Pressure (mm Hg) | | Fetal Placental Resistance [mm Hg/ (ml/min x kg fetus)] | | | | | |
|---|---|---|---|---|---|---|---|---|
| | | | Cotyledons | | Fetal Placental Membranes | | Fetal Kidneys | |
| | C | T | C | T | C | T | C | T |
| 19a | 32 | 35 | 0.294 | 0.530 | 8.00 | 17.50 | 4.00 | 4.37 |
| 19b | 24 | 28 | 0.289 | 0.400 | 2.67 | 5.60 | 4.00 | 7.00 |
| 20a | 36 | 35 | 0.424 | 5.000 | 6.00 | 35.00 | 6.00 | 17.50 |
| 20b | 35 | 35 | 0.340 | 0.603 | 3.18 | 5.00 | 5.00 | 5.83 |
| 22a | 48 | 49 | 0.274 | 1.324 | 12.00 | 24.50 | 16.00 | 16.33 |
| 22b | 35 | 35 | 0.229 | 0.530 | 5.83 | 11.67 | 8.75 | 8.75 |
| 23a | 41 | 53 | 0.192 | 0.353 | 5.86 | 8.83 | 13.67 | 17.67 |
| 23b | 42 | 53 | 0.280 | 0.445 | 6.00 | 7.57 | 14.00 | 17.67 |
| 25  | 43 | 58 | 0.250 | 0.784 | 8.60 | 29.00 | 10.75 | 11.60 |
| 26a | 50 | 55 | 0.298 | 0.579 | 4.55 | 11.00 | 25.00 | 13.70 |
| 26b | 43 | 52 | 0.295 | 1.405 | 2.26 | 5.20 | 14.33 | 10.40 |
| MEAN | 39 | 44 | 0.288 | 1.087 | 5.90 | 14.62 | 11.05 | 11.89 |
| | P < .004 | | P < .001 | | P < .001 | | NS | |

Table 4

Fetal Arterial-Venous Pressure Changes (Blood Pressure) and Vasomotor Responses of Several Fetal Organs. Data Are the Ratio of Observations Made After the Injection of 50 μg Prostaglandin $E_2$ to the Equivalent Observations Made Before the Injection. Data Were Obtained Without Pretreatment (Control), After the Administration of Phenoxybenzamine to the Mother and Fetus (Alpha Block), After the Administration of Angiotensin Inhibitor to the Mother and Fetus (Angiotensin Block), and Maternal and Fetal Alpha Blockade and Angiotensin Inhibition (Alpha and Angiotensin Block)

|  | Control | Alpha Block | Angiotensin Block | Alpha and Angiotensin Block |
|---|---|---|---|---|
| Cotyledons | 3.03 | 3.87 | 3.36 | 3.23 |
| Membranes | 3.03 | 2.39 | 2.35 | 2.21 |
| Kidneys | 1.68 | 2.03 | 1.09* | 1.15* |
| Heart[a] | 0.43 | 0.81 | 0.40 | 0.37 |
| Brain[a] | 0.72 | 0.55 | 0.49 | 0.67 |
| Blood pressure | 1.40 | 1.23 | 1.15 | 1.13 |
| Number | 6 | 9 | 9 | 11 |

[a] Apparent resistances.
* Not significantly different from unity.

fore follows that the vasomotor responses that we observed when prostaglandin $E_2$ was injected into these animals were not due to the release of those endogenous agents.

We have included data pertaining to the fetal brain and heart in Table 4. The data are based on apparent resistances because the integrated arterial blood sample was drawn from the fetal femoral artery and is not representative of blood flowing to the fetal heart and brain. The apparent vasodilatation seen in these organs after the injection of prostaglandin $E_2$ into the fetal circulation was not affected by alpha receptor blockade or angiotensin inhibition. These data are compatible with several possible hypotheses including vasodilatation in the organs themselves and redistribution of inferior caval flow from the upper regions of the fetus perhaps by dilatation of the ductus arteriosus. The experimental protocols used in these studies did not permit further explanation of these observations.

The similarity between the placenta and the lung is quite striking. In both organs there is a need to maintain the perfusion-perfusion ratios at the local level and prostaglandin $E_2$ is shown to have opposite actions on each side of the exchanging area in the adult lung and the placenta. Local mechanisms for the control of the ventilation-perfusion ratios have been observed in the lung. If analogous mechanisms exist in the placenta then they should be demonstrable. The most relevant experiment in which such a demonstration was attempted was performed by Raye et al. (20), who killed fetal sheep and looked for changes in the uterine blood flow. They did not see significant changes in the uterine blood flow for a period of 2 hr after fetal death. This observation would appear to demonstrate that local, interdependent, flow-regulating mechanisms did not exist in the placenta, but in a recent publication Grant et al. (8) have shown that, in the lung, the increase in pulmonary vascular resistance secondary to alveolar hypoxia is governed by a mechanism with a variable gain. This system has a maximum gain in the normal range of ventilation-perfusion ratios and has an extremely low gain outside the normal range. In the experiments of Raye et al., fetal death would indicate that the perfusion-perfusion ratios of the placenta were in an extremely abnormal range. If an analogous mechanism exists in the placenta then it would not be observed in the experimental series of Raye et al.

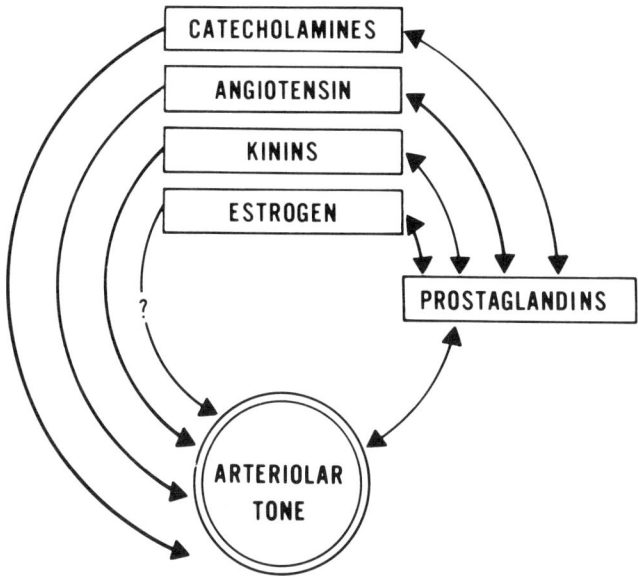

Fig. 2. Schema of the interrelationship between various systemic vasoactive agents, local prostaglandins, and arteriolar tone. The interactions between the various systems are referenced in the text. The prostaglandins are depicted as local modulators of other vasoactive systems.

The role that prostaglandins play in the regulation of perfusion-perfusion ratios in the near-term placenta will not be defined without a great deal more investigation. The matter is extremely complex because prostaglandins interact with most of the other vasoactive systems that have been observed to affect the placental circulation. The interaction of locally produced prostaglandins with exogenous systems is presented in diagramatic form in Figure 2. Evidence is available to support the interaction between prostaglandins and the renin-angiotensin system, the vasodilating effects of estrogen, the effects of catecholamines, and the effects of kinins (4, 7, 11, 21, 25). All these systems are plastic and capable of responding to internal and external changes. The effect of multicontrol systems such as these renders the interpretation of experiments in which only one parameter is controlled to be extremely difficult. It is probable that locally produced prostaglandins are a form of modulator used by the tissue to perhaps directly affect the vascular resistance and to modify the effects of external systems upon the vascular resistance. This concept of the role that prostaglandins play in the maintenance of vascular homoestasis is concordant with ideas recently expressed by Armstrong et al. (2).

In the placenta, it appears that prostaglandin $E_2$ may be the primary modulator which protects the perfusion-perfusion ratios from the effects of exogenous agents. In disease states such as preeclampsia, where there appears to be a defect in the mechanism whereby the flows and their ratios are regulated, there is a change in the ability of the placental tissue to synthesize prostaglandins of the E series (1). Talledo et al. (23) have shown that preeclamptic patients have an increased sensitivity to circulating norepinephrine and angiotensin II. This type of hypertension is described by Armstrong et al. (2), who find that it is compatible with the concept of an abnormality of the modulating (prostaglandin) system. It is possible that the description of the actions of prostaglandins on the placental vasculature will enable us to protect the fetus from defects in the mechanisms whereby the supply of nutrients are assured.

## ACKNOWLEDGEMENTS

The author wishes to acknowledge the kindness of Dr. John Pike of the Upjohn Company in supplying the prostaglandins for these experiments. I also wish to acknowledge the assistance of Mr. Terrance Phernetton in the execution of the experiments, as well as the help of Ms. Jeannine Hedberg, Mr. William Schneider, and Mrs. Ruth Ledin in the experimental procedure and preparation of the manuscript.

Supported by grant HD06736.

## REFERENCES

1. Alam, N. A.; Clary, P.; and Russell, P. T. 1973. Depressed placental prostaglandin $E_1$ metabolism in toxemia of pregnancy. *Prostaglandins* 4:363-370.

2. Armstrong, J. M.; Blackwell, G. J.; Flower, R. J.; McGiff, J. C.; Mullane, K. M.; and Vane, J. R. 1976. Genetic hypertension in rats is accompanied by a defect in renal prostaglandin catabolism. *Nature* 260: 582-586.

3. Bravo, E. L.; Kohsla, M. C.; and Bumpus, F. M. 1975. Vascular and adrenocortical responses to a specific antagonist of angiotensin II. *Am. J. Physiol.* 228: 110-114.

4. Brody, M. J. and Kadowitz, P. J. 1974. Prostaglandins as modulators of the autonomic nervous system. *Fed. Proc.* 33:48-60.

5. Broughton-Pipkin, F.; Kirkpatrick, S. M. L.; Lumbers, E. R.; and Mott, J. C. 1974. Renin and angiotensin-like levels in foetal, new-born and adult sheep. *J. Physiol.* 241:575-588.

6. Faber, J. J. 1969. Application of the theory of heat exchangers to the transfer of inert materials in placentas. *Circ. Res.* 24:221-234.

7. Gimbrone, M. A. and Alexander, R. W. 1975. Angiotensin II stimulation of prostaglandin production in cultured human vascular endothelium. *Science* 189: 219-220.

8. Grant, B. J. B.; Davies, E. E.; Jones, H. A.; and Hughes, J. M. B. 1976. Local regulation of pulmonary blood flow and ventilation-perfusion ratios in the coatimundi. *J. Appl. Physiol.* 40:216-228.

9. Kadowitz, P. J.; Joiner, P. D.; and Hyman, A. L. 1975. Effect of prostaglandin $E_2$ on pulmonary vascular resistance in intact dog, swine and lamb. *Europ. J. Pharmacol.* 31:72-80.

10. Lonigro, A. J.; Itskovitz, H. D.; Crowshaw, K.; and McGiff, J. C. 1973. Dependency of renal blood flow on prostaglandin synthesis in the dog. *Circ. Res.* 32:712-717.

11. McGiff, J. C.; Itskovitz, H. D.; Terragno, A.; and Wong, P. Y-K. 1976. Modulation and mediation of the action of the renal kallikrein-kinin system by prostaglandins. *Fed. Proc.* 35:175-180.

12. Novy, M. J.; Piasecki, G.; and Jackson, B. T. 1974. Effects of prostaglandins $E_2$ and $F_{2alpha}$ on umbilical blood flow and fetal hemodynamics. *Prostaglandins* 5:543-555.

13. Orehek, J.; Douglas, J. S.; Lewis, A. J.; and Bouhuys, A. 1973. Prostaglandin regulation of airway smooth muscle tone. *Nature, New Biol.* 245:84-85.

14. Rankin, J.; Meschia, G.; Makowski, E. L.; and Battaglia, F. C. 1970. Macroscopic distribution of blood flow in the sheep placenta. *Am. J. Physiol.* 219:9-16.

15. Rankin, J. H. G. 1976. A role for prostaglandins in the regulation of the placental blood flows. *Prostaglandins* 11:343-353.

16. Rankin, J. H. G. and Phernetton, T. M. 1976. Circulatory responses of the near-term sheep fetus to prostaglandin $E_2$. *Am. J. Physiol.* 231. (Submitted for publication.)

17. ----. 1976. Effect of alpha and angiotensin receptor blockade on the fetal circulatory response to prostaglandin $E_2$. *Am. J. Physiol.* (In review.)

18. ----. 1976. Effect of prostaglandin $E_2$ on ovine maternal placental blood flow. *Am. J. Physiol.* 231. (To be pub.)

19. Rankin, J. H. G. and Schneider, J. M. 1975. Effectt of surgical stress on the distribution of placental blood flows. *Resp. Physiol.* 24:373-383.

20. Raye, J. R.; Killam, A. P.; Battaglia, F. C.; Makowski, E. L.; and Meschia, G. 1971. Uterine blood flow and $O_2$ consumption following fetal death in sheep. *Am. J. Obstet.-Gynec.* 111:917-924.

21. Ryan, M. J.; Clark, K. E.; Van Orden, D. E.; Farley, D.; Edvinsson, L.; Sjoberg, N. O.; Van Orden, L. S., III; and Brody, M. J. 1974. Role of prostaglandins in estrogen induced uterine hyperemia. *Prostaglandins* 5:257-268.

22. Said, S. I.; Yoshida, T.; Kitamura, S.; and Vreim, C. 1974. Pulmonary alveolar hypoxia: Release of prostaglandins and other humoral mediators. *Science* 185:1181-1183.

23. Talledo, O. E.; Chesley, L. C.; and Zuspan, F. P. 1968. Renin-angiotensin system in normal and toxemic pregnancies. III. Differential sensitivity to angiotensin II and norepinephrine in toxemia of pregnancy. *Am. J. Obstet.-Gynec.* 100:218-221.

24. Terragno, N. A.; Terragno, A.; and McGiff, J. C. 1974. Prostaglandin E - angiotensin II interaction in the gravid uterus. *Acta Physiol. Lat. Am.* 24:550-554.

25. Terragno, N. A.; Terragno, D. A.; Pacholczyk, D.; and McGiff, J. C. 1974. Prostaglandins and the regulation of uterine blood flow in pregnancy. *Nature* 249:57-58.

26. Trimper, C. E. and Lumbers, E. R. 1972. The renin--angiotensin system in foetal lambs. *Pflugers Arc.* 336:1-10.

27. Tucker, A.; Weir, E. K.; Reeves, J. T.; and Grover, R. F. 1976. Pulmonary microembolism: Attenuated pulmonary vasoconstriction with prostaglandin inhibitors and antihistamines. *Prostaglandins* 11:31-41.

28. Venuto, R. C.; O'Dorisio, T.; Stein, J. H.; and Ferris, T. F. 1975. Uterine prostaglandin E secretion and uterine blood flow in the pregnant rabbit. *J. Clin. Invest.* 55:193-197.

29. Yam, J. and Roberts, R. J. 1976. Modification of alveolar hyperoxia induced pulmonary vasodilatation by indomethacin. *Prostaglandins* 11:679-689.

# Mathematical Models of the Developing Circulatory System

# 21

## Multicomponent Analysis of the Fetal System

James M. Cameron, Jr., Daniel D. Reneau, and Eric J. Guilbeau

Department of Biomedical Engineering
Louisiana Tech University
Ruston, Louisiana

INTRODUCTION

Under normal conditions, the fetus lives in a quiet, well-controlled environment, but the environment can quickly change and often becomes hostile. During labor and birth, the fetus is subjected to large pressure variations and the fetal oxygen supply is diminished and possibly stopped. A fundamental question arises: Is the fetus completely at the mercy of its environment, or does the fetus, through control mechanisms, respond in an attempt to insure its survival?

Fetal physiology can be divided into two branches. The first is steady-state analysis, which includes determination of the normal state of the fetus and sensitivity analysis of physiological parameters. This approach can yield information concerning cause and effect relations in regard to the pathophysiology of the fetus. The second branch is unsteady-state analysis, which is the study of the fetal transient response to stimuli. The transient analysis of the fetal system includes study of gas exchange, cardiovascular dynamics and control systems. The study of the fetus during transient response could be termed "fetal dynamics."

In addition to laboratory research of fetal dynamics, mathematical modeling can be used to gain further insight concerning fetal control mechanisms and the response of the total fetal system to selected disturbances. Contrasted with experimental data, the model can be used to determine cause-and-effect relationships of fetal control. This method requires a systems approach, since the dynamics of the system are dependent upon the interrelationship of the individual components. A systems analysis uses engineering techniques to study the overall behavior.

The purpose of this chapter is to present a systems analysis of fetal dynamics based on mathematical modeling. Since most available fetal data is for sheep, the following review and mathematical developments are for the fetal lamb. However, all of the models developed during the course of this research can, with proper adjustment of parameters, be applied to the human fetus.

## BACKGROUND

References to quantitative analysis of the fetal system are extremely rare. A pioneering report by Bekey, et al. (6) concerning hemodynamics of the fetal circulation was published in 1963. Morris, et al. (41) published an article in 1965 in which Bekey's model was contrasted with experimental results.

Mathematical modeling of gas transport in the fetal system seems to be an offshoot from earlier placental modeling (21, 27, 30, 31, 43, 44, 45). In 1972 and 1973, Guilbeau and Reneau (22, 23) presented a lumped distributed parameter model of a combined counter-current and concurrent flow geometry for the placenta but only a single compartment for the fetal body. This made it possible to evaluate the effect of recirculation of fetal blood, but did not give an adequate indication of fetal oxygenation. An analysis was made to determine the effect of variation of the maternal blood flow rate to placental oxygen transfer. Previously the modelers had considered only the effect of the fetus on transfer across the placenta rather than the effect of placental transfer on the well-being of the fetus.

Moll (40) in 1973 reported on a model of the fetal-placental system in which rudimentary divisions placed fetal tissue in parallel with the placenta. The model was used to study the effect of uterine contractions on fetal oxygenation by varying maternal blood flow rate through the placenta.

The effect of reducing fetal cardiac output during labor also was studied. The results show that fetal arterial oxygen values are a poor indication of fetal oxygenation because loss of cardiac output can possibly raise the arterial oxygen concentration.

Perhaps the first model in which the fetus was the major component rather than the placenta was that of Cameron, Reneau, and Guilbeau (11) in 1974. In this model, the placenta was one of several fetal body divisions. A lumped parameter model was developed by dividing each fetal organ into two compartments: one for the capillary bed and one for tissue. The placenta was divided into a lump for the maternal flow stream and another for the fetal stream. Additional volume lumps were used to represent major blood vessels which provided the proper transport delays.

Steady-state solutions of the model were used to study the sensitivity of fetal oxygenation to the maternal blood flow rate through the placenta, maternal arterial oxygen concentration, and fetal cardiac output (holding the distribution constant). The effect of labor on fetal oxygenation was simulated with nonsteady-state solutions of the model. The maternal blood flow rate through the placenta was varied proportional to the intrauterine pressure. The effect of strength of contraction and frequency of contraction presented a systematic analysis of oxygen transport in the fetal-placental system. Models were presented which were used to study the mass transfer characteristics of the placenta and to study the total fetal system.

Hill, Power, and Longo (27) in 1973 published a model of simultaneous transfer of oxygen and carbon dioxide in the placenta. Using this model, it was possible to determine the effect of carbon dioxide exchange on oxygen transport. It was concluded that carbon dioxide transfer improves oxygen transport through the combined Bohr and Haldane effects by as much as eight percent.

## FETAL CIRCULATION

The most prominent feature of the fetus is its dependency upon the mother for the supply of all nutrients, including oxygen, and the removal of all wastes, including carbon dioxide. The only link between fetus and mother is the placenta, an organ that provides a means of exchange for both mass and energy. Exchange is accomplished without an intermixing of maternal and fetal blood and takes place by diffusive transport through the placental membranes which separate the fetal and maternal circulation systems. Since the fetus is dependent upon the placenta and the lungs are inactive, a special circulatory system exists.

There are three specialized blood pathways in the fetus: (1) the ductus venosus, which shunts blood around the liver; (2) the ductus arteriosus, which carries blood from the right ventricle directly to the aorta, bypassing both the lung and left heart; and (3) the foramen ovale, which allows blood from the inferior vena cava to flow directly into the left atrium. The foramen ovale and ductus arteriosus allow the right and left hearts to operate in parallel rather than in series as in the adult. The fetal cardiac output, when compared to the adult cardiac output (measured relative to body weight), is two to three times greater due to the parallel pumping arrangement of the heart (14).

## FETAL CONTROL

Fetal control mechanisms develop throughout gestation, many of which are fully operational at birth. Barocontrol and chemocontrol appear to be functioning in the term (near the end of gestation) fetal lamb and Rudolph and Heymann (58, 59) have described the viable control mechanisms present in the term fetus. The following discussion owes much to these reviews.

Existence of baroreflex activity has been demonstrated in term fetal lambs. Shinebourne et al. (61) conducted chronic experiments with an implanted balloon catheter placed in the fetal descending aorta. Inflation of the balloon caused a rise in arterial pressure, and the fetal heart rate diminished in response. This response could be abolished by cutting the carotid sinus nerve endings and stripping the aorta, confirming, according to Rudolph and Heymann, that the response is due to the baroreceptor reflex.

Evidence for central control of the fetal circulation is somewhat obscure. Rudolph and Heymann (58, 59) refer to earlier work where the hypothalmus of fetal lambs was stimulated. In these chronic preparations, it was shown that hypothalamic stimulation resulted in cardiovascular responses including a rise in fetal heart rate. Whether or not central control is normally active in the term fetal lamb remains unanswered. On the other hand, the work of Purves and James (47) has shown that autoregulation, at least in the brain, is present in the term fetal lamb.

While many studies conducted on acute preparations have demonstrated that stimulation of aortic chemoreceptors causes elevation of the heart rate, the chronic experiments of Cohn et al. (12) show different behavior. The more recent chronic experiments conducted on fetal lambs by Boddy et al. (7) tend to confirm the results of Cohn and associates. In the work of Cohn, both the carotid and aortic chemoreceptors were stimulated; both resulted in bradycardia. The research of both groups show that hypoxia and/or acidemia cause an immediate drop in fetal heart rate. Blood flow to the placenta was preserved as was blood flow to the brain and myocardium. There was, however, a reduction of cardiac output accompanied by peripheral vasoconstriction and a rise in arterial pressure. Fetal response to hypoxemia consists of control of cardiac output and redistribution of blood flow.

Control of both maternal and fetal blood supply to the placenta is minimal if at all existent. In fact, control of the fetal blood supply, other than passive, has never been demonstrated. Meschia and Battaglia (38) have conducted experiments demonstrating that uterine blood flow is insensitive to variations of arterial oxygen supply. Similarly, Assali and Brinkman (1) report that uterine blood flow is a passive function of arterial pressure; the only control demonstrated in their studies involves a shutdown of blood supply to the uterus during times of extreme maternal stress (8).

## QUANTITATIVE FETAL PHYSIOLOGY

The accuracy of measurements of physiological parameters should usually be viewed with a certain degree of caution. Many procedures involve anesthetizing the animal and extensive surgery before data are collected. Fetal experiments involve additional trauma, as extensive surgery must be performed on the mother to expose the fetus; and during the course of the experiment, further surgery is performed on the fetus. Finally, physiological measurements are often made with the fetus in an artificial position.

In 1967, Heymann and Rudolph (26) demonstrated that exteriorization of the fetus has a considerable effect on results. Since that time, additional techniques have been developed which allow gathering of data from fetuses *in vivo* by using chronic preparations. In this section, experimental data necessary for the development of a mathematical model of the fetal system are presented, and emphasis is placed on data collected from *in vivo*, chronic preparations.

### Fetal Circulation

The cardiac output of the fetal lamb (141-150 days gestation) was measured by Rudolph and Heymann (59) as 548 ml/min/kg of fetal body weight. Additional experiments by Rudolph and Heymann (57, 58) were used to construct Table 1, which shows the distribution of the fetal lamb cardiac output and blood flow rates to individual organs of the fetal lamb. For this research, the blood flow rates at various locations in the fetal lamb were determined by multiplying the total cardiac output times the distribution of cardiac output. These and other flow rates, determined by simple mass balances, are also given in Table 1.

By pacing the heart, Rudolph and Heymann (58) were able to construct a plot showing the relation between cardiac output and heart rate. The normal fetal heart rate is near that which provides the maximum cardiac output,

Table 1

Distribution of Cardiac Output of Fetal Lamb*

|  | Percent of Total Cardiac Output | Blood Flow to Individual Organs (ml/min--100 g Organ Weight) | Calculated Blood Flow to Organs[a] (ml/min--100 g Fetal Body) |
|---|---|---|---|
| Placenta | 41.0 | . . . | 224.5 |
| Lungs | 7.0 | 126.0 | 38.5 |
| Kidneys | 1.9 | 173.0 | 10.5 |
| Gut | 5.5 | 69.0 | 30.0 |
| Spleen | 0.7 | 240.0 | 4.0 |
| Myocardium | 3.6 | 291.0 | 19.5 |
| Lower carcass | 21.3 | 26.0 | 116.5 |
| Upper carcass | 16.5 | 26.0 | 90.5 |
| Brain | 3.0 | 132.0 | 16.5 |
| Inferior vena cava | 70.0 | -- | 383.5 |
| Superior vena cava | 19.5 | -- | 107.0 |
| Right heart | 66.0 | -- | 361.5 |
| Left heart | 34.0 | -- | 186.5 |
| Ductus arteriosus | -- | -- | 323.5 |
| Foramen ovale | -- | -- | 148.0 |
| Ductus venosus | -- | -- | 130.5 |

Source: Data of Rudolph and Heymann (57, 59)
[a] Based on a total cardiac output of 548 ml/min-kg fetal body weight.

while in the adult the normal heart rate is approximately half that which corresponds to maximum cardiac output. In another communication, Heymann and Rudolph (25) reported measurements of normal pressures in the fetal lamb circulation system. These pressures were measured relative to intrauterine pressure.

In order to complete the description of the fetal lamb circulation, it is necessary to estimate the distribution of the blood volume. Pipkin and Kilpatrick (42) measured the total blood volume of the term fetal lamb and placenta to be 121 ml/kg of fetal body weight. The capillary blood volume of each organ was estimated by use of an empirical relation for packed beds, a relation that Bischoff (5) used successfully in pharmacokinetic models. Additional blood contained in the venules was determined by ratio from the capillary volume. The total blood volume for each organ and major body division is given in Table 2.

### Fetal Gas Transport

The distribution and transport of oxygen and carbon dioxide are dependent upon several factors, such as (1) solubilities, (2) metabolic reactions, (3) chemistry of gas transport in blood, (4) the rate of supply to an organ, and (5) the blood volume of an organ. In dynamic analysis the size of the organs, which is one major indication of diffusional lag, and normal steady-state concentrations must be known. While some of these factors such as solubilities are easily located in handbooks, other data are not as readily accessible. The purpose of this subsection is to present a sampling of data necessary for a dynamic analysis of gas transport in the fetal lamb.

Division of the blood flow rate to an organ (per unit weight of fetal body) by the blood flow rate to the same organ (per mass of organ weight) yields the mass of the organ per weight of fetal body. The values in Table 3 were calculated by assuming, for the sake of simplicity, a density of one and rounding off to the nearest half gram.

According to Fick's law (10), diffusion of oxygen and carbon dioxide between organ tissue and blood is proportional to the partial pressure difference between the compartments. The partial pressure of a gas at low pres-

Table 2

Blood Volume of Major Body Divisions of Fetal Lamb Calculated by Use of an Empirical Correlation for Packed Beds

|  | Organ Blood Volume (ml/kg Fetal Body Weight) |
|---|---|
| Brain | 0.496 |
| Myocardium | 0.471 |
| Lungs | 1.557 |
| Liver | 5.449 |
| G. I. tract | 1.567 |
| Upper body | 8.246 |
| Lower body | 12.694 |
| Placenta (active) | 10.504 |
| Placenta (shunt) | 1.370 |

Table 3

Volume of the Tissue Portion of Major Body Divisions of the Fetal Lamb Estimated from Experimental Data

|  | Tissue Volume (ml/kg Fetal Body Weight) |
|---|---|
| Brain | 12.5 |
| Myocardium | 6.5 |
| G. I. tract | 45.0 |
| Lungs | 30.5 |
| Lower body | 454.0 |
| Upper body | 348.0 |
| Liver | 34.5 |
| Placenta (fetal) | 99.0 |
| Placenta (maternal) | 78.0 |
| Uterus | 166.5 |

sure is proportional to the concentration of the gas, as stated by Henry's law (10). While Henry's law holds for freely dissolved gases, the partial pressure is not a direct indication of total concentration when the gas is present in other chemical forms.

Oxygen concentration in blood is dependent not only on the oxygen partial pressure ($PO_2$), but also on the hemoglobin concentration. The oxygen capacity, a measure of the hemoglobin content of blood, is about 17 ml $O_2$/100 ml of blood in the fetal lamb and 15 ml $O_2$/100 ml blood in the ewe (39). The fractional saturation ($\Psi$) of hemoglobin with oxygen multiplied times the oxygen capacity yields the total concentration of oxygen in blood.

In this research, equilibrium between oxygen and oxyhemoglobin is assumed to exist and an empirical oxygen dissociation curve,

$$\log (PO_2) = k_1 + k_2(7.4 - pH) + k_3 \log \left(\frac{\Psi}{1 - \Psi}\right) \quad (1)$$

which includes the Bohr effect, was used. The constants for the curve were measured by Battaglia et al. (4), and the difference between the binding capacities of maternal and fetal blood is reflected through the values of the constants.

Carbon dioxide content of blood is also characterized by a dissociation curve; however, the rates of carbon dioxide reactions are not as fast as the oxygen and hemoglobin reactions, and it is not always safe to assume equilibrium exists for carbon dioxide in dynamic analysis. The reactions of carbon dioxide in blood, though not completely understood, have been thoroughly researched by Ross-Bernardi and Roughton (54) and Sirs (62). The mathematical derivation of the kinetics has been expanded by Hill et al. (27) and applied to fetal physiology through placental modeling. A similar development is given here. Table 4 stochiometrically shows the various reactions in which carbon dioxide participates. These reactions represent, in addition to carbon dioxide kinetics, the ionization of hemoglobin and the effect of oxygenation of hemoglobin on carbon dioxide transport.

By the law of mass action, the rate of formation of carbon dioxide by reaction A is

## Table 4
## List of Reactions Relevant to
## Carbon Dioxide Transport in Blood

A. $\quad CO_2 + H_2O \underset{k_2}{\overset{k_1}{\rightleftharpoons}} H_2CO_3$

B. $\quad H_2CO_3 \overset{K}{\rightleftharpoons} H^+ + HCO_3^-$

C. $\quad HbNH_3^+ \overset{K''}{\rightleftharpoons} HbNH_2 + H^+$

D. $\quad CO_2 + HbNH_2 \underset{k_4}{\overset{k_3}{\rightleftharpoons}} HbNHCOOH$

E. $\quad HbNHCOOH \overset{K'}{\rightleftharpoons} HbNHCOO^- + H^+$

F. $\quad O_2HbNH_3^+ \overset{K'''}{\rightleftharpoons} O_2HbNH_2 + H^+$

G. $\quad CO_2 + O_2HbNH_2 \underset{k_4}{\overset{k_3}{\rightleftharpoons}} O_2HbNHCOOH$

H. $\quad O_2HbNHCOOH \overset{K'}{\rightleftharpoons} O_2HbNHCOO^- + H^+$

$$r_A = k_2[H_2CO_3] - k_1[CO_2][H_2O] \tag{2}$$

A complete list of nomenclature is given in Appendix A, at the end of the chapter. The faster ionization reaction B is assumed to be in equilibrium and by definition

$$K = \frac{[H^+][HCO_3^-]}{[H_2CO_3]} \tag{3}$$

Combining these two equations, and recognizing that water is in excess, yields

$$r_A = \frac{k_2}{K}[H^+][HCO_3^-] - \bar{k}_1[CO_2] \tag{4}$$

which gives the rate of formation of carbon dioxide from decomposition of carbonic acid.

The rate of formation of carbon dioxide by reaction D, which is the reaction of carbon dioxide with reduced hemoglobin to form carbamino-bound carbon dioxide, is, according to the law of mass action.

$$r_D = k_4[HbNHCOOH] - k_3[CO_2][HbNH_2] \quad (5)$$

The ionization reaction C is assumed to be in equilibrium,

$$K'' = \frac{[HbNH_2][H^+]}{[HbNH_3^+]} \quad (6)$$

as is the ionization reaction E,

$$K' = \frac{[HbNHCOO^-][H^+]}{[HbNHCOOH]}. \quad (7)$$

Combining equations 6 and 7 into the rate equation (5) yields

$$r_D = \frac{k_4}{K'}[HbNHCOO^-][H^+] - \frac{k_3 K''[CO_2][HbNH_3^+]}{[H^+]} \quad (8)$$

The total reduced hemoglobin is

$$[Hb] = [HbNH_2] + [HbNH_3^+] + [HbNHCOO^-] + [HbNHCOOH] \quad (9)$$

which, together with equations 6 and 7, results in

$$[Hb] = \frac{K''[HbNH_3^+]}{[H^+]} + [HbNH_3] + [HbNHCOO^-] + \frac{[HbNHCOO^-][H^+]}{K'} \quad (10)$$

Solving this relation for $[HbNH_3^+]$ and substituting into equation 8 gives the rate expression

$$r_D = \frac{k_4}{K'}[HbNHCOO^-][H^+]$$
$$- k_3[CO_2]\left(\frac{[Hb] - 1 + [H^+]/K'[HbNHCOO^-]}{1 + [H^+]/K''}\right) \quad (11)$$

The rate of formation of carbon dioxide from carbamino-bound carbon dioxide of oxygenated hemoglobin in reaction G is given by

$$r_G = k_4 \, [O_2HbNHCOOH] - k_3 \, [CO_2] \, [O_2HbHN_2] \quad (12)$$

The ionization reactions for oxyhemoglobin F and H yield

$$K''' = \frac{[O_2HbNH_2]\,[H^+]}{[O_2HbNH_3^+]} \quad (13)$$

$$K' = \frac{[O_2HbNHCOO^-]\,[H^+]}{[O_2HbNHCOOH]} \quad (14)$$

Equations 12, 13, and 14 are analogous to equations 5, 6, and 7 for reduced hemoglobin. A similar development yields the rate of formation of carbon dioxide by reaction G which is

$$r_G = \frac{k_4}{K'} [O_2HbNHCOO^-][H^+]$$
$$- k_3 \, [CO_2] \left( \frac{[HbO_2] - (1 + [H^+]/K')\,[O_2HbNHCOO^-]}{1 + [H^+]/K'''} \right) \quad (15)$$

Unfortunately, the ionization constant K' is not what has been measured. An equivalent reaction for reactions D and E or G and H can be written. Using D and E as an example, an equivalent expression is

$$CO_2 + HbNH_2 \underset{}{\overset{K_I}{\rightleftharpoons}} HbNHCOO^- + H^+ \quad (16)$$

where $K_I$ is the equivalent ionization constant and is equal to

$$K_I = \frac{[HbNHCOO^-]\,[H^+]}{[CO_2]\,[HbNH_2]} = \frac{K'k_3}{k_4} \quad (17)$$

and according to E. Hill (27) the term $[H^+]/K'$ is small and can be neglected.

The total rate of production of carbon dioxide from carbamino-bound forms is

$$r_{D+G} = r_D + r_G \tag{18}$$

Since total hemoglobin is always constant,

$$[Hb]_t = \Psi[HbO_2] + [1 - \Psi][Hb] \tag{19}$$

and the ionized carbamino-bound carbon dioxide can be summed to equal

$$[HbNHCOO^-]_t = [HbNHCOO^-] + [O_2HbNHCOO^-] \tag{20}$$

the number of variables can be reduced.

Combining equations 11 and 25 for determinations of the total reaction rate yields in terms of $[Hb]_t$ and $[HbNHCOO^-]_t$

$$r_{D+G} = \frac{k_3}{K_I}[H^+][HbNHCOO^-]_t + \left([HbNHCOO^-]_t - [Hb]_t\right) k_3[CO_2] \times \left(\frac{1-\Psi}{1+[H^+]/K''} + \frac{\Psi}{1+[H^+]/K'''}\right) \tag{21}$$

The formation of carbonic acid from carbon dioxide and the following ionization reaction results in the formation of one mole of hydrogen ions per mole of carbon dioxide. In other words, the rate of formation of acid due to reaction A is equal to, but opposite in sign of the rate of carbon dioxide formation. According to Hill et al. (27), the average acid production rate dur to the combined reaction rate of carbamino-bound carbon dioxide is approximately 1.5 moles of hydrogen ions produced per mole of carbon dioxide consumed; also, there are approximately 0.6 mole of hydrogen ions produced per equivalent of oxyhemoglobin formed. The total rate of acid production is given by

$$r_{acid} = -r_A - \frac{3}{2}r_{D+G} + 0.6 \text{ Hb}\frac{d\Psi}{dt} \tag{22}$$

where $d\Psi/dt$ is the time rate of change of fractional saturation of hemoglobin with oxygen.

Acid produced by these reactions is buffered in blood. The actual production of free hydrogen ions, which is an indication of blood pH, is related to acid production by the buffer curve (56).

Measurement of oxygen and carbon dioxide partial pressures and pH in the fetus is a difficult task. As shown in Table 5, different researchers often get different values for similar measurements. The variability of procedures and date of gestation probably accounts for some of the differences.

Included in gas transport is the metabolic utilization or production of chemical components. Normal oxygen consumption in the fetus is zero order; i.e., independent of oxygen concentration. Carbon dioxide production is proportional to oxygen consumption and, in the fetal lamb, about 0.94 moles of carbon dioxide are produced for each mole of oxygen consumed (28). As tissue oxygen concentration decreases, the reaction rate is believed to become a function of oxygen concentration. When cellular oxygen concentration falls below the critical value, which is the point where the reaction becomes a function of tissue oxygen concentration, lactate is produced by glycolysis in an effort to satisfy energy demands. Anaerobic glycolysis is less effective than aerobic glycolysis, and lactate accumulates, which lowers pH and thereby threatens tissue damage.

## FETAL MODEL

In biomedical simulation a system must be defined anatomically and physiologically. The fetal system in this research was comprised of the fetus and the placenta. The environmental interactions with the system were maternal blood supply to the placenta, quality of maternal arterial blood, and direct pressure on the fetal system due to uterine activity.

Specific anatomical locations in the system were represented by a compartment (Fig. 1). Each compartment coincided with a major blood vessel, organ, or larger body

Table 5

Measurements of Oxygen, Carbon Dioxide, and pH in Fetal Lambs

|  | Percent Saturation | $PO_2$ (mm Hg) | $PCO_2$ (mm Hg) | pH | Reference |
|---|---|---|---|---|---|
| Umbilical vein | 75-80 ... ... | 28-30 30±2 34.8 | 37-38 38±3 41.5 | 7.36-7.38 7.35±0.01 7.391 | 25 57 13 |
| Umbilical artery | ... ... | 21±1 22.8 | 45±1 47.5 | 7.33±0.01 7.340 | 57 13 |
| Maternal artery | ... ... | 73±3 99.0 | 34±2 33.5 | 7.44±0.02 7.478 | 57 13 |
| Uterine vein | ... | 53.0 | 38.0 | 7.440 | 13 |
| Inferior vena cava | 70 | ... | ... | ... | 25 |
| Jugular | ... | 14.0 | 47.0 | 7.33 | 48 |
| Ascending aorta | ... ... | 21.4±1 18.0 | 53.8±1 44.0 | 7.35±0.004 7.35 | 34 48 |
| Descending aorta | 60 | 19-23 | 42-47 | 7.32-7.34 | 25 |
| Carotid | 65 | 23-26 | 39-42 | 7.34-7.36 | 25 |
| Superior vena cava | 40 | ... | ... | ... | 25 |
| Pulmonary trunk | 55 | 17-20 | ... | ... | 25 |
| Ductus arteriosus | 55 | ... | ... | ... | 25 |
| Foramen ovale | 70 | ... | ... | ... | 25 |

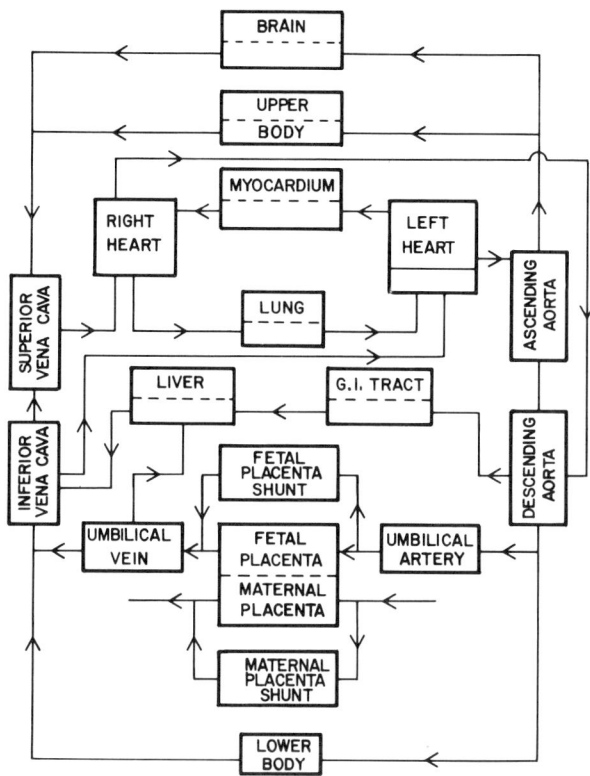

Fig. 1. Division of fetal circulation for oxygen model.

division. The organs were further subdivided into lumps representing tissue and local circulation. A direct anatomical relation was maintained between model divisions and the actual system. The physiological state of the fetus was assumed to be described by the oxygen distribution within the fetus and by cardiovascular function. Figure 1 also serves as a flow diagram indicating the directions of blood flow and the paths of mass transfer. The solid lines represent paths of transport by flowing blood (the arrows indicate the direction of flow) and the dotted lines within the organ divisions represent barriers to transport by flow but across which transport can occur by diffusion. Inefficiencies of the placenta are considered by the inclusion of both fetal and maternal shunts which consist of both intracotelydonary and intercotelydonary shunts and tissue oxygen consumption.

Since the compartments are basically alike, the mathematical expressions describing the compartments were obtained by deriving the equation in a general form and modifying it to fit each compartment. These individualized sets of equations were combined to form the overall models. Two basic models of oxygen transport within the fetal system were derived. The first model neglected all control systems; flow rates were considered constant. A second model, which was later modified, was used for simulation of oxygen transport in the controlled fetal system. The basic difference was that in the second model, blood flow rates, local blood volumes, blood pressures, and the fetal heart rate were variables. The third, most complex model, was developed by adding equations for carbon dioxide and lactate transport to the controlled oxygen model. Since the equations are basically the same for all applications, the derivation is divided into oxygen transport, hemodynamics, and multicomponent transport. The oxygen model can stand alone for calculations assuming constant cardiovascular parameters, or be combined with the hemodynamic model. The following derivations demonstrate the development of equations for tissue, capillary, and blood vessel compartments for cardiovascular oxygen, and multicomponent calculations.

## Cardiovascular Model

Disregarding for the moment the tissue portions of the fetal divisions, a model of the cardiovascular system was derived. Since this model was to be used in conjunction with the oxygen transport model, it was not necessary to calculate instantaneous blood flow rates. The cardiovascular model needed only to provide the average blood flow rates as a function of fetal heart rate, local blood pressures, and local blood flow resistances. The following is a brief description of that model.

The derivation of the model is based on the circulation divisions shown in Figure 1. These divisions were classified into two categories: the first representing major blood vessels, arteries and veins; and the second representing the fetal organs and larger body divisions, such as the forequarters. The major blood vessels were each assumed to consist of a single compartment. Figure 2 shows a schematic illustration of the vasculature of an organ lump; the circulation is divided into sections representing arterioles, capillaries, and venules. In order to simplify the organ circulation, it was assumed that the blood volume of the microcirculation could be assigned to the venules and that determination of the venule volume would sufficiently describe the capacitance of the system. This simplification results in a one lump model for the organs, but maintains the concept of arteriole and capillary resistances which are useful for control relations.

The volume of a compartment was determined from a total mass balance.

$$\begin{pmatrix} \text{rate of change} \\ \text{of mass in a} \\ \text{compartment} \end{pmatrix} = \begin{pmatrix} \text{rate of mass} \\ \text{flow into the} \\ \text{compartment} \end{pmatrix} - \begin{pmatrix} \text{rate of mass} \\ \text{flow out of} \\ \text{the compartment} \end{pmatrix}$$

(23)

The density of blood remains nearly constant, which allows substitution of the volumetric flow rate for the mass flow rate and calculation of the volume of the compartment from the mass balance. The volume of a compartment at any time can be determined by integration of the differential equation

$$\frac{dv}{dt} = Q_A - Q_V \qquad (24)$$

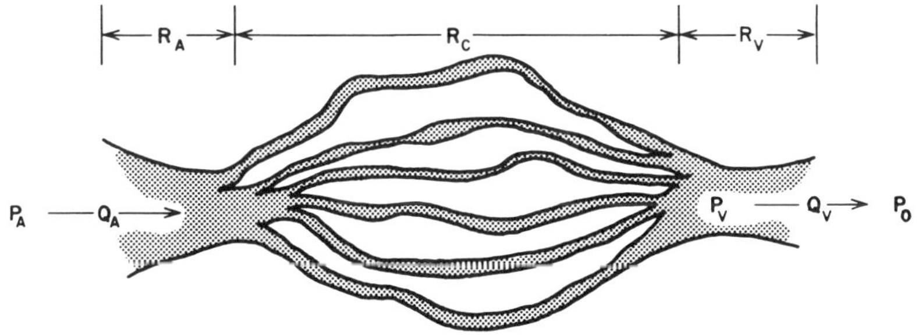

Fig. 2. Capillary bed configuration used for development of local control equations.

where differentiation is with respect to time, and

v = volume

t = time

Q = volumetric flow rate

As is the practice in hemodynamic calculations (37), the volumetric flow rate between compartments was assumed to be linearly related to the pressure difference between compartments. For instance, the arterial blood flow rate into an organ capillary division was given by

$$Q_A = (P_A - P_V) / R_a \tag{25}$$

and the venous blood flow rate by

$$Q_V = (P_V - P_o) / R_V \tag{26}$$

where R is the resistance to blood flow and the pressures correspond to those shown in Figure 2. This is the Hagen-Poiseuille equation, valid for laminar flow of a Newtonian fluid.

For the capillary beds, the resistance to arterial blood flow was divided into arteriole resistance and capillary resistance, that is,

$$R_a = R_A + R_C \tag{27}$$

This was done to facilitate inclusion of control mechanisms following the method of Granger et al.(20).

Major blood vessels were modeled as simple lumps and the volume calculated using mass balance of equation (23). The blood flow rates into and out of the vessels were calculated in the same manner as those for the microcirculation, except that arterial resistance was a single value, independent of control mechanisms, as was the venous resistance.

Blood pressure of a compartment was determined, using a simple pressure volume relation, as a function of the volume of the compartment. The pressure was related to the volume by the bilinear function

$$P = \begin{array}{l} 1/C(V-V_u); \quad V > V_u \\ 0 \quad\quad\quad\quad\quad ; \quad V \leq V_u \end{array} \qquad (28)$$

where V represents the volume of the compartment and $V_u$ represents the unstressed volume. The unstressed volume is the volume at which the blood vessel is not stretched, there is no elastic effect, and pressure drops to zero.

The heart chambers were modeled somewhat differently. The flow rates in and out of the heart chambers were assumed to be known and a function of the fetal heart rate; this made it unnecessary to determine an average pressure for these chambers. Similarly, it was not necessary to calculate the volume of a chamber; instead, an average volume was assigned. The cardiac output of these chambers was determined from the fetal heart rate using experimental data of Rudolf and Heymann (58). This data, which was gathered for the fetal lamb by pacing the heart, correlates the fetal heart rate and right ventricular output. It was assumed that the left ventricle had a similarly shaped curve, but that at a given heart rate the left ventricular cardiac output was less than that of the right ventricle.

### Oxygen Model

There were two oxygen models used; the first assumed constant fetal cardiac output and blood flow distribution; the second model was essentially the same as the first, except control mechanisms were included and the cardiovascular model used to calculate blood flow variations resulting from controller responses to stimuli. Figure 1 schematically shows the flow of the fetal system used for the oxygen model.

Compartments representing organs and major body divisions were subdivided into lumps representing local circulation and tissue. The system equations were derived using the following simple mass balance:

$$\begin{pmatrix} \text{rate of} \\ \text{accumulation} \\ \text{of oxygen} \end{pmatrix} = \begin{pmatrix} \text{net rate of} \\ \text{oxygen in} \\ \text{by convection} \end{pmatrix}$$
$$+ \begin{pmatrix} \text{net rate of} \\ \text{oxygen in} \\ \text{by diffusion} \end{pmatrix} + \begin{pmatrix} \text{net rate of} \\ \text{oxygen gained} \\ \text{in reactions} \end{pmatrix} \qquad (29)$$

Each division including blood and tissue lumps was represented by an oxygen mass balance with some of the terms being zero.

The oxygen balance for the blood portion of an organ was

$$\frac{d\, V[O_2]}{dt} = Q_A [O_2]_A - Q_V [O_2]_V + D \qquad (30)$$

where

$V$ = volume

$[O_2]$ = oxygen concentration

$Q$ = volumetric flow rate

$D$ = diffusion of oxygen from tissue to blood

It was assumed that the concentration of oxygen in blood can be related to the fractional saturation of hemoglobin through the oxygen capacity by

$$[O_2] = \Psi M \qquad (31)$$

where

$\Psi$ = fractional saturation of hemoglobin

$M$ = oxygen capacity

which gave

$$\frac{d(V\Psi)}{dt} = Q_A \Psi_A - Q_V \Psi_V + D/M \qquad (32)$$

The assumption that a compartment behaves like an ideal stirred tank, that is, the concentration is homogeneous throughout, implied that

$$\Psi_V = \Psi \tag{33}$$

and resulted in

$$\frac{d(V\Psi)}{dt} = Q_A \Psi_A - Q_V \Psi + D/M \tag{34}$$

For the model without control, the volume was assumed constant, which meant that

$$Q_A = Q_V = Q \tag{35}$$

and yielded

$$\frac{d\Psi}{dt} = (Q(\Psi_A - \Psi) + D/M)/V \tag{36}$$

where D is zero for blood lumps which do not exchange oxygen by diffusion. The oxygen equation for a major blood vessel was similar to equations 34 or 36 except the diffusion term was not included.

The derivation of oxygen equations for tissue compartments is similar except there is no bulk flow; that is, Q is zero. Another difference is that the oxygen concentration was related to the partial pressure by

$$[O_2] = \alpha_{O_2} PO \tag{37}$$

where

$P_{O_2}$ = oxygen partial pressure
$\alpha_{O_2}$ = solubility

Using this relation yielded

$$\frac{dP_{O_2}}{dt} = D/\alpha_{O_2} V + R/\alpha_{O_2} \tag{38}$$

The mass transfer rate of oxygen by diffusion was calculated from

$$D = K\Delta PO_2 \tag{39}$$

where K was the transfer coefficient. The diffusional driving force $\Delta P_{O_2}$ is the difference between blood and tissue oxygen partial pressure, or in the case of the placenta, the difference between maternal and fetal blood oxygen partial pressures. The oxygen partial pressure of the blood compartments was determined as a function of the fractional saturation using the oxygen dissociation curve.

The oxygen consumption rate is usually assumed to be of zero order, which, with a slight modification, was the procedure followed here. It was assumed that as the oxygen concentration approached zero, a critical point (brain and heart--1 mm Hg; other tissue--5 mm Hg) was reached where the reaction rate became first order with respect to oxygen. The reaction rate is given by

$$R = \begin{cases} A \text{ (constant)} ; & PO_2 > PO_{2_{critical}} \\ kPO_2 ; & PO_2 \leq PO_{2_{critical}} \end{cases} \tag{40}$$

When this oxygen model was used in conjunction with the hemodynamic model to study oxygen control, the volume of blood compartments is not constant, but is a function of time determined from a total mass balance. Additionally, the flow rates $Q_A$ and $Q_V$ are independent of one another and variables.

## Multicomponent Model

The derivation of the reaction rate expressions in the introduction indicated that in blood five individual component mass balances were necessary to describe carbon dioxide transport; in tissues four were necessary. If hypoxia or anoxia were to be included in the simulations, anaerobic metabolic pathways must be represented in some form. Therefore, the components modeled in the multicomponent system were (1) oxygen, (2) carbon dioxide, (3) bicarbonate, (4) carbamino-bound carbon dioxide, (5) hydrogen ions, and (6) lactate, where the carbamino-bound carbon dioxide was not determined for the tissue lump.

The equations were derived with a simple mass balance similar to that used for the formulation of the oxygen model. For the blood lump, the oxygen equation remained the same as equation (32).

The carbon dioxide balance resulted in

$$\frac{dv[CO_2]}{dt} = Q_{IN}[CO_2]_{IN} - Q_{OUT}[CO_2] + Vr_A + Vr_{D+G} + D \qquad (41)$$

where $r_A$ and $r_{D+G}$ represented the rate equations from the introduction. Application of Henry's law allowed direct calculation of the carbon dioxide partial pressure from the differential equation

$$\frac{d(vPCO_2)}{dt} = Q_{IN}PCO_{[IN]} - Q_{OUT}PCO_2 + \frac{(Vr_A + Vr_{D+G} + D)}{\alpha_{CO_2}} \qquad (42)$$

Combination of carbon dioxide with hemoglobin to form carbamino-bound carbon dioxide was determined by

$$\frac{dV[HBNCOO^-]_t}{dt} = Q_{IN}[HBNCOO^-]_{IN} - Q_{OUT}[HBNCOO^-]_t - Vr_{D+G} \qquad (43)$$

where there was no diffusion term because hemoglobin is confined to the circulatory system. Bicarbonate concentration was determined from

$$\frac{dV[HCO_3^-]}{dt} = Q_{IN}[HCO_3^-]_{IN} - Q_{OUT}[HCO_3^-] - Vr_A \qquad (44)$$

According to Hill et al. (27) (see Introduction), acid production is related to $r_A$, $r_{D+G}$, and the rate of oxygenation of hemoglobin, and during anaerobic metabolism to the rate of production of lactate, which gave

$$r_{acid} = -r_A - 1.5 r_{D+G} + 0.6 [Hb]_t \frac{d\Psi}{dt} + \frac{d[L]}{dt} \qquad (45)$$

The total accumulation of acid in a compartment was

$$\frac{dv[acid]}{dt} = Q_{IN}[H^+]_{IN} - Q_{OUT}[H^+] + Vr_{acid} \qquad (46)$$

and the accumulation of free hydrogen ions was related to acid production by the buffer curve of blood. The slope of the curve

$$\frac{d[acid]}{dpH} = \text{slope of buffer curve} \qquad (47)$$

allowed application of the chain rule which gave

$$\frac{d[acid]}{d[H^+]} = \frac{-\text{slope}}{[H^+]2.303} \qquad (48)$$

and hydrogen ion balance was then given by

$$\frac{dV[H^+]}{dt} = \frac{-[H^+](2.303)}{\text{slope}} \left( Q_{IN}[H^+]_{IN} - Q_{OUT}[H^+] + Vr_{acid} \right) \qquad (49)$$

In tissue, the flow rates are zero and hemoglobin is not present; the tissue equations were determined to be

$$\frac{dPCO_2}{dt} = (Vr_A - D)/\alpha_{CO_2} V \qquad (50)$$

$$\frac{d[HCO_3^-]}{dt} = -r_A \qquad (51)$$

$$\frac{d[H^+]}{dt} = \frac{-[H^+](2.303)r_{acid}}{\text{slope}} \qquad (52)$$

where the tissue buffer curve slope is different than that for blood.

Diffusion of all species other than oxygen, carbon dioxide, and lactate was neglected. Like oxygen, the rate of diffusion of carbon dioxide was determined from the partial pressure difference between blood and tissue by

$$D = K\Delta PCO_2 \qquad (53)$$

where K was a transfer coefficient.

Lactate was assumed to exist in two forms, either dissociated or not dissociated. In tissue the mass balance for lactate was

$$\frac{d[LH]V}{dt} = Vr_{LH} - Vr_i \tag{54}$$

where $r_{LH}$ was the rate of lactate production by metabolism and $r_i$ was the rate of dissociation of lactate into the lactate ion for which the mass balance was

$$\frac{d[L^-]V}{dt} = Vr_i \tag{55}$$

Summing equations 54 and 55 yields a total lactate balance given by

$$\frac{dV([LH] + [L^-])}{dt} = \frac{dV[L]}{dt} = Vr_{LH} \tag{56}$$

where L represented total lactate. The equilibrium relation for lactate

$$K = \frac{[L^-][H^+]}{[LH]} \tag{57}$$

indicated that $\frac{d[L]}{dt} \sim \frac{d[L^-]}{dt}$ since

$$\frac{[L^-]}{[LH]} = \frac{K}{[H^+]} \ggg 1 \tag{58}$$

and therefore, the hydrogen production was assumed equal to the lactate production rate. Lactate is known to diffuse from tissue to blood; therefore, diffusion was added to the mass balance yielding

$$\frac{dV[L]}{dt} = Vr_{LH} - D \tag{59}$$

The blood compartment mass balance for lactate was

$$\frac{dV[L]}{dt} = Q_{IN}[L]_{IN} - Q_{OUT}[L] + D \tag{60}$$

where lactate diffused proportional to the concentration difference between blood and tissue lumps according to

$$D = K \Delta[L] \tag{61}$$

where K is a transfer coefficient.

Production of lactate by metabolism was assumed to be zero during normal conditions. As the oxygen concentration dropped below the critical value, consumption of oxygen became a function of oxygen concentration and decreased. At this time, lactate production was started proportional to the loss of oxygen consumption. A proportionality constant was determined from the energy production rate of aerobic metabolism relative to anaerobic metabolism which for glucose is 38:2 (24).

### Control Systems

The control systems studied in the fetus were (1) local, (2) baroreceptor, and (3) chemoreceptor. All controllers were assumed to be of the integral type; that is, the controlled variable is related to the integral of the error. The error is the difference between the desired value, or set point, and the actual value of the controlled parameter.

The baroreceptor reflex was assumed to control the arterial pressure by adjusting the fetal heart rate and by causing vasoconstriction or vasodilation. The pressure in the ascending aorta was monitored for this purpose. Similarly, the oxygen partial pressure of the ascending aorta was monitored for chemocontrol which acted through manipulation of the heart rate and vasoconstriction. Local control of oxygen within the microcirculation was assumed to be active in all organs of the fetus and to operate as a function of local tissue oxygen partial pressure.

There are three types of error signals for each capillary compartment except the brain and myocardium, which were assumed to be independent of chemo and baro control. The first two error signals represent the deviation of the arterial pressure and arterial oxygen partial pressure from the set points, calculated for the ascending aorta. The fetal heart rate was calculated as a function of these two error signals, using the equation

$$FHR = FHR_0 + \int (k_1 e_1 + k_2 e_2) \, dt \qquad (62)$$

The error signal for brain and myocardium is the deviation of the tissue oxygen partial pressure from the set point while the error signal for other organs and body divisions consist of contributions from the baro and chemo error signals in addition to the deviation of the tissue oxygen partial pressure from the set point. The total error for a typical organ was

$$e = k_3 e_3 + k_4 e_1 + k_5 e_2 \tag{63}$$

where $k_4$ and $k_5$ are zero for brain and myocardium.

It was assumed that regulation of blood flow through an organ is accomplished by variation of the local arterial resistance and that this resistance is made up of two parts; one representing arteriole resistance and the other capillary resistance. The equations derived for modifying the local resistances are similar to those of Granger and Shepard (20). A capillary is thought to be either completely open or closed by the pre-capillary sphincter. As a result of this, the total resistance of the capillary bed is a function of the number of open capillaries. Additionally, since oxygen exchange occurs within the capillary bed, the mass transfer distance is a function of the number of open capillaries, and the mass transfer coefficient must be calculated from the number of open capillaries. The number of open capillaries was calculated from the total error signal for each compartment, using

$$N = N_0 + k_b \int e \, dt \tag{64}$$

and from this, the capillary resistance and mass transfer coefficients were determined by

$$R_c = R_{co} N_0 / N \tag{65}$$

and

$$K = N/N_0 \left[ \frac{K}{B(N/N_0)^{\frac{1}{2}} - 1} \right] \tag{66}$$

The arteriole resistance is controlled, presumably as a function of local oxygen partial pressure and by nervous intervention, by smooth muscle present in the vessel walls. Therefore arteriole resistance was determined directly from the total error signal for each compartment by

$$R_A = R_{Ao} + k_7 \int e \, dt \tag{67}$$

and the total blood flow resistance for each compartment was the sum of capillary and arteriole resistance.

## Combined Model

The preceding derivations describe the equations needed to model a single compartment. The various mathematical models of the fetal system were formed by combining the equations for the individual compartments. These sets of equations were interrelated by blood flow and diffusion connections; that is, the blood leaving a compartment, with an oxygen concentration representing that compartment, is the input for the downstream compartment.

Each box shown in Figure 1 is represented by an oxygen balance, and the mass balance equations for carbon dioxide, hydrogen ion, and lactate. Diffusion of oxygen across the dashed lines of Figure 1 is calculated for each organ using the oxygen partial pressure difference between the two compartments. For blood compartments, the oxygen partial pressure was determined from the fractional saturation of hemoglobin with oxygen using the oxygen dissociation curve presented earlier.

In the control models, the blood flow rates and blood compartment volumes are not constants, and additional equations must be added to the model. The blood flow rates between compartments were calculated using equations (25) or (26) where the pressures appearing in the equations were the pressures determined for each blood compartment using equation (28). The volume of a blood compartment is determined using equation (24). The fetal heart rate and the blood flow resistance and mass transfer coefficient of each capillary bed were calculated using the control equations, (62) through (67).

Since flow expressions and oxygen concentrations connect the compartments, the equations for each compartment could not be solved separately. The total set of differential and algebraic equations must be solved simultaneously; and since the set was large and the equations nonlinear, solution by analytical means was precluded. The models were solved using CSMP III (44) on an IBM 370/145.

## RESULTS

### Oxygen Control

The oxygen model was used to validate the control equations by formulating the model in several steps beginning with a model having a fixed cardiac output and no control and resulting in a model with several control mechanisms keying on oxygen concentration. Data of maternal anoxia and recovery was used to evaluate the oxygen model at each state of development. In this experiment, the mother goat was spontaneously breathing pure oxygen. The goat was made anoxic by having the goat respire nitrogen, which was maintained until the maternal goat became visibly distressed. Data was tabulated for maternal arterial oxygen partial pressure during the anoxic period. More importantly, the oxygen partial pressure in the brain of the fetal goat was measured with an oxygen microelectrode (see Fig. 3). Though this was an acute experiment with the animals under anesthesia, it was unique in that the fetal response to a maternal variation was measured. The data of the maternal arterial oxygen partial pressure was used as input for simulation of maternal anoxia.

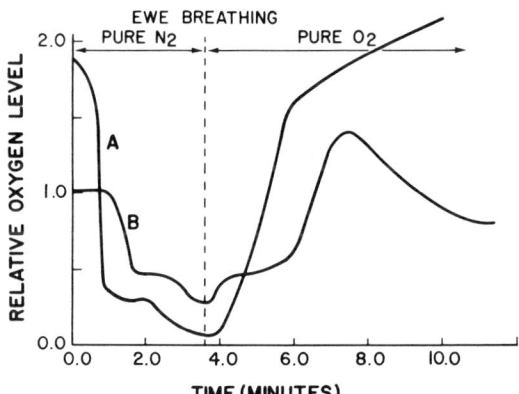

Fig. 3. Relative oxygen level during hyperoxia and anoxia. Line A shows the maternal arterial oxygen level measured in the femoral artery (normal approximately $PO_2$ 100 mm Hg) and line B shows the oxygen level in fetal brain tissue (normal $PO_2$ 7.5 mm Hg) measured with a microelectrode.

Simulation using the oxygen model of the uncontrolled fetal system did not reproduce the experimental curve (Fig. 3) for the brain tissue response. The calculated brain tissue oxygen partial pressure (Fig. 4) did not exhibit the overshoot which followed the anoxic period in the experimental results. This was the expected result since the overshoot phenomena is indicative of control.

A premise of the study was that local control is perhaps the simplest most primitive control system, requiring no central intervention or extensive innervation. Therefore, the second version of the oxygen models included local control of average tissue oxygen partial pressure. All organs and major body divisions responded to variation of tissue oxygen partial pressure by altering the arteriole resistance and the number of open capillaries. The addition of local control produced an overshoot in the brain tissue oxygen partial pressure response, but there was a substantial decrease of arterial blood pressure which does not agree with the experimental evidence discussed previously.

In an effort to maintain arterial pressure during simulation of anoxia, a simple pressure control mechanism was included in the model; the fetal heart rate was ad-

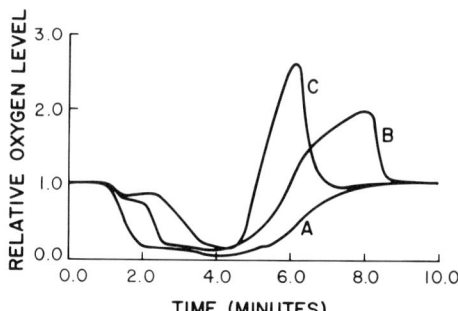

Fig. 4. Relative oxygen level calculated for fetal sheep brain tissue during maternal hyperoxia and anoxia. Line A was determined using the model with no control; line B was calculated using the model with local oxygen and baroreceptor control; line C was calculated using the model with local and chemoreceptor oxygen control and baroreceptor control.

justed by the controller in order to maintain arterial pressure. When this version of the model was solved, it was found that arterial pressure still lagged far below normal. The addition of this simple baroreceptor reflex accentuated the overshoot phenomena, because not only was the local resistance lowered to allow increased flow, but also the cardiac output rose in response to lowered arterial pressure; but, lowered arterial pressure resulted from the lack of redistribution of blood flow because all organs had equally effective local control mechanism. A further departure of the model solution from experimental results was the fact that extensive vasodilatation throughout the fetus caused a drop in umbilical blood flow due to a drop of the arterial pressure.

The oxygen model was further modified to include vasoconstriction as part of the baroreceptor reflex. It was assumed that all organs, except the brain and heart, would undergo vasocontrol for maintenance of blood pressure. This addition was to help provide blood flow redistribution. Figure 4 shows the brain tissue oxygen response calculated for maternal anoxia by this version of the oxygen model. The curve corresponds well to the experimental curve. However, the model was still not reproducing the body responses discussed in the introduction (see "Fetal Control"). There was not an immediate drop of fetal heart rate, nor was there a rise of arterial pressure (arterial pressure still lagged below normal, though to a lesser extent than previously calculated); umbilical flow to the placenta was not being maintained.

The results to this point indicated that a controller sensitive to oxygen partial pressure was needed which would cause redistribution of blood flow during anoxia. Since Rudolph and Heymann (58, 59) contend that the chemoreceptor is active in the fetus and sensitive to arterial oxygen pressure, the chemoreceptor was added to the oxygen model. The controller was assumed to vary the fetal heart rate and to cause vasoconstriction relative to the oxygen concentration measured in the aorta. Again maternal anoxia was simulated using yet another version of the oxygen model.

Simulation of maternal anoxia and recovery with the newly revised oxygen model produced the brain tissue response shown in Figure 4. Though the curve did not match the data exactly, there was general agreement with experimental trends. (No attempt was made to fit the model to

the experimental results.) Not only did the results of
this simulation agree with the brain tissue response, but
the results for the rest of the model corresponded with
the experimental fetal responses discussed in the introduction. The blood supply to the brain increased (Fig. 5),
and the fetal heart rate dropped (Fig. 6). The fetal heart
rate underwent a slight initial rise before falling sharply.
This rise of fetal heart rate indicated an attempt to maintain oxygen levels by local control, which of course fails
during severe anoxia, and the fetal heart rate drops when
the chemoreceptor causes flow redistribution. Other calculated results that agree with what has been noted in the
fetal lamb are (1) elevation of arterial blood pressure
(Fig. 6), (2) maintenance of umbilical blood flow (Fig. 5),
and (3) a loss of blood supply to the hindquarters (Fig. 5).
The overshoot of the brain tissue response corresponds

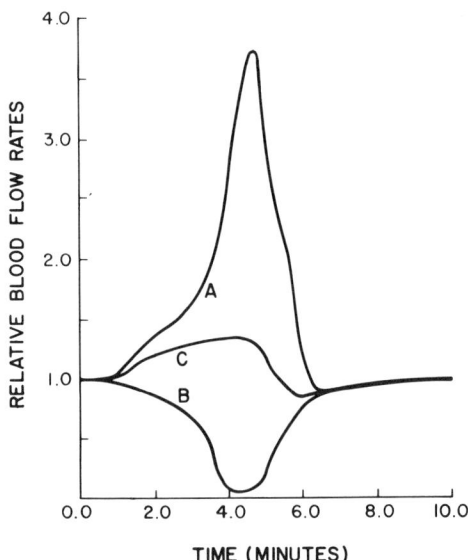

Fig. 5. Relative blood flow rates calculated
for fetal lamb during maternal anoxia using
final model. Line A shows the blood flow rate
to fetal brain; line B shows the blood flow
rate to fetal hindquarter; line C shows the
blood flow rate through the fetal side of the
placenta.

with additional overshoots calculated for the arterial
pressure and for the umbilical flow rate. The return of
the system to the normal state following maternal anoxia
causes an overshoot of blood pressure to a value slightly
less than normal, which causes a drop of blood supply to
the placenta which has no control mechanisms.

## Fetal System During Labor

Labor was simulated by varying the maternal blood
flow rate proportional to the curve shown in Figure 7. If
a series of contractions was desired, the curve was used
repeatedly with a delay between contractions. The severity
of a contraction and the length of a contraction were not
varied during this study.

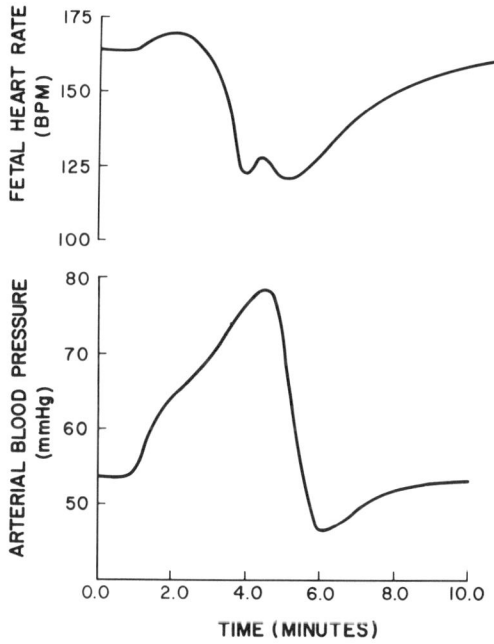

Fig. 6. Fetal lamb heart rate and arterial
pressure during maternal anoxia. Calculated
for fetal lamb using final oxygen control
model.

The effect of a single uterine contraction was simulated with the multicomponent model (Fig. 8). Oxygen partial pressure calculated for brain tissue remained essentially constant while the hindquarter oxygen partial pressure dropped to about 65% of normal. The control mechanisms caused the oxygen partial pressure in the fetal brain to remain higher than previously simulated by reducing the oxygen supply to the hindquarters (11). Furthermore, since hypoxia was minimal, there was no anaerobic metabolism and, therefore, the other chemical components remained essentially constant.

## 1. Normal Labor

A series of 3 one-minute contractions spaced one minute apart was used with the oxygen model of the controlled fetal system. (The first contraction began at time

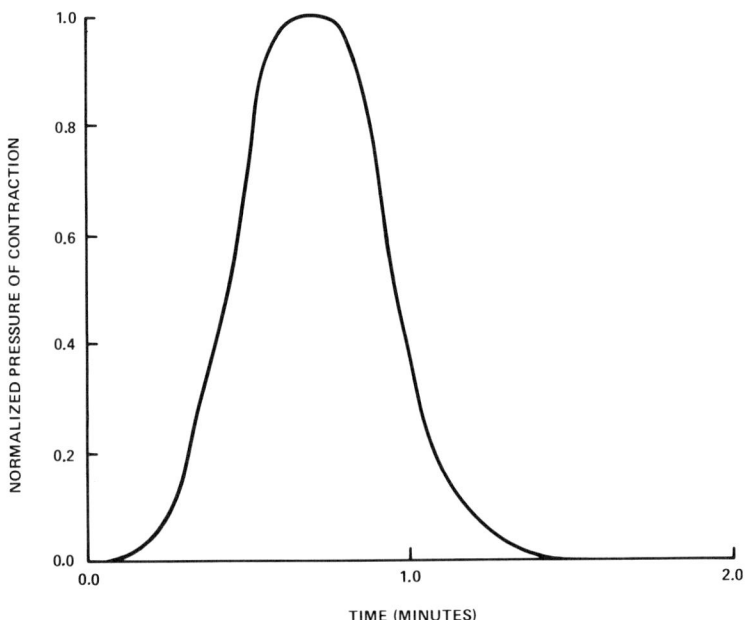

Fig. 7. Curve used to simulate the increase of intrauterine pressure during labor contractions.

zero.) During the episode, the fetal brain tissue oxygen partial pressure barely dropped below normal (Fig. 9); and the hindquarter oxygen concentration, which reflected the oxygen loss to the system due to inadequate oxygen transport in the placenta, dropped as much as 70% below normal. The fetal heart rate rose slowly during the episode. This is consistent with normal labor since a slight acceleration and variability of the fetal heart rate is expected during contractions.

## 2. Abnormal Labor

The multicomponent model was used to simulate umbilical cord occlusion. It was assumed that the blood flow resistance through the umbilical cord varied proportional to the intrauterine pressure curve (Fig. 7). Partial occlusion of the cord during a single contraction is shown in Figures 10, 11, and 12. Blood flow through the fetal placenta dropped to 35 percent of normal during the contraction. Constriction of the umbilical cord caused an

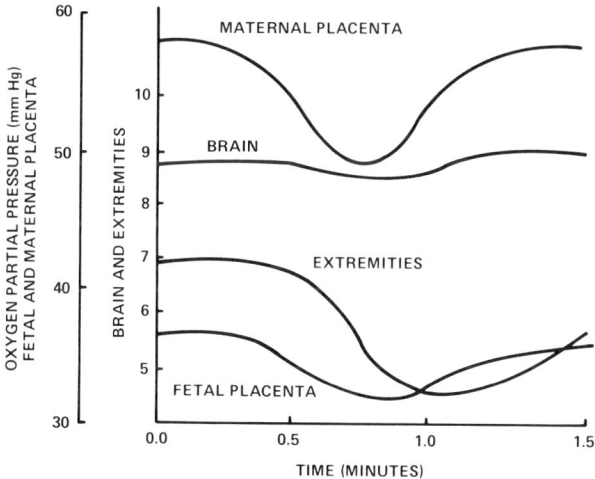

Fig. 8. Oxygen partial pressure in the fetal system during a single uterine contraction. The maternal blood flow rate to the placenta was varied proportional to the intrauterine pressure (minimum was 52% of normal).

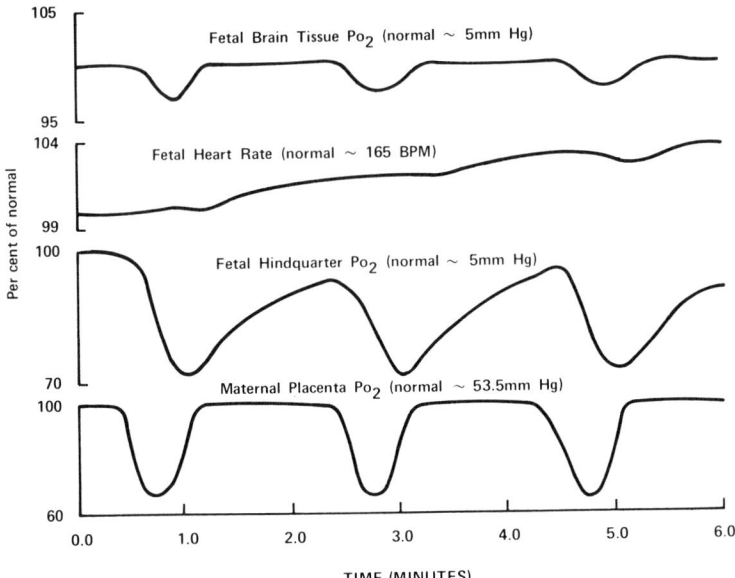

Fig. 9. The effect of uterine contractions on fetal oxygen levels and the fetal heart rate. Three contractions one minute apart were simulated. The maternal blood supply was varied proportional to the intrauterine pressure. Simulated for the fetal lamb with the oxygen model of the controlled fetal system.

increase of the blood supply to the hindquarters, and during the initial phase of the contraction the hindquarter oxygen partial pressure rose, after which the oxygen partial pressure fell to 45 percent of normal (Fig. 10).

Since the blood flow rate through the placenta fell during the contraction, the fetal blood in the placenta had time to become oxygenated above normal, but there was less oxygen transferred as indicated by the simultaneous increase of the oxygen concentration in the maternal side of the placenta and by the decrease of the oxygen concentration in the fetal aorta. Anaerobic metabolism in the hindquarters caused a slight elevation of the hydrogen ion concentration in blood which, in turn, caused a rise of

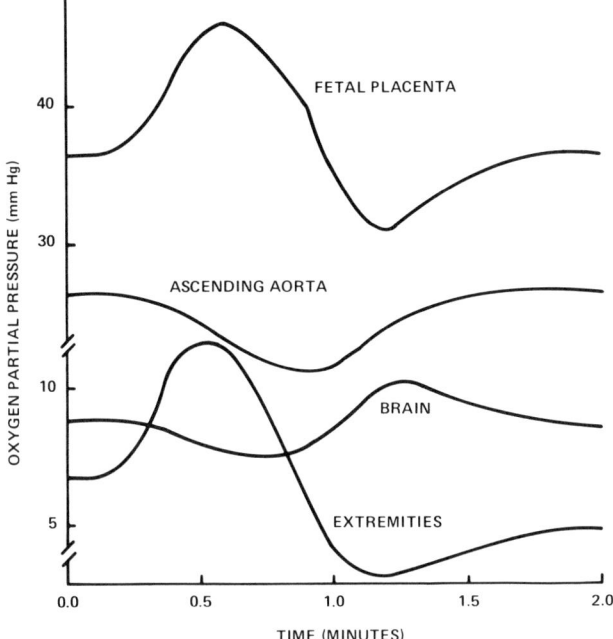

Fig. 10. Oxygen partial pressure in the fetal system during a single uterine contraction. The umbilical cord is occluded and the maternal blood flow rate to the placenta is varied proportional to the intrauterine pressure.

the carbon dioxide partial pressure. In response to elevated arterial pressure, the fetal heart rate (Fig. 11) fell during the contraction, bottoming out at the end of the contraction. Thus, the model predicts late deceleration as a result of partial occlusion of the umbilical cord, which is consistent with current thoughts on the subject.

A series of three contractions, 1 minute apart, with complete occlusion of the umbilical cord, was simulated, using the oxygen model (Fig. 1). The hindquarter of the fetus remained essentially without oxygen during the simulation (Fig. 13); and though the brain tissue oxygen partial pressure oscillated through a wide range of values, the brain remained fairly well oxygenated. It is interest-

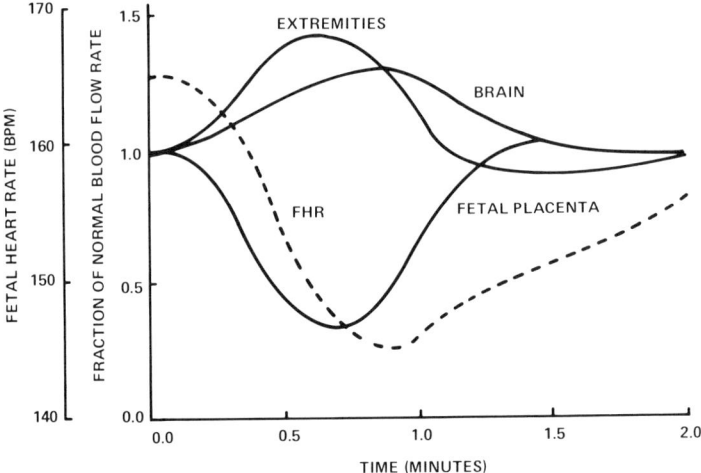

Fig. 11. Blood flow rates and fetal heart rate during a single uterine contraction. The umbilical cord is partially occluded.

ing that the maternal oxygen concentration of blood leaving the placenta oscillated with a frequency twice that of labor (Fig. 13).

The fetal heart rate (Fig. 14) fell during the initial contraction and remained at an overall low level. The fetal heart rate curve takes on a somewhat jagged appearance similar to late deceleration degenerating into variable deceleration. The low level of oxygen in the hindquarters is an indication of the overall hypoxic state of the fetus. Figure 14 also shows the blood flow rates to the brain, hindquarters, and fetal placenta.

## Maternal Death

Maternal death was simulated by stopping the maternal blood supply to the placenta. Results of the multicomponent model indicated that the brain oxygen partial pressure remained nearly constant during the first two minutes while the oxygen partial pressure calculated for the extremities quickly fell to zero (see Fig. 15).

Additional parameters, calculated using the multi-component model, indicate a rise of the carbon dioxide partial pressure and hydrogen ion concentration. It is interesting that the initial loss of oxygen causes a redistribution of carbon dioxide through the Haldane effect. This is evidenced by a drop of carbon dioxide partial pressure at the onset of maternal death (Fig. 16).

During the simulation, the fetal heart rate rose and then began to fall as flow to the extremities was slowed by control mechanisms (Fig. 17). The blood flow rate to the brain increased due to not only a decrease of local blood flow resistance, but also because of rising arterial pressure. Blood flow through the fetal placenta increased due solely to increased arterial pressure.

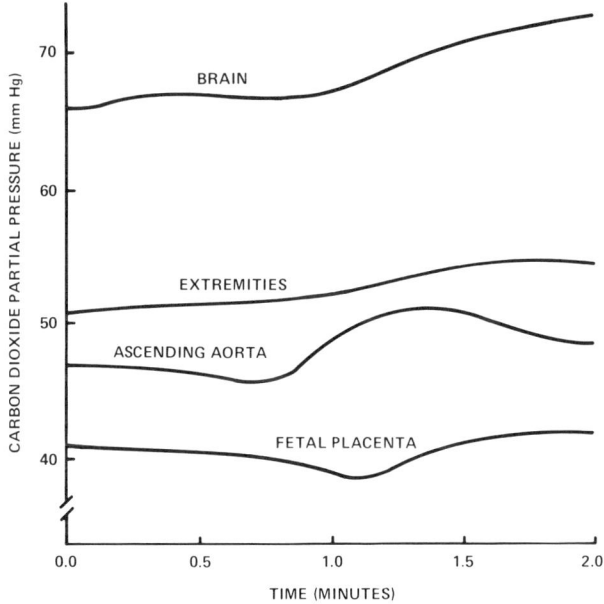

Fig. 12. Carbon dioxide partial pressure in the fetal system during a single contraction. The umbilical cord is occluded and the maternal blood flow rate to the placenta is varied proportional to the intrauterine pressure.

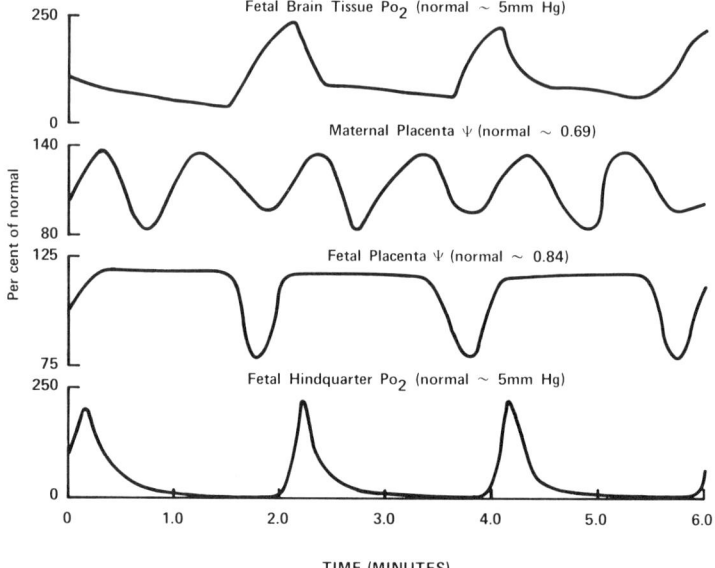

Fig. 13. The effect on fetal oxygenation of total occlusion of the umbilical cord during each contraction of a series of contractions. Simulated for the fetal lamb with the oxygen model of the controlled fetal system.

## SIMULATION EVALUATION

Experimental results indicate that the term fetus is a controlled system (40, 41). The oxygen and multicomponent models of the controlled fetal system gave results consistent with measured fetal physiology. During maternal anoxia, the model predicts (1) an increase of blood supply to the brain, (2) a loss of blood supply to the extremities, (3) a rise of arterial pressure, (4) a drop of fetal heart rate, and (5) maintenance of umbilical blood flow. This is in agreement with the data presented in the introduction. Though the simulated brain tissue oxygen response in Figure 4 did not match the single experimental curve exactly, there was a general agreement of the response.

The key to obtaining these results was the inclusion of oxygen sensitive chemoreceptors in the aorta. The con-

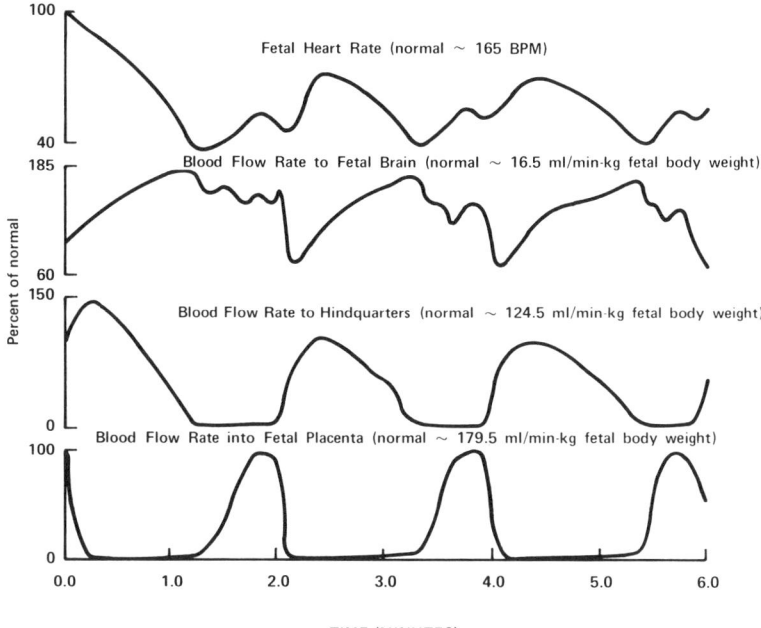

Fig. 14. The effect of total occlusion of the umbilical cord on fetal cardiovascular parameters during each contraction of a series of contractions. Simulated for the fetal lamb with the oxygen model of the controlled fetal system.

trol mechanisms included in the final oxygen model were: (1) chemoreceptor, (2) local oxygen control, and (3) baroreceptor reflex. All of the controllers were integral. Baro and chemocontrol varied the fetal heart rate and local resistance in order to maintain the desired set point, and local control adjusted the local blood flow resistance.

Simulation of labor provides interesting results. It appears that if the main influence of labor upon the fetus is variation of the maternal blood supply to the placenta, the fetus can tolerate one-minute contractions spaced one minute apart (Fig. 8). The fetal heart rate remained essentially constant exhibiting baseline variability which is consistent with normal labor. Complete occlusion of the umbilical cord was simulated with the oxygen model

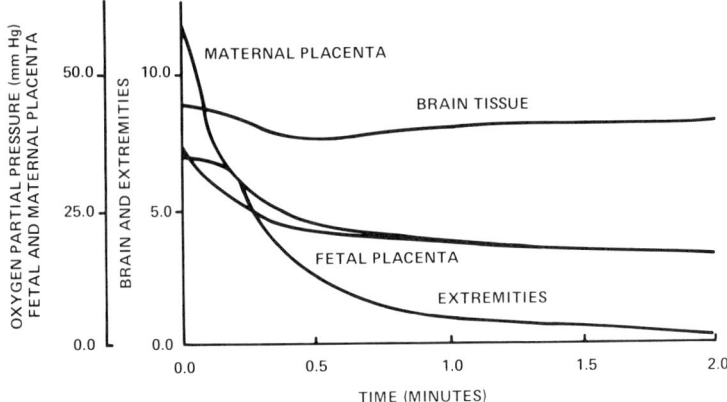

Fig. 15. Oxygen partial pressure in the fetal system following cessation of maternal blood flow to the placenta.

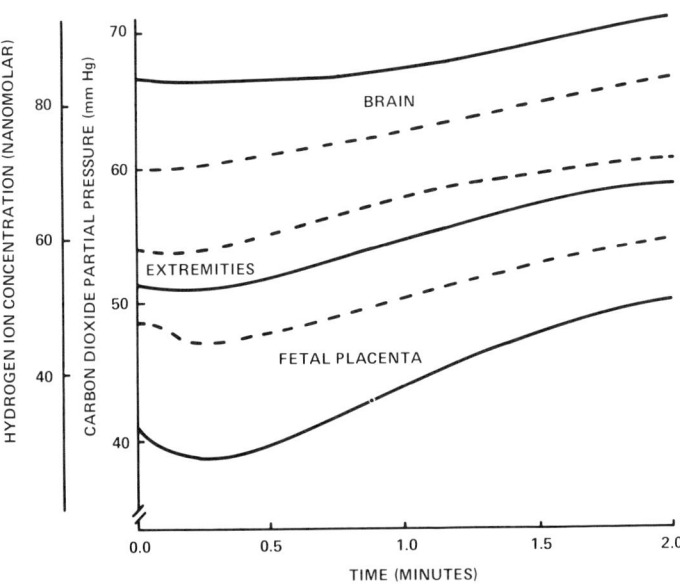

Fig. 16. Carbon dioxide partial pressure (solid lines) and hydrogen ion concentration in the fetal system following cessation of maternal blood supply to the placenta.

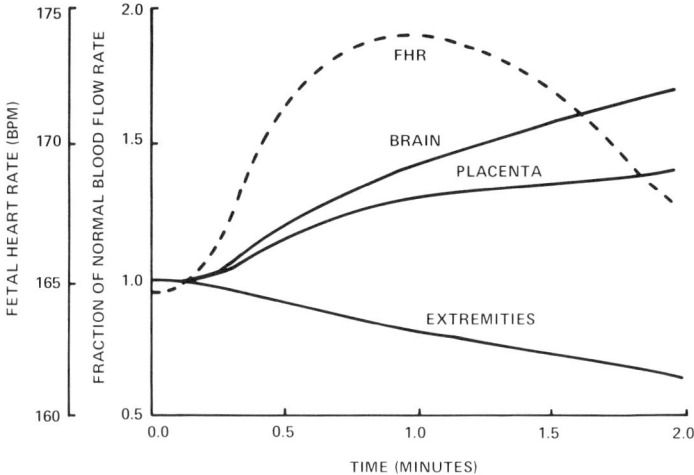

Fig. 17. Blood flow distribution in the fetal system following cessation of maternal blood supply to the placenta.

(Figs. 13 and 14) during which the fetal heart rate exhibited late deceleration and possibly variable deceleration, which is currently accepted as an indication of cord occlusion. The results indicate that cord occlusion causes a considerable loss of oxygen supply to the fetus and that an oxygen debt was forming. Unfortunately, the oxygen model cannot determine the extent of an oxygen debt since lactate and other metabolites are not included in the model.

The multicomponent model was used to simulate a single uterine contraction for both normal and abnormal labor. Results showed that during partial occlusion of the umbilical cord a rise of hydrogen ion concentration and carbon dioxide partial pressure results; and, since the values did not immediately return to normal, an accumulative effect would be expected during a series of contractions.

These results only hint at the great potential of mathematical modeling as a research tool in fetal physiology. These models appear capable of reproducing conditions ranging from maternal anoxia to cord occlusion during labor, and further evaluation of the models should serve only to increase their potential.

ACKNOWLEDGMENT

This research was supported in part by the Frost Foundation, Shreveport, Louisiana.

APPENDIX A

Definition of Symbols

A--constant oxygen consumption rate, ml $O_2$/ml--min

c--compliance, ml/mm Hg-kg fetal body weight

$[CO_2]$--concentration of carbon dioxide, mMolar

D--rate of transfer by diffusion, mMoles/min

e--error signal for controller

$[H^+]$--concentration of hydrogen ions, mMolar

$[Hb]$, $[HbNH_2]$, $[HbNH_3^+]$--concentration of reduced, carbon dioxide free hemoglobin, mEquiv.

$[HbNHCOOH]$, $[HbNHCOO^-]$-- concentration of reduced hemoglobin in combination with carbon dioxide, mEquiv.

$[HbO_2]$--concentration of oxygenated hemoglobin

$[H_2CO_3]$--concentration of carbonic acid, mMolar

$[HCO_3-]$--concentration of carbonic acid anion, mMolar

K--equilibrium constant

k--transfer coefficient, mass/time-driving force

k--reaction rate constant, $min^{-1}$

$[L]$, $[L^-]$, $[LH]$--lactate concentration, mMolar

M--oxygen capacity, mMolar

N--number of open capillaries

$[O_2]$--oxygen concentraion, mMolar

$[O_2HbNH_2]$, $[O_2HbNH_3^+]$--concentration of oxygenated, carbon dioxide free hemoglobin, mEquiv.

$[O_2HbNHCOOH]$, $[O_2HbNHCOO^-]$--concentration of oxygenated hemoglobin in combination with carbon dioxide, mEquiv.

P--pressure, mm Hg

$PCO_2$--partial pressure of carbon dioxide, mm Hg

$PO_2$--partial pressure of oxygen, mm Hg

Q--volumetric flow rate, ml/min-kg fetal body weight

R--flow resistance, mm Hg-min-kg fetal body weight/ml

r--reaction rate, mMoles/liter-min

S--percent saturation of hemoglobin with oxygen

t--time, minutes

v--volume, ml/kg fetal body weight

## Greek Letters

$\alpha$--solubility, mMolar/mm Hg

$\beta$--constant

$\Psi$--fractional saturation of hemoglobin with oxygen

$\rho$--density, mass/volume

## Subscripts

A--arterial

A--G--pertaining to that reaction in Table 6

$CO_2$--carbon dioxide

f--fetal

m--maternal

0--normal state, venous

$O_2$--oxygen

ut--uterus

v--venous, venule

## REFERENCES

1. Assali, N. S. and Brinkman, C. R., III. 1972. The uterine circulation and its control. Respiratory gas exchange and blood flow in the placenta: proceedings of symposium at the 25th International Congress of Physiological Sciences, Hanover, Germany, U. S. Department HEW, pp. 570.

2. Bartels, H.; Moll, W.; and Metcalfe, J. 1962. Physiology of gas exchange in the human placenta. *Am. J. Obstet. Gynecol.* 84:1714-30.

3. Bartels, H. and Moll, W. 1964. Passage of inert substances and oxygen in the human placenta. *Pflugers Archiv.* 280:165-77.

4. Battaglia, F. C.; McGaughey, H.; Makowski, E. L.; and Meschia, G. 1970. Postnatal changes in oxygen affinity of sheep red cells: A dual role of diphosphoglyceric acid. *Am. J. Physiol.* 219:217-21.

5. Bischoff, K. B. and Brown, R. C. 1965. Drug distribution in mammals, *Chemical engineering in medicine*, Chemical Engineering Progress Symposium Series, American Institute of Chemical Engineers, New York, 62:32-45.

6. Bekey, G. A.; Darms, D. A.; Manson, W. A., and Assali, N. S. Analog computer simulation of the cardiovascular system of the fetal lamb. EAI General Purpose Analog Computation Biomedical Application Study: 4.4.3a.

7. Boddy, K.; Dawes, G. S.; Fisher, R.; Pinter, S.; and Robinson, J. S. 1974. Foetal respiratory movements, electrocortical and cardiovascular response to hypoxaemia and hypercapnia in sheep. *J. Physiol.* 243: 599-618. London.

8. Brinkman, C. R.; Mofid, M.; and Assali, N. S. 1974. Circulatory shock in pregnant sheep. *Am. J. Obstet. Gynecol.* 118:77-90.

9. Brown, W. E. L. and Hill, A. V. 1922. The oxygen dissociation curve of blood and its thermodynamic basis. *Proc. R. Soc.* B94:297-334. London.

10. Bird, B. R.; Stewart, W. E.; and Lightfoot, E. N. 1960. *Transport Phenomena.* New York: Wiley.

11. Cameron, J. M.; Reneau, D. D.; and Guilbeau, E. J. *A mathematical analysis of oxygen transport in the combined fetal-placental system*, Proceedings of the Summer Simulation Conference, Houston, Texas, July 1975.

12. Cohn, H. E.; Sacks, E. J.; Heymann, M. A.; and Rudolph, A. M. 1974. Cardiovascular responses to hypoxemia and acidemia in fetal lambs. *Am. J. Obstet. Gynecol.* 120:817-824.

13. Comline, R. S. and Silver, M. 1970. Daily changes in foetal and maternal blood of conscious pregnant ewes, with catheters in umbilical and uterine vessels. *J. Physiol.* 209:567-586. London.

14. Dawes, G. S. 1968. *Foetal and neonatal physiology.* Chicago: Year Book Medical Pub.

15. Diemer, K. 1963. Eine Verbesserte Modellvorstellung zur Sauerstaffversorgung des Gehiras. *Naturwissenschaften* 50:617-18.

16. ----. 1964. Ueber die Entwicklung der Gefassrersoryung des Gehirns im Säugling sultur. *Monatsehr Kinderheilk.* 112:240.

17. ----. 1965. Ueber die Sauerstoffdiffusion im Gehirn. Mitteilung I. Raunliche Vorstellung and Berechnung der Sauerstoffdiffusion. *Pflugers Archiv.* 285:99-108.

18. ----. 1965. Ueber die Sauerstoffdiffusion im Gehirm. II Mitteilung. Der Sauerstoffdiffusion bei $O_2$-Mingelzustaden. *Pflugers Archiv*. 285:109-118.

19. Downing, S. E.; Lee, J. C.; Taylor, J. F. N.; and Halloran, K. 1973. Influence of norepinephrine and digitalis on myocardial oxygen consumption in the newborn lamb. *Circ. Res.* 32:471-79.

20. **Granger, H. J.** and Shepard, A. P., Jr. 1973. Intrinsic microvascular control of tissue oxygen delivery. *Microvasc. Res.* 5:49-72.

21. Guilbeau, G. J. and Reneau, D. D. A **Mathematical** simulation of oxygen transfer across the human placenta. *157th American Chemical Society National Meeting*, Abstracts of Papers, Minneapolis, Minnesota, April 1969.

22. ----. Mathematical Analysis of Combined Placental-Fetal Oxygen Transport. *International Symposium on Oxygen Transport to Tissue*, Charleston-Clemson, South Carolina, April 1973.

23. Guilbeau, G. J.; Reneau, D. D.; and Chew, W. W. A transient analysis of oxygen transport in the placenta-fetus system, *164th American Chemical Society National Meeting*, Abstracts of Papers, New York, August 1972.

24. Harper, H. A. 1971. *Review of physiological chemistry*. Los Altos, Calif.: Lange Medical Publications.

25. Heymann, M. A., and Rudolph, A. M. 1972. Effects of congenital heart disease on fetal and neonatal circulations. *Prog. Cardiovas. Dis.* 15:115-143.

26. ----. 1967. Effects of exteriorization of the sheep fetus on its cardiovascular function. *Circ. Res.* 21: 741-745.

27. Hill, E. V.; Power, G. G.; and Longo, L. D. 1973. A mathematical model of carbon dioxide transfer in the placenta and its interaction with oxygen. *J. Appl. Physiol.* 224, no. 2:283-299.

28. James, E. J.; Raye, J. R.; Gresham, E. L.; Makowski, E. J.; Meschia, G.; and Battaglia, F. C. 1972. Fetal oxygen consumption, carbon dioxide production and glucose uptake in a chronic sheep preparation. *Pediatrics* 50:361-71.

29. Lamport, H. 1959. The transport of oxygen in the sheep's placenta: The diffusion constant of the placenta. *Yale J. Biol. Med.* 27:26-34.

30. Longo, L. D. and Power, G. G. 1969. Analysis of $PO_2$ and $PCO_2$ difference between maternal and fetal blood in the placenta. *J. Appl. Physiol.* 26, no. 1:48-55.

31. Longo, L. D.; Hill, E. P.; and Power, G. G. 1972. Theoretical analysis of factors affecting placental $O_2$ transfer. *J. Appl. Physiol.* 222, no. 3:730-739.

32. Lucas, W.; Kirschbaum, T.; and Assali, N. S. 1966. Cephalic circulation and oxygen consumption before and after birth. *Am. J. Physiol.* 210:287-292.

33. Makowski, E. L.; Meschia, G.; Droegmueller, W.; and Battaglia, F. C. 1968. Distribution of uterine blood flow in the pregnant sheep. *Am. J. Obstet. Gynecol.* 101:409-412.

34. Makowski, E. L.; Schneider, J. M.; Tsoulos, N. G.; Colwill, J. R.; Battaglia, F. C.; and Meschia, G. 1972. Cerebral blood flow, oxygen consumption, and glucose utilization of fetal lambs *in utero*. *Am. J. Obstet. Gynecol.* 114:292-303.

35. Mann, L. I. 1970. Effects of hypoxia on umbilical circulation and fetal metabolism. *Am. J. Physiol.* 218:1453-1458.

36. Mann, L. I. 1970. Fetal brain metabolism and function. *Clin. Obstet. Gynecol.* 13:638-651.

37. McLeod, J. 1966. PHYSBE...A physiological simulation benchmark experiment. *Simulation* 7, no. 6:324-329.

38. Meschia, G. and Battaglia, F. C. 1973. Acute changes of oxygen pressure and the regulation of uterine blood flow. *Foetal and neonatal physiology*, Sir

Joseph Barcroft Centenary Symposium, Cambridge: Cambridge Unitersity Press.

39. Metcalfe, J.; Dhindsa, D. S.; and Novy, M. J. 1972. General aspects of oxygen transport in maternal and fetal blood. Respiratory gas exchange and blood flow in the placenta: *Proceedings of a Symposium at the 25th International Congress of Physiological Sciences*, Hanover, Germany, U. S. Dept. HEW.

40. Moll, W. 1973. Placental function and oxygenation in the fetus. *International Symposium on Oxygen Transport to Tissue*, Medical University of South Carolina and Clemson University, New York: Plenum Press.

41. Morris, J. A.; Bekey, G. A.; Assali, N. S.; and Beck, R. 1965. Dynamics of blood flow in the ductus arteriosus. *Am. J. Physiol.* 208:471-473.

42. Pipkin, F. B., and Kilpatrick, S. M. L. 1973. The blood volumes of fetal and newborn sheep. *Quart. J. Exp. Physiol.* 58:181-188.

43. Power, G. G.; Butler, L. A.; and Longo, L. D. 164th American Chemical Society National Meeting, Abstracts of Papers, New York, August 1972.

44. Power, G. G.; Hill, E. P.; and Longo, L. D. 1972. Analysis of uneven distribution of diffusing capacity and blood flow in the placenta. *Am. J. Physiol.* 222, no. 3:740-746.

45. Power, G. G., and Longo, L. D. 1969. Graphic analysis of maternal and fetal exchange of $O_2$ and $CO_2$. *J. Appl. Physiol.* 26, no. 1:38-41.

46. ----. (In press) Sluice flow in the placenta: maternal vascular pressure effect on the fetal circulation.

47. Purves, M. J. and James, I. M. 1969. Observations on the control of cerebral blood flow in the sheep fetus and newborn lamb. *Circ. Res.* 25:651-667.

48. Quilligan, E. J.; Hon, E. H.; Anderson, G. G.; and Yeh, S. V. 1968. Fetal cephalic metabolism in sheep. *Am. J. Obstet. Gynecol.* 102:716-726.

49. Rankin, J. H. G. 1972. The effects of shunted and unevenly distributed blood flows on crosscurrent exchange in the sheep placenta: *Proceedings of a symposium at the 25th International Congress of Physiological Sciences*, Hanover, Germany, U. S. Dept. HEW.

50. Reneau, D. D.; Bruley, D. F.; and Knisely, M. H. 1967. A mathematical simulation of oxygen release, diffusion, and consumption in the capillaries and tissue of the human brain. *Chemical Engineering in Medicine and Biology*, ed., D. Hershey. New York: Plenum Press.

51. Reneau, D. D.; Cameron, J. M.; Melton, P.; Guilbeau, E. J.; and Chew, W. W. A systems analysis of oxygen transport in the infant brain. Abstracts of Papers, 164th National Meeting of the American Chemical Society, New York, 1972.

52. Reneau, D. D.; Guilbeau, E. J.; and Cameron, J. M. 1974. A theoretical analysis of the dynamics of oxygen transport and exchange in the placental-fetal system. *Microvasc. Res.* 8:346-361.

53. Rideout, V. C. and Dick, D. E. 1976. Difference-differential equations for fluid flow in distensible tubes. *IEEE Trans. Biomed. Eng.* BME-14, no. 3.

54. Rossi-Bernardi, L. and Roughton, F. J. W. 1967. The specific influence of carbon dioxide and carbomate compounds on the buffer power and Bohr effects in human hemoglobin solutions. *J. Physiol.* 189:1-29. London.

55. Roughton, F. J. W. 1964. Transport of oxygen and carbon dioxide. *Handbook of physiology*, American Physiological Society, Washington, D. C., sec. 3, 1.

56. Ruch, T. C., and Patton, H. C. 1965. *Physiology and biophysics*. Philadelphia: W. B. Saunders Co.

57. Rudolph, A. M. 1970. Circulatory changes during growth in the fetal lamb. *Circ. Res.* 26:289.

58. Rudolph, A. M. and Heymann, M. A. 1973. Control of the foetal circulation. *Foetal and neonatal physio-*

59. ----. 1974. Fetal and neonatal circulation and respiration. *Annual Review of Physiology* 36:187-207.

60. Shier, R. W. and Dilts, P. V., Jr. 1972. Comparison of fetal scalp, carotid, and jugular blood gas values and oxygen consumption in the lamb. *Am. J. Obstet. Gynecol.* 112:397-403.

61. Shinebourne, E. A.; Vapaavuori, E. K.; Williams, R. L.; Heymann, M. A.; and Rudolph, A. M. 1972. Development of baroreflex activity in unanesthetized fetal and neonatal lambs. *Circ. Res.* 31:710-718.

62. Sirs, J. A. 1973. The interaction of carbon dioxide transfer in the placenta and its interaction with oxygen. *J. Appl. Physiol.* 224, no. 2:283-299.

63. Smith, J. M. 1970. *Chemical Engineering Kinetics*, 2nd ed., New York: McGraw-Hill.

64. Thews, G.; Fischer; and Vogel, H. R. 1965. Den Gasaustauisch in der Menschlichen Placenta unter Berucksichtigung der Wechselseitigen Abhangigkert von $O_2$- und-$CO_2$ Transport. *Pfluger Archiv.* 286:257-274.

# 22

## Fetal Oxygen Responses to Hypoxia:
## A Mathematical Model

William W. Allen, Gordon G. Power, and
Lawrence D. Longo

Division of Perinatal Biology
Departments of Physiology and Biomathematics
Loma Linda University
School of Medicine
Loma Linda, California

To predict the extent and severity of fetal hypoxia during labor and also in states of fetal anemia, based on existing experimental data and plausible assumptions, we constructed a mathematical model of the major fetal organs and the vascular system supporting them. Figure 1 represents the organs and the vascular system chosen for the model as well as the blood flows (expressed as percentages of total cardiac output) and the normal oxygen consumption of each organ (expressed as percentages of the $O_2$ crossing the placenta). The model relates oxygen transfer across capillary walls to vascular oxygen tension levels, relates the oxygen content of blood to $PO_2$ via the oxyhemoglobin dissociation curve, and characterizes the vascular transfer of oxygen across organs by their distributions of transit times.

In the model, oxygen transfer across capillary walls is a function of vascular $PO_2$ values. Previous theoretical

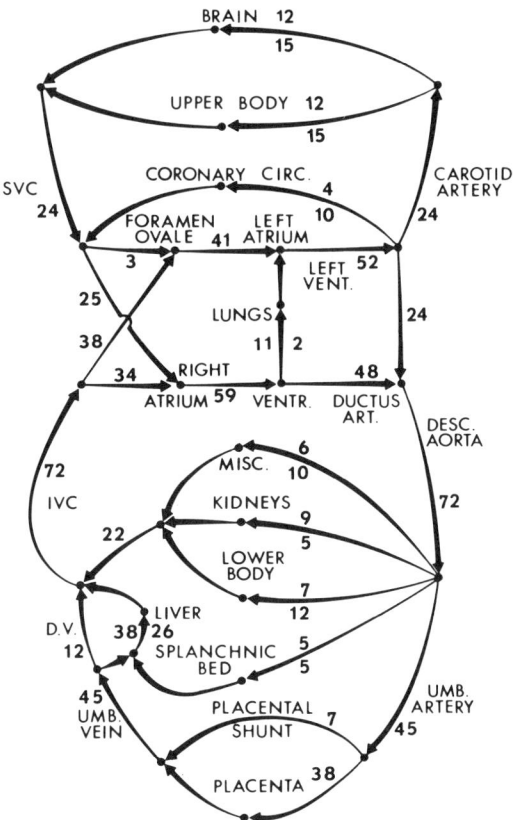

Fig. 1. The fetal model: organs, and vascular system connecting them. Numerical values are blood flows, as percentages of total cardiac output; and (where values are paired) oxygen consumptions, as percentages of placental $O_2$ transfer.

(3) and experimental (6) studies in our laboratory have delineated a dependence of placental $O_2$ transfer on umbilical arterial $PO_2$ as shown in Figure 2, the lower panel. The work of Stainsby and Otis on adult skeletal muscle (9) and our studies of $O_2$ consumption for the fetus as a whole (7) led us to posit a relation between fetal $O_2$ consumption and arterial $PO_2$ as shown in Figure 2, the upper panel. In the absence of unique relations for each organ, we employ the

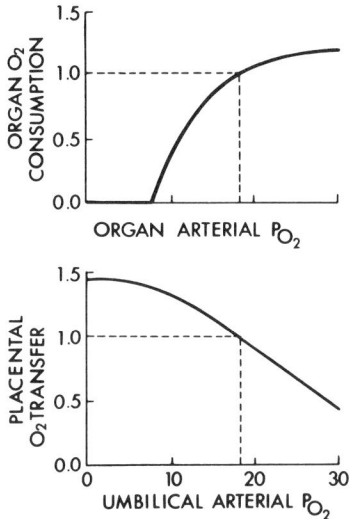

Fig. 2. Oxygen consumption and transfer as functions of arterial $PO_2$ values, expressed as fractions of normal consumption and transfer values.

same relation for all the fetal organs, scaling the $PO_2$ axis for each so that at its normal oxygen tension, the $O_2$ consumption fraction is unity.

For the relation between blood $O_2$ content and $PO_2$, we use the Hill equation with empirical constants from Hellegers and Schruefer's data (2).

At mixing points, such as the convergence of large vessels, mixed $O_2$ contents are calculated as the sum of the inflowing $O_2$ contents ($C_i$) weighted by their respective flows ($\dot{Q}_i$): $C_m = (\sum_i C_i \dot{Q}_i)/\sum_i \dot{Q}_i$

The model characterizes vascular $O_2$ transfer through capillary beds by distributions of blood transit times. These were derived from experimental fetal dye dilution curves (8). Where the injection and measurement points of the experimental curves did not coincide with the inflows and outflows of particular organs, the curves were judiciously modified. A typical distribution of transit times has unit area, shows no outflow of indicator prior to the elapse of an "appearance time," $t_a$, and has substantially no more indicator remaining to be discharged in the outflow by time $t_m$.

Given the inflow $O_2$ contents for an organ and its distribution of transit times, its present (time $t$) outflow $O_2$ content can be calculated in the following way (Fig. 3). Consider an inflow $O_2$ content, $C_{in}$, which entered organ seconds ago, at time $t-\tau$, $C_{in}(t-\tau)$. From the distribution of transit times, $h(t)$, we know that the fraction $h(\tau)d\tau$, of that inflow content, will now be appearing at the outflow. Thus we know that the total $O_2$ outflow content is

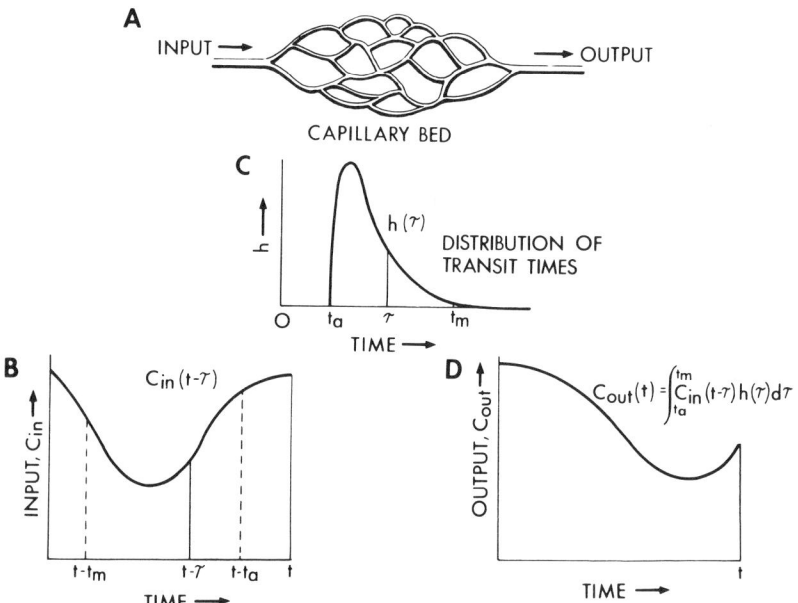

Fig. 3. Outflow oxygen calculations, the convolution integral. A represents a vascular bed. B is an arbitrary "history" of the $O_2$ contents which have flowed into the vascular bed. C represents the distribution of transit items characterizing in detail the transfer of substances across the bascular bed. D represents the calculated outflow $O_2$ contents, each point of which is the convolution integral of the history of inflowing $O_2$ contents and the distribution of transit times.

the "sum" of all such inflow contributions, the integral of $C_{in}(t-\tau)h(\tau)d\tau$, over the appropriate limits, from $t_a$ to $t_m$:

$$C_{out}(t) = \int_{t_a}^{t_m} C_{in}(t-\tau)h(\tau)d\tau$$

The model was constructed to calculate this integral for each major fetal organ, to augment or diminish the calculated outflow $O_2$ content by the amount of trans-capillary $O_2$ transferred, based on the local vascular $PO_2$, and to determine the $O_2$ content at all major mixing points throughout the fetus. A computer performed these calculations iteratively, beginning from a set of $O_2$ values throughout the fetus calculated to be consistent with local blood flow rates and $O_2$ consumptions. (For details of the process, and the derivation of iterative forms of the mixing equation and the outflow $O_2$ integral, see Ref. 1.)

To investigate the $PO_2$ levels likely to result during uterine contractions, we imposed Gaussian-shaped decreases in $O_2$ transfer on the placenta. In the standard case, representing relatively intense labor, the duration of the Gaussian-shaped $O_2$ transfer decrease was one minute, the extent of the maximum decrease was 50%, and the interval between contraction peaks was three minutes. Figure 4 illustrates a series of five such contractions.

One of the most striking results was that except for the liver, oxygen tensions in fetal organs fell only 2 or 3 mm Hg after the first contraction, and that following successive contractions, they fell to minimum levels only a fraction of a mm Hg lower than those reached after the first contraction, rather than continuing steadily to fall. For the brain and the heart, the venous $PO_2$ levels after the first contraction fell 2 and 1.5 torr, and following the fifth contraction, they fell only 2.4 and 1.9 torr below their baseline values. The oxygen consumption deficits incurred by these two organs during the five contractions were 2.54 and 1.68 ml, respectively. The umbilical venous $PO_2$ fell 10.3 mm Hg during the first contraction. Consequently, the liver, the first organ to which the umbilical venous blood returns, experienced not only the greatest $PO_2$ decrease, 4.5 mm Hg, but also the greatest oxygen consumption deficit (since, like all the other fetal

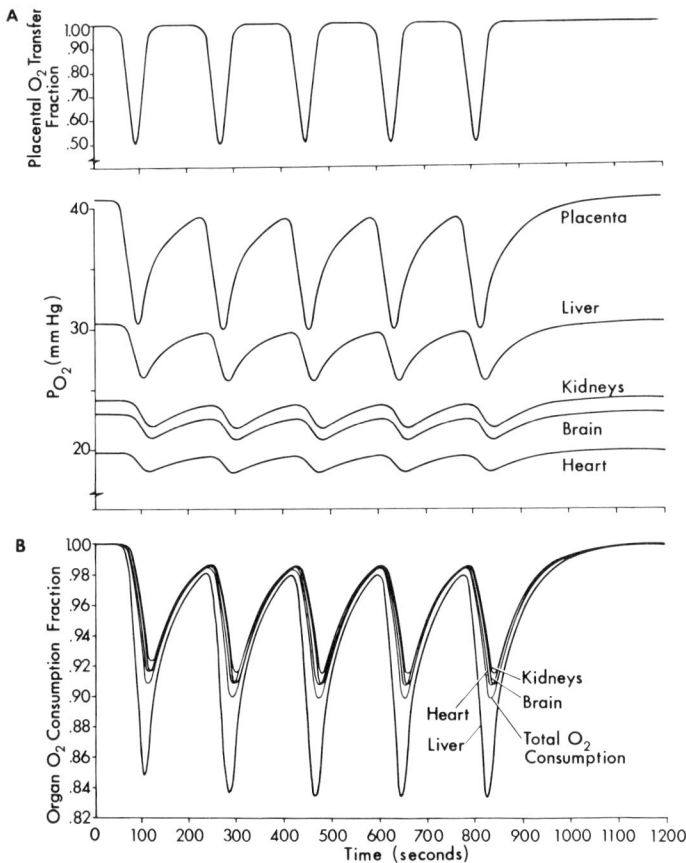

Fig. 4. A sequence of five simulated uterine contractions, and some of the resulting $PO_2$ levels and $O_2$ consumption decreases. Representing an advanced stage of labor, this is the standard case, with Gaussian-shaped decreases in placental $O_2$ transfer to 50% of its normal value. The duration of these decreases was one minute, and their peak-to-peak period was three minutes. Oxygen consumption for each (illustrated) organ is expressed as a fraction of its normal value.

organs in the model, its $O_2$ consumption diminishes with decreases in its arterial $PO_2$), after five contractions, 6.6 ml.

We wished to examine the relative effects of contraction intensity, duration and period (peak-to-peak), and of blood $O_2$ capacity on the impairment of fetal oxygenation. To do so, we varied each of these parameters about its

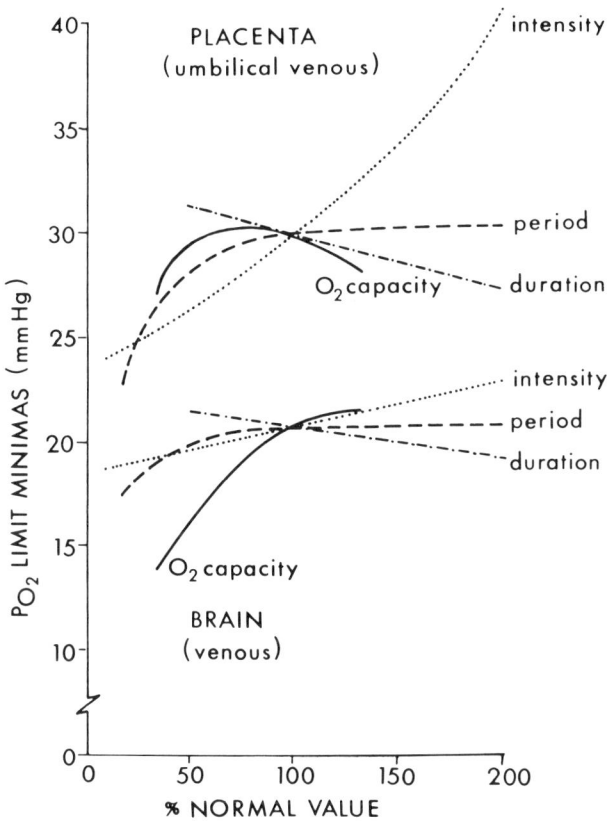

Fig. 5. A comparison of four factors affecting fetal oxygenation, based on the extent to which they alter the minimum $PO_2$ values (as in Figure 4) that would be reached after an extended unbroken chain of contractions, illustrated for the venous outflows of the placenta and the brain.

normal value (intensity, 50% decrease in placental $O_2$ transfer at the peak of a contraction; duration, 60 seconds; period, 3 minutes peak-to-peak; and blood $O_2$ capacity, 0.21 ml $O_2$/ml blood) and compared the minimum $PO_2$ values attained as measures of impairment. Figure 5 illustrates the results for the venous outflows of the brain and the placenta. For the brain, sensitivity to contraction period is slight until the period has become short enough that one contraction begins to overlap the next. Its sensitivity to contraction intensity and to contraction duration are of about equal magnitude. Brain minimum $PO_2$ values are more sensitive to changes in blood $O_2$ capacity than to any of the other three factors. At 50% of normal capacity, minimum $PO_2$ values fall 5 mm Hg below those reached at normal $O_2$ capacity. The corresponding parameter variations result in similar, although more pronounced, effects on placental venous blood. The lesser magnitudes of the effects elsewhere can be explained at least in part by the admixture of placental venous with fetal venous blood, moderating the $O_2$ changes experienced by the placenta.

This version of the model does not permit changes in the volume or distribution of blood flows. The model predicts small decreases in systemic $PO_2$ levels succeeding large decreases in umbilical venous $PO_2$ levels simulating those that should accompany the final stages of labor. This analysis must be considered a worst-case prediction because the fetus probably defends it vital organs against severe hypoxia by distributing blood to them preferentially during hypoxic crises. That fetal $O_2$ levels would be maintained as well as predicted by the model, even without compensatory blood flow redistribution, is noteworthy.

The prediction of $O_2$ levels in the organs of a fetus subjected to hypoxic stresses, considering the redistribution of fetal blood flow, is benefiting from studies (Ref. 4 and 5, and Vol. 2, ch. 16) indicating that blood flow redistributions can be divided into three categories based on the character of the organ perfused. Vital organs whose $O_2$ supply must not be interrupted, such as the brain and the heart, increase their perfusion (hyperbolically) to maintain constant $O_2$ extraction. Lung blood flow decreases roughly exponentially with decreasing $O_2$ levels. Blood flows to other organs initially increase slightly with decreases in $O_2$ levels but finally fall precipitously.

A model incorporating vascular resistances based on such $O_2$ responses is being prepared. It expresses the output of each ventricle in terms of inflow (atrial) and outflow (afterload) pressures, and recasts the convolution integral $O_2$ outflow calculations, basing them not on elapsed times, but on elapsed volumes.

## REFERENCES

1. Allen, W. W.; Power, G. G.; and Longo, L. D. 1977. Fetal $O_2$ changes in response to hypoxic stress: a mathematical model. *J. Appl. Physiol.* 42:179-190.

2. Hellegers, A. E. and Shrueffer, J. J. 1961. Nomograms and empirical equations relating oxygen tension, percentage saturation, and pH in maternal and fetal blood. *Am. J. Obstet. Gynecol.* 81:377-384.

3. Hill, E. P.; Power, G. G.; and Longo, L. D. 1972. A mathematical model of placental $O_2$ transfer with consideration of hemoglobin reaction rates. *Am. J. Physiol.* 222:721-729.

4. Longo, L. D.; Wyatt, J.; and Gilbert, R. D. (In press) A comparison of the effects of hypoxic hypoxia and carbon monoxide hypoxia on the fetal circulation. *Am. J. Physiol.*

5. Peeters, Louis L. 1976. Personal communication.

6. Power, G. G. and Jenkins, F. 1975. Factors affecting $O_2$ transfer in sheep and rabbit placenta perfused in situ. *Am. J. Physiol.* 229:1147-1153.

7. Power, G. G. and Longo, L. D. 1975. Placental $O_2$ transfer and fetal consumption at varying fetal arterial $PO_2$. (Abs.) *Fed. Proc.* 34:451.

8. ------. 1975. Fetal circulation times and their implications for tissue oxygenation. *Gynecol. Invest.* 6:342-355.

9. Stainsby, W. N., and Otis, A. B. 1964. Blood flow, blood oxygen tension, oxygen uptake, and oxygen transport in skeletal muscle. *Am. J. Physiol.* 206:858-866.

# 23
## A Model of Placental Parameters During Labor Contractions

L. Allan Butler, Lawrence D. Longo, and Gordon G. Power

Departments of Anesthesiology, Physiology and Obstetrics and Gynecology
Loma Linda University
School of Medicine
Loma Linda, California

INTRODUCTION

In defining a science as an organized body of knowledge we imply, firstly, that meaningful data exist, and, secondly, that relationships between data elements have been formulated. The data are generally obtained empirically from the real universe using experimental techniques. Relationships however, are normally formulated abstractly to correlate classes of data. When such correlations are of sufficient generality they are termed the "laws" of the particular science. Science normally advances on both the empirical and theoretical fronts simultaneously. Theoretical correlations must take into account all verifiable empirical data, and the experimenter gains point and direction for his work from the theoretical framework of his science. This chapter is an example of theoretical deduction extending the empirical data on placental conditions during uterine contractions. As a fetus grows, oxygen and nutrients must be continually transferred across the placenta. The rates of maternal and fetal blood flows in the placenta are among the important factors affecting the rate of oxygen transfer. In turn, a factor known to affect these flows is the uterine contractions during labor. Uterine blood flow can be so profoundly affected under some conditions that fetal viability is threatened. Because of ethical and technical difficulties, only limited experimental data are available on such events in humans.

This chapter describes modeling techniques used to characterize placental blood flows and pressures for the maternal and fetal circulations. Time lags are built into the model to allow for circulation of blood from the placenta to the fetal tissues and then back to the placenta.

Such a theoretical study can be useful to the understanding of the role of various factors affecting maternal and fetal placental blood flows and hence, placental oxygen transfer during contractions.

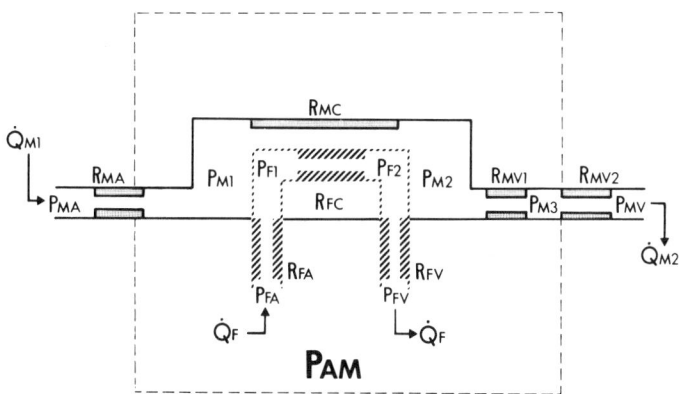

Fig. 1. Placental idealization where pressures, flows, and resistances are labeled with symbols beginning with $P$, $Q$, and $R$. $M$ = maternal, $F$ = fetal, $A$ = arterial, $C$ = cotyledon, $V$ = venous. Exception: $PAM$ = pressure of amniotic fluid.

## DESCRIPTION OF MODEL

It is assumed that maternal blood pressures and arterial blood gas values remain constant during uterine contractions; in other words, the model excludes maternal extrauterine effects. The uterus, shown diagramatically in Figure 1, is considered to be a closed container of amniotic fluid whose hydrostatic pressure (PAM) increases during uterine contractions. Since the fetal circulation is wholly contained within the uterus, hydrostatic pressure changes result in exactly equal changes of pressure in the various parts of the fetal circulation and in the maternal circulation inside the uterus. Secondary compensations of the fetal circulation (e. g., baroreceptors) are ignored.

Increases in amniotic fluid pressure cause increases in the maternal blood pressures at the entrance and exit (PM1 and PM2) of the exchange units of the placenta (Figure 1). This alters maternal blood flows into (QM1) and out of (QM2) the placenta (17). This is because the maternal arterial pressure (PMA) and venous pressure (PMV) are assumed constant as are resistances to flow (RMA, RMC, RMV1, and RMV2). Values for all of these parameters are initially chosen which lead to assumed normal blood flows and pressures. Fortunately, the computation is fairly insensitive to changes in these parameters (e. g., decreasing RMA 20% alters the oxygen exchange rate by only 5%). Despite this, the simplifying assumption of constant vascular resistance might be a shortcoming of the model. Borell, Fernstrom, Ohlson, and Wiqvist (2) showed that lumen size of many branches of uterine artery decreased during uterine contraction. In a few instances the size increased. These and other workers (17) have also shown a marked decrease in the intensity of the uterine venous pattern during a contraction. Resistances in these vessels (RMA and RMV in the model) could have been varied. This further variability was not introduced for lack of quantitative data on either the magnitude of the changes or their change in relation to other resistances.

Since the pressure differential between the maternal artery and the exchange unit decreases, we can expect QM1 to decrease and QM2 to increase. Thus, the volume (VM) of maternal blood in the placental intervillous space will decrease until the consequent drop in pressures (PM1 and PM2) stabilizes the blood flows (QM1 and QM2) at a new level.

Nonrigid placental tissues which permit increases in hydrostatic pressure to affect the maternal and fetal blood flows imply a further complication. If the pressure within some part of a given blood vessel does not exceed the surrounding pressure, collapse will take place. The pressure inside the vessel will then build up because of the impeded flow. Thus, maternal flow will slow during a contraction but not stop completely because maternal arterial pressure exceeds hydrostatic pressure even at the peak of the contraction. QM1 will depend on the difference between maternal arterial pressure and PM1 even at the peak of the contraction. QM2 will depend on the difference between PM2 and the surrounding pressure during collapse, rather than the inflow minus outflow pressure difference. This is termed "sluice" or "waterfall" flow, as only upstream conditions affect the flow over a waterfall, not the height of the fall. Flow through a real tube (i. e., having resistance) causes a pressure drop. As one proceeds along the course of the blood vessel the intravascular pressure decreases. Thus partial collapse will occur first in the maternal circulation where the uterine veins exit from the uterus. Fetal placental sluice will occur first where the umbilical circulation exits from the exchange unit. Experimental evidence for fetal sluice flow in the sheep was reported recently by Power and Longo (14), and Bissonnette and Farrell (1).

It is also essential to consider the changes in distending (transmural) pressure on the volume (VM) of maternal blood in the intervillous space. This relation apparently has not been experimentally determined. In resting conditions the intervillous space volume has been estimated to be about 110 ml from morphometric studies (10). Measurement of intervillous space pressure has suffered from the problem of ascertaining the exact location of pressure sensing tips inside an intact placenta. Values have ranged from 30 to 70 mm Hg. We elected to assume a resting pressure of 50 mm Hg and a resting volume of 100 ml and constructed an arbitrary compliance curve. When pressure is low and increases the volume is likely to increase rapidly. When the pressure is high and increases still further, the increase in volume is not likely to be as great due to fiber recruitment. Thus, the intervillous space compliance varies from a high value at low pressures to a low value at high pressures. Some assumption is essential to permit the development of the model. We assumed that the curve for compliance followed an exponential form, such as that of the charging of a capacitance through a resistor. It is convenient to specify the "full distension" as a parameter. Fivefold variation in the pressure required to distend fully the intervillous space does not modify any of the results quantitatively.

## DERIVATION OF EQUATIONS

The resistance to blood flow is the ratio of driving pressure to flow. Maternal flow at each step in the placental circulation may be expressed as follows:

$$QM1 = (PMA - PM1)/RMA \tag{1}$$

$$(QM1 + QM2)/2 = (PM1 - PM2)/RMC \tag{2}$$

$$QM2 = (PM2 - PMV)/(RMV1 + RMV2) \text{ if } PM3 > PAM$$
$$\text{(non-sluice)} \tag{3}$$

$$= (PM2 - PAM)/RMV1 \text{ if } PAM > PM3$$
$$\text{(non-sluice)}$$

where $PM3 = PMV + QM2 \; RMV2$

Note that equation (3) takes two forms, depending on whether sluice flow exists in the maternal circuit. Since VM equals the sum of all blood which has ever flowed into the intervillous space less the sum of all which has flowed out,

$$V = \int_{-\infty}^{t} (QM1 - QM2) \, dt \tag{4}$$

Using the usual pressure vs. volume equation for the intervillous lake volume,

$$VM = C \left[1 - \exp\left(-K(PM1 + PM2 - 2PAM)\right)\right] \tag{5}$$

where C and K are arbitrary constants whose values establish the resting compliance and volume of the intervillous space. Also, the fetal flow may or may not be in sluice.

$$QF = (PFA - PFV)/RFA + RFC + RFV) \quad \text{if } PF2 > PM2$$
$$\text{(non-sluice)}$$

$$= (PFA - PM2)/RFA + RFC \quad \text{if PM2} > \text{PF2 (sluice)}$$
where PF2 = PFV + QF RFV  (6)

Solutions of equations (1) through (6) are time dependent, due to the time integral of equation (4). Blood flows, pressures, and volume are dependent on past as well as present conditons.

In the strict sense, the term "model" applies to the foregoing set of assumptions and equations derived from them. The model is exercised by solving the equations for sets of values for each of the independent variables. Comparisons with the real universe constitute tests of the validity of the model. Such exercises may point to hitherto unobserved experimental situations and indicate areas in which experimental evidence is lacking.

## SOLUTION OF EQUATIONS

It can be seen that analytic solution of equations (1) through (6) is not possible. Numerical solution of the equations is therefore required. Equation (4) is deliberately set up as an integral rather than a differential equation. Solution of the corresponding differential equation could be accomplished using standard Runge-Kutta or other techniques. However, the step size required for such solutions is relatively small. In comparable work by Hill, Power, and Longo (9), the step size was 10 milliseconds. Use of such a small step size is dictated by the requirement for solution stability. Such a step size requires large amounts of computer time. With the formulation of equation (4) in integral form a much larger step size is permissible. (See Appendix II). In the present case stability of solutions is obtained with a one-second step size requiring but 60 steps to cover a one-minute contraction. The saving in computer time for comparable accuracy is considerable.

Since the equations are nonlinear, we have used a least-squares minimization technique for solution. We insert initial guesses for the independent variables (QM1, QM2, PM1, and PM2) into each of the equations and calculate for each equation an imbalance. The sum of squares of the imbalances is considered to be a function F(QM1, QM2, PM1, PM2) of the independent variables. The Fletcher-Powell minimization technique is used to adjust the independent variables such that F is minimized. The solution was considered adequate when F was below $10^{-6}$. A series of such computations yields estimates for maternal and fetal placental flows and their changes with time as the amniotic fluid pressure rises and falls in a pattern simulating labor. Each minute of contraction simulation requires approximately 15 seconds of computer time on an IBM 370/158.

## PLACENTAL OXYGEN TRANSFER

Given the computed blood flows of the fetal and maternal circuits, the next step is to estimate the placental oxygen transfer. This transfer is a function of several factors, including four quite rapidly changing variables: (1) The maternal placental blood flow (QM). (2) The fetal blood flow (QF). (3) The maternal arterial oxygen tension. (4) The umbilical arterial oxygen tension. We chose representative values for these variables spanning the physiologic range, and then calculated the placental oxygen transfer using the placental simulation described by Hill, Power, and Longo (9) for all permutations of the selected values, while recognizing that other models such as that described by Guilbeau would have been equally applicable. These results were stored in a four-dimensional data array. In making these calculations other variables that change slowly but can still affect the oxygen transfer, such as the placental diffusing capacity and maternal and fetal hemoglobin concentrations, were assumed to remain constant at normal levels over the time span of a contraction.

Having computed the blood flows using the equations above, and assigning a suitable value to the maternal arterial oxygen tension, there is one rapidly changing variable remaining, i. e., the inflowing fetal umbilical arterial oxygen tension. This tension depends on a balance between the rates at which oxygen is transferred across the placenta and the rate at which it is consumed by the fetus.

The umbilical arterial oxygen pressure is determined by an iterative procedure. First we assume a starting value for the oxygen tension and then interpolate the umbilical venous oxygen tension from the data array, using the known flows. We subtract an estimate for the oxygen used by the fetal tissues (12). We assume a constant drop in oxygen saturation in a normal range of oxygen tensions, but then a progressively smaller drop at lower tensions where oxygen utilization presumably diminishes, assuming a relationship similar to that in adults:

$$SA = SV - 30\ (1 - \exp(-SV/5)) \tag{7}$$

where SV is fetal oxygen saturation and SA is fetal arterial oxygen saturation. The saturation of the blood is then used to make a new estimate for the umbilical arterial oxygen pressure, using the empirical relation

$$PA = 21.71\ [SA/\ (100 - SA)]^{0.389} \tag{8}$$

where PA is fetal arterial oxygen tension. With this new value, umbilical venous oxygen tension is again interpolated from the array, and so on. Within five iterations the umbilical oxygen tensions converge to stable values, irrespective of the initial assumed value.

In addition to time dependences due to filling and emptying of the intervillous space, the time required for circulation through the fetal body also introduces a time lag. For example, the umbilical arterial $PO_2$ will remain low for some time after a contraction. The time delay for the fetal circulation will not be the same for all elements of blood because of their different routes through the fetal body. We used the injection of a dye bolus into the blood leaving the placenta, with sampling from blood returning to the placenta, to estimate the probability of blood presently in the umbilical vein returning to the placenta in a certain time period (13). This was found to be zero for some time--normally 4 seconds. The probability rises to a maximum nine seconds thereafter, a rise we idealized by the left half of a Gaussian curve. From this time on the probability of blood returning decreases to zero; we assumed the fall to follow the right half of a Gaussian curve with a longer decay time of 22 seconds. Thus oxygen saturation of blood returning to the placenta at a given instant was computed as a weighted average of that which left the placenta some 4 to 35 seconds previously, less consumed oxygen.

To summarize, the input variables are: maternal and fetal blood pressures, resistances to flows, and maternal oxygen tension, placental vascular compliance, and the pressure-time profile for uterine contractions. Time is the independent variable. Intermediate variables calcu-

*Placental Modeling* 569

lated in the course of the computation are maternal placental inflow and outflow, the intervillous space volume, and umbilical blood flow. The placental oxygen transfer rate and the partial pressures of oxygen in the fetal blood entering and leaving the placenta are calculated.

NORMAL CASE

Figure 2 has six panels that show the changes occuring during a uterine contraction.

The upper left panel shows the amniotic fluid pressure assumed for a contraction of moderate intensity. Pressure rises from a resting tonus of 10 mm Hg to a peak of 50 mm Hg and then falls to the resting tonus, the whole process taking one minute. This pressure-time profile was chosen to be representative of human labor (3). Although

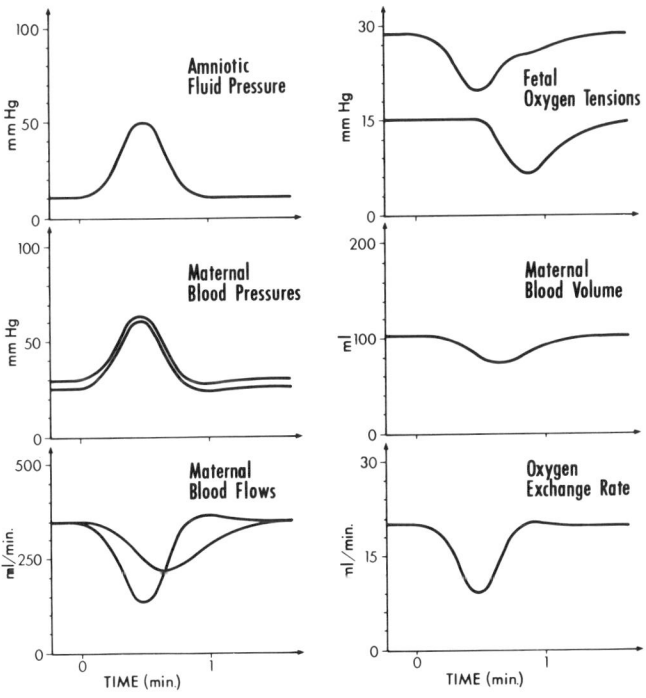

Fig. 2. Exercise of the model for assumed normal contraction.

not shown here, contractions of other intensities and durations gave similar patterns of response, although of course the effects varied quantitatively.

The left center panel shows maternal blood pressures. Blood enters the intervillous space, upper curve, at a slightly higher pressure than it exits, in the lower curve. Due to the pliant walls of the placental vessels, both pressures are elevated during a contraction.

The left lower panel shows maternal flows into and out of the intervillous space. As can be expected the rate of maternal inflow falls sharply during a contraction. Perhaps surprisingly, the rate of outflow is also somewhat reduced in spite of the greater pressure difference between the intervillous space and the maternal vein. This occurs because the myometrium squeezes on the maternal veins causing them to collapse partially as they exit from the uterus. As the contraction subsides there is an inrush of fresh blood while the outflow remains less than normal. About 30 seconds after the contraction subsides, inflows and outflows are identical and there are no further changes.

Fetal umbilical blood flow is predicted to remain essentially constant during and between contractions, a result not shown in the figure. Fetal capillary pressure is predicted to exceed the pressure in the maternal intervillous lakes. Fetal flow in such conditions is not influenced by surrounding intervillous lake pressure but depends solely on the inflow minus outflow pressure. Since the fetus is enclosed entirely within the uterus, the inflow and outflow pressures rise and fall together during the contraction and flow remains constant. It is interesting to note, however, that had a lesser share of the total resistance to maternal flow been ascribed to the spiral arteries (PMA) then fetal vessels would have been predicted to be partially collapsed between contractions (sluice flow). Then, as fetal pressures rise during a contraction, the fetal pressures would exceed surrounding intervillous lake pressure. Fetal vessels would open and fetal flow would increase during the contraction compared with resting levels. Experimental measurement of umbilical flow changes during labor would be an interesting test of the model's assumptions and predictions. The right center panel shows the volume of the intervillous space. Because the maternal inflow is less than outflow early in the contraction, the volume decreases. The fall is only to about 80% of

its resting level and is not greater because the collapse
of maternal veins occurs downstream from the lake. Relative maintenance of this volume is advantageous to the
fetus because an oxygen supply is provided during the contraction. Experimental evidence points to that maintenance
of the intervillous volume during contractions in rhesus
monkeys (17).

The right upper panel shows fetal oxygen tensions.
The upper curve shows the oxygen tension in blood leaving
the placenta in the umbilical veins, while the lower curve
shows the tension in blood supplying the placenta in the
umbilical artery. The time lag in the fetal circulation
is apparent. Primarily as a result of reduced maternal
flow the umbilical venous tension falls from about 28 to
18 mm Hg. After leaving the placenta and perfusing fetal
tissues this blood with lower oxygen tension returns to
the placenta with the result that placental inflow and outflow tensions remain depressed even as the contraction is
subsiding. The right lower panel shows the placental oxygen exchange rate. Early in a contraction the umbilical
venous oxygen tension is reduced without any change in the
arterial oxygen tension. This corresponds to a reduced
exchange rate from 20 to 12 ml/min. Near the end of the
contraction maternal inflow increases while fetal blood
remains hypoxemic. Both of these factors lead to oxygen
transfer slightly higher than normal, shown in the diagram
by a slight elevation above the baseline.

In order to compare the effects of different labor
patterns it is useful to estimate the oxygen deprivation
during a contraction. The difficulty is deciding from
what baseline the exchange rate is reduced. We assume
that the fetus is accustomed to precontraction conditions
("assumed normal" or otherwise). This avoids calculating
all the deficits from a "normal" exchange rate and thereby
introducing constant deficits for the whole time duration
of computation. Such deficits would be unrelated to the
contraction effects. If the oxygen exchange rate is reduced by 10 ml/min for 1 min, the fetus is deprived of 10
ml of oxygen. Generalizing, we may sum the oxygen deprivations second by second over the duration of a contraction
and arrive at a total oxygen deficit for that contraction.
To calculate oxygen deprivation over longer period of time,
it is only necessary to multiply the number of contractions
which take place by the deficit for each contraction. If

desired, the steady-state deficit over the whole period due to abnormal conditions could be added at this point.

## EXERCISE OF THE MODEL

The effect of isolated changes in the various parameters for oxygen deficit caused by a contraction depends upon the parameter changed. The umbilical venous oxygen tension is most sensitive to the intensity of uterine contractions and to the maternal mean-arterial blood pressure. Changes in the maternal arterial oxygen tension have less effect. Perhaps most interesting is the finding that changes in duration of the contraction have virtually no effect on the minimum oxygen tension. If a contraction lasts longer than about 25 seconds, the model predicts that adjustments related to the circulation of hypoxic blood in the fetus essentially lead to a new steady state. Therefore, prolonging the contraction beyond 25 seconds does not lead to any further diminution in umbilical oxygen tension. There is, of course, an accumulating oxygen deficit with longer contractions. When two or more factors change simultaneously, the combined effect is not generally the same as the sum of the individual effects. That is, there is interaction. For example, this is found when both the intensity and the duration of contractions are changed simultaneously from the assumed normal values. After studying a number of cases it was found possible to express the interaction with a simple empirical equation,

$$DEF = 0.33 + DUR (PEAK/490 - 0.03) \qquad (9)$$

where DEF is oxygen deficit in ml, DUR is contraction duration in seconds, and PEAK is maximum hydrostatic pressure during the contraction. This relation applies for any value of PEAK between 20 and 60 mm Hg, and any value of **DUR from 30 to 120 seconds. The maximum error in DEF** for any combination of parameters within these ranges is 0.4 ml. With an intensity of 50 mm and a duration of 60 seconds the oxygen deficit is 4.7 ml. This deficit occurring over a minute constitutes about a fourth of the normal oxygen consumption of a term human fetus. During more intense, longer contractions, the oxygen delivery may fall to less than half of normal. The minimum oxygen tension in the umbilical vein can be estimated with an empirical relation:

$$POMIN = 26.7 - 0.291 \ PEAK \qquad (10)$$

where POMIN is in mm Hg. The relation (10) applies within the same ranges as above. Maximum error in applying (10) within the range of validity is 1 mm Hg. If the maternal blood pressure is low, contractions cause an especially large oxygen deficit. In these instances, maternal flow through the uterus may well cease entirely. The oxygen deficit is greater still when the maternal arterial oxygen tension is low. Such a combination is especially devastating for the fetus. Although oxygen administered to the mother may have only modest benefit under normal conditions, it may be expected to be much more beneficial if the mother is hypotensive.

## COMPARISON WITH OTHER MODELS AND RESULTS

Guilbeau and co-workers (7, 18) were the first theoretically to consider the effects of uterine contractions as placental blood flow was modelled with a sinusoidally varying waveform. The period varied from 15 to 45 seconds. Flow pattern was a combination of concurrent and countercurrent types, formulated in lumped parameter differential equations. Sluice flow was not included. Their conclusion that the oxygen tension values in fetal placental end capillary blood decreased as a function of maternal placental flow is similar to the results of the present study, as well as previous work from this group (11).

There are few experimental data with which to compare the predictions of the model. Several reasons account for this. It has not been feasible to study effects of uterine contractions on the fetal circulation with the uterus open. Although catheters and flowmeters can now be implanted and left in a closed uterus, the effects of labor have not been reported in such chronic animals. Even with animal studies, the results may not apply to humans since the magnitude of changes in intrauterine pressure and the vascular geometry is quite different in the different species. The effect of uterine contractions on maternal placental flow has been studied using cineradioangiography in monkeys (15, 16) and man (2). Greiss (5) used electromagnetic flowmeters to show that uterine blood flow decreased with increased uterine resting tonus and also with amniotic fluid pressure, and with both the increased frequency and duration of contraction (6). The direction of these changes agrees with the predictions of the model. Dawes, Jacobson, Mott, and Shelley (4) demonstrated that fetal femoral oxyhemoglobin saturation decreased from 1.5 to 9% during a uterine contraction in rhesus monkeys. Since the intrauterine pressure was not measured, however, no quantitative relation between amniotic fluid pressure and fetal oxygen tensions may be determined. Studies of the relation of scalp blood oxygen tension to uterine activity in the human fetus (19, 20), while of interest, yield little quantitative relation between oxygen tension and intrauterine conditions because of the problems inherent in scalp samplings during labor.

In conclusion, this study has estimated the effects of different patterns of uterine contraction and oxygen delivery to be fetus *in utero*. Oxygen deprivation during human labor has not yet been measured directly and it remains to test the predictions of the model experimentally. Oxygen deprivation may result in neurological damage and other sequelae, depending upon its severity. Using the predictions reported here, one may assess the effects of different labor patterns, and insofar as the type of labor can be modified, minimize the consequences of fetal hypoxia.

## ACKNOWLEDGMENTS

This study was supported by Public Health Service grants HD 03807, HD 04394 and HL 15655 and the United Cerebral Palsy Foundation. Computation was performed on an IBM 370/158, Data Processing Center, Loma Linda University.

L. D. Longo was the recipient of Public Health Service Research Career development Award 2-K4 HD 23676.

G. G. Power was the recipient of Public Health Service Research Career Development Award 1-K4 HD 20253.

## REFERENCES

1. Bissonnette, J. M. and Farrell, R. D. 1973. *J. Appl. Physiol.* 35:355.

2. Borell, U.; Fernstrom, I.; Ohlson, L.; and Wiqvist, N. 1965. *Am. J. Obstet. Gynec.* 93:44.

3. Caldeyro-Barcia, R. and Poseiro, J. J. 1960. *Clin. Obstet. Gynec.* 3:386.

4. Dawes, G. S.; Jacobson, J. N.; Mott, J. C.; and Shelley, H. J. 1960. *J. Physiol.* 152:171.

5. Greiss, F. C., Jr. 1965. *Am. J. Obstet. Gynec.* 93:917.

6. Greiss, F. C., Jr. and Anderson, S. G. 1968. *Clin. Obstet. Gynec.* 11:96.

7. Guilbeau, J. J. and Reneau, D. D. 1973. In *Oxygen transport to tissue*, vol. 2, *pharmacology, mathematical studies and neonatology*, eds., D. F. Bruley and H. I. Bicher, pp. 1007-1016. New York: Plenum Press.

8. Hildebrand, F. B. 1956. *Introduction to numerical analysis.* New York: McGraw-Hill.

9. Hill, E. P.; Power, G. G.; and Longo, L. D. 1972. *Am. J. Physiol.* 222:721.

10. Laga, E. M.; Driscoll, S. G.; and Munro, H. N. 1973. *Biol. Neonate* 23:231-259.

11. Longo, L. D.; Hill, E. P.; and Power, G. G. 1972. *Am. J. Physiol.* 222:730.

12. Meschia, G.; Cotter, J. R.; Makowski, E. L.; and Barron, D. H. 1967. *Quart. J. Exp. Physiol.* 52:2.

13. Power, G. G., and Longo, L. C. 1973. *Fed. Am. Socs. Exp. Biol.* 32:440.

14. ----. 1973. *Am. J. Physiol.* 225:1490.

15. Ramsey, E. M. 1968. *Clin. Obstet. Gynec.* 11:78; *Obstet. Gynec.* 86.

16. Ramsey, E. M.; Corner, G. W., Jr.; and Donner, M. W. 1953. *Am. J.* 213.

17. Ramsey, E. M.; Martin, C. B., Jr.; McGaughey, H. S., Jr.; Kaiser, I. H.; and Donner, N. W. 1966. *Am. J. Obstet. Gynec.* 95:948.

18. Reneau, D. D.; Guilbeau, E. J.; and Cameron, J. M. 1974. *Microvasc. Res.* 8:346.

19. Renou, P.; Newman, W.; Lumley, J.; and Wood, C. 1968. J. Obstet. Gynec. Br. Comm. 75:629.

20. Saling, E. 1964. Ztsch. Geburtsch. Gynak. 161:262.

## APPENDIX I

### Evaluation of Integrals

Since for a well-behaved function

$$\int_b^a f(x)\, dx = \int_b^c f(x)\, dx + \int_c^a f(x)\, dx \tag{11}$$

VM at time t + h is the sum of VM at time t plus an integral term:

$$VM_{t+h} = VM_t + \int_t^{t+h} (QM1 - QM2)\, dt \tag{12}$$

Similarly, VM at later times t + 2h, t + 3h, . . . , is obtained by adding further integral terms. The time interval h between computed values of VM is the "step size" for progressing through the derivation of a uterine contraction.

Even for well-behaved $f(x)$ (i.e., $f(x)$ and its derivatives exist and are continuous in the interval of integration, the integral can not be evaluated exactly when $f(x)$ is known only for x = 0, h, 2h, . . . , nh.

Frequently a polynomial fitting the known values of $f(x)$ is used to approximate $f(x)$ (8). The integral of this approximating polynomial is used as an approximation to the desired integral. For example:

$$\int_x^{x+h} f(x)\, dx = h[f(x) + f(x+h)]/2 - h^3 f''(X) \tag{13}$$

Placental Modeling 577

where x < X < x + h. The integral is approximated by ignoring the "error" term involving X, since X cannot be determined. Adequate accuracy is obtained only if h is sufficiently small and f(x) is sufficiently linear, i.e., has small higher derivatives. Rather than forcing small h, additional ordinates may be used in the integration:

$$\int_x^{x+2h} f(x)\, dx = h\left[f(x) + 4(f(x + h) + f(x + 2h)\right]/3$$

$$-h^5 f''''(X)/90 \qquad (14)$$

where x < X < x + 2h. For many "smooth" functions the fourth derivative is much smaller than the second. Also, the higher power of h permits use of larger h (i.e., step size) for comparable accuracy. A net saving of computation is realized despite the extra terms in (14) compared with (13). More complex formulas of similar form may be obtained (8) with error terms involving higher derivatives and smaller factions of higher powers of h.

An alternate approach to reducing computation involves use of the form of f(x) to obtain more accurate approximations. Earlier work (9) indicates that in the placental system functions f(x) approach asymptotic values in an exponential fashion after a perturbation. (This expectation was confirmed with exercises of the present model.) We therefore seek an alternate approximation to f(x) which is suitable for use with large h:

$$f(x) = c + a\, \exp bx + R \qquad (15)$$

Unfortunately, the error term R is not as amenable to analysis as the terms in polynomial approximations. Further, numerical exercise of approximations is of limited generality. Testing with the present model shows that $|R| < 0.005$ when h = 1 second.

We require R = 0 for x = 0, h, 2h in equation (15).

Let:

$$A = f(h) - f(0) \qquad (16)$$
$$B = (2h) - (h) \qquad (17)$$

$$g(x) = p \exp qx \qquad (18)$$
$$g(0) = A \qquad (19)$$
$$g(h) = B \qquad (20)$$

Clearly, $p = A$ and hence $\exp qh = B/A$ i.e., $qh = \ln (B/A)$. We seek to identify $q$ with $b$ and hence determine $a$ and $c$:

$$f(0) = c + a \qquad (21)$$
$$f(h) = c + a \exp (\ln B/A) = c + a\, B/A \qquad (22)$$

Hence:

$$a = A^2/(B - A) \qquad (23)$$
$$c = f(0) \qquad (24)$$

We have used data only for $f(0)$ and $f(h)$ with the conjecture that $b = q$. Proof rests on the identity for $f(2h)$:

$$\begin{aligned}
f(2h) &= c + a \exp 2bh \\
&= c + a \exp (2\ln B/A) \\
&= c + a\, (B/A)^2 \\
&= f(0) + (B^2 - A^2)/(B - A) \\
&= f(0) + A + B \\
&= f(0) + f(h) - f(0) + f(2h) - f(h) \\
&= f(2h) \qquad (25)
\end{aligned}$$

Tabulated values of QM1 and QM2 from three successive points in time permit calculation of $a$, $b$, and $c$, and hence an approximation of the integral to compute VM.

## APPENDIX II

Comparison of Differential and Integral Formulations

Numerical solution of differential equations requires estimating derivatives from tabulated points. The derivative formula corresponding to equation (14) is (8):

$$f'(x) = [f(x + h) - f(x - h)]/2h - h^2 f''(X)/6 \qquad (26)$$

where $x - h < X < x + h$. The error $h^2 f''(X)/6$ forces use of smaller h (i.e., more steps for solution) than $h^5 f''''(X)/90$ from (14). Similar savings in computation may be made with other approximations whenever an integral rather than a differential formulation is available.

# 24

## The Role of the Placental Vascular Bed in the Fetal Response to Cord Occlusion

A.F.L. Veth and J.H. van Bemmel
Department of Medical Informatics
Free University
Faculty of Medicine
Amsterdam, The Netherlands

The fetal heart rate (FHR), in combination with the intrauterine pressure, is recorded frequently in clinical obstetrics. During labor the fetal heart rate often varies, especially with uterine contractions, and considerable attention has been devoted to the temporary decreases in heart rate (the so-called decelerations or dips) and to the beat-to-beat variations in this rate. Several classification schemes correlate specific fetal heart rate patterns with certain pathological conditions of the fetus (6, 14, 15, 21).

We have analyzed the fetal circulatory response to temporary occlusion of the umbilical cord, as it may occur during uterine contractions or as a result of problems, such as entanglement or compression. This choice was prompted by several considerations: (1) Cord problems, mostly entanglements, are reported to occur in about 30% of deliveries (11, 16). (2) According to Hon (14) and Gabert and

Stenchever (11), cord compression accounts for about 90% of all abnormal heart rate patterns and the resultant variable deceleration is the commonest specific type of fetal heart rate pattern. This has been confirmed by Rey et al. (18). (3) Repeated occlusion of the umbilical cord may damage the fetus, because the fetal circulation not only is disturbed hemodynamically, but placental gas exchange may be decreased during the occlusion. (4) We expected the fetal circulatory response to be determined primarily by hemodynamic changes, assuming that the duration of the occlusion was short enough to prevent major metabolic changes. Because numerous investigators have modeled the adult human circulation, these results might be used to explore the fetal circulatory response.

In order to study the behavior of the fetal circulation, especially its response to temporary umbilical cord occlusion, we developed a mathematical model based on physiological knowledge (20). We described the fetal circulation mathematically and used a digital computer for the computations. The data for such a description were derived from the literature and from the results of our own experiments on the exteriorized fetal lamb.

The ultimate goal of this investigation was the definition of the physiologic parameters that determine the fetal heart rate patterns observed clinically. Hopefully, this would allow one to relate these heart rate patterns to changes in the fetal condition during birth.

## THE FETO-PLACENTAL UNIT

More than just the exchange organ between the mother and fetus, to a great extent the placenta determines fetal hemodynamics. About 40 to 65% of the fetal cardiac output flows through the placenta. The placenta also contains a large amount of blood. Creasy et al. (8) computed a placental blood volume of sheep of between 10 to 65% of the total fetal blood volume. Bissonnette and Farrell (4) reported a volume of 32 ml/kg fetal weight in the perfused sheep placenta.

The umbilical arterial pressure almost equals the arterial pressure, but the umbilical venous pressure is higher than the central venous pressure. Reynolds and Paul (19) found an umbilical vein pressure of 20 to 35 mm Hg, while Assali et al. (2) and Dawes (9) reported pressures of about 15 mm Hg. Bissonnette and Farrell (4), and Power and Longo (17) described the placental vascular bed as a Starling resistor. This means that the placenta is described as an elastic tube with an input pressure, $P_{UA}$, and an input resistance, $R_{UA}$, an output pressure, $P_{UV}$, and an output resistance, $R_{UV}$, and a surrounding pressure, $R_{SURR}$.

Depending on the magnitude of the pressures, the following relations hold (Fig. 1):

(a)  If $P_{SURR} > P_{UA} > P_{UV}$, then $F = 0$ (1)
(b)  If $P_{UA} > P_{SURR} > P_{UV}$, then $F \approx (P_{UA}-P_{SURR})/R_{UA}$ (2)
(c)  If $P_{UA} > P_{UV} > P_{SURR}$, then $F = (P_{UA}-P_{UV})/(R_{UA}+R_{UV})$ (3)

These relations explain the influence of the maternal circulation on the fetal placental flow. By increasing the venous outflow pressure until the arterial inflow pressure also starts to increase, Bissonnette and Farrell (4) showed that the maternal inferior vena caval pressure almost equals the surrounding pressure. These workers, as well as Power and Longo (17), reported surrounding pressures of 15 mm Hg and demonstrated changes in the fetal placental flow by changing the uterine arterial or venous pressure.

Apparently, the placental vascular bed is relatively insensitive to changes in fetal blood gas values. Dawes (9) found a slight vasoconstriction as the $PO_2$ was decreased from 25 to 8 mm Hg. During fetal hypercapnia ($PCO_2$ up to 80 torr), the umbilical resistance increased only minimally.

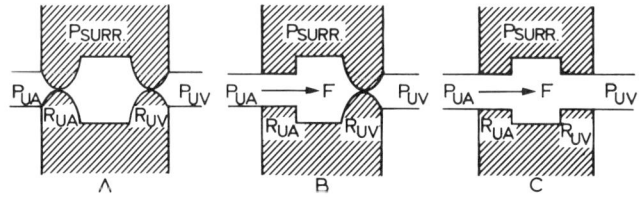

Fig. 1. The placental vascular bed, described as a Starling resistor (conditions as indicated by equations 1 to 3).

Whether the placental vascular bed is controlled by the autonomic nervous sytem remains controversial. Fox and Jacobson (10) demonstrated nerves in the abdominal and extra-abdominal part of the umbilical arteries; however, changes in the vascular tone by sympathetic stimulation have never been shown in an experiment. Dawes (9) reported vasoconstriction in the placental vascular bed after infusion of adrenaline or noradrenaline. This vasoconstriction occurred in the umbilical arteries and veins, as well as in the placental cotyledons; while no vasoconstriction was observed in the ductus venosus or the liver vessels. As Comline and Silver (7) demonstrated, cathecholamine excretion increased during fetal asphyxia. The placental vascular resistance may increase during this condition.

It is of interest to ponder to what extent changes in the placental vascular bed may occur during cord occlusion. From the available data, it is unlikely that these changes result from reflex stimulation during the occlusion. After releasing the cord, spasm may occur, as we sometimes observed during our experiments. If catecholamines were excreted during the cord occlusion, placental vasoconstriction could occur after release depending upon the degree of asphyxia induced by the occlusion.

We mentioned above that the placental vascular bed is an important determinant of fetal hemodynamics. Not only changes in placental vascular resistance, but also changes in placental compliance, will influence the fetal circulation. This may be illustrated by the difference between the response of the fetal circulation to separate occlusions of the umbilical arteries and veins, measured for example by Assali et al. (1, 2). Occluding the umbilical arteries for 3 min caused a bradycardia, an increase in arterial pressure and carotid artery flow, and a slight increase in femoral artery flow. In general, the blood pressure decreased prior to release of the occlusion. In contrast, occlusion of the umbilical veins caused a bradycardia with decreased arterial pressure and carotid and femoral blood flow. Gasps and efforts to breathe accompanied the umbilical arterial occlusion.

This never was seen when occluding the umbilical veins. The difference between the fetal response to separate occlusions of the umbilical arteries and veins results from the placental vascular bed characteristics. If the placental vessels should behave like a rigid tube, no difference would occur. In that case, umbilical flow would cease immediately after the occlusion. However, the placenta acts as a reservoir and stores a certain amount of blood. After umbilical arterial occlusion, blood will flow from the placenta to the venous system until the placental vascular pressure equals the venous pressure. For reasons of simplicity, we assume no limiting factors, such as the effect of surrounding pressures, that may stop the placental outflow before the placental vascular pressure equals the venous pressure. As long as blood flows out of the placental vessels, the venous return and the mean systemic pressure (13) of the fetal circulation will increase. The effect on the cardiac output will be diminished by the increase in arterial pressure and by the degree of asphyxia reached during the cord occlusion. After umbilical vein occlusion, the response will be reversed.

Now, blood will continue to flow into the placenta until the placental vascular pressure equals the arterial pressure. This will decrease venous return to the heart and the mean systemic pressure will decrease. If we neglect the effect of asphyxia, the steady-state pressure reached during cord occlusion will be determined by the total amount of blood shifted into or out of the placental vessels. The placental compliance, therefore, causes the difference in response of the fetal circulation to separate occlusion of the umbilical arteries and veins. A striking feature of the experiments of Assali et al. (2) was the decreased fetal heart rate during both types of occlusion. If this bradycardia results from baroreceptor activity, one would expect an increase in the fetal heart rate during occlusion of the umbilical vein. We observed the same type of response after we pharmacologically blocked the autonomic nervous system. Figure 2 shows the responses of separate occlusions of the umbilical arteries and veins of a lamb fetus whose nervous system has been blocked. The bradycardia that occurs during the occlusion may be caused by myocardial hypoxia, as will be discussed later. Although Assali et al. (2) did not mention a blockade of the nervous system, it appears from their results that this system was inactive.

## MODELING THE PLACENTAL VASCULAR BED

The peripheral circulation is represented as arterial and venous reservoirs which behave as compliances around

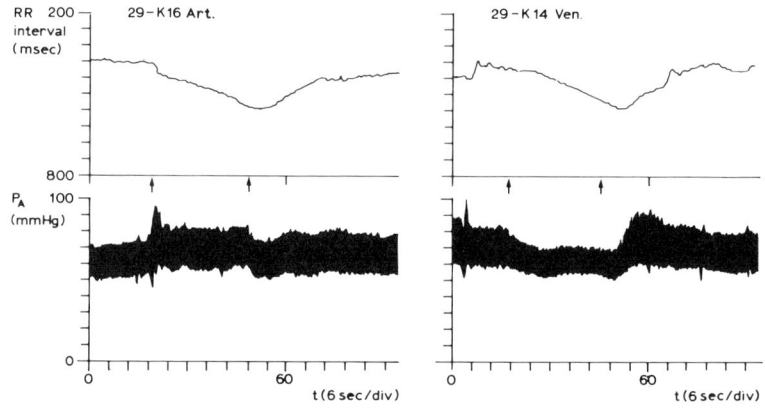

Fig. 2. Measured response of the RR interval (T) and the arterial pressure ($P_A$) of the lamb to separate occlusions of the umbilical arteries (*left*) and veins (*right*). (Occlusion between arrows.)

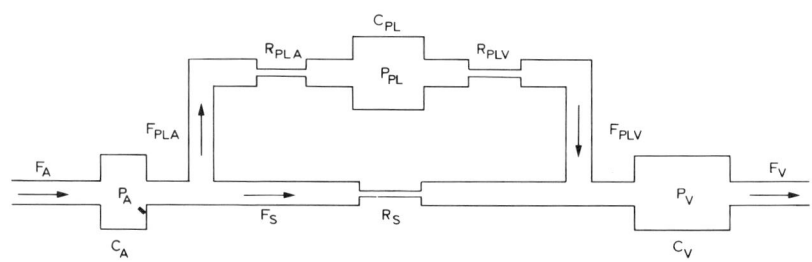

Fig. 3. The hydraulic equivalent of the fetal peripheral circulation, including the placental vascular bed.

the setpoints of the circulation, and are interconnected via a resistance. The placental vascular bed is represented as one reservoir with an arterial inflow and a venous outflow resistance. Figure 3 shows a hydraulic analog of this system. The compliances have been described according to Beneken (3), who defined the so-called unstressed volume, below which no pressure can be built up within the reservoir.

As discussed earlier, experiments (4, 5, 17) indicate that the placental vascular bed behaves like a Starling resistor. We determined certain parameter values from the results of these studies. Bissonnette (5) measured the blood flow and vascular volume of the sheep placenta as a function of the inflow pressure ($P_{FA}$) and outflow pressure ($P_{FV}$). From his results, we computed the total placental vascular resistance ($R_{PT}$).

Figure 4 shows the changes in total placental vascular resistance as a function of inflow pressure and outflow pressure, respectively. Since the weights of the lambs differed considerably, we normalized the total placental vascular resistance by expressing the flow in units per kg of lamb body weight. As can be seen in Figure 4, total placental vascular resistance decreases with increasing inflow pressure until the inflow pressure reaches 35 mm Hg. This decrease in total placental vascular resistance may be caused by passive distension of the placental vascular bed. If inflow pressure exceeds 35 mm Hg, the placental resistance remains almost constant. In terms of a Starling resistor, we may say that the surrounding pressure on the arterial side of the placenta equals approximately 35 mm Hg. When the outflow pressure is increased to 15 mm Hg,

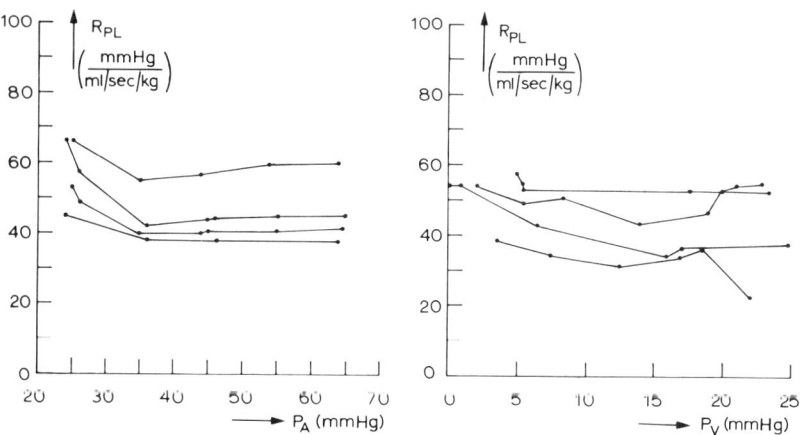

Fig. 4. (*Left*) Placental vascular resistance ($R_{PL}$) as a function of the arterial pressure ($P_A$). (*Right*) As a function of the venous pressure ($P_V$), computed from the data of Bissonnette (5).

total placental vascular resistance decreases (Fig. 4) indicating a surrounding pressure of 15 mm Hg at the venous side of the placenta. This value was also found by Power and Longo (17). The inconsistent changes occurring in total placental vascular resistance and outflow pressure of more than 15 mm Hg may be explained by the differences in inflow pressure during these experiments. From these data, we conclude that the placental vascular resistance is independent of the arterial inflow pressure, if this pressure 35 mm Hg. The effect of the venous outflow pressure on this resistance remains unclear. Bissonnette (5) also measured the change in placental vascular volume ($\Delta Q_{PL}$) as a function of changes in the outflow pressure, ($\Delta P_{VA}$), and outflow pressure ($\Delta P_{FV}$). He found:

$$\left[\frac{\Delta Q_{PL}}{\Delta Q_{FA}}\right]_{P_{FV}} = \frac{\partial Q_{PL}}{\partial P_{FA}} = 1.0 \text{ ml/mm Hg for } P_{FA} > 35 \text{ mm Hg} \quad (4)$$

and

$$\left[\frac{\Delta Q_{PL}}{\Delta P_{FV}}\right]_{P_{FA}} = \frac{\partial Q_{PL}}{\partial P_{FV}} = 0.5 \text{ ml/mm Hg for } P_{FV} > 15 \text{ mm Hg} \quad (5)$$

In order to translate these data to model parameters, we describe the placental pressure of the model as follows:

$$P_{PL} = \frac{R_{PLA}}{R_{PLT}} P_V + \frac{R_{PLV}}{R_{PLT}} P_A \quad (6)$$

where:

$$R_{PLT} = R_{PLA} + R_{PLV}$$

It can be derived that:

$$C_{PL} = \frac{\partial Q_{PL}}{\partial P_A} + \frac{\partial Q_{PL}}{\partial P_V} \quad (7)$$

After substituting the values as found in equations (4) and (5), we find

$$C_{PL} = 1.5 \text{ ml/mm Hg}$$

Using the assumption that the placenta contains 100 ml of blood at a pressure of 25 mm Hg, we compute the placental unstressed volume as:

$$Q_{PLU} = 62.5 \text{ ml}$$

From the data of Bissonnette (5), and assuming that the model truly represents the placental vascular bed, we can also compute the ratio of the arterial inflow and venous outflow resistances of the placenta.

$$\frac{\Delta Q_{PL}}{\Delta P_A} \frac{\Delta P_V}{\Delta Q_{PL}} = \frac{R_{PLV}}{R_{PLA}} = 2 \tag{8}$$

Given our normal values of the umbilical arterial pressure of 55 mm Hg and venous pressure of 5 mm Hg, this would result in a mean placental vascular pressure of 38 mm Hg, which is rather high as compared to our estimated value of 25 mm Hg. We do not know the extent to which the results of Bissonnette were influenced by his experimental conditions. In an attempt to measure both the vascular and extravascular volumes of the placenta, the umbilical vessels were perfused with a Dextran-Ringers solution while the maternal side of the placenta was left intact. It would be interesting to repeat these experiments for placentas still connected to the fetus. Because placental compliance determines to a great extent the differences in response of the fetal circulation to separate occlusion of the umbilical arteries and veins, it is of interest to analyze the responses of the complete fetal circulatory model when we include the estimated placental compliance. The model of the fetal heart used for these simulations has been described elsewhere (20).

The changes in the arterial pressure ($\Delta P_A$) and the placental vascular volume ($\Delta Q_{PL}$) caused by separate occlusion of the umbilical arteries and veins have been represented as a function of the placental compliance in Figure 5. For a placental compliance ($C_{PL}$) of 1.5 ml/mm Hg, we calculate that the arterial pressure increases 50 mm Hg after occlusion of the umbilical arteries and decreases 13 mm Hg after occlusion of the umbilical veins. Furthermore, the placental blood volume increases by 25 ml and decreases by 26 ml, respectively. If the placental compliance equals zero, no difference occurs between separate occlusions of the umbilical arteries or veins. In both cases the arterial pressure increases 18 mm Hg.

In our animal experiments, we found an increase in the arterial pressure of 10 to 15 mm Hg after separate occlusion of the umbilical arteries, and an equal decrease after occlusion of the umbilical veins. To investigate the effect of a deterioration in the fetal heart function during the occlusion period, we computed the responses for a cardiac feedback parameter of 70% of its original value. This would decrease the cardiac output by 30%, assuming constant arterial and venous pressure. These responses are shown in Figure 5 by the broken lines. In this case, there is a smaller increase in the arterial pressure after separate occlusion of the umbilical arteries, as well as after total cord occlusion. The effect of deterioration on the

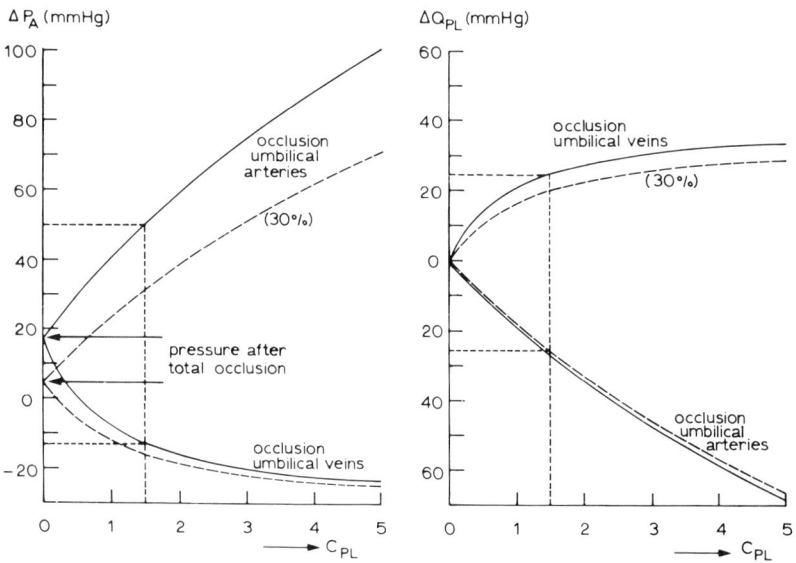

Fig. 5. Predicted change in the arterial pressure ($\Delta P_A$) and the placental blood volume ($\Delta Q_{PL}$) as a function of the placental compliance ($C_{PL}$) after separate occlusion of the umbilical arteries or veins. The broken lines represent these changes with decreased fetal cardiac output.

placental vascular pressure and volume responses to separate occlusion of the umbilical veins is much less.

Without assuming a change in the fetal cardiac output, the separate occlusions of the umbilical arteries or veins decrease or increase, respectively, the placental blood volume by about 25 ml. Thus, the fetal blood volume changes about 7%. Deterioration in the fetal cardiac output only slightly affects this volume shift. The greater sensitivity of the circulation to changes in blood volume also has been found in the human newborn by Young and Cottom (22). After withdrawal of 20 ml of blood from the inferior vena cava, they found a decrease in the arterial pressure of 10 to 20 mm Hg, in agreement with the change predicted by our model.

The model predicts a much larger response of the arterial pressure to separate occlusion of the umbilical vessels, as compared to that of the fetal lambs during our experiments. From Figure 5 we can see that this discrepancy may be partly explained as a deterioration in the fetal cardiac output. As will be discussed later, another explanation could be a limitation in the blood volume that can leave the placenta.

## STEADY-STATE ANALYSIS

### Variation of the Systemic Vascular Resistance ($R_S$)

Figure 6 shows the effect of systemic vascular resistance ($R_S$) on the distribution of the blood volume among the three compartments. The logarithm of the normalized resistance, $R_S/R_S(0)$, is indicated on horizontal scale. Although this function may vary over a wide range, a tenfold increase of the systemic vascular resistance only doubles the total peripheral resistance because the placental vascular resistance parallels the systemic vascular resistance. Figure 6 demonstrates that the arterial and placental blood volumes, $Q_A$ and $Q_{PL}$ respectively, increase at the expense of the umbilical venous volume. The pressures within the vascular compartments change according to the volume changes. The compartment having the smallest compliance, in this case the arteries, shows the greatest pressure increase (Fig. 6). Increasing the systemic vascular resistance ten times increases the arterial pressure ($P_A$) only 16 mm Hg.

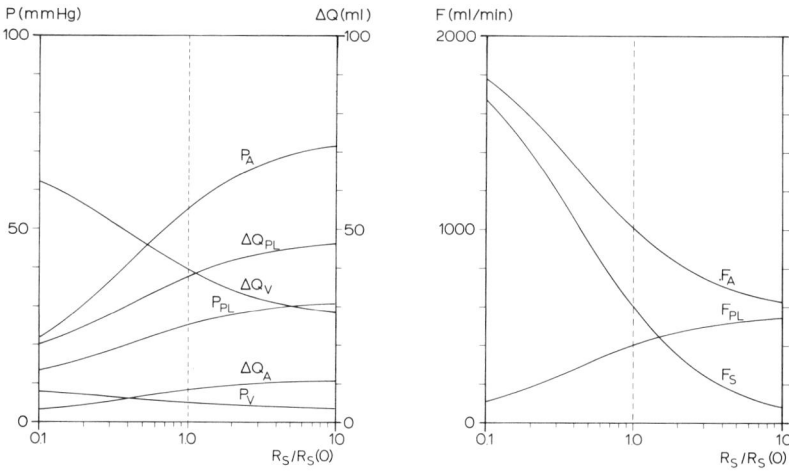

Fig. 6. Predicted steady-state relations as a function of the systemic vascular resistance ($R_S$).

The relation between the systemic resistance and the blood flow through each of the compartments also is shown in Figure 6 (right). As the systemic resistance increases, the cardiac output ($F_A$) and systemic flow ($F_S$) decrease, while the placental blood flow ($F_{PL}$) increases. This decrease in cardiac output results from both the increase in arterial pressure and the decrease in venous pressure. As the systemic resistance increases, a greater portion of the cardiac output flows to the placenta. A tenfold increase in systemic vascular resistance diminishes the cardiac output by 37%. This value compares to that of the adult circulation as Guyton (13) showed a 30% decrease in the dog cardiac output. This value also compares with the simulations of Beneken (3), while Grodins (12) reported a 20% decrease. This comparison does not hold completely, because our model of the fetal circulation includes the placental vascular bed. From this analysis, it appears that a temporary increase in the systemic vascular resistance may be an advantageous fetal defense mechanism. If the increase results from vasoconstriction of "less important" vascular beds such as those of the skeleton, skeletal muscles and intestines, two purposes can be served. The placental blood flow increases, resulting in a better respiratory gas exchange, and the flow through vital organs, such as the brain and the coronary vessels, increases.

## Variation of the Placental Resistance

As discussed earlier, the effect of changes in the placental arterial inflow resistance on the fetal circulation differs markedly from those effects caused by changes in the venous outflow resistance. This is illustrated by the model predictions (Fig. 5) and by the animal experiments (Fig. 2). Because the pressure within the umbilical vein is lower than that within the umbilical artery, we predict that during birth or cord compression most change will occur in the venous outflow resistance.

The effect of changes in the umbilical venous outflow resistance, $R_{PLV}$, on the steady-state pressure and the blood volume of the three compartments is shown in Figure 7. The horizontal axis is the logarithm of the normalized umbilical venous outflow resistance. A tenfold increase in umbilical venous outflow resistance results in a 45% increase in the total peripheral resistance. With increasing umbilical venous outflow resistance, placental vascular

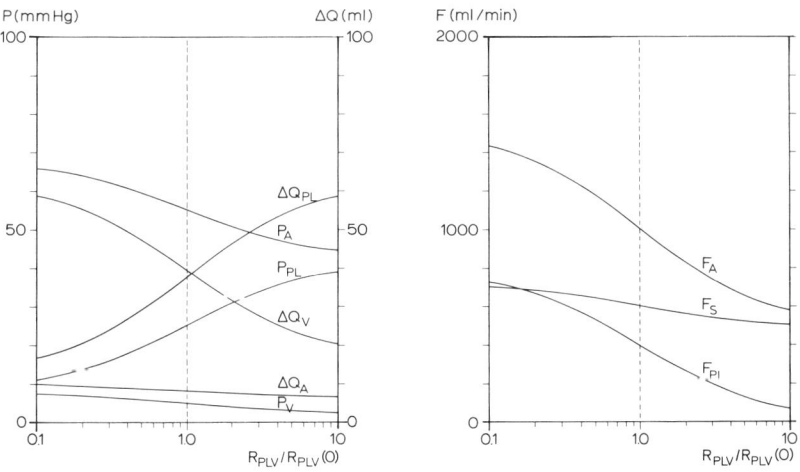

Fig. 7. Predicted steady-state relations as a function of the placental outflow resistance ($R_{PLV}$).

volume and pressure increase at the expense of the blood volume within the arterial and venous compartments. These shifts are accompanied by decreasing arterial and venous blood pressures. As can be seen from Figure 7, the placental compartment is the most sensitive to changes in umbilical venous outflow resistance. Figure 7 also shows that increases in placental outflow resistance decreases the blood flow through the compartments. The cardiac output and the placental blood flow are the most sensitive to these variations.

The relation between the placental arterial inflow resistance ($R_{PLA}$) and the pressure and blood volume within the compartments is shown in Figure 8. In this case, the arterial and venous pressure increase with rising placental arterial inflow resistance as blood shifts away from the placental vessels to the arterial and venous compartments. Remarkable is the increase in the arterial pressure, although the total peripheral resistance increases only 50% with a tenfold increase in placental arterial inflow resistance. This results in a rise of 66% in arterial pressure while an equal increase in umbilical venous outflow resistance decreases the arterial pressure by only 10%. As mentioned before, this difference is due mainly to the properties of the placental vascular bed. Figure 8 also shows

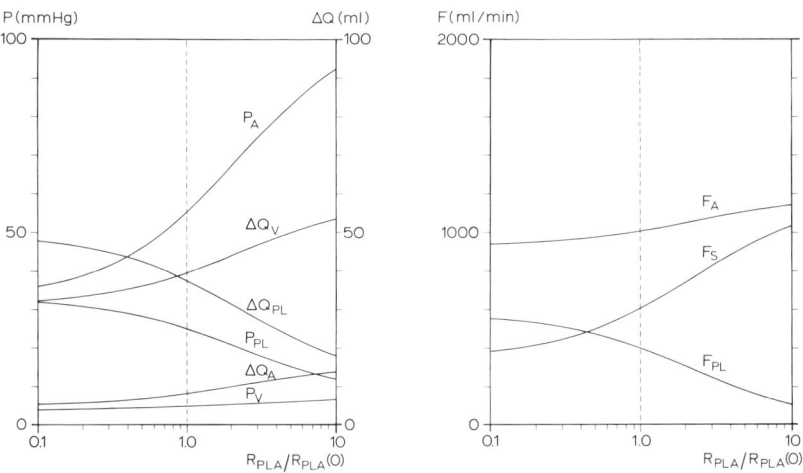

Fig. 8. Predicted steady-state relations as a function of placental inflow resistance ($R_{PLA}$).

the relation between placental arterial inflow resistance and the flows through the compartments. A rising placental arterial inflow resistance results in a major increase in the systemic flow with a minor increase in cardiac output.

In summary, changes in the placental vascular resistances are accompanied by a redistribution of blood between the three compartments. Increase in the venous outflow resistance results in a shift of blood from the fetal circulation to the placental bed, accompanied by a decrease in the cardiac output and in the arterial and venous pressures. Increasing the placental arterial inflow resistance shifts blood from the placental vascular bed to the fetal circulation. This causes a minor increase in cardiac output and in venous pressures and a major rise in arterial pressure.

## DYNAMIC ANALYSIS

### The Fetal Circulatory Response to Umbilical Arterial Occlusion

The main difference between occlusion of the whole umbilical cord or just the umbilical arteries lies in the shift of blood from the placenta to the fetus. Following occlusion the placental vascular pressure ($P_{PL}$) will equal the venous pressure ($P_V$). Figure 9 presents a detailed response of the predicted pressures and flows during a 20-second occlusion of the umbilical arteries. It takes about 14 seconds for the system to reach a new steady state. The duration of the transient part of the response is determined mainly by the venous compliance, parallel with the placental compliance ($C_{PL}$), and the placental outflow resistance, resulting in a time constant of 4 seconds. The shift of blood to the fetal circulation increases the mean systemic pressure which results in an increased cardiac output. After releasing the arterial occlusion, the cardiac output shows an overshoot, due to the rapid decrease in the arterial pressure and the relatively slow decrease in the venous pressure. Figure 10 shows the predicted response of the fetal circulation to occlusion of the umbilical arteries under several conditions, namely: (1) an uncontrolled circulation, i.e., no changes in the systemic resistance nor in the cardiac rate or output; (2) changes in the RR interval during the occlusion; (3) changes in the systemic resistance, due to the autoregulation of blood flow; and (4) changes in both the RR interval and the systemic vascular resistance.

From the figure it is evident that the changes in the RR interval or in the systemic vascular resistance tend to decrease the cardiac output during the occlusion. However, the variation in the RR interval decreases the arterial pressure, while a rise in the systemic vascular resistance results in a further increase of arterial pressure.

During the animal experiments, we observed a much smaller increase in arterial pressure after occluding the umbilical arteries than predicted by the model. Because

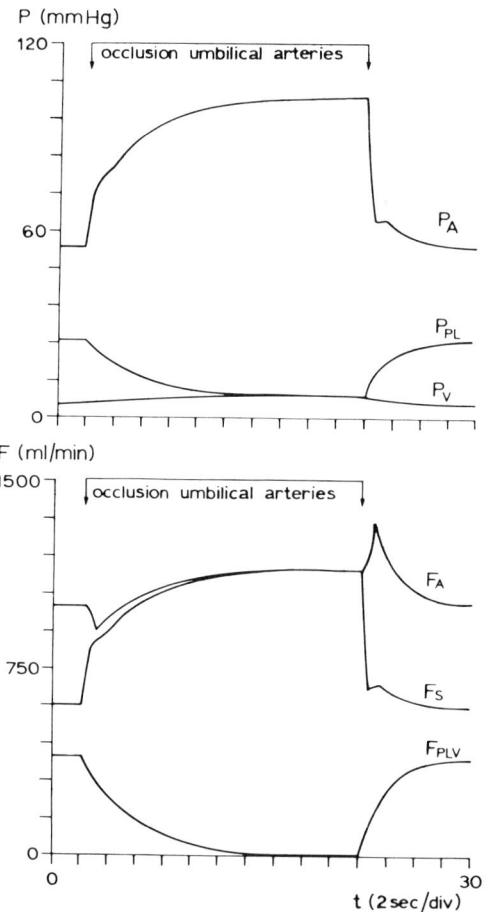

Fig. 9. Predicted response to occlusion of the umbilical arteries.

the increase in arterial pressure is determined to a great extent by the shift of blood from the placenta to the fetus, we expect that this shift is limited by other factors. If the placental vascular bed behaves as a Starling resis-

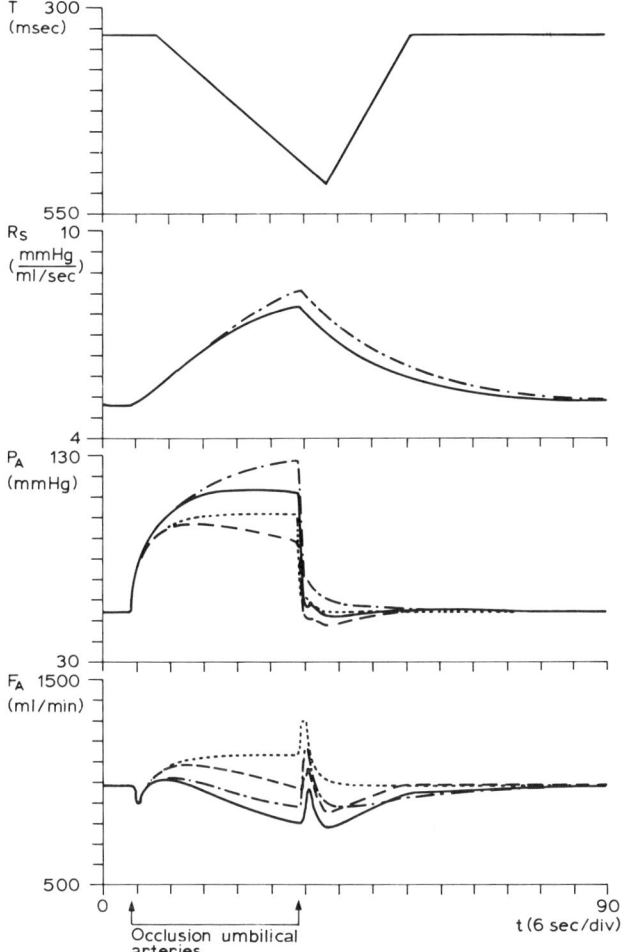

Fig. 10. Predicted response to occlusion of the umbilical arteries: without any regulatory systems (-----); with a changing RR interval (─ ─ ─); with autoregulation of blood flow (·······); with autoregulation of blood flow and a changing RR interval (───).

tor, the shift of blood may be limited by an occlusion of
the placental outflow vessels. This will occur if the
surrounding pressure exceeds the placental outflow pressure;
and will prevent the placental vascular pressure from
equalizing with the venous pressure, thus diminishing the
shift of blood.

The Fetal Circulatory Response to Umbilical Venous Occlusion

After occluding the umbilical veins, the placental
vascular pressure will equal the arterial pressure, with
a shift of blood from the fetus to the placenta. Figure
11 illustrates the predicted results in response to a 20-
second occlusion. The transient part of this response
lasts about 2 seconds less than the response to umbilical
arterial occlusion. The arterial pressure decreases about
13 mm Hg. Since the arterial pressure increases about 18
mm Hg after occlusion of the whole umbilical cord, the arterial pressure decreases about 31 mm Hg as a result of the
shift of blood out of the fetus.

We modeled the fetal circulatory response to occlusion of the umbilical veins alone, using the four assumptions of the changes in the RR interval and systemic vascular resistance as described previously. The results are
shown in Figure 12. As may be seen from this figure, the
RR interval variation results in a decrease in both the
arterial pressure and cardiac output.

The autoregulation of blood flow, resulting in this
case in a decrease in the systemic resistance, increases
the mean arterial pressure and the cardiac output. However,
the changes are small as compared to those caused by variation of the RR interval.

Because the increase in total peripheral resistance
following occlusion of the umbilical veins is partly compensated for by the shift of blood from the fetus to the
placenta, the additional effects of flow regulation and RR
interval increase have little effect on the arterial pressure. This holds, of course, in comparison with the effect
seen during occlusion of the umbilical arteries.

## DISCUSSION

The placental vascular bed determines to a great extent the hemodynamic responses of the fetal circulation. This is illustrated by the differences in the response of the fetal circulation to separate occlusion of the umbilical arteries and veins. The model of the uncontrolled fetal circulation revealed that after umbilical arterial occlusion blood flows from the placenta to the fetus for about 14 seconds. After occluding the umbilical veins,

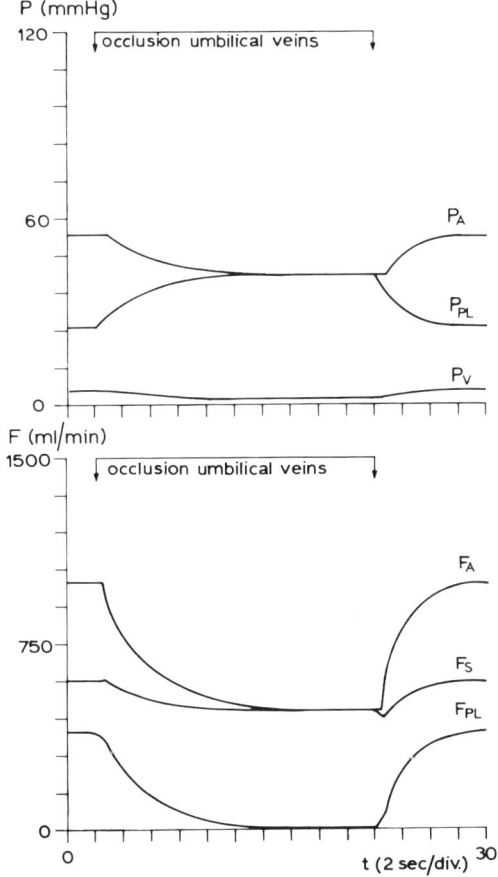

Fig. 11. Predicted response to occlusion of the umbilical veins.

blood flows into the placenta for about 12 seconds. These predictions suggest that the flow of blood out of the placental vessels after umbilical arterial occlusion is limited.

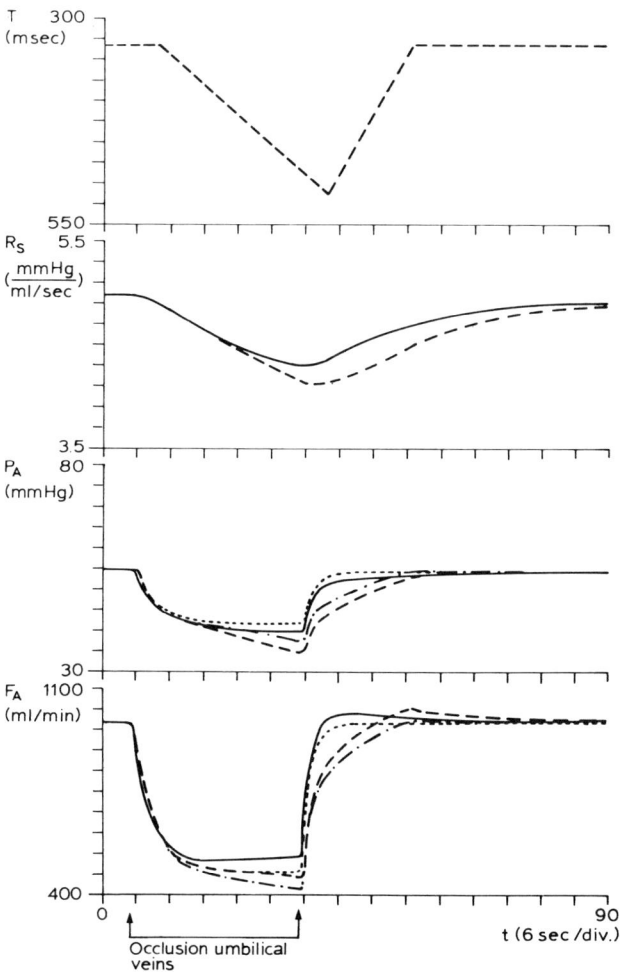

Fig. 12. Predicted response to occlusion of the umbilical veins: without any regulatory systems (-----); with a changing RR interval (-..-..-); with autoregulation of blood flow (———); with autoregulation of blood flow and a changing RR interval (——).

This means that the flow ceases before the placental vascular pressure equals the central venous pressure. If we assume the placental vascular bed acts as a Starling-resistor, this effect results from an external pressure that is higher than the central venous pressure. Modeling the controlled fetal circulation reveals that the response to umbilical arterial occlusion is mediated mainly via the fetal heart rate. After umbilical venous occlusion the fetal heart rate increases as a result of the baroreceptor action not increasing the cardiac output. In this case, the response is mediated primarily via the peripheral resistance.

## Clinical Implications

When monitoring the fetus during labor, one observes certain fetal heart rate patterns that may be explained by occlusion of the umbilical cord or by an impediment of umbilical blood flow during a uterine contraction. The predictions from our model allow for an alternative analysis of these patterns which may be illustrated by the following statements:

A temporary acceleration of the fetal heart rate preceding the deceleration can be explained by a preliminary occlusion of the umbilical vein. Since the pressure within the umbilical vein is lower than that within the umbilical artery, the vein will be occluded first, if the intra-uterine pressure increases. We simulated this effect on the computer and the results are shown in Figure 13. The response of the fetal heart rate and the arterial pressure to simultaneous occlusion of the umbilical arteries and veins is depicted by (————). If the umbilical arteries are occluded for 2.5 seconds after umbilical venous occlusion and are released 2.5 seconds before the release of the umbilical venous occlusion, a response as given by (- - -) is seen. If we increase this time delay between arterial and venous occlusion to 5 seconds, a response occurs as given by (-·-). The dramatic effect of a delayed occlusion of the umbilical arteries is evident from this figure.

During the response of the fetal circulation to umbilical venous occlusion about 25 ml of blood shifts from the fetus to the placenta during about 12 seconds. This implies that: (1) umbilical arterial occlusion 12 seconds after the umbilical venous occlusion does not affect the fetal circulation, because at that time the placental blood has already ceased flowing; and (2) the arterial pressure decreases during the umbilical venous occlusion as a result of a decreased venous return and mean systemic pressure.

Thus, the shift of blood to the placenta attenuates the arterial pressure response and results in a lessened

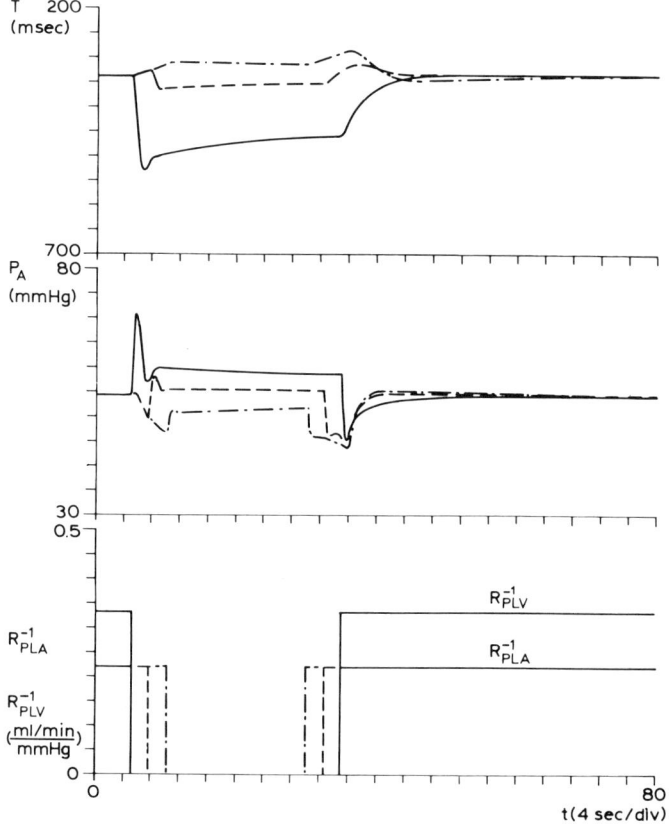

Fig. 13. Predicted response to asynchronous occlusion of the umbilical arteries and veins, mediated by baroreceptor control of both heart period and systemic resistance, for three different times delays ($\Delta T$): $\Delta T = 0$ (———); $\Delta T = 2.5$ sec (– – –); $\Delta T = 5$ sec (·—··—·).

deceleration of the fetal heart rate. If the time delay is large enough, e.g. 5 seconds, the arterial pressure does not rise above its control level, and the fetal heart rate can even accelerate.

Thus, we may expect a fetal heart rate deceleration during labor to become less pronounced if it is preceded by an acceleration. Alternatively, if a fetal heart rate deceleration becomes less when preceded by an acceleration, we may expect an occlusion of the umbilical cord. Further clinical research should examine this question, but a preliminary investigation of our patient material seems to support this hypothesis.

If a fetal heart rate pattern results from complete umbilical cord occlusion during a contraction, there will be no relation between the shape of the fetal heart rate response and the uterine contraction. However, such a relation will exist if the placental flow is impeded partially during the contraction.

In this instance the fetal heart rate response will depend on the vessel being occluded. An increased arterial inflow resistance will result in deceleration of the fetal heart rate, while an increased venous outflow resistance will result in acceleration. A change in placental vascular compliance during a contraction may also result in a fetal heart rate change. A decreasing compliance will result in a fetal heart rate deceleration resulting from an increased placental blood volume. The venous return and the arterial pressure will increase, with resultant fetal heart rate decreases from the baroreceptor reflex.

## ACKNOWLEDGMENTS

This research was performed during a cooperative project between the Institute of Medical Physics at Utrecht and the Department of Obstetrics and Gynecology of the Free University in Amsterdam. It was supported financially by the Dutch Ministry of Public Health and Environmental Hygiene and by the Foundation for Preventive Medicine.

REFERENCES

1. Assali, N. S.; Holm, L. W.; and Seghal, N. 1962. Hemodynamic changes in fetal lamb *in utero* in response to asphyxia, hypoxia and hypercapnia. *Circ. Res.* 11: 423.

2. Assali, N. S.; Bekey, G. A.; and Morrison, L. W. 1968. Fetal and neonatal circulation. In *Biology of gestation*, vol. 2, ed., N. S. Assali. New York and London: Academic Press.

3. Beneken, J. E. W. 1965. *A mathematical approach to cardiovascular function: the uncontrolled human system.* Report Inst. of Med. Physics TNO.

4. Bissonnette, J. M. and Farrell, R. C. 1973. Pressure-flow and pressure-volume relationships in the fetal placental circulation. *J. Appl. Phys.* 35(3): 355.

5. Bissonnette, J. M. 1975. Control of vascular volume in the sheep umbilical circulation. *J. Appl. Phys.* 38(6):1057.

6. Caldeyro-Barcia, R.; Méndez-Bauer, C.; Poseiro, J. J.; Escarcena, L. A.; Pose, S. V.; Bieniarz, J.; Arnt, I.; Gulin, L.; and Althabe, O. 1966. Control of human fetal heart rate during labor. In *The heart and circulation in the newborn and infant*, ed., D. E. Cassels. New York: Grune and Stratton.

7. Comline, R. S. and Silver, M. 1961. The release of adrenaline and noradrenaline from the adrenal glands of the foetal sheep. *J. Phys.* 156:424.

8. Creasy, R. K.; Drost, M.; Green, M. V.; and Morris, J. A. 1970. Determination of fetal, placental and neonatal blood volumes in the sheep. *Circ. Res.* 27 (4): 487.

9. Dawes, G. S. 1968. *Fetal and neonatal physiology*. Chicago: Year Book Medical Publishers.

10. Fox, H., and Jacobson, H. J. 1969. Innervation of the human umbilical cord and umbilical vessels. *Am. J. Obstet. Gynecol.* 103(3):384.

11. Gabert, H. A. and Stenchever, M. A. 1973. Electronic fetal monitoring in association with paracervical blocks. *Am. J. Obstet. Gynecol.* 116(8):1143.

12. Grodins, F. S. 1963. *Control theory and biological systems*. New York and London: Columbia University Press.

13. Guyton, A. C.; Jones, C. E.; and Coleman, T. G. 1973. *Circulatory physiology: cardiac output and its regulation*. Philadelphia: Saunders.

14. Hon, E. H. 1968. *An atlas of fetal heart rate patterns*. New Haven: Hartley Press.

15. Maeda, K. 1969. *Pathophysiology of the fetus: Report 21st Ann. Meet. Jap Obstet. Gyn. Soc.* Kanazawa.

16. O'Gureck, J. E.; Roux, J. F.; and Neuman, M. R. 1972. Neonatal depression and fetal heart rate patterns during labor. *Obstet. Gynecol.* 40(3): 347.

17. Power, G. G., and Longo, L. D. 1973. Sluice flow in placenta: maternal vascular pressure effects on fetal circulation. *Am. J. Phys.* 225(6):1490.

18. Rey, H. R.; Bowe, E. T.; and James, L. S. 1974. *Impact of fetal heart rate monitoring and blood sampling on infant morbidity: ongoing studies. Pediatr. Res.* 8:450.

19. Reynolds, S. R. M., and Paul, W. M. 1958. Pressures in umbilical arteries and veins of the fetal lamb in utero. *Am. J. Phys.* 193(2): 257.

20. Veth, A. F. L. 1976. *Modeling the foetal circulation: simulation of the response to umbilical cord occlusion*. Ph.D. dissertation, Free University, Amsterdam.

21. Wood, C.; Newman, W.; Lumley, J.; and Hammond, J. 1969. Classification of fetal heart rate in relation to fetal scalp, blood measurements and Apgar score. *Am. J. Obstet. Gynec.* 105:942.

22. Young, M., and Cottom, D. 1966. An investigation of baroreceptor responses in the newborn infant. In *The heart and circulation in the newborn infant*, ed., D. E. Cassels. New York and London: Grune and Stratton.

# 25

## A Model of the Fetal Circulation During Labor

### D.T. Gibbons and J.T.M. Wright
Bioengineering Unit/University of Liverpool
Liverpool, England

### F. Johnson
Medical Physics Department/University Hospital
Nottingham, England

Computerized models of physiological situations have now been used for some time. An example is the model of Cerasi et al. (1), which successfully describes the glucose induced insulin release in man. Use of a mathematical model allows the prediction of the body's response to treatment, thus providing a test of a proposed regime of treatment. Internal parameters, which may be difficult to assess in a patient, can be easily examined if the physiological basis of the model is sufficiently accurate. Thus in the case of the model described above, it is possible to make some postulates about the moment-to-moment behavior of the glucose response of the body.

The purpose of the model we describe, however, is to attempt to derive clinically useful data. During the last decade electronic fetal monitoring has become a routine labor ward technique. Maternal intrauterine pressure (IUP) and fetal heart rate (FHR) signals are continuously recorded on a paper chart. These recordings are used by the clinicians to assess the fetal condition. Our model is consistent with this situation in that it has a similar graphical output of the two parameters, fetal heart rate and intrauterine pressure. This consistency provides the possibility of modelling a particular patient situation on a convenient bedside computer and running the model faster than real time to predict the effect of drug administration.

Our modeling philosophy has been to use only a restricted set of parameters with some of these specified without rigor because of lack of physiological data. The model is constructed from as much known data as is possible to incorporate, then the performance of the model is compared with reality. Divergence in performance from the true events is then corrected by adjusting the unknown parameters of the model (the guesses) until a reasonable fit is obtained. A full treatment of this philosophy of modeling is given in Harmon and Lewis (2).

Thus far we have been unable to emulate the characteristic beat-to-beat variation of the fetal heart rate. This has been considered to originate from several sources which include fetal breathing movements, hypoxia, hypercapnia and gestational age (3). None of these are modeled, and it is felt that rather than being sources of the variation, these only form modifying effects. The real origin of the variation remains a subject for speculation.

Further input of physiological data is being made to improve consideration of placental exchange, and explicit recognition of $CO_2$ and pH may then be possible.

## FETAL HEMODYNAMICS

The simplified form of the fetal circulation which is used in the model is shown in Figure 1. The fetal components are considered as lumped parameters interconnected by the major blood vessels. The fetal circulation differs from the adult in that there exist two shunts in the fetus: the foramen ovale (F.O.) and the ductus arteriosus (D.A.). These shunts mean that both sides of the heart work in parallel. The foramen ovale allows oxygenated blood coming from the placenta, via the inferior vena cava, to pass directly into the left heart (4) and not between the two ventricles, as generally assumed. The lungs are redundant in

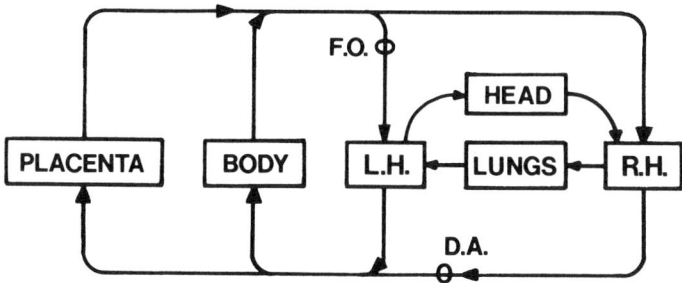

Fig. 1. Simplified fetal circulation used to build the model.

fetal life and appear in this model simply as a blood vessel. The ductus arteriosus allows blood from the right heart to bypass the lungs. The head is fed with blood via the carotid artery which separates from the descending aorta before the ductus arteriosus combines with this aorta.

## THE MODEL

This simplified concept of the fetal circulation forms the basis of the model. The model simulates the blood flows in each of the vessels indicated in Figure 1 as a linear function of fetal heart rate. The oxygen content in each blood stream is calculated, being related to input oxygenation, flow rate and oxygen consumption, or, in the case of the placenta, oxygen gain.

The blood flow rate $Q_G$ in the general vessel G is given by

$$Q_G = K_G \times FHR$$

where $K_G$ is a constant for the vessel. The blood oxygen content of any vessel is a weighted function of the blood flow rate for that vessel. The oxygen content of the blood leaving the placenta is given by

$$O_{PLO} = \frac{O_{MP} \cdot \dot{Q}_{PL}}{140} \times 1 - [ (1 - O_{PLI}/O_{MP}) \dot{x} \exp(- 1.05/QPL)]$$

where
> $O_{MP}$ is the maternal placental site oxygenation
> $Q_{PL}$ is the foetal placental blood flow
> $O_{PLI}$ is the oxygenation of the blood entering the placenta

The first term provides the blood flow weighting function which is normalized to the mean heart rate of 140 beats per minute (b.p.m.).

The oxygen content $O_G$ of any blood stream G is determined by a method of mixtures:

$$O_G = \frac{\sum_{1}^{n} Q_i O_i - U_G}{Q_G}$$

where
> $Q_i$ represents the flow rates of blood streams entering the vessel
> $O_i$ represents the respective oxygenations of these streams
> n is the total number of streams converging on one vessel
> $U_G$ is the oxygen usage in the vessel
> $Q_G$ is the blood flow in the vessel

Time delays representative of blood transit times in the simulated vessels are incorporated within the scheme.

The basic blood and oxygen flow pattern is then combined with other physiological features. Figure 2 illustrates the complete flowchart of the model.

The mean arterial blood pressure is calculated as being proportional to the fetal heart rate (the heart stroke volume and peripheral resistance being assumed constant under all conditions in this model). From the blood pressure a parasympathetic index is derived. This assumes an exponential relationship between the carotid arterial pressoreceptor output and the blood pressure.

The blood pressure is given by

$$BP = K \times FHR$$

and the parasympathetic index by

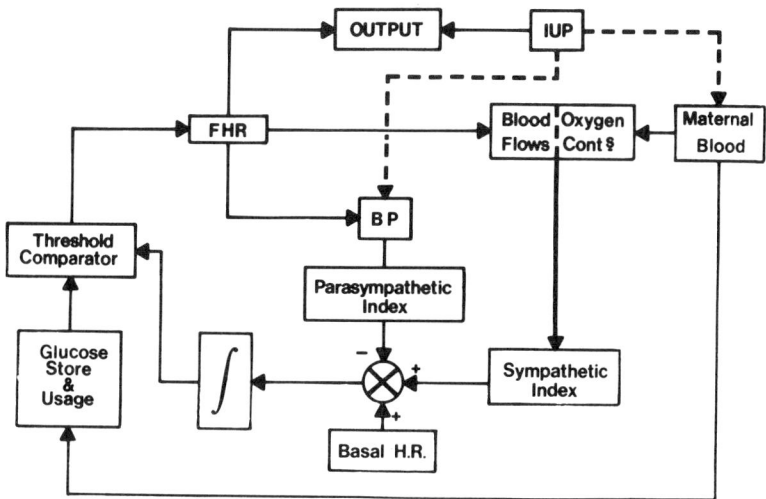

Fig. 2. Components of the model described.

$$PS = 2^{(BP/41.6 + PRES/53.5)}$$

where PRES is the intrauterine pressure above a 15 mm Hg threshold level.

The intrauterine pressure is a raised sine curve placed on a 5 mm Hg basal tone and reaching a maximum of 100 mm Hg with a width at the 5 mm Hg level of 1 min.

The circulatory blood and oxygen flows are calculated from the fetal heart rate and the maternal placental site oxygenation. A sympathetic index arising from the medullary chemoreceptor output is derived as an exponential function of the head oxygenation. The parasympathetic index is subtracted from the sum of the basal heart rate index and the sympathetic indices.

The parasympathetic index is given by
$$S = 2^{[0.25(102.6 - O_{LH})]}$$

where $O_{LH}$ is the oxygenation of the blood leaving the left heart, i.e., that entering the head.

The final index before integration is given by

$$I = 32.3S - 21.6 \, PS + 140$$

where 140 is the basal heart rate.

The result of this operation is integrated and compared with a generator potential threshold value, to determine whether the pacemaker cell has fired. This cell is modeled in the manner of the neuron model of Stein et al. (5), with a perfect integrator. The threshold being used here is shown in Figure 3, an exponentially decaying value following a 150 ms refractory period. The dotted lines indicate the normal working point, given a mean heart rate of 140 beats per min.

## CLASSIFICATION OF FETAL HEART RATE DIP PATTERNS

The classification of the dip patterns we are using in this model is that proposed by Caldeyro-Barcia (6) (see

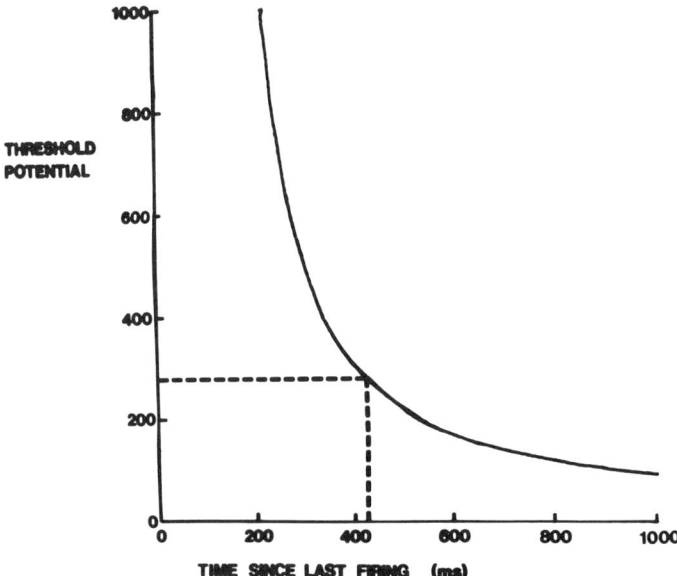

Fig. 3. Time course of pacemaker cell firing threshold.

Fig. 4). In essence all decelerations in the fetal heart rate, which are maternal contraction originated, are separated into two groups.

Type I dips are characterized by the fetal heart rate falling from a baseline level in the range of 120 to 160 beats per min by about 30 beats per min. This fall is in time with the contraction and is almost a reflection of the intrauterine pressure.

Type II dips are characterized by the dip in fetal heart rate occuring between 20 and 40 seconds after the contraction. The dip is also 'U' shaped instead of the 'V' which characterized the type I dip. The fall is generally in the order of 50 beats per min.

## RESULTS FROM THE MODEL

Figure 5 shows the model output of a simulated type I dip. This is produced by increasing head pressure during a contraction. This in turn increases the parasym-

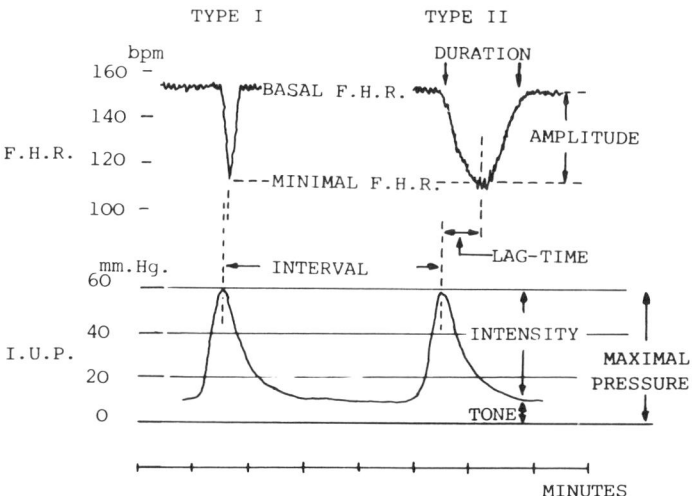

Fig. 4. Type I dip (*left*) and Type II dip (*right*) showing the commonly used parameters of classification. (Modified from Caldeyro-Barcia.)

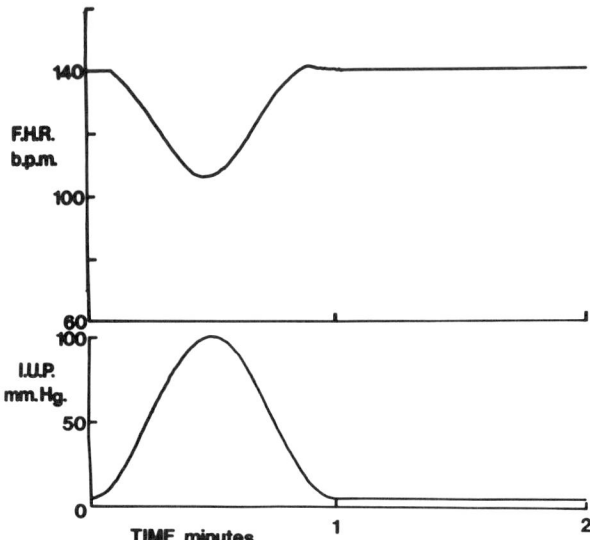

Fig. 5.  Output of model for Type I dip.

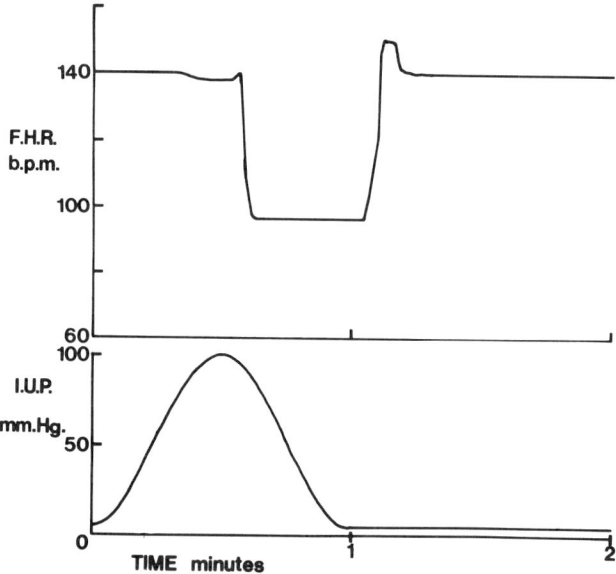

Fig. 6.  Output of model for Type II dip.

pathetic index which leads to a lowering of the heart rate. The mechanism for this is taken from Hon (7).

Figure 6 illustrates the output of the model for a type II dip. The mechanism for this type of dip relies on the energy saving characteristic of the heart. The model has a glucose store which is filled by the "glucose index." This is numerically the same as the blood oxygenation index. Glucose is removed from the store as a function of heart rate. During a contraction the store receives little glucose as the maternal placental site blood flow decreases and placental exchange is impaired. Thus the store becomes depleted and, at a certain threshold level, the heart rate drops to conserve glucose. The heart rate rises when the glucose level is again above its threshold value. The dip lagtime is thus related to the glucose level and will vary with consecutive contractions. The basis for this dip mechanism is taken from Caldeyro-Barcia (6), and Hon (7).

## CONCLUSION

The model is being run remotely on a large computer using a Fortran program, and interactively in real time, or faster than real time on a microprocessor. The graphical output is consistent with the routine recordings made of fetal heart rate and intrauterine pressure on labor ward instrumentation. The parameters of the model are easily changed, and this may be done to fit the model performance to the case in hand if the computer being used is 'on-line.'

There is therefore the possibility of using the model for predictive monitoring. Although this is the initial goal of the project, much work has yet to be done to improve the fit of the model before it is capable of the therapeutic applications that have been made with the pancreas model of Cerasi (1). As it stands at present, however, the model has value in allowing investigation of the mechanisms behind the heart rate patterns. Even this facility has heretofore been unattainable without use of expensive animal experiments, and thus constitutes a useful contribution to the science of engineering in obstetrics.

REFERENCES

1. Cerasi, E.; Fick, G.; and Rudemo, M. 1974. A mathematical model for the glucose induced insulin release in man. *Europ. J. Clin. Invest.* 4:267-278.

2. Harmon, L.D. and Lewis, E. R. 1966. Neural modeling. *Physiol. Rev.* 46:513-591.

3. Dawes, G. S.; Dalton, K. J.; and Patrick, J. *Measurement of foetal heart rate variation*. Blair Bell Meeting, Oxford, October 1975.

4. Dawes, G. S. 1968. *Foetal and neonatal physiology*. Chicago: Year Book Medical Pub.

5. Stein, R.; French, A. S.; and Holden, A. V. 1972. The frequency response, coherence, and information capacity of two neural models. *Biophys. J.* 12:295-322.

6. Caldeyro-Barcia, R. et al. 1966. In *The heart and circulation in the newborn and infant*. Ed., D. E. Cassels. New York: Grune and Stratton.

7. Hon, E. H. 1968. *An atlas of foetal heart rate patterns*. New Haven, Conn.: Harty Press.

# Index

absorption theory of respiration, 20
acetylcholine, 72-75, 80, 81, 83, 86, 103, 170, 181, 187, 202, 223, 225, 370, 440
 and arterial pressure, 80, 81
 cardiovascular response of fetus, 74
 dose-response curve, 75
 and heart rate, 80
 in lamb, fetal, 76
 neonate, 79
action potential, 195, 208, 211
acylcarnitine transferase, 266
adenosine, 167
adenosine monophosphate, cyclic (cyclic AMP), 215, 217, 225, 264, 282, 285, 290
 in embryonic chick heart, 215, 217, 226, 229, 230
adenosine triphosphatase (ATPase), 202-204
adenosine triphosphate (ATP), 278, 281, 283, 289
adenylate cyclase, 264
adrenal cortex, 424
adrenal gland, fetal, 83
adrenaline, 416, 431

adrenergic
 alpha, agonist, 153-166
  gestational changes, mechanism of, 172-175
 beta, blockade, 54, 56
  stimulation, 71
 blockade, 58, 59, 83
 stimulation
  fetal, 71
  transmural nerve, 161, 173
afferent interaction, somatic-visceral
 in piglet, 128-135
age dependency of right to left ventricle effect, 408, 409, 411
agonist
 alpha-adrenergic, 153-166
  and cardiovascular response, fetal, 153-166
 autonomic in sheep, response of
  in adult, 66
  in fetus, 66, 85-86
  in neonate, 66
amniotic fluid pressure, 563, 569
anemia, fetal, 55
anesthesia, 418, 424-425
 spinal, 306, 309, 311, 322
 see separate anesthetic agents
angiotensin, 167, 416, 418, 486
 II, 416, 422, 423, 429-431, 485, 486

converting enzyme,
  421-424
correlation of fetal
  and maternal, 423
in lamb delivered by
  Caesarean section,
  429
by vagina under farm
  conditions, 429
plasma concentration
  in mature fetal
  lamb, 422
metabolism, 424
in plasma, fetal, 417,
  422
pressor action, 416
radioimmunoassay, 418
receptor blockade, 485-
  492
-renin system, fetal,
  415-438
see renin
anoxia
  and arterial pressure,
    531
  and blood flow, rate of,
    529, 530
  and heart rate, 531
  and oxygen level, 527,
    528
antihistamine, 441
aorta, thoracic, 181
  chemoreceptor, 501
  pressure, 116, 136,
    137, 311
arachidonic acid, 456,
  457, 459
  precursor, 456
arterial blood pressure,
  58, 80, 81, 585
  in anoxia, 531
  in general anesthesia,
    418
  mean, 608
  placental, 323-328
  resting in sheep, 58
    adult, 58
    fetus, 58
    neonate, 58

arterial flow response, 136,
  137
arteriole, resistance of, 525
artery
  carotid, occlusion of, 136,
    137
  mesenteric, anterior, 181
  pulmonary, 59
    and acetylcholine, 81
    and blood pressure, 410
  umbilical, occlusion of,
    594, 595, 600
aspirin inhibiting prostaglandin synthetase, 459
atrial
  beat, 379
  cell of young heart, 202
  output, 379
  pacemaker site, 379
  pacing, 270, 375, 379
  pressure, right, 301
  stretching, right, 64
atropine, 54, 73
automaticity, 204-205
  major requirement for, 204
  of ventricular cells, 204
autonomic nervous system, 84
autoregulation of blood flow,
  596
azotemia
  transient appearance of, 473

Barcroft, Joseph, xxviii-xxxii,
  xxxv, 36-37
Barensprung, E., 25-26
baroreceptor
  afferent, 129-135
  reflex, 524
baroreflex activity in fetal
  lamb, 501
Barron, D.H., 35-36
  an appreciation, xxvii-xxxvi
  list of publications, xxxii-
    xxxiv
beat, cardiac
  dependent decay, 397
  per minute, 379
  potentiation of the follow-
    ing, 395

# Index

beating cells in induced heart, 193
Bezold-Jarish reflex, 60-63, 65
bicarbonate, 330-334, 521
biomedical engineering, 44
birth and resting pressure
 after, 57
 before, 57
Bischoff, T.L.W., 14-17
blastoderm, 192-194
bleeding of infant rabbit, 416
 superior ability to withstand, 416
blockade
 alpha-adrenergic, 58
 cholinergic, 54, 56
 ganglionic, 58, 59
blood
 buffer curve, 522
 carbon dioxide of, 506, 507, 520
 see carbon dioxide
 circulatory, 619
 fetal, pathway of, 500
 flow
  in anoxia, 529, 530
  autoregulation, 596
  between compartments, 526
  fetal, 536, 540, 541
  in lung, 439, 440
  maternal, 570
  model of, 562
  placental, 567
  and prostaglandin, 479-496
  rate, 49, 607
  resistance, 526, 565
  umbilical, fetal, 571
 measurement, 346
 oxygen balance, 518
 oxygen concentration, 506
 oxygen content, 607, 608

pressure, 57, 139, 140, 160, 410, 484, 487-489, 516
vessel, fetal
 constrictor response, 176-178
 model of, 516
 and norepinephrine, 175
 reactivity, 175-181
 and serotonin, 175
volume, fetal, 352-354, 357, 589
whole, 355-356
Bohr effect, 506
botulinum toxin, 441
brachial plexus, 128
 stimulation, 128
bradykinin, 440
brain stem in piglet, 102
 diagram of coronal sections, 102
brain tissue oxygen, 528
 partial pressure, 528
 response, 529
bretylium tosylate, 171
bromine, radioactive, 347

calcium, 225-228, 389, 392, 398
capillary
 bed, configuration of, 515
 number of open, 525
 resistance, 525
carbon dioxide, 329-334, 337-340, 506, 507, 520, 521
 balance, 521
 carbonic acid formation from, 510
 content in blood, 506
 diffusion, rate of, 522
 exerts osmotic force, 329
 formation, rate of, 506-509
 hypothesis, 322
 measurement, 512
 partial pressure, 521, 537, 540
 transport in blood, 507, 520
  reactions occurring in, 507

water
  flux, induced, 333
  movement in response to, 330
  transfer, placental, 317-344
carbonic acid, 510
cardiac
  action potential, 234
  beat, 193, 379, 395, 397
  cell, embryonic
    and acetylcholine, 202
    action potential, rate of rise in, 211
    adenosine triphosphatase, 202-204
    automaticity, 204-205
    calcium channel, flow in, 225-228
    of chick, 190-236
    cultured, 221-225
    hyperpolarizing afterpotentials, 205
    ion content, 205-213
      of potassium, 208
      of sodium, 205-208
    K+ sensitivity, 201-202
    membrane
      channel, voltage-dependent, 208-213
      properties, 213-215
      resistance, 200-201
    mucopolysaccharide jelly, 206
    properties
      electrical, 190-236
      K+ permeability, 194, 200, 202
      resting potential, 194-200
    and tetrodotoxin, 211
  differentiation, electrical, 221
  function curve for normal cells, 301
  myocyte in sheep
    fetal, 369
    postnatal, 369
  output, 345, 503
    control, XX, 299-316, 345, 378
    normal values, 308
    physiologic, 345
    determination by indicator-dilution method, 360
    as a function of blood volume in young lamb, 357
    and indicator-dilution technique, 348
  venous return, 299-316
  ventricular muscle, 369, 370
    fetal, 370
    glycolytic pathway, 272
    mass action ratio, 291
    in rhesus monkey, fetal, 271-297
    velocity curve in extracts, 286
cardiovascular
  controlling system, XX, 138, 140
    postnatal maturation, 138
  functions, 50, 75, 371, 374
    autonomic control of, 47-91
      resting heart rate, 52-56
        neural control of, 52-56
      in sheep, 47-91
        adult, 51-52
        fetal, 49-51, 371
        neonatal, 51-52
  homeostasis, 385
  mechanism
    adrenergic, 119-127
    peripheral, 119-127
  response
    alpha-adrenergic agonist, 153-166
    and central nervous system stimulation, direct, 103-118
    and high frequency stimulation, 108, 112-113

Index                                                                                   619

and reflex, 127-137
system, regulatory, central, neural, 94-152
postnatal maturation, 94-152
carnitine transferase, 266
carotid artery of lamb, 172-173
  contractile response, 172
  occlusion, 136, 137
carotid sinus
  inhibition, 97
  stimulation, 97, 134
catecholamine, 126, 225, 226, 263, 485
  cardiac effect, 120, 370
  peripheral vascular effect, 124
catechol-O-methyltransferase, 169, 174, 183
cat heart, 392, 395-397
  force-frequency relation, 389-398
cation pump capability, 204
cell, see separate cells
central
  controlling system, 140
  nervous system
    interaction, 127-137
    stimulation, 103-118
  neural cardiovascular regulatory system, 94-152
  vasoactive site stimulation, 135
chemical theory of respiration, 20
chick embryo heart, 208-209
  and acetylcholine, 202
  action potential, rise in, 210
  and adenosine monophosphate, cyclic, 215, 217, 226, 229, 230

atrial cell, 200
blastoderm and precardiac area, 192-194
cardiac cell, 190-236
  see cardiac cell
cell
  atrial, 200
  cardiac, 190-236
  myocardial, 234-255
    activity, electrical, 234-255
    cultured, 246
    electrical activity, 234-255
    and ouabain, 244, 246
    potassium, see potassium below
    sodium content, intracellular, 246
    transmembrane potassium flux, 238-242
    transport, active, 242-246
  ultrastructure in situ, 218
development of heart, 193-255
and isoproterenol, 217
membrane
  fluidity in development, 214
  resistance, 200-201
metabolism, 216-219
organ culture, 217, 220-225
potassium
  content, 208, 246
  permeability, 194-200, 202
  sensitivity, 201-202
  transmembrane flux in myocardial cell, 238-242
  uptake, 244
rate, 194
resting potential, 194-200
sodium content, intracellular, 246
tetrodotoxin, sensitivity to, 209
tissue electrolytes, 205
transmembrane potential, 247, 248

ultrastructure, 216-219
ventricular muscle, 211
chloralose, 424, 425, 441
chloride space, 359
chromium, radioactive, 347
chronaxie of young heart, 200
circulation, fetal, xviii, xxi, xxii, 41, 47-50, 57, 482-483, 501, 513, 568, 579, 597, 599, 605-614
cistern, subsarcolemmal, 219
citrate, 282, 285, 290
cocaine hydrochloride, 171
compartment volume, 515
compliance, vascular, 305, 309, 312, 314
computerized models of physiological situations, 605
contraction, uterine, 533-539
  and blood flow rate, 536
  and carbon dioxide partial pressure, 537
  -excitation coupling, 389
  heart rate, fetal, 376, 534, 536, 538
  intensity, 557
  and oxygen partial pressure, 533, 535
  response, 170
  and umbilical cord occlusion, 538, 539
  see labor
control systems, 524-536
  baroreceptor, 524
  chemoreceptor, 524
  local, 524
  parasympathetic, 82-85
  sympathetic, 82-85
converting enzyme
  of angiotensin II, 421-424
  pulmonary, 422

convolution integral, 554
cord
  compression, 580
  occlusion and the placental-fetal vascular bed, 579-604
coronal section, diagram of diencephalon of piglet, 110
cotyledon, vascular resistance, 484-489
creatininemia, transient appearance of, 473
creatinine sulfate, 171
cycloheximide, 229

D-*600*, 213
Darwin, Erasmus
  fetus *in utero* respires via placenta, 4
  *Zoonomia* (1794/96), title page, 5
decongestative therapy, 468
denervation, pharmacologic, 55-56
diabetes, 291
diencephalon of piglet, 110
  coronal sections, diagram of, 110
  site, 118
  stimulation, 109
dihomo-$\gamma$-linolenic acid, 456-459
  precursor, 456
dilatation by prostaglandins of a constricted ductus arteriosus, 465
distal resistance, 452
dog, adult
  chronically instrumented, 377
drug receptor interaction, 154
ductus arteriosus, 51, 70, 74, 76, 500
  dilatation, 467
  and prostaglandin, 465, 472
  shunt of, 465
ductus venosus, 500
dynamic analysis, 593-596
  fetal, 498

Index

ear artery, central, 161
  isolated helical strips of, 159
  and norepinephrine, 161
  and transmural stimulation, 161, 173
Earle's salt solution (medium 199), 258
effector-receptor system, 163-164
  development of, 154
electrical activity of cell, myocardial, 234-255
electrical stimulation, transmural, 163
electron micrograph of papillary muscle, 393, 394
endoperoxide circulation, 457
ephedrine, 154-158, 160
  administration, 155, 156
  increase in blood pressure, 160
  pressor amine, 162
epinephrine, 66, 70, 119, 122, 126, 310-312
erythrocyte
  chromium, radioactive, 347
  volume, following infusion of Tyrode's solution, 352-354
estrogen, 481
exchange, placental, XXIII
excitability, 208
extracellular
  fluid, 352-354
    volume, 346
  space, 359

Fairbairn, J., 37
fatty acid
  uptake by the heart, 258, 263
fetal
  circulation, 300

component, major
  heart, 300
  vasculature, peripheral, 300
  model, mathematical, 605-614
  schema, simplified, 607
  components, 609
control, 498
heart of lamb, 371
  instrumentation, 371
  monitoring, 52
  rate, 524, 536, 579, 599, 601, 605
hemodynamics, 582, 606-607
hypoxia during labor, 552
instrumented, chronically, experimental animal
dog, 377
lamb, 371
  atrial stretching, 64
  baroflex activity, 501
  cardiac output, 502
  chronic preparation, 465
  circulatory change, 63
  ganglionic stimulation with DMPP, 64, 65
  immature, 60-64, 67
  isotope disappearance in, 351
  maternal blood injection, 428
  mature, 60-64, 67
  nephrectomy, bilateral, 419, 432-433
  norepinephrine injection, 67-69
  parturition, 429-431
  plasma potassium, 428-430
  plasma sodium, 428-430
  potassium in plasma, 428-430
  premature, 60-64, 67
  renin activity in plasma, 419, 421, 425, 426
    endogenous velocity, 417

maternal anesthesia during operation, 425
sodium infusion, prolonged, 428
sodium in plasma, 428-430
veratridine, 60-63
membrane, placental, 484, 487, 488
model, 511-526, 552
  cardiovascular, 514-517
  combined model, 526
  control system, 524-526
  multicomponent, 520-524
  oxygen, 517-520
nephrectomy, 432
  cardiovascular consequences, long-term, 432
oxygen tension, 571
oxygenation effect on labor, 499, 557
physiology, 541
  brances of, 497-498
  conferences relating to, 44
  steady state analysis, 497, 589-600
  predictions, 590-600
  unsteady state analysis, 498
placental blood flow, 581
respiration, 383-386
  history, 1-32
  question and Zweifel's answer (1651, 1876), 1-32
studies, 49-51
  methodology, brief account of, 49
  vascular function, 49
system
  circulation, 500, 502-511
  control, 500-502
  during labor, 531-538

model, mathematical, of gas transport, 498, 514
multicomponent analysis, 497-550
physiology, quantitative, 502-511
water accumulation, 320-321
  during fetal growth, 321
  hypothesis for, 320
fetologists and international cooperative research, 42
feto-placental unit, 580-583
fetus, 58, 499, 502-511
  blood pathways in, 500
  control systems in, 524, 538
  immature, 58, 75
  premature, 75
  the term, 58, 538
  see fetal
fibroblast-like cell, 241, 249
Fletcher-Powell minimization technique, 566
flow
  regional, 107
  response, 106, 107
foramen ovale, XVIII, 405-410, 500
Frank-Starling mechanism, 380, 384, 385
  ank-Starling relation, 380-387
frog heart, 392-394
fructose-6-phosphate, 278-280, 290
furosemide, 425

Galen
  describes the foramen ovale, xviii
ganglionic blockade, 58
gas transport in fetus, mathematical modeling of, 498, 514
glucagon, 262-264, 266
glucose
  index, 613
  and insulin release in man, 605

uptake, 258, 260, 262, 263
  see glycolysis
glycogen, 219, 263
glycolysis, 272, 274-277, 289
  control of, 271-297
  nonequilibrium reactions in, 289
  see glucose
goat, 527
Goldman's constant field equation, 198
  equation, 334
  hypothesis, 334
Gusserow, A., 24

Hagen-Poiseuille equation, 516
halothane, 100-101, 104, 108, 112, 130-133, 272
Harvey, Wm.
  De Generatione, title page, 2
  describes circulation of fetus, xvii, 1
heart
  automaticity, 204-205
  chambers, model of, 517
  denervation, pharmacologic, 55, 56
  developing, mechanical properties of, 369
  left ventricular
    cardiac output, 385
    cell, automaticity in, 204-205
    changes, 400
    effect on right ventricular output, 403
    end diastolic diameter, 382
      pressure, 383, 406-407
    hypertrophy, 399, 411
  -lung bypass, 302-304
  muscle, embryonic, 238, 242

membrane
  differentiation, 248
  resting potential, 242
pacing, 502
rate, 56, 80, 98, 194, 374, 534, 536, 538
  of adult sheep, nonpregnant, 53
  in anoxia, 531
  blockade
    beta-adrenergic, 56
    cholinergic, 56
  of chick embryo heart, 376, 534, 536, 538
  denervation, pharmacologic, 55, 56
  immature fetal lamb, 53
  mature fetal lamb, 53
  premature fetal lamb, 53
  and propranolol, 53
  resting (beats per minute), 52-56, 386
hemodynamic equation, 516
  response, 582, 597, 606-617
hemoglobin, 26
hemorrhage, 117, 118, 425-527
Henry's law, 506
Hill equation, 553
histamine, 229, 230, 440
histofluorescence, 179, 181, 185
  photographs of, 179, 181, 185
Hoboken nodes, 35
Hoppe-Seyler, 26
hormone and fetal mouse heart, 261-264
Huggett, A., XVIII, 37-43
Hunter, J., 35
Hunter, W., 35
hydrostatic force, 319, 329
5-hydroxytryptamine, 155, 159
hypercapnia, in piglet, 115-117
hyperoxia, 527-528
hyperpolarizing current pulse, 208
hypothalamus, 112, 116
  stimulation, 109, 114, 137
  vasoactive site, 109

hypoxia, XXIII, 427, 452-454, 552
  fetal oxygen response to, 551-559
    a mathematical model for, 551-559

index, parasympathetic, 609
indicator-dilution technique, 348
indocyanine green dye for cardiac output, 348
indomethacin, 452-454, 459, 466-468, 473
infant, premature
  pharmacological closure of patent ductus arteriosus, 465-478
  respiratory distress syndrome, 467, 472, 474
inflow resistance, placental, 592
inotropic agent, 225-228
insulin, 261, 266
  glucose-induced release, 605
  space, 208
intrauterine pressure, maternal, 605, 609
inulin
  measurement, 357
  space, 208
  technecium-labeled, 347
isoproterenol, 66, 70-73, 78-79, 119, 123, 126, 127, 153, 159, 215-217, 440
  in fetal lamb, 72, 73
  in neonatal lamb, 79
  in sheep, 73
isotope disappearance in fetal lamb, 351

ketalar, 272
kidney, 484, 487, 488

kinetics, 239-243
  one-compartment, 242
  two-compartment, 242
kitten heart, neonatal, 394
  similar to frog heart, 394

labor
  and fetal system, 531-538
    abnormal, 533-536
    death of mother, 533-536
    heart rate decceleration during, 601
    hypoxia, 551
    normal, 532-533
    intrauterine pressure during, 532
  monitoring the fetus during, 599
  pressure, intrauterine, 532
  simulation, 539
  uterine, 531-538, 562
lactic acid, 521, 523
  release, 260, 262
lamb, fetal
  blood volume in organs, 505
  carbon dioxide measurement, 512
  carotid artery response, 172, 173
  oxygen measurement, 512
  tissue volume in organs, 505
lamb, neonatal
  baseline control values, 350
  blood volume, 345-367
    whole blood infusion and, 355-357
  cardiac output, 345-367, 503
  control values, 350
  determinant, myocardial, 369-389
  determination, 360
  Frank-Starling relation, 380-387
  heart rate and potentiation, 373-377
  instrumentation, 372
  myocardial determinant, 369-389

Index

pacemaker location, 378-380
potentiation and heart rate, 373-377
experiment *in vivo*, 155
extracellular fluid volume, 345-367
measurement of
 blood-gas, 348-349
 blood pressure, 348-349
 blood volume, 347-348, 358, 360, 361
 fluid volume, 347-348, 358, 360, 361
 radioactive compounds, 347, 350-351, 357
 Tyrode's solution infusion, 351-354
 whole blood infusion, 355-357
Laplace's law, 400
law
 Henry's, 506
 of Laplace, 400
 of mass action, 508
least squares minimization technique, 566
left ventricular heart, see heart
lidocaine, 169
Liebig, J. von, 15-16
liver with oxygen consumption deficit, 555
lung, 479
 blood flow-through, 439
 similarity to placenta, 490
 ventilation-perfusion ratio, 479
lysine vasopressin, 155, 159

*Macaca mulatta*, see rhesus monkey
Magnus, G., 17-18
mammal

blood pressure, progressive increase, in neonatal, 139
heart innervated at birth, 138, 139, 389
mass
 action ratio, 291
 balance, 517
 transfer coefficient, 525
 transfer rate of oxygen, 520
mathematical models, 334, 335, 541, 551, 561-578, 580
Mathews, B., 36
meclofenate
 inhibitor of prostaglandin synthetase, 459
medium-*199*, see Earle's salt solution
medulla and high-frequency stimulation, 108, 112-113
medullary pressor
 area, 136
 response, 107
 stimulation, 106, 107
membrane, 213-215
 fluidity (microviscosity), 214, 215
 pores, 325
 resistance, 200-201
mesencephalic stimulation, 115
mesencephalon, 112
mesenteric artery, 181
mesoderm, 193
miniature swine, see swine
minimization technique, 566
mitochondrion, 219
model, 511-526, 552, 551-559
 of fetal circulation, 605-614
 of fetal oxygen response to hypoxia, 551-559
 of gas transport, 498, 514
 of placental parameters during labor contractions, 561-578
 equations, 565-567
 predictions, 573, 613

model, computerized, 605
  of vascular bed, 583–589
monoamine oxidase, 169, 171, 174, 183
  assay, 171
mouse heart, fetal
  cultured, 260
  and fatty acid, effect of, 258, 263
  glucose uptake, 260, 262, 263
  and hormone, response to, 261–264
  and insulin, effect of, 263
  lactic acid release, 260, 262
  metabolic maturation, 257–270
  octanoate uptake, 263
mucoplosaccharide jelly, cardiac, 206
Mueller, J., 9–14
  *De Respiratione foetus: commentario physiologica*, title page, 10
multicomponent
  model, 541
  system, 520
myelination, 106
  postnatal, 109
myocardial cell, embryonic, chicken
  coat, 249
  cultured, 246
  electrical activity, 234–255
  potassium content, intracellular, 246
  sodium content, intracellular, 246
myocardium, 247
  early, 247
  inotropic state, 387
myocyte, cardiac, 369
myofibril, 219
myofilament, 216

neonate, *see* separate mammals
nephrectomy, fetal, bilateral, 419, 432–433
Nernst equation, 198
nerve
  autonomic system, 84, 167
  presynaptic development, 154
  sciatic, stimulation, 96, 128, 135, 141
  somatic, stimulation, 98
  transmural, stimulation, 173
nervous system, autonomic, 84, 167
neuroeffector transmission, 167
neurohumoral control, 59
neuron model, 610
noradrenalin, 416
**norepinephrine**, 86
l-norepinephrine, 50, 66–71, 76–78, 119, 122, 126, 127, 154, 155, 159, 160, 161, 163, 164, 167, 169–179, 182–187, 229, 230, 263, 265, 370, 486

octanoate, 260, 262, 263
oliguria, 471, 473
organ culture, 217, 220–225, 258
osmotic
  buffering, 327
  force, 319, 329, 333, 335
ouabain, 244, 246
outflow resistance, placental, 591
ovine fetus, *see* lamb
oxygen
  and anoxia, *see* anoxia
  balance of blood, 518
  of brain tissue, 528
  capacity, 518
  concentration in blood, 506, 617, 618
  consumption
    deficit in liver, 555
    rate, 520
  control, 527–531
    in anoxia, 527, 528
    in brain of sheep, 528

Index

in hyperoxia, 527, 528
in uterine contraction, 533, 534
deprivation, 574
during labor, 571, 572, 574
dissociation curve, 506
equilibrium between oxyhemoglobin and oxygen, 506
flow, 609
and hypoxia, see hypoxia
measurement, 512
model, fetal, 517-520
outflow, 554
partial pressure, 533, 535
pressure, 567, 568
partial, 533, 535
tension, 571
transfer, 567-569
oxygenation, fetal, factors affecting, 499, 557
oxyhemoglobin, 26, 506

pacemaker, 378-380
atrial site, 379
cell, 610
and firing threshold, 610
potential in embryonic chick heart, 205
pacing, 270, 375, 379, 502
parasympathetic index, 609
patent ductus arteriosus
closure, pharmacologic, 465-478
in premature infant, 465-478
prostaglandin infusion to dilate, 465
Paul, W., 40-41
pentobarbital, 104, 120, 124
pentobarbitone, 425, 427
pentose shunt pathway, 219
perfusion-perfusion ratio
normal, 481
placental, 480
prostaglandin role in regulation of, 491
perinatal research
history, 33-45
progress, 33-45
peripheral circulation, 583
fetal-hydraulic equivalent, 584
Pflueger, E., 22-23
phenoxybenzamine, 58, 62, 68, 69, 83, 484-489
phentolamine, 158, 159, 162
phenylephrine, 156-160, 162, 163
phosphofructokinase, 273, 277-285, 289
phosphogluconate pathway, 219
physiology, fetal, quantitative, 502-511
pig heart, 401, 402, 404
piglet, 100-102, 104, 108, 112, 120, 124, 128-135
neonatal, 399-413
placenta, 3, 500, 580, 583, 584, 587
bicarbonate, 330-334
blood flow and prostaglandin, 479-496
capillary bed, filtration coefficient, 327
carbon dioxide, 329-334, 339
as fetal lung, 3
filtration coefficient of capillary bed, 327
first definitive account of, 35
idealization, 562
inflow resistance, 602
mathematical model, 561-578
normal case, 569-572
oxygen transfer, 567-569
vascular bed, 579-604
cord occlusion, 579-604
modeling of, 583-589
water transfer across, 317-344

and arterial pressure,
    fetal, 323-328
and cardiac output,
    fetal, 317-344
    factors, 334-340
venous pressure, fetal,
    323-328
placental
  compliance, 587, 588
  filtration coefficient,
    326
  membrane, 318, 333
  outflow resistance, 591
  oxygen transfer, factors
    of, 567-569
  perfusion-perfusion
    ratio, 480
  pressure, 586
  resistance, 591-593
  tissue, nonrigid, 564
  vascular bed, 581
    resistance, 585, 593
    as Starling resistor,
      581, 585, 595, 599
    volume, 582, 586
  vasculature, 492
    and prostaglandin
      action, 492
  water movement across,
    317-344
plasma, 352-355
  volume following infu-
    sion, 352-355
polyribosome, free cyto-
    plasmic, 216
polystrand preparation,
    243, 245
pontocaine, 169
potassium, 194-202, 204,
    208, 238-244, 246,
    248, 428-430
Potter's syndrome, 432
predictive monitoring,
    model for, 613
preeclampsia, 492
pressor action, 416
pressure
  atrial, right, 301
  of blood, see blood pressure
  response, 106, 107
  systemic, 304, 312, 314
Priestley's discovery of de-
    phlogisticated air, 3
propranolol, 53, 54, 62, 69,
    82, 83, 127, 377
prostaglandin
  activation, site of, 450
  aerosolized, 447-449
  and blood flow, placental,
    479-496
  ductus arteriosus dilated by,
    465, 467, 472
  E series, 441-453, 458, 465,
    481-483, 495, 490
  F series, 441, 454, 459
    as vasoconstrictor, 459,
      472
  inactivation, site of, 450
  labor-inducing, 454
  in lung, 455
  precursor of, 455-457
  pulmonary circulation, 439-
    464
  synthesis, inhibition of,
    472
  vasoconstriction by, 454-455,
    459
  vasodilatation, 442, 458,
    465, 467, 472
prostaglandin synthetase,
    inhibitor of
  aspirin, 459
  indomethacin, 459
  meclofenamate, 459
pulmonary
  arterial pressure, 451
  artery, 59, 81, 410
  blood flow, 440
  circulation, 439-464
    fetal and prostaglandin,
      439-464
    hypoxia and indomethacin,
      452-454
    neonatal and prostaglandin,
      439-464

perinatal, 439-441
  and drugs, 440-441
  and prostaglandin, 452
  and prostaglandin aerosol, 447-449
  vascular resistance and prostaglandin, 441-447, 449-452
  vasoconstriction and prostaglandin, 454-455
valve stenosis, congenital, 399
vascular bed, 85
vasculature, 449
  flow-through, 449
  resistance, 445, 447, 454, 459
venous pressure, 451
pyruvate kinase, 273, 284-288, 291

rabbit, 416
radioimmunoassay, 418
radioiodinated serum albumin (RISA), 347
radionuclide, 357
  validity of methods with, 357
rat heart, 192
receptor blockade, 485-492
reflex response, 127-137
  and afferent stimulation, 96-103
  development, 60-66
renin
  -angiotensin system, fetal, 415-438
    see angiotensin
  in fetal lamb, unanesthetized, 419-421
  in fetal plasma, 417, 423, 428, 430, 431
    factors affecting, 424-427
      anesthetics, 424-425
      furosemide, 425

hemorrhage, 425-427
hypoxemia, 427
renal, 418
as substrate, 423-424
resistance, proximal, 452
respiration, 383-386
  history, 1-32
  theory
    absorption, 20
    chemical, 20
    tissue, 21
respiratory distress syndrome, 467, 472, 474
resting potential, 194-200
reticulum, sarcoplasmic, 219
rhesus monkey, fetal
  cardiac muscle, 271-297
    cofactors, 276
    glycolysis, 274-277
    glycolytic control, 271-297
    mass action ratio, 274-277
    phosphofructokinase activity, 277-284
      kinetic data, 282
    pyruvate kinase activity, 284-288
      kinetic data, 287
    rate-limiting reactions, 274-277
right ventricular heart
  end diastolic pressure, 406, 407
  to left effect, mechanism of, 400
RISA, see radioiodinated serum albumin

saphenous vein, 181
sarcolemma, 215
  cholesterol/phospholipid ratio, 215
sarcoplasmic reticulum, 389, 392
Scheel, P., 7
Schuetz, G. F., 7
  Experimenta circa calorem foetus et sanguinem ipsius instituta (1799), title page, 8

Schwann, T., 13, 17–20
  *Mikroskopische Untersuchingen ueber die Uebereinstimmung in der Struktur und dem Wachstum* (1839), title page, 19
Schwartz, H., 21–23
sciatic nerve stimulation, 100–101
science, 561
  "laws" of, 561
serotonin, 173–177, 182–187
Severinus, XXIV
sheep, see lamb
  adult, 51–52, 58, 66, 155
sluice, see waterfall flow
sodium, 201, 205–208, 212–214, 220, 225, 226, 229, 246, 428–430, 473
sodium pentobarbitone, see pentobarbitone
spinal anesthesia, see anesthesia
Starling's law, 318, 322, 325
Starling's resistor, 449, 450, 581, 585, 595, 599
steady-state analysis, 589–593
stenosis, subvalvular, 411
stimulation, 103–118, 134
  adrenergic, 71
  afferent and reflex response, 96–103
  sympathetic nerve, 167–190
  transmural, 161, 173
subsarcolemmal cistern, 219
subvalvular stenosis, 411
swine
  mature, 104
  miniature, 120, 124
  see pig, piglet
symbols, definition of, 542–544
sympathetic nerve, stimulation of, 167–190
systemic pressure, mean, 304, 312, 314

technecium-labeled inulin, see inulin
tetrodotoxin (TTX), 208–211, 220–223, 225, 226, 229, 248, 249
thiopentone, 424–425
tissue compartment
  oxygen equation for, 519
transmembrane potential, 247, 248
transmural stimulation, 161, 173
transplacental water flux, 324, 329
transport, active, 242–246
trimethaphan, 58–60
tropomyosin-troponin complex, 219
trypsin, 193
tyramine, 154
Tyrode's solution, 351–354

umbilical
  arterial pressure, 580, 587
  artery occlusion, 594, 595, 600
  cord occlusion, 533, 535, 538, 539, 579–604
  vein
    occlusion, 599–600
    oxygen tension, minimum, 572
uterus, 563
  contraction, 533–539, 573
  stimulation, 556

vagal nerve traction
  and circulatory effect, 57, 87

## Index

vagotomy, bilateral, in neonate piglet, 104
vascular
  compliance, 305, 309, 312, 314
  homeostasis, 492
    and prostaglandin, 492
  oxygen transfer, 553
  resistance, 442, 452, 484-489
    pulmonary, 452
    systemic, 589-591
  smooth muscle, fetal, reactivity to
    sympathetic nerve stimulation, 167-190
    vasoactive agent, 167-190
  system, 551
vascular bed, fetal placental, 579-604
  and cord occlusion, 579-604
  modeling of, 583-589
vasoactive agent, 167-190
  see prostaglandin

vein, umbilical
  occlusion, 609-610
venous return curve, 299, 301, 304, 306, 307, 309, 311, 313
ventricular, see heart
verapamil, 213
veratridine, 60-63
Vesalius, xviii, 35

water
  accumulation, fetal, 320-321
  flux, transplacental, 324, 329-334
    see carbon dioxide
  transfer across
    intestine, 327
    lung, 327
    muscle, 327
    placenta, 317-344
waterfall flow (sluice), 564
Wurster, G., 25-26

zona incerta, 112
Zweifel, P., 1, 26-29